The Advanced Very High Resolution Radiometer (AVHRR)

Arthur P. Cracknell graduated with a degree in Physics from Cambridge University in 1961 and then completed a DPhil at Oxford University in 1964 in the Department of Metallurgy, working on band-theory calculations for metals. From 1964–1967 he worked as a lecturer in physics at the University of Singapore (now the National University of Singapore) and from 1967–1970 as a lecturer in physics at the University of Essex. He moved to Dundee University in 1970 where he was successively a senior lecturer and reader before being appointed to a personal chair in 1978. He was recently transferred to the established Carnegie Chair of Physics at Dundee.

Until about 1978 his research work was concerned with the Fermi surfaces of metals, with the use of group theory in solid state physics and, especially, with the theory of magnetic point groups and space groups and their relation to the magnetic structure and properties of solids. From about 1978 he turned his attention to remote sensing, beginning with work on the Coastal Zone Colour Scanner and the development of atmospheric correction and chlorophyll extraction algorithms. Subsequently he has been involved in research projects involving data from a wide variety of Earth-orbiting satellites as well as from airborne instruments. For about twelve years he has been editor of the International Journal of Remote Sensing and he has been responsible for the editing of several editions of the Remote Sensing Yearbook. He is the author of over 20 books and of over 200 research papers. Forthcoming with Taylor & Francis is his book co-authored with Professor K. Ya. Kondratyev, *Observing Global Climate Change*.

In recent years, because of the internationally-famous Dundee satellite data receiving station, Cracknell has developed a special interest in the Advanced Very High Resolution Radiometer (AVHRR) which is flown on the NOAA series of polar-orbiting meteorological satellites. This instrument is the subject of the present book.

The Advanced Very High Resolution Radiometer (AVHRR)

ARTHUR P. CRACKNELL

Department of Applied Physics and Electronic and Mechanical Engineering,
University of Dundee, Dundee DD1 4HN, Scotland, UK

CRC Press
Taylor & Francis Group
Boca Raton   London   New York

CRC Press is an imprint of the
Taylor & Francis Group, an **informa** business

UK   Taylor & Francis Ltd, 1 Gunpowder Square, London EC4A 3DE
USA Taylor & Francis Inc., 1900 Frost Road, Suite 101, Bristol, PA 19007

British Library Cataloguing in Publication Data

A catalogue record for this book is available from the British Library.
ISBN 0-7484-0209-8

Library of Congress Cataloging in Publication data are available

Cover design by Amanda Barragry

Contents

Preface

I had wanted to give this book a slightly unconventional title such as 'The exciting and totally unanticipated success of the AVHRR in applications for which it was never intended'. But I was persuaded, from the marketing point of view, that since this was intended to be a serious-minded book or monograph, such a title, while accurate and trying to convey some of my excitement and enthusiasm about the AVHRR, would not be a very good idea.

There have been quite a number of satellite remote sensing systems launched over the last 20–30 years. Mostly they have been of limited duration and most of them have met with some measure of success. However, their useful lifetimes in space have been relatively short and the usefulness of the data generated has been restricted to a rather small and specialised scientific community. I think of projects like Seasat or HCMM, for example. However, the AVHRR is different. It is part of a system which has been operational for over 15 years and which looks like being operational beyond the end of the present century. The AVHRR was designed as an instrument to be flown on a polar-orbiting satellite for meteorological purposes. The data gathered by the AVHRR have come to be used for a whole range of purposes, not just in meteorology and oceanography for which it was planned but in a whole range of applications that were never envisaged when the system was originally specified and designed. The most spectacular and successful of these has been the use of vegetation indices which are now routinely produced and very widely used, but there are many other environmental applications of the data which were never envisaged when the AVHRR was originally designed in the mid-1970s. It is really because of this extraordinary success and undoubted importance of the system that I felt that it would be a good idea to write this book.

The main purpose of this book is to describe the enormously varied non-meteorological applications of AVHRR data. As a prerequisite to this, it was felt to be necessary to discuss the development and operation of the system, the data distribution arrangements and various aspects of the calibration of the AVHRR instruments. Much of this material is only available scattered around in various NOAA Technical Memoranda and Reports.

When it came to writing the text, I had one particular problem to face. This was concerned with whether I should attempt to include any significant discussion of the meteorological uses of AVHRR data, which were the originally intended use of the data. In the end I decided not to include that material for the following reasons.

1   Polar-orbiting satellites are complementary to geostationary (geosynchronous) satellites. Given the timescales on which meteorological systems develop and the repeat frequencies of geostationary and polar-orbiting satellite data acquisition, the polar-orbiting satellites are probably subsidiary to the geostationary ones in terms of their usefulness in weather forecasting. To attempt to discuss the role in weather forecasting of only one of the instruments on one polar-orbiting satellite system would be silly.

2   The use of polar-orbiting satellite data in meteorology is now firmly established and textbooks on meteorology make use of the data (mainly as images) as and when required. An excellent account of the integrated use of geostationary and polar-orbiting satellite data in meteorology is given in section 7 of the book by Rao *et al.* (1990) (see the references list); that section covers 140 pages. A very good pictorial overview of the applications of meteorological satellite data (from both Meteosat and AVHRR) will be found in the publication *EUMETSAT Directory of Meteorological Satellite Applications* (Darmstadt: EUMETSAT, 1993). There is also a complete book by Scorer (1986) (see references list) which is almost entirely devoted to meteorological interpretations of images of clouds obtained from the AVHRR. To include a discussion of meteorological applications of AVHRR data in the present book would only increase its length very substantially without contributing much that is not already available elsewhere in the literature.

3   I know very little about meteorology anyway and would simply be copying from existing materials.

It should, perhaps, be made clear that the AVHRR was never designed specifically or primarily for quantitative radiometric work. It was designed for use as an imaging radiometer, primarily for meteorological purposes. However, with great foresight it was very wisely decided to arrange for its calibration in absolute units because quantitative applications (particularly in connection with sea surface temperatures) were foreseen. This was actually very fortunate because as new applications of AVHRR data were found, then in some of these applications there has been far more use of the data in a quantitative radiometric manner than could ever have been envisaged when the system was originally being planned 20 years ago.

Scientific knowledge and technological capability, along with access to relevant data, and therefore the ability to generate potentially useful results from remote sensing are, however, no guarantee that environmental benefits will actually be obtained for mankind. There must be many examples to illustrate this statement; I quote one in particular which is discussed in some detail in section 6.2.3. Five remote sensing surveys were made over a period of two decades to study deforestation in the Philippines, but this work took place in a politically repressive environment. In the Philippines political considerations from 1965 to 1986 consistently overrode the value of any data provided by remote sensing. The first four surveys were ignored by the Government and had no impact on forest management.

I suppose that, like many other books, this one arises originally out of a set of lectures. In Dundee University I have, for a number of years, been giving a set of lectures to postgraduate students in our MSc course on *Remote sensing, image processing and applications*. These originated as a set of lectures on the use of data from

the thermal infrared channels of the AVHRR for determining sea surface temperatures and covering the calibration of the satellite data, the consideration of
atmospheric corrections and the question of the validation of the satellite-derived
sea surface temperatures. Over the years the scope of these lectures expanded until
they became a set of lectures on as many applications of AVHRR data as I could
think of. I became conscious of the fact that there was no book or monograph
which I could recommend to the students for coverage of all this material. It was all
scattered around in NOAA Technical Memoranda and individual research papers.
Therefore I thought it would be a good idea to try to bring it all together. I am
particularly grateful to Dr. Michael Matson of NOAA NESDIS who, during April
1993, arranged for me to see a number of key people within NOAA itself. This
enabled me to talk to a large number of people who were able to provide various
details that had escaped inclusion in the various published NOAA materials. Dr. C.
R. Nagaraja Rao, of NOAA NESDIS, provided me with considerable help on the
question of the calibration of the AVHRR (both pre-launch and post-launch). Mr.
William G. Knorr of ITT Aerospace/Communications Division, the builders of the
AVHRR, provided me with several photographs and diagrams of the instrument
which I have used in chapters 1 and 2. Sources of these and all other figures are
acknowledged *in situ*. I have also received a considerable amount of help from a
number of other individual research workers who have provided me with references,
reprints or images related to their own particular applications of AVHRR data. I
should also like to acknowledge the help I have received over the last 16 or so years
from staff and students in Dundee who have worked with AVHRR, especially the
staff of the NERC-supported ground station in this Department. I understand that
this was one of the very first ground stations to be set up to receive NOAA data
outside North America; it was set up in the pre-AVHRR days of the VHRR on
NOAA-5. I can recall the station director, Mr. Peter E. Baylis, giving a seminar in
the Physics Department on almost exactly the day on which TIROS-N was
launched. The Dundee ground station started receiving AVHRR data from
TIROS-N as soon as the satellite started operating and has been receiving and
archiving AVHRR data ever since then. I do not think any of us realised then quite
how important AVHRR was to become.

In the conventional manner, however, I myself accept the blame for any errors
that may be present in this book. If anyone detects such errors I would be most
grateful to hear about them. I am also slightly fearful that there may be some very
important successful application of AVHRR data which I have omitted; again I
shall be very happy to hear from anyone wishing to point out any such omission. If
the book goes to a second edition then corrections can be made to cover errors and
omissions.

In these days of word-processing one does a great deal of the typing and revising
of a text oneself. However, the initial keyboarding of raw handwritten text and of
(physical) cut and pasted photocopied material has been done by others; Mrs Pat
Cunningham keyboarded some of the earlier sections while Miss Nikki Millar keyboarded a very large number of the later sections. Figures are always also a problem
and I have had considerable help from various people in obtaining material in a
form suitable for use by the printer.

ARTHUR P. CRACKNELL
Dundee

CHAPTER ONE

The spacecraft and instruments

1.1 METEOROLOGICAL SATELLITES

1.1.1 Early meteorological satellites

We can ask ourselves what are the really successful commercial applications of space technology. The answer of course is telecommunications satellites; telephones, telex, TV relay and satellite TV broadcasting are all heavily dependent on space technology and are commercially successful and widely used throughout the whole world.

Remote sensing (Earth observation) from satellites comes in a completely different category. There is considerable discussion about commercial applications of remote sensing. There was, for instance, a whole special issue of the *International Journal of Remote Sensing* (volume 10, number 2, February 1989) devoted to the commercialisation of remote sensing and the subject has been discussed extensively at various conferences. But people are still talking about it; conferences are being held to address the problem. There is no case of a commercial project which consists of designing, building and launching a remote sensing satellite specifically for the purpose of that project. Commercialisation – in the sense of a fully commercial system – is at present out of the question. But there are some *operational* remote sensing programmes; this is, perhaps, the next best thing to a commercial system. These operational systems are primarily meteorological satellites. The first meteorological satellite, TIROS-1 (Television and Infrared Observation Satellite), was launched from Cape Canaveral in Florida on 1 April 1960. Thereafter a large number of meteorological satellites have followed, involving several different countries, and there are now two main components of the international meteorological satellite programme. This is an operational programme because

- the US NOAA (National Oceanic and Atmospheric Administration) has committed itself to an on-going operational programme of polar-orbiting satellites (the TIROS-N, NOAA-6, NOAA-7, etc. series), and
- the international meteorological community has committed itself to a series of geostationary satellites for meteorological purposes.

These two systems are the backbone of the meteorological satellite systems. This book is concerned with the data generated from one instrument, the Advanced Very High Resolution Radiometer (AVHRR), which is flown on the NOAA polar-orbiting meteorological satellite system. The operational nature of the NOAA polar-orbiting series of satellites means that there is already an extensive archive of historical AVHRR data and that one can be assured of new data from the AVHRR for a number of years to come. The system has been operational for over 15 years and it is planned to continue being operational beyond the end of the present century. The AVHRR was designed in the mid-1970s as an instrument to be flown on a polar-orbiting satellite for meteorological purposes. However, the data gathered by the AVHRR has come to be used for a whole range of purposes, not just in meteorology and oceanography for which it was planned, but in a whole range of applications that were never envisaged when the system was originally specified and designed. The most spectacular and successful of these has been the use of vegetation indices which are now routinely produced and very widely used, but there are many other environmental applications of the data which were never envisaged when the AVHRR was originally designed. It is really because of this extraordinary success and undoubted importance of the system that it seemed to be a good idea to write this book. We shall concentrate on the non-meteorological applications of AVHRR data since there is an excellent account of the integrated use of geostationary and polar-orbiting satellite data in section 7 of the book by Rao *et al.* (1990) and there is also a complete book by Scorer (1986) which is almost entirely devoted to meteorological interpretations of images of clouds obtained from the AVHRR.

Let us recall briefly the distinction between polar-orbiting and geostationary satellites. A geostationary, or geosynchronous, satellite is in an equatorial orbit and it travels in the same sense as the Earth's rotation. Its height above the Earth is arranged so that the satellite's period in its orbit is exactly 24 hours and the satellite (ideally) remains directly above the same point on the equator all the time. Most communications satellites are in geostationary orbits. The operational meteorological geostationary satellites are arranged around the Earth at intervals of about 70° longitude from one another. Polar-orbiting satellites circle the Earth at an altitude low enough to complete several orbits each day. Depending on the inclination of the orbit (relative to the equatorial plane), the intersection of the orbit with the equatorial plane (the nodes) will precess slightly; however, during a time period of one day, the Earth rotates beneath the orbital track. Sun-synchronous polar-orbiting satellites have an orbital inclination such that the rate of precession compensates for the motion of the Earth around the Sun (i.e. without precession, a node of local noon, i.e. with the Sun directly overhead, would have moved 90° (to 6:00 a.m. local time) after three months (Figure 1.1)). Thus, a Sun-synchronous satellite will continue to cross the equator at approximately the same local time each day.

The two meteorological satellite systems, polar-orbiting and geostationary, are complementary, each having been designed to meet mission requirements for which it is uniquely suited. For example, the monitoring of the development of rapidly developing weather systems is largely done with data from the geostationary satellite system with its frequent imaging of an area at 30-minute intervals. On the other hand, the determination of the vertical temperature profile of the atmosphere is obtained by using data from the sounding instruments on the polar-orbiting satellites.

The series of NOAA polar-orbiting satellites began with the launch of TIROS-1

(a)

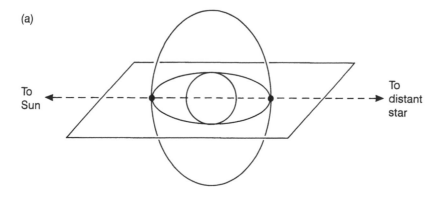

To
Sun

To
distant
star

(b)

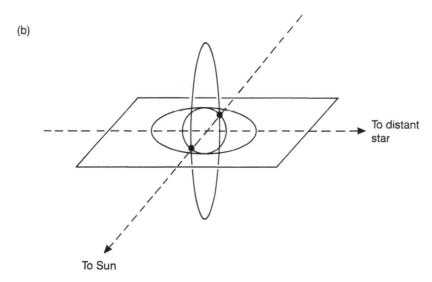

To distant
star

To Sun

Figure 1.1 Sun-synchronous orbit (rate of precession approximately 1° per day), (a) at time t_0 and (b) at time $t_0 + 3$ months.

in 1960 and reached a third generation with the launch of TIROS-N in 1978, see Table 1.1. In the three decades since 1960 the spacecraft, the sensors, the communication links, the data and the data uses of the weather satellites have evolved so that they now bear little resemblance to what they were at the time of that first pioneering launch of TIROS-1. For example, the satellite weight has grown from about 100 kg to nearly a tonne, vidicon cameras have given way to scanning radiometers and hand-drawn analyses of the data have been replaced by computer-generated products. Analogue data have been replaced by digital data and the simple generation of images has been augmented by the use of systems to determine geophysical parameters of the atmosphere or Earth's surface quantitatively. What was initiated by the USA has now become an activity that is shared by a number of other nations as well. Polar-orbiting weather satellites have now also been launched by the former Soviet Union and the People's Republic of China, while geostationary satellites have

Table 1.1   Polar-orbiting satellites (NOAA)

Name	Launch date	Dates of data archived by SDSD	Form	Instrument data archived
TIROS-1	1 April 1960	1 April 1960–14 June 1960	I	Vidicon, (TV Camera System)
TIROS-2	23 November 1960	23 November 1960–27 September 1961	I	Vidicon, IR Radiometer
TIROS-3	12 July 1961	12 July 1961–23 January 1962	I	Vidicon
TIROS-4	8 February 1962	8 February 1962–18 June 1962	I	Vidicon, IR Radiometer
TIROS-5	19 June 1962	19 June 1962–14 May 1963	I	Vidicon
TIROS-6	18 September 1962	18 September 1962–21 October 1963	I	Vidicon
TIROS-7	19 June 1963	19 June 1963–26 February 1966	I	Vidicon, IR Radiometer
TIROS-8	21 December 1963	21 December 1963–12 February 1966	I	Vidicon
TIROS-9	22 January 1965	23 January 1965–9 September 1966	I	Vidicon
TIROS-10	2 July 1965	2 July 1965–2 April 1966	I	Vidicon
ESSA-1	3 February 1966	4 February 1966–6 October 1966	I	Advanced Vidicon Camera System (AVCS)
ESSA-3	2 October 1966	4 October 1966–1 June 1967	I/D	AVCS and Low Resolution, IR Radiometer (LRIR)
ESSA-5	20 April 1967	1 June 1967–3 December 1968	I/D	AVCS, LRIR
ESSA-7	16 April 1968	16 April 1968–31 March 1969	I/D	AVCS, LRIR
ESSA-9	26 February 1969	1 April 1969–15 November 1972	I/D	AVCS, LRIR
ITOS-1	23 January 1970	28 April 1970–17 June 1971	I/D	Scanning Radiometer (SR)
NOAA-1	11 December 1970	26 April 1971–20 June 1971	I/D	SR
NOAA-2	15 October 1972	16 November 1972–19 March 1974	I/D	SR, VHRR, VTPR (see below)
NOAA-3	6 November 1973	26 March 1974–17 December 1974	I/D	SR, VHRR, VTPR
NOAA-4	15 November 1974	17 December 1974–15 September 1976	I/D	SR, VHRR, VTPR
NOAA-5	29 July 1976	15 September 1976–16 March 1978	I/D	SR, VHRR, VTPR
TIROS-N	13 October 1978	30 October 1978–1 November 1980	I/D	AVHRR, TOVS (See below)
NOAA-6	27 June 1979	27 June 1979–20 June 1983	I/D	AVHRR, TOVS
NOAA-7	23 June 1981	23 June 1981–25 February 1985	I/D	AVHRR, TOVS
NOAA-8	28 March 1983	20 June 1983–31 October 1985	I/D	AVHRR, TOVS
NOAA-9	12 December 1984	25 February 1985–7 November 1988	I/D	AVHRR, TOVS, SBUV
NOAA-10	17 September 1986	17 November 1986–16 September 1991	I/D	AVHRR, TOVS
NOAA-11	24 September 1988	8 November 1988–Present	I/D	AVHRR, TOVS, SBUV
NOAA-12	14 May 1991	1 September 1991–Present	I/D	AVHRR, TOVS

Form: I – Image; D – Digital
VHRR – Very High Resolution Radiometer; AVHRR – Advanced VHRR
VTPR – Vertical Temperature Profile Radiometer
TOVS – TIROS Operational Vertical Sounder (three sensors; see description)

been launched by the USA, the European Space Agency (ESA), Japan and India. Further weather satellites are under development in other countries too.

The origins of the weather satellite programme were two-fold. The first was the increasing awareness of the inadequacy of conventional methods of gathering meteorological data, particularly as in the 1950s numerical weather prediction models began to come into existence. Such numerical models need to have meteorological input data on a regular grid and at regular intervals. Conventional observations are only available on a highly irregular distribution over the globe; they are very concentrated over land areas in advanced industrialised countries, they are very sparse over some of the large-area developing countries and they are extremely sparse (almost non-existent) over the polar regions and over large areas of the world's oceans. The second was the realisation from balloon observations and from observations from high-flying aircraft that artificial satellites could be used as weather observing platforms in space. The enabling technology was available as a result of the developments of military equipment and systems during the later part of the Second World War and in the following decade. The history of the development of civilian weather satellites is described in chapter 2 of Rao *et al.* (1990). Many of the important characteristics of the various satellite systems, the instruments flown on them and the data generated by them are described in several of the later chapters of the book by Rao *et al.* (1990).

The first generation in the series of polar-orbiting weather satellites consisted of the TIROS series of satellites, of which the first was TIROS-1 (launched on 1 April 1960) and the last, TIROS-10, was launched on 2 July 1965. These were largely research and development satellites and carried a vidicon camera with an array of lenses. The vidicon was essentially a television camera which provided visible data at a maximum spatial resolution of 3.8 km. Initially the orbital inclination was approximately 64° but the later satellites in the series had higher inclinations (currently 98.9°) and were therefore able to image more of the Earth's surface.

The second series of polar-orbiting satellites were those of the Environmental Science Services Administration (ESSA). These satellites are considered the first generation of operational polar orbiters, the TIROS Operational System (TOS). Also placed in near Sun-synchronous orbits, satellites in this series worked in pairs so that two were usually operational at one time. Odd-numbered satellites were equipped with an Advanced Vidicon Camera System (AVCS), which provided visible data at a maximum spatial resolution of 2.2 km, and a Low Resolution Infrared Radiometer (LRIR) which provided infrared measurements at varying spatial resolutions. Even-numbered satellites were equipped with an Automatic Picture Transmission (APT) system which provided local imagery (at spatial resolutions of 3.8–7.4 km) to suitably equipped ground stations. The orbits were designed to give morning and afternoon coverage from each satellite over a given area; odd-numbered satellites were in afternoon orbits and even-numbered ones were in morning orbits. The ESSA series began with the launch of ESSA-1 on 3 February 1966 and ended with ESSA-9 which was launched on 26 February 1969.

1.1.2 The NOAA series of satellites

The second generation of operational polar orbiters began on 23 January 1970 with the launch of ITOS-1 (Improved TIROS Operational System) and continued with

NOAA-1 to NOAA-5. NOAA-5 was launched 29 July 1976. These satellites were also placed in near Sun-synchronous orbits with nodes of 0900 Greenwich Mean Time (GMT, descending) and 2100 GMT (ascending). The sensors on ITOS-1 and NOAA-1 included an AVCS and a Scanning Radiometer (SR). A Very High Resolution Radiometer (VHRR) and a Vertical Temperature Profile Radiometer (VTPR) were part of the payload for NOAA-2 to NOAA-5. The SR provided global visible and infrared data at 4 and 8 km spatial resolutions, respectively. The VHRR was a two-channel radiometer that provided visible and infrared data for specific Earth locations at a resolution near 1 km. The VTPR was an eight-channel radiometer that provided infrared measurements at 68 km resolution.

The latest generation of polar-orbiting satellites was initiated with the launch of TIROS-N on 13 October 1978; this has been followed by NOAA-6 to NOAA-14 and further satellites in the series are planned for operation through the 1990s. Some characteristics of these satellites are given in Table 1.2(a). The major advance introduced with this series was the shift from an analogue to a fully digital system; the data are digitised on board before transmission to Earth. A great deal of information about the TIROS-N series and the Advanced TIROS-N series of spacecraft is given in two NOAA Technical Memoranda (Schwalb 1978, 1982) and in two

Table 1.2   (a) TIROS-N/NOAA A–G summary sheet (Schwalb 1978)

Spacecraft	Total weight – 737 kg (excludes expendables)
Payload	Weight including tape recorders – 194 kg Reserved for growth – 36.4 kg
Instrument Complement	Advanced Very High Resolution Radiometer (AVHRR) High Resolution Infrared Radiation Sounder (HIRS/2) Stratospheric Sounder Unit (SSU) Microwave Sounder Unit (MSU) Data Collection System – ARGOS (DCS) Space Environment Monitor (SEM)
Spacecraft Size	3.71 m in length 1.38 m in diameter
Solar Array	2.37 m × 4.91 m: 11.6 m² 420 W, end of life, at worst solar angle
Power Requirement	Full operation – 330 W Reserved for growth – 90 W
Attitude Control System	0.2° all axes 0.14° determination
Communications	
Command Link	148.56 MHz
Beacon	136.77; 137.77 MHz
S-Band	1698; 1702.5; 1707 MHz
APT	137.50; 137.62 MHz
DCS (uplink)	401.65 MHz
Data Processing	All digital (APT; analogue)
Orbit	833; 870 km nominal
Launch Vehicle	Atlas E/F
Lifetime	2 years planned

editions of a data user's guide (Kidwell 1984, 1991). The TIROS-N series of space-craft, like their predecessors, operate in nearly Sun-synchronous orbits. They have an inclination of 98.9° and orbital period of approximately 102 minutes. Consecutive equatorial crossings are separated by about 25° latitude. This produces 14.1 orbits per day. Each satellite crosses the equatorial plane at a fixed (local solar) time each day; these are either 1500 hours (ascending node) and 0300 hours (descending node) or 1930 hours and 0730. These times are approximate as there is, in fact, a slow precession and the orbital tracks do not repeat on a daily basis; however, similar equatorial nodes are experienced every eight days. There is also long-term drift (over a period of a year or two) in some spacecraft in the series (Price 1991, Brest and Rossow 1992).

Table 1.2   (b) Advanced TIROS-N/NOAA E–J summary sheet (Schwalb 1982)

Spacecraft	Total weight – 1009 kg
	(excludes expendibles)
Payload	Weight, including tape recorders – 386 kg
Instrument Complement	Advanced Very High Resolution Radiometer (AVHRR/2)
	High Resolution Infrared Radiation Sounder (HIRS/2)
	Stratospheric Sounder Unit (SSU)
	Microwave Sounder Unit (MSU)
	Data Collection System – ARGOS (DCS)
	Space Environment Monitor (SEM)
	Search and Rescue (SAR) Satellite Aided Tracking (SARSAT)
	Solar Backscatter UltraViolet Radiometer (SBUV/2) – NOAA F and on Earth Radiation Budget Experiment (ERBE) – NOAA F and G only
Spacecraft Size	3.71 m in length
	1.88 m in diameter
Solar Array	2.37 m × 4.91 m: 11.6 m²
	515 W, end of life at worst solar angle
	(violet, high efficiency solar cells)
Power Requirement	Full operation – 475 W
	Reserved – 40 W
Attitude Control System	0.2° all axes
	0.14° determination
Communications	Command Link – 148.56 MHz
	Beacon – 136.77; 137.77 MHz
	S-Band – 1698; 1702.5; 1707 MHz
	APT – 137.50; 137.62 MHz
	DCS (uplink) – 401.65 MHz
	SAR – 1544.5 MHz
	SAR (uplink) – 121.5; 243.0; 406 MHz
Data Processing	All digital (APT translated to Analogue)
Orbit	833; 870 km nominal, Sun-synchronous
Launch Vehicle	Atlas E/F
Lifetime	2 years planned

The instruments originally flown on board all these spacecraft include the following:

- the Advanced Very High Resolution Radiometer (AVHRR)
- the TIROS Operational Vertical Sounder (TOVS)
- the ARGOS data collection and location system
- the Space Environment Monitor (SEM).

The instruments flown on TIROS-N and its successors are successors of earlier instruments, scanning radiometers and sounders, etc., which were flown on earlier generations of polar-orbiting satellites. Although most of the system was a NOAA/NASA development in the USA, there were contributions towards the TOVS from the UK Meteorological Office and of the ARGOS system from the French Centre Nationale d'Etudes Spatiales (CNES).

Table 1.2(a) was produced before any of these spacecraft were launched and in fact this description only applies to the earlier spacecraft in the series. Significant modifications, described as the Advanced TIROS-N (ATN) programme, were made to NOAA-E, -F and -G and three new spacecraft, NOAA-H, -I and -J were added to the series, see Table 1.2(b). The main changes were (a) an increased payload capability which allowed extra instruments to be flown, (b) an extra power specification and (c) the addition of a new downlink capability for the search and rescue system which was added to the system. An Earth Radiation Budget Experiment (ERBE) has been flown on NOAA-9 and NOAA-10. A Solar Backscatter Ultraviolet radiometer (SBUV) had been flown on Nimbus-7, an experimental satellite which was launched in 1978, to monitor the distribution of ozone in the atmosphere. An improved version of this instrument has been developed and flown on NOAA-9 and NOAA-11.

The instruments on board the TIROS-N series of satellites will be described later in this chapter; it is one of these instruments in particular, the Advanced Very High Resolution Radiometer (AVHRR), which forms the subject of this book. It should, perhaps, be made clear that the AVHRR was never designed specifically or primarily for quantitative radiometric work. It was designed for use as an imaging radiometer, primarily for meteorological purposes. However, with great foresight it was very wisely decided to arrange for its calibration in absolute units because quantitative applications (particularly in connection with sea surface temperatures) were foreseen. This was actually very fortunate because, as new applications of AVHRR data were found, then in some of these applications there has been far more use of the data in a quantitative radiometric manner than could ever have been envisaged when the system was originally being planned 20 years ago. Figure 1.2 shows a drawing of the TIROS-N spacecraft and Figure 1.3 shows a photograph of one of the advanced series before launch. Figure 1.4 shows an artist's impression of one of the spacecraft in flight; the AVHRR is at the left-hand side. The direction of flight of the spacecraft is approximately perpendicular to its length and to the local vertical.

In keeping with the general philosophy of making use of previously developed space components, it was decided that the TIROS-N programme would make use of applicable parts of the Defense Meteorological Satellite Program (DMSP) Block 5D spacecraft for the flight of the payload specified by NOAA. The TIROS-N spacecraft is an adaptation of the spacecraft built by the RCA Corporation and first

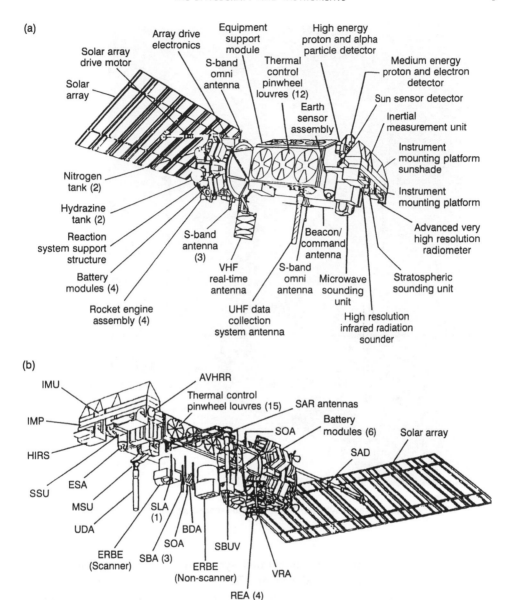

Figure 1.2 Sketch of (a) TIROS-N spacecraft (Schwalb 1978) and (b) Advanced TIROS-N spacecraft (ITT).

launched in 1976. RCA was also responsible for building the TIROS-N/NOAA-A, ..., -G spacecraft bus and testing the fully integrated satellite (including the instruments). Ground rules used during the design were to:

- meet the NOAA mission requirements
- make maximum use of existing Block 5D designs where applicable (excluding the attitude control system)
- change 5D only where necessary or where cost advantages can be shown for the change.

AVHRR
and
HIRS

Figure 1.3 Photograph of an Advanced TIROS-N series spacecraft before launch (ITT).

The TIROS-N satellite is an integrated system designed to provide for, and control injection into, a nominal 833 or 870 km circular Sun-synchronous orbit after separation from the Atlas-E/F launch vehicle. The spacecraft structure consists of four components: (1) the Reaction System Support (RSS) structure, (2) the Equipment Support Module (ESM), (3) the Instrument Mounting Platform (IMP) and (4) the Solar Array (SA). Instruments are located on both the IMP and the ESM. With the exception of the SEM, all instruments face the Earth when the satellite is in mission orientation.

Spacecraft power is provided by a direct energy transfer system whose primary source is a single-axis oriented solar array; the secondary source is a pair of nickel cadmium batteries. The solar array is made up of eight panels of solar cells; these panels are very prominent in Figures 1.3 and 1.4. A solar array drive system causes the array to rotate once per orbit so that the array continuously faces towards the Sun. Current supplied to the satellite through slip rings during daylight portions of the orbit is used to operate the satellite and to charge the batteries. These batteries supply spacecraft power during dark portions of the orbit and augment the array

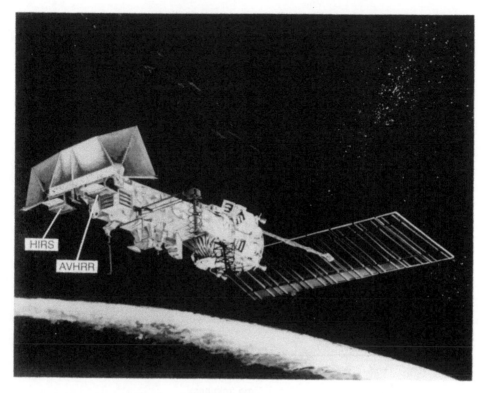

Figure 1.4 Artist's impression of an Advanced TIROS-N series spacecraft in flight (ITT).

during daylight peak load conditions. The total orbit average load capacity for the system is expected to be 420 W at the end of two years in orbit at a worst-case Sun angle.

The spacecraft of the TIROS-N series are three-axis stabilised spacecraft. This is achieved with the on-orbit Attitude Determination and Control Subsystem (ADACS) which provides three-axis pointing control for the satellite. The ADACS maintains system pointing by controlling torque in three mutually orthogonal wheels (a fourth skewed wheel is available in the event of failure of one of these three). The torque is determined by analysis of spacecraft orientation in space. Input to these computations is acquired from the Earth Sensor Assembly (ESA) for pitch and roll, and an inertial reference with Sun sensor updates for yaw. The ADACS is required to control spacecraft attitude so that orientation of the three axes is maintained to within $\pm 0.2°$ (3σ) of the local geographic reference. An indication of the stability actually achieved is given by Figure 1.5, see also Baldwin and Emery (1995). Information to permit computation of yaw, pitch and roll to within $0.1°$ by computer processing on the ground, after the fact, is also available.

The thermal design of the satellite involves accurate temperature control of both the spacecraft structure and the instrument payload. Both active and passive elements are used for this purpose. Passive control is effected by the appropriate use of multilayer insulation blankets, aluminised teflon thermal shielding, special finishes and thermal conduction control materials. The two major elements of the thermal control system are heaters and louvre-controlled cooling radiators. There are two

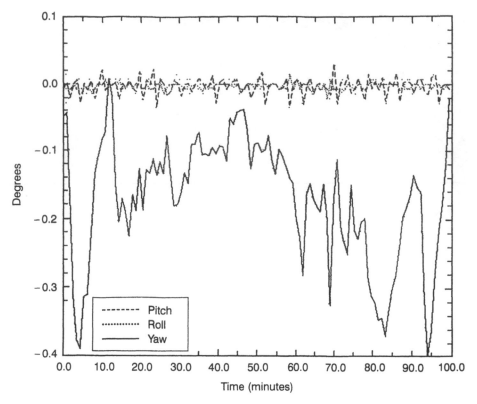

Figure 1.5   Roll, pitch and yaw for one orbit of NOAA-11 on 24 March 1992 (Krasnopolsky and Breaker 1994).

types of louvres, vane and pin-wheel. Control of both types is maintained thermostatically by the Thermal Control Electronics (TCE) unit.

The TIROS-N data handling subsystem consists of four primary components:

- the TIROS Information Processor (TIP) – a low data-rate processor
- the Manipulated Information Rate Processor (MIRP) – a high data-rate processor
- digital tape recorders (DTRs)
- a Cross Strap Unit (XSU).

All data available for transmission to the ground through the lifetime of the mission are processed by some or all of these components, as will be seen in chapter 2.

The spacecraft NOAA-A, NOAA-B, etc., only acquired numbers after their launches; this is to allow for the fact that there was no guarantee in advance either that they would all be launched successfully or that they would be launched in alphabetical order. In fact the numbers acquired are summarised in Table 1.3. We summarise a few of the main features of the launch and operation of each of these spacecraft. The aim is to have two spacecraft operational at a time. Each satellite crosses the equatorial plane at an almost fixed (local solar) time each day; these are 1500 hours (ascending node) and 0300 hours (descending node) for one satellite, and

Table 1.3 Pre-launch and post-launch names of spacecraft

Pre-launch name	Post-launch name	Four- or five-channel AVHRR	Spacecraft ID
TIROS-N	TIROS-N	4	1
NOAA-A	NOAA-6	4	2
NOAA-B*	—	—	—
NOAA-C	NOAA-7	5	4
NOAA-D	NOAA-12	4	5
NOAA-E	NOAA-8	5	6
NOAA-F	NOAA-9	4	7
NOAA-G	NOAA-10	5	8
NOAA-H	NOAA-11	5	1
NOAA-I*	—	—	—
NOAA-J			
NOAA-K			
NOAA-L			
NOAA-M			

1930 hours and 0730 hours for the other. As already mentioned, these times are approximate as there is a slight precession of the orbits from one day to the next.

TIROS-N was launched at 11:23 Z on 13 October 1978 and was operational from 19 October 1978 until 30 January 1980.

NOAA-A (NOAA-6) was launched on 27 June 1979 at 15:52 Z. The satellite greatly exceeded its two-year lifetime and was deactivated on 31 March 1987 after nearly eight years of operational service.

NOAA-B was launched on 29 May 1980 at 10:53 Z and failed to achieve a usable orbit because of a booster engine anomaly.

NOAA-C (NOAA-7) was launched on 23 June 1981 at 10:53 Z and was deactivated in June 1986 following failure of the power system.

NOAA-E (NOAA-8) was launched on 28 March 1983 at 15:52 Z. It was the first of the ATN spacecraft and included a stretched structure, which provided additional capability, and it also included the first SAR[1] (search and rescue) package. The redundant crystal oscillator (RXO) failed after 14 months in orbit. The RXO recovered from its failure, finally locking up on the RXO backup side in May 1985. The spacecraft was stabilised and declared operational by NOAA on 1 July 1985. The satellite was finally lost on 29 December 1985, following a thermal runaway which destroyed the battery.

NOAA-F (NOAA-9) was launched on 12 December 1984 at 10:42 Z and is currently in standby operation. Digital Tape Recorder (DTR) 1A/1B failed two months after launch. The Earth Radiation Budget Experiment (ERBE) Scanner stopped outputting science data in January 1987. The Advanced Very High Resolution Radiometer (AVHRR) has at times exhibited anomalous behaviour in its synchronisation with the Manipulated Information Rate Processor (MIRP), Microwave Sounding Unit (MSU) channels 2 and 3 have failed and the power system is degraded. A Solar Backscatter UltraViolet (SBUV/2) instrument is also on board and is operating

---

[1] Note that in this context SAR refers to search and rescue and not to synthetic aperture radar (as it commonly does elsewhere in remote sensing).

satisfactorily. The satellite is collecting, processing and distributing SBUV/2 and ERBE-Nonscanner (NS) data. It is also providing real-time search and rescue data.

NOAA-G (NOAA-10) was launched on 17 September 1986 at 15:52 Z into a morning orbit and it is currently transmitting data for local weather analysis directly to users around the world. All instruments and subsystems are performing well except the ERBE-Scanner, which has exhibited a scan sticking anomaly that is apparently generic to the instrument, and the Search and Rescue Processor (SARP) 406 MHz receiver, which has failed.

NOAA-H (NOAA-11) was launched on 24 September 1988 at 10:02 Z into an afternoon orbit with a 1:40 p.m. ascending node crossing time. It is currently trans-mitting data for local weather analysis directly to users around the world. All instru-ments are operational. The NOAA-H had been modified for a 0° to 80° Sun angle and includes fixed and deployable sunshades on the Instrument Mounting Platform (IMP) and the capability for a deployable Medium Energy Proton and Electron Detector (MEPED). The increase of maximum Sun angle from 68° to 80° allows an afternoon nodal crossing closer to noon to enhance data collection. Two gyros have failed, however, and attitude control is being maintained through the use of new reduced gyro flight software. In addition, before the NOAA-D launch, a gyroless flight software package was installed on NOAA-11 which will provide attitude control, at expected reduced accuracy, should the X gyro fail.

NOAA-D (NOAA-12) was launched into a morning orbit on 14 May 1991 at 15:52 Z and is functioning well. It replaced NOAA-10 in orbit. However, it does not contain a search and rescue package.

For a discussion of NOAA-I see section 1.5.

1.2 THE ADVANCED VERY HIGH RESOLUTION RADIOMETER (AVHRR)

1.2.1 Origins of the AVHRR

The AVHRR has been produced in various versions, a four-channel version flown on TIROS-N, a slightly different four-channel version flown on NOAA-6, -9 and -12, and a five-channel version flown on NOAA-7, -8, -10 and -11 ... (see Table 1.3). The approximate wavelength ranges of the spectral channels of the AVHRR are indi-cated in Table 1.4. Channels 1 and 2 are designed, and calibrated before launch, to provide direct quasi-linear conversion between the 10-bit digital numbers and the albedo. In addition the thermal channels are designed, and calibrated before launch as well as in space, to provide direct quasi-linear conversion between the digital numbers and the temperature in degrees Celsius (or Kelvin). As the thermal infrared channels were optimised for measuring the skin temperature of the sea surface, their range is approximately -25 to $+49°C$ for channel 3, -100 to $+57°C$ for channel 4 and -105 to $+50°C$ for channel 5. The main reason for the introduction of the five-channel version is in connection with atmospheric corrections in the determi-nation of sea surface temperatures (see section 4.2.3). A very brief account of the AVHRR is given by Hastings and Emery (1992). They point out that it has been the policy of NOAA to meet operational requirements with instruments whose potential has been proved in space. Predecessor instruments were flown on experimental satellites before they were accepted and flown on operational monitoring satellites. These instruments were redesigned to meet both the scientific and technical require-

Table 1.4 Channels and band widths on AVHRR (μm)

Channel	TIROS-N	NOAA-6, -9, -12	NOAA-7, -8, -10, -11, -I, -J
1	0.55–0.90	0.58–0.68	0.58–0.68
2	0.725–1.10	0.725–1.10	0.725–1.10
3	3.55–3.93	3.55–3.93	3.55–3.93
4	10.50–11.50	10.50–11.50	10.30–11.30
5	Channel 4 repeated in both		11.50–12.50

ments of the mission; the aim was to improve the reliability of the instruments and the quality of the data without changing the previously proved measurement concepts. This philosophy brings both benefits and challenges to the user. The benefits are centred around relative reliability, conservative technology, continuity of access and application of the data compared to other satellite systems. The challenges include desires to use the system beyond its original design and conflicting desires by users for greater support for their own particular scientific disciplines with more advanced sensors and more sophisticated customer support. A regular feature of information about the AVHRR and examples of the use of AVHRR data, edited by David Hastings, Michael Matson and Andrew H. Horvitz, was started in the journal *Photogrammetric Engineering and Remote Sensing* in January 1988 in volume 54.

There were two radiometers, the Scanning Radiometer (SR) and the Very High Resolution Radiometer (VHRR), on the satellites of the second series of polar-orbiting meteorological satellites (see section 1.1). Each of these instruments was a two-channel instrument operating in the visible and thermal infrared wavelength ranges and the output was downlinked as analogue data. In the mid-1970s it was decided, as a joint NASA/NOAA initiative, to commission a new and better scanner. This scanner, which became known as the Advanced Very High Resolution Radiometer (AVHRR), was designed and built by ITT Aerospace. In about 1976 ITT produced an engineering model leading to the launch of the first instrument in the series into space on TIROS-N in October 1978. Subsequent instruments in the series have been built by ITT Aerospace under contract to NASA which, in turn, procured the instruments on behalf of NOAA. There are two series of AVHRR instruments. The AVHRR/1 is a four-channel, filter-wheel spectrometer/radiometer while the AVHRR/2, built in the early 1980s, is identical except for the addition of channel 5. Calibration of each AVHRR instrument was carried out by ITT (see section 2.2.2). It should, however, perhaps be made clear that the AVHRR was originally designed as an imaging radiometer, which happened to be rather well calibrated; it was not really designed or intended to be used as an absolute radiometer. That the AVHRR is coming more and more to be used for absolute radiometric work can be regarded as a fortunate accident or as a largely unforeseen consequence of good design.

1.2.2 Construction of the instrument

Details of the construction of the AVHRR are given in various contractor reports by ITT to NASA but reasonably short summaries are given by Schwalb (1978), Lauritson *et al.* (1979) and Foote and Draper (1980). The instrument has overall

Figure 1.6   Photograph of an AVHRR instrument (ITT).

dimensions of approximately 25 cm by 36 cm by 76 cm, weighs 27 kg and consumes about 25 W. The photograph in Figure 1.6 shows one of the AVHRR instruments and Figure 1.7 illustrates the component modules diagrammatically. For orientation purposes, when the instrument is on an orbiting spacecraft, its nadir view would be up in this figure. The AVHRR is mounted at the left-hand end of the spacecraft as seen in Figure 1.4.

The instrument scans in the east–west direction with a continuously rotating scan mirror whose surface makes a 45° angle with the telescope axis. This is towards the left-hand end of Figure 1.6 or Figure 1.7. The AVHRR achieves a daily observation cycle with far off-nadir views. Other satellite remote sensing systems, which were designed for land studies, have very limited scan swaths. The Landsat MSS only sweeps a 12–15° swath (6–7.5° from nadir), the Thematic Mapper sweeps 15° (7.5° from nadir) and the SPOT HRV sensor, in nadir mode, only views a swath within 2.5° of nadir. The AVHRR sensor views a swath of over 110°, viewing over 55° off nadir. The AVHRR instrument's spatial resolution is approximately 1.1 km when the view is at nadir. Scanning to 55° (= 68° look zenith angle relative to the Earth's surface) off nadir produces an effective ground resolution of over 2.4 km by 6.5 km at the maximum off-nadir position. There is also significant overlap between adjacent off-nadir pixels which causes the off-nadir observations to be highly redundant (see Figure 3.8). The rotation of the scan mirror causes the scene below the spacecraft to be observed in a continuous line from horizon to horizon. The scan rate and

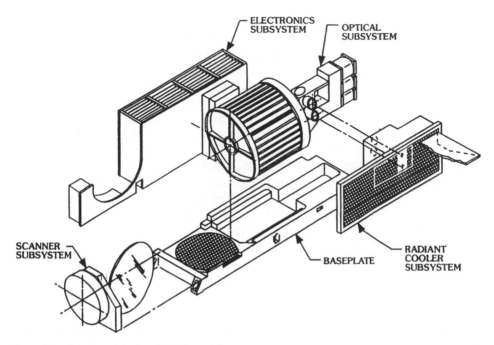

Figure 1.7 Diagram showing AVHRR modules.

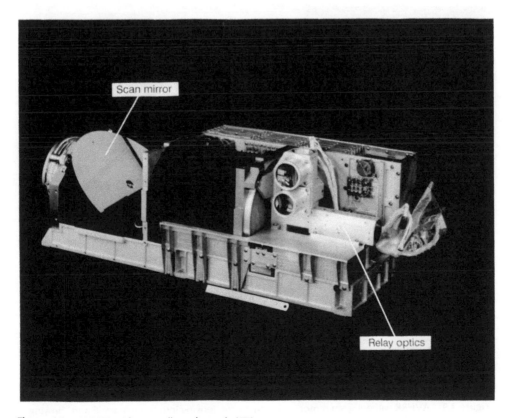

Figure 1.8 AVHRR without walls and panels (ITT).

the instrument field of view are chosen to ensure that consecutive scan lines are contiguous at the subsatellite point; adjacent scans increasingly overlap towards the horizon. The north–south component of the scan pattern is accomplished by the spacecraft motion along the orbit track. The mirror can be seen more clearly in the photograph in Figure 1.8; the black cylinder in the centre is the telescope, while Figure 1.9 shows the instrument without the scan motor and scan mirror.

The signal from the instrument during each rotation of the scan mirror is as shown in Figure 1.10. The sequence begins at the 'line-synchronisation signal'. This signal is produced by a magnetic pickup each time the scan mirror passes this physical position and is sent to the spacecraft central processor to initiate each scan sequence. Next, the scan mirror position causes the detector to look into deep space. During this time, the detector electronics are clamped to the radiance input as a reference level. While the scan mirror rotates through the next portion of the scan, a ramp calibration signal is inserted into the data stream. This ramp increases 6.25 mV each scan line so that it goes from zero to the maximum extent of the radiometer output in 1024 scan lines. It then returns to zero and begins the ramp again. This system is a useful diagnostic tool for examining the health of the electronics if data problems occur. During the Earth scan region the Earth observation data are obtained; then in the back-scan region the housekeeping telemetry is multiplexed into the data stream. Finally, the scan mirror looks at the internal calibration target region (see Figure 1.9) and then sweeps on to begin the sequence again.

Figure 1.9   AVHRR without scan motor and scan mirror, looking into the telescope and showing the onboard blackbody (internal calibration target) for the thermal infrared channels at the front right.

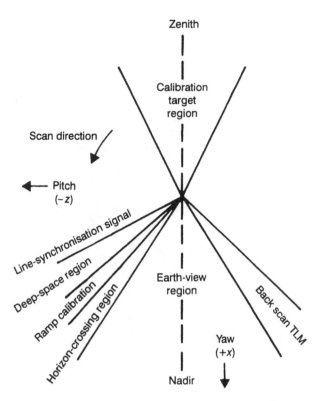

Figure 1.10 Signal position as a function of scan mirror angle (Foote and Draper 1980).

The exploded view of the AVHRR (Figure 1.7) shows that the instrument has a modular construction that permits the interchanging of modules during construction if any serious problems are encountered. The instrument consists of a baseplate, an optical subsystem, a radiant cooler, a scanner assembly and an electronics unit.

The baseplate is essentially an optical bench to which everything else is attached. The darkened area on the bottom of the scan cavity is the onboard temperature target, which the radiometer views during each scan line. Four precision platinum resistance thermometers are mounted in this target and their outputs are multiplexed into the data stream during each scan in the back-scan telemetry region referred to previously.

The scanner assembly consists of an 80-pole hysteresis synchronous motor that uses a set of duplex bearings. The scanner operates at 360 rpm and exceeds the jitter requirements, which call for less than 17 μs of jitter from scan line to scan line. The scan mirror is made from beryllium with a waffle construction for lightness and high rigidity and is attached to the motor shaft. This entire assembly is dynamically balanced to 706 dyne-cm at 800 rpm. The mirror blank is nickel plated after machining to provide a good polished surface. After polishing, the mirror is aluminised and overcoated for high reflective characteristics.

The optical subsystem of the five-channel version, AVHRR/2, minus the scan mirror, is shown schematically in Figure 1.11. The energy from the scene is collected by a 20 cm diameter clear aperture afocal telescope, the primary (M1) being the entrance aperture. Dichroic 1 (D1) transmits channels 1 and 2 (visible and near-infrared channels) and reflects channels 3, 4 and 5 (infrared channels). The beam that

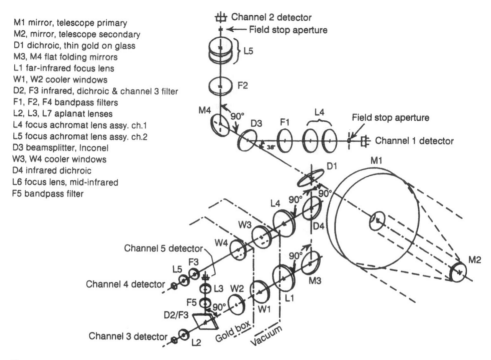

M1 mirror, telescope primary
M2, mirror, telescope secondary
D1 dichroic, thin gold on glass
M3, M4 flat folding mirrors
L1 far-infrared focus lens
W1, W2 cooler windows
D2, F3 infrared, dichroic & channel 3 filter
F1, F2, F4 bandpass filters
L2, L3, L7 aplanat lenses
L4 focus achromat lens assy. ch.1
L5 focus achromat lens assy. ch.2
D3 beamsplitter, Inconel
W3, W4 cooler windows
D4 infrared dichroic
L6 focus lens, mid-infrared
F5 bandpass filter

Figure 1.11   AVHRR optical subsystem five-channel version, AVHRR/2.

is transmitted by dichroic 1 is subsequently separated by the beam splitter (D3), which reflects channel 1 and transmits channel 2. The angle of reflection at D3 was chosen to minimise the effects of polarisation in channel 1. Elements F1 and F2 are absorbing spectral filters for channels 1 and 2, respectively, and lens assemblies L4 and L5 are focusing elements that form images at the field-stop aperture of each detector. The field-stop apertures are in plates located in front of silicon pin photo diodes that are used in both solar channels. The beam that is reflected at dichroic 1 (D1) is then separated at dichroic 4 (D4) into one reflected beam (channel 4) and one transmitted beam (channels 3 and 5). Channels 3 and 5 are then separated by dichroic 2 (D2). The various bandpass filters (F3, F4 and F5) are of multilayer interference-type construction. Channel 3 is the 3.5–3.9 μm channel, which uses an indium antimonide photo-diode detector, while channels 4 and 5 in the 10.3–12.5 μm range use biased mercury cadmium telluride detectors.

The instantaneous field of view of each channel is a square, 1.3 milliradians on a side, and the channel-to-channel registration is within 0.1 milliradians after all environmental testing is complete. The signal-to-noise ratio of the solar channels is specified as 3 to 1 at 0.5% albedo and the noise equivalent temperature differences (NEΔT) of the infrared channels are specified as 0.12 K at 300 K.

The detectors for the three infrared channels are operated at 107 K and this is part of the reason for their physical separation from the detectors for channels 1 and 2 (see Figure 1.11). This temperature is achieved by the radiant cooler which is at the front of the instrument in Figure 1.6. The cooler construction is shown detached in Figure 1.12(a) and an exploded view is given in Figure 1.12(b). The radiant cooler consists of four separate components, (a) the cooler housing, (b) the first stage

(a)

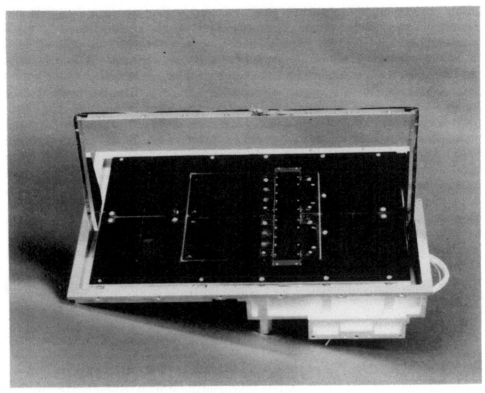

(b)

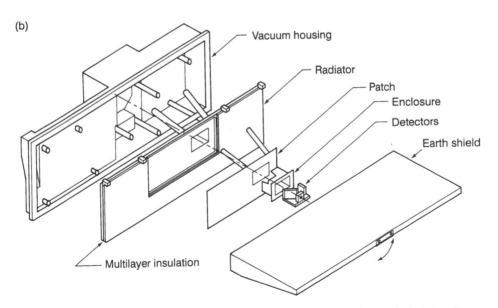

Figure 1.12   Radiant cooler (a) photograph (without Earth shield) (ITT) and (b) exploded view (Foote and Draper 1980).

radiator (or radiator), (c) the second stage radiator (or patch) and (d) the Earth shield. The vacuum housing is physically attached to the instrument baseplate and is normally at 15°C. The radiator is attached to the vacuum housing by 13 synthane support tubes that have very poor thermal conductance characteristics. Multilayer insulation blankets minimise the thermal radiation between the vacuum housing and the radiator. The radiator and the patch have a 2π steradian clear view to space, except for the solid angle subtended by the Earth. The Earth radiation is shaded from the cooler by the Earth shield (at the top in Figure 1.6). The Earth shield is also used as a protective cover before it is deployed to keep the Sun from impinging on the cooler during spacecraft acquisition manoeuvres. The Earth shield is then deployed on command after orbit is attained. The radiator and the Earth shield, which is attached to the radiator, normally run at 107 K. The patch, which includes the detector/filter assembly, is attached to the radiator by four synthane support tubes. The radiation between the patch and the radiator is minimised by low-emissivity finishes on the surfaces of both structures. The portion of the patch that views space has a highly emissive honeycomb surface. The patch temperature is controlled through the system being designed so that the patch's temperature would fall below the required operating temperature and then have sufficient energy supplied by a heater to maintain the patch at the required operating temperature of 107 K.

1.2.3 Data collection and transmission

The spacecraft provides 28 V, a clock and commands to the instrument. The instrument electronics module then provides the command storage, power conversion and regulation, timing and control signal generation, signal amplification and the analogue-to-digital (A/D) conversion necessary for performing the instrument functions.

The detector signals in the various channels are amplified, filtered and then applied to the 10-bit A/D converter, which samples all five channels simultaneously every 25 µs. The converter operates in the track mode for 9 µs before the simultaneous sampling occurs and then converts each of the five channels sequentially with a basic 3 µs conversion time. The time allowed for the channel 5 conversion is increased to 4 µs to minimise the data storage and gating circuits required. The A/D converter provides for six input channels: five data channels and a sixth channel that switches in telemetry data at appropriate times in the scan line. Because the A/D converter design anticipated the five-channel AVHRR from the beginning, the four-channel AVHRR samples channel 4 twice to fill all the appropriate data words. The instrument has 26 separate command functions and 20 analogue telemetry points for housekeeping. It also has a digital telemetry feature that continuously indicates the status of all the commands.

The 10-bit-resolution digital data from the AVHRR is processed in the spacecraft's Manipulated Information Rate Processor (MIRP), see Figure 1.13, to produce the following final four AVHRR products:

- direct readout to ground stations of High Resolution Picture Transmission (HRPT) worldwide
- direct readout to ground stations of Automatic Picture Transmission (APT) worldwide

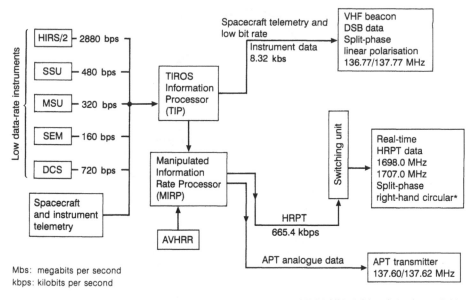

Figure 1.13   TIROS-N/NOAA satellites real-time systems data flow (Lauritson *et al.* 1979).

- Global Area Coverage (GAC) at relatively low resolution (4 km) for central processing

- Local Area Coverage (LAC) from selected portions of each orbit at high resolution (1 km) for central processing.

The spacecraft carries digital tape recorders to record some of these data. The tape-recorded data can then be downlinked when the spacecraft is near to one of the North American Command and Data Acquisition (CDA) ground stations operated by NOAA, see Figure 1.14 (see also section 2.3). Each of the four AVHRR products will now be described in a little more detail.

- HRPT. The High Resolution Picture Transmission is used for real-time transmission by an S-band transmitter. This product contains the data as they are produced from the AVHRR along with some other housekeeping telemetry. The spatial resolution of the image data is about 1 km.

- APT. The Automatic Picture Transmission selects any two AVHRR channels and is used for real-time transmission by a VHF transmitter. This product uses only one out of every three scan lines from the AVHRR and averages the data along the scan line to reduce the resolution and correct the geometric distortion in the raw data. The resolution of the data is reduced from 1 km to about 4 km for this product.

- GAC. The Global Area Coverage is data from an entire orbit that is stored on a digital tape recorder for transmission to a central data station. This product uses only one out of every three AVHRR scan lines and averages four adjacent samples to produce one data point. The resolution of these data is approximately 4 km.

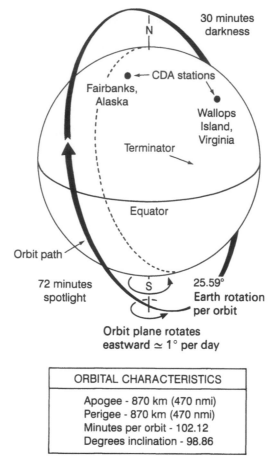

Figure 1.14 Typical NOAA satellite orbit, showing the Command and Data Acquisition (CDA) stations (ITT).

- **LAC.** The Local Area Coverage is essentially HRPT data recorded for about 10 minutes anywhere in the orbit for transmission to a central data station.

Complete global recovery of GAC data from the on-board tape recorders is available, but complete recovery of HRPT/LAC data is only possible by a combination of the use of the on-board tape recorders and a global distribution of direct readout ground stations (see sections 2.5 and 7.9.6).

We have mentioned already that there are 14 orbits of the spacecraft per day, leading to an approximate separation of equator crossings of about 26° as the Earth rotates beneath the (essentially) stationary satellite orbit. Given the large scan angle of the AVHRR, this leads to a slight overlap at the equator between the swath (the area scanned) for one orbit and the swath for the next orbit. Each satellite therefore gives global coverage once every 24 hours. At higher latitudes the overlap between the swaths for adjacent orbits increases until at the poles there is a very large overlap and the poles are scanned in every orbit. With two satellites in operation, global coverage is available for any point on the Earth at least once every 12 hours and considerably more frequently at higher latitudes.

It can be seen from Tables 1.3 and 1.4 that the four-channel and five-channel versions of the AVHRR have been flown in an irregular manner on the spacecraft NOAA-6, -7, -8, -9, -10, -11 and -12. What happened was that a number of four-channel versions of the instrument were procured. The five-channel version, which is sometimes described as AVHRR/2, was developed and ready to be flown before all the four-channel versions had been flown. Because of the very great importance of the five-channel version in connection with making atmospheric corrections to sea surface temperature data (see section 4.2.3), it was decided to fly the five-channel version before all the four-channel versions had been flown. However, because of the very substantial cost of the instruments it was decided not to scrap the unflown four-channel instruments but to fly them alternately with the five-channel versions on successively launched spacecraft.

There have recently been some further developments in the AVHRR and a new version of the instrument will be flown on NOAA-K, -L and -M. This will be a six-channel instrument, the AVHRR/3, having the same five channels as the present version together with one additional short wavelength infrared channel (1.3–1.9 µm), see Figure 1.15. However, it was decided to impose a minimum dislocation both to existing receiving and archiving systems and to existing software for processing AVHRR data. This is to be achieved by designating the channels

- centred at 1.6 µm as channel 3A

- centred at 3.7 µm as channel 3B (= existing channel 3)

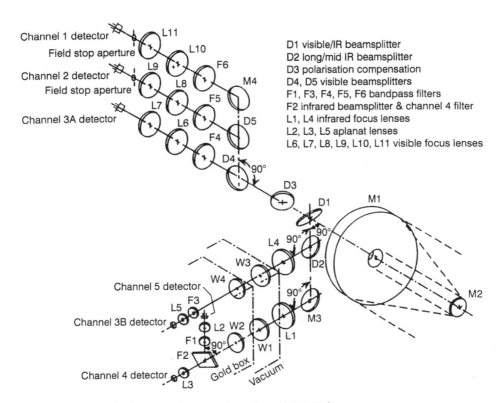

Figure 1.15 Optical subsystem of proposed six-channel AVHRR/3.

and making the data from these two channels share the position occupied by channel 3 data in the present data structure. At any one time *either* channel 3A data *or* channel 3B data will be transmitted and it will occupy the position presently occupied by channel 3 data in the transmitted data stream (see section 2.1); a flag will be set in the header, using a slot that presently is spare, to indicate whether it is channel 3A or channel 3B data that follows later in that frame. It is intended that the switching between the transmission of channel 3A and channel 3B data will be regular and automatic. Channel 3B will be transmitted during the night-time (because of its importance at night in relation to sea surface temperature determination, see section 4.2.3) and during the hours of daylight in the morning. The new channel, channel 3A, had been developed primarily for studying aerosols, and the light levels are rather low for this in the morning. Thus, the channel 3A data will only be transmitted during the hours of daylight in the afternoon. The decision to accommodate this development without a major disruption to the format of the data stream does mean that some data will be lost. However, it is a recognition of the massive investment that has been made worldwide in terms of data reception and archival and of processing software.

There is one other change that will also be made to the AVHRR instrument for NOAA-K, -L and -M. This will be to introduce a split gain for channels 1, 2 and 3A; this will enable good use to be made of the 1024-level resolution of the 10-bit data for low intensities of radiation while reducing the risk of saturation for high intensities of radiation. An indicator (flag) will be added to the header to indicate which gain applies to the data. Although this technique of variable gain has been used previously with other multispectral scanners, e.g. the Coastal Zone Colour Scanner (CZCS) on Nimbus-7, it has not been used on the first or second version of the AVHRR.

These developments take us up to and including NOAA-M and to the end of the century. Discussion of developments from 2000 and covering NOAA-N and -N' will be postponed until the end of section 1.5.

1.2.4   Use of AVHRR data

We have already dismissed the meteorological uses of AVHRR as far as this book is concerned. We do aim, however, to cover the main present non-meteorological applications of AVHRR data in this book but we shall not attempt to introduce or summarise these here. We would, however, just mention at this stage two general recent reviews of some uses of AVHRR data by Hastings and Emery (1992) and Ehrlich *et al.* (1994) and make two other general points at this stage.

First, a number of AVHRR data users' conferences have been held. The papers from an early one in Europe were published as a special double issue of the *International Journal of Remote Sensing* (volume 10, numbers 4 and 5, 1989). Papers from several subsequent conferences (Rothenburg, Germany, 5–8 September 1989; Tromsø, Norway, 25–28 June 1991; Belgirate, Italy, 29 June–2 July 1993 and Winchester, UK, 4–8 September 1995) have been published by EUMETSAT (Am Elfengrund 45, D-6100 Darmstadt, Germany). The first Australian AVHRR Conference was in Perth, WA, 22–24 October 1986 (proceedings from Dr. Fred Prata, CSIRO Division of Groundwater Research, Private Bag, P.O., Wembley, WA 6014, Australia) and the first North American NOAA Polar Orbiter Users Group meeting was in Boulder, Colorado, 14–16 July 1987 (proceedings from National Geophysical Data Center, NOAA, Code E/GC4, 325 Broadway, Boulder, Co, 80303, USA).

Secondly, a number of AVHRR data user groups have been formed in various parts of the world. NOAA/NESDIS in Washington, DC, attempts to maintain an up-to-date list of contact names and addresses for these (contact: Mary F. Hughes, Data Collection and Direct Broadcast Branch, NOAA/NESDIS, World Weather Building, Washington, DC 20233, USA).

1.3 THE ARGOS DATA COLLECTION AND PLATFORM LOCATION SYSTEM

We now turn to a description of the ARGOS data collection and platform location system which is flown on each of the satellites of the TIROS-N/NOAA series. An overview of the system is given in Figure 1.16.

Although the major part of the data transmitted on the communications link from a meteorological satellite back to Earth consists of the data that the instruments on the satellite have gathered, some of these satellites also fulfil a communications role. For example, the geostationary satellite Meteosat serves as a communications satellite to transmit processed Meteosat data and other geostationary satellite data to users from the European Space Operations Centre (ESOC) at Darmstadt in Germany. Another aspect of the use of remote sensing satellites for communications purposes that is of particular relevance to environmental scientists and engineers is that some satellites carry data collection systems. Such systems enable them to collect data from instruments situated in difficult or inaccessible locations on the land or sea surface. Such instruments may be at sea on a moored or

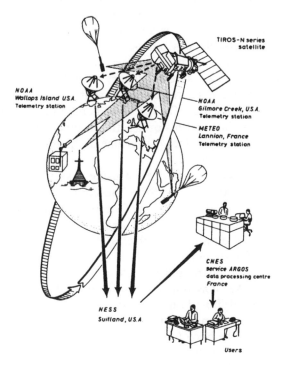

Figure 1.16 Overview of the ARGOS data collection and platform location system (System ARGOS).

drifting buoy or on a weather station in a hostile or otherwise inaccessible environment, in the Arctic, or in deserts, and so on.

There are several options for recording and retrieving data from an unmanned data gathering station such as a buoy or an isolated weather station or hydrological station; these include using

- cassette tape recorders with occasional visits to change the tapes
- a direct radio link to a receiving station conveniently situated on the ground
- a radio link via a satellite.

The first option may be satisfactory if the amount of data received is relatively small, but if there is a large quantity of data and one is only able to visit the site occasionally this may not be very suitable. It is not at all suitable if the data are needed in near-real time. The second option may be satisfactory over short distances but becomes progressively more difficult over longer distances. The third option has some attractions and is worth a little further consideration here. There are two satellite-based data collection systems which are of importance. One involves the use of a geostationary satellite such as Meteosat, while the other is the ARGOS system. The Meteosat system differs from the ARGOS system in one very important way. This is that the platform location capability, which is available with the ARGOS system, does not exist with the Meteosat system. This is because the location facility relies on a Doppler shift of the frequency of the carrier radio wave and for a geostationary satellite there is no relative motion of the satellite and the Earth and therefore there is no Doppler shift.

To some extent the Meteosat and ARGOS data collection systems are complementary. A data collection system using a geostationary satellite, such as Meteosat, has the advantage that the satellite is always overhead and therefore always available, in principle, to receive data. A polar-orbiting satellite will, however, be out of sight of the data collection platform a great deal of the time. However, a data collection system cannot be used on a geostationary satellite if the data collection platform is situated in extreme polar regions because it will be out of the line of sight from the satellite, see Figure 1.17. On the other hand a data collection system that uses a polar-orbiting satellite will perform better in polar regions because the satellite will be in sight of a platform that is near one of the poles much more frequently than a platform near the equator.

There are several possible advantages in using a satellite and not just using a direct radio transmission from the platform housing the data-collecting instruments to the user's own radio receiving station. One of these is simply convenience. It saves on the cost of reception equipment and of operating staff for a receiving station of one's own; it also simplifies problems of frequency allocations. There may, however, be the more fundamental problem of distance. If the satellite is orbiting, it can store the messages on board and play them back later – perhaps on the other side of the Earth. The ARGOS system accordingly enables someone in Europe to receive data from buoys drifting in the Pacific Ocean or in Antarctica, for example. As well as recovering data from a drifting buoy the ARGOS system can also be used to locate the position of the buoy (or of any other moving platform). The platform location facility is described, for instance, by Cracknell and Hayes (1991). The ARGOS system is particularly useful for gathering *in situ* data for calibrating algorithms for the extraction of geophysical parameters (e.g. sea surface temperature, see

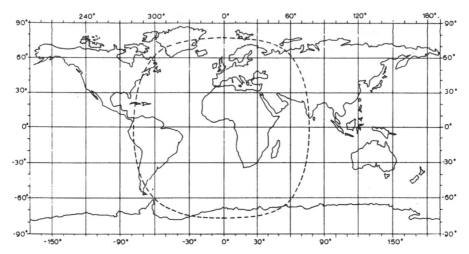

Figure 1.17 Meteosat reception area (European Space Agency).

section 4.2) from AVHRR data; this is because, being flown on the same satellite as the AVHRR, it is guaranteed to gather the *in situ* data simultaneously with the acquisition of the AVHRR data.

The ARGOS platform location and data collection system is the result of a cooperative programme between the French Centre Nationale d'Etudes Spatiales (CNES) and two US organisations, the National Aeronautics and Space Administration (NASA) and the National Oceanic and Atmospheric Administration (NOAA). The ARGOS system's main mission is to provide an operational environmental data collection service for the entire duration of the NOAA TIROS-N programme, i.e. from 1979 to 2000 at least. After several years of operational service the efficiency and reliability of the ARGOS system has been demonstrated very successfully.

The ARGOS system can be considered as comprising four segments:

- the set of all users' platforms (buoys, balloons, fixed or offshore stations, animals, etc.), each being equipped with a PTT (Platform Transmitter Terminal)
- the onboard Data Collection System (DCS) flown on each satellite of the series TIROS-N, NOAA-6, -7, -8, etc.
- the ARGOS data-processing centre in Toulouse
- the distribution system for results.

As an alternative to the official distribution system for the results one can also extract the ARGOS data from the direct readout HRPT data if one has access to a ground receiving station.

1.4 THE TIROS OPERATIONAL VERTICAL SOUNDER (TOVS) SYSTEM

1.4.1 The TOVS package

The TIROS Operational Vertical Sounder (TOVS) was a successor to the sounding instruments that were flown on the previous generations of polar-orbiting meteorological satellites. Although the TOVS and atmospheric sounding are not central to the theme of this book we shall, nevertheless, consider this subject briefly in this section because the TOVS data are sometimes used for the atmospheric correction of AVHRR data. The TOVS system consists of three separate and independent instruments and the data from them are combined for the computation of atmospheric profiles. The three instruments are

- the High resolution Infrared Radiation Sounder (HIRS/2)
- the Stratospheric Sounding Unit (SSU)
- the Microwave Sounding Unit (MSU)

and they are described in some detail by Schwalb (1978) and Lauritson et al. (1979).

The TOVS was designed for the determination of (i) atmospheric profiles, that is curves of temperature, T, as a function of pressure, P, from the surface to 10 mb, (ii) the water vapour content at three levels in the atmosphere and (iii) total ozone content. This information was traditionally obtained from radiosondes. Radiosondes have the advantage that they measure the atmospheric profiles of the pressure, temperature and humidity (and sometimes even the ozone concentration) directly. However, the spatial coverage of the Earth's surface is irregular. There are large numbers of radiosonde balloon launches in the densely populated and industrialised countries, relatively fewer in developing countries and almost none in the polar regions and over the oceans. A satellite sounding system has the advantage of giving a much more regularly spaced coverage and this is particularly important if one is seeking to supply data for a regular grid of points for a numerical weather-forecasting or climate-prediction model. The disadvantage of the satellite sounding instruments is that they do not measure the required meteorological parameters directly. They are basically low-resolution scanners and the meteorological parameters have to be retrieved by a complicated inversion procedure that is very demanding in terms of computing resources. The atmospheric sounders on the previous generation of polar-orbiting satellites generated atmospheric temperature profiles that varied from radiosonde measurements by about 2.5°C r.m.s; the TOVS was intended to reduce this error to about 1.5°C r.m.s.

1.4.2 The HIRS/2

The High resolution Infrared Radiation Sounder (HIRS/2) is a development from the HIRS/1 instrument that had earlier been flown on Nimbus-6. The HIRS/2 was also built by the Optical Division of ITT Aerospace. One can think of the HIRS/2 as a scanner of very much lower spatial resolution than the AVHRR but with many more spectral channels, with all except one of them in the infrared. Its characteristics as a scanner are summarised in Table 1.5. It is instructive to compare these characteristics with those of the AVHRR. The scan angle of the HIRS/2 is $\pm 49.5°$, which is only slightly less than that of the AVHRR ($\pm 55°$), while the number of steps (pixels) for the HIRS/2 along one scan line is only 56 whereas there are 2048 steps (pixels) along one scan line of the AVHRR. The HIRS/2 is a 20-channel instrument that measures radiation in 19 regions of the infrared and one region of the visible parts of the spectrum. The spectral characteristics are shown in Table 1.6. The radiance in each of the spectral channels of the HIRS/2 can be determined from the calibration of the instrument. This calibration is very similar to that of the

Table 1.5 HIRS/2 system parameters (Schwalb 1978)

Parameter	Value
Calibration	Stable blackbodies (2) and space background
Cross-track scan	$\pm 49.5°$ (± 1120 km)
Scan time	6.4 s
Number of steps	56
Optical FOV	1.25°
Step angle	1.8°
Step time	100 ms
Ground IFOV (nadir)	17.4 km diameter
Ground IFOV (end of scan)	58.5 km cross-track by 29.9 km along-track
Distance between IFOVs	42 km along-track
Data rate	2800 bits/s

Table 1.6 HIRS/2 instrument specifications (Schwalb 1978)

Channel	Central wave number (cm^{-1})	Half power band width (cm^{-1})	Wavelength (μm)	Specified NEΔN $(mW\ m^{-2}\ sr^{-1}\ cm)$
1	688.5 ± 1.3	$3.0^{+1}_{-0.5}$	14.96	0.80*
2	680.0 ± 1.8	10.0^{+4}_{-1}	14.71	0.27
3	690.0 ± 1.8	12.0^{+6}_{-0}	14.49	0.27
4	703.0 ± 1.8	16.0^{+4}_{-2}	14.22	0.22
5	716.0 ± 1.8	16.0^{+4}_{-2}	13.97	0.22
6	733.0 ± 1.8	16.0^{+4}_{-2}	13.64	0.22
7	749.0 ± 1.8	16.0^{+4}_{-2}	13.35	0.22
8	900.0 ± 2.7	35.0 ± 5	11.11	0.11
9	1030.0 ± 4	25.0 ± 3	9.71	0.16
10	1225.0 ± 4	60.0^{+10}_{-3}	8.16	0.16
11	1365.0 ± 5	40.0 ± 5	7.32	0.22
12	1488.0 ± 4.7	80.0^{+15}_{-4}	6.72	0.11
13	2190.0 ± 4.4	23.0 ± 3	4.56	0.002
14	2210.0 ± 4.4	23.0 ± 3	4.52	0.002
15	2240.0 ± 4.4	23.0 ± 3	4.46	0.002
16	2270.0 ± 4.7	23.0 ± 3	4.41	0.002
17	2360.0 ± 4.7	23.0 ± 3	4.24	0.002
18	2515.0 ± 5	35.0 ± 5	3.98	0.002
19	2660.0 ± 9.5	100.0 ± 15	3.76	0.001
20	$14\,500.0 \pm 20$	1000.0 ± 15	0.69	0.15 Albedo

* 1.70 most likely achievable

thermal infrared channels of the AVHRR (see section 2.2) but the HIRS/2 calibration will not be described in this book; there is an appropriate NOAA Technical
Memorandum describing the calibration (Lauritson *et al.* 1979, Planet 1988).

1.4.3   The SSU

The SSU was designed by the UK Meteorological Office; it employs a selective
absorption technique to make measurements in three channels. These three channels
all work at the same wavelength but they involve absorption in cells of CO_2 at

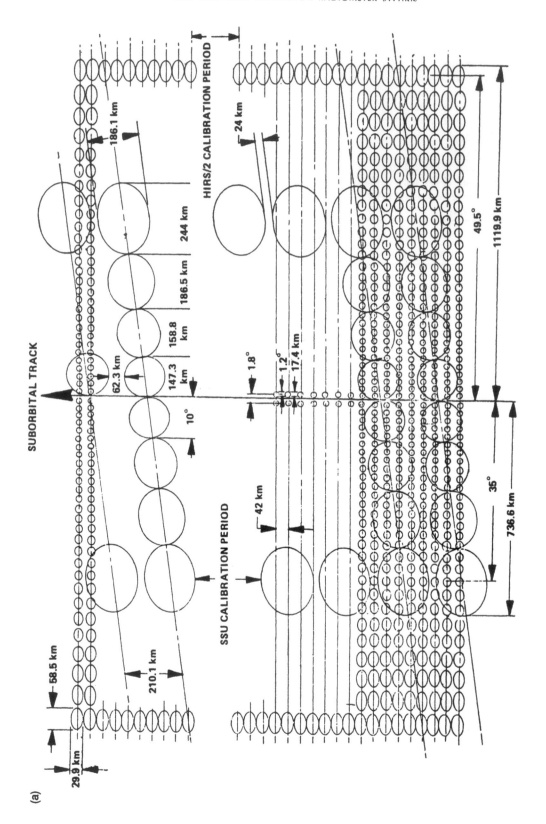

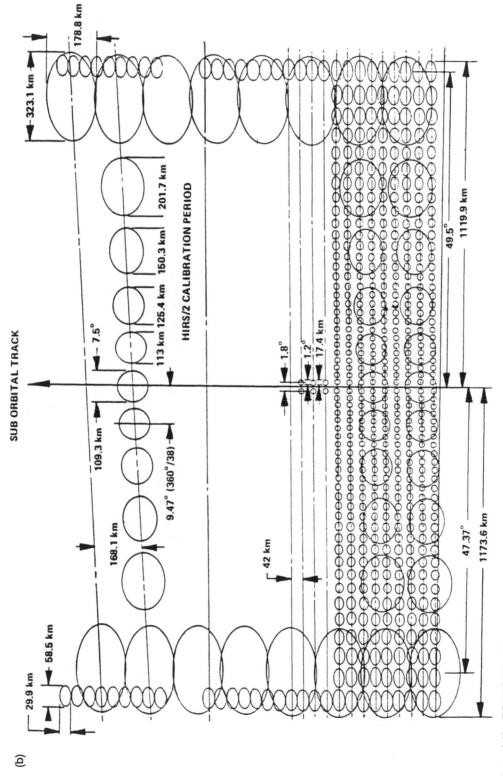

Figure 1.18 TOVS field of view patterns projected on Earth for (a) HIRS/2 and SSU and (b) HIRS/2 and MSU (Schwalb 1978).

Table 1.7   SSU characteristics (Schwalb 1978)

Channel Number	Central wave number (cm^{-1})	Cell pressure (mb)	Pressure of weighting function peak	
			mb	km
1	668	100	15	29
2	668	35	5	37
3	668	10	1.5	45
Calibration		Stable blackbody and space		
Angular field-of-view		$10°$		
Number of Earth views/line		8		
Time interval between steps		4 s		
Total scan angle		$\pm 40°$ from nadir		
Scan time		32 s		
Data rate		480 bits/s		

different pressures, see Table 1.7. The spectral characteristics are determined by the pressure of CO_2 in the cell. The SSU can, like the HIRS/2, be thought of as a very low spatial resolution scanner. The scan angle is smaller than that of the HIRS/2, $\pm 40°$ instead of $\pm 49.5°$ and the number of samples (fields of view) along a scan line is only 8 instead of the 56 of the HIRS/2. This is illustrated in Figure 1.18(a). The swath width shown in Figure 1.18(a) is somewhat narrower than that of the AVHRR.

The SSU makes use of a pressure modulation technique to measure the radiation emitted from CO_2 in the stratosphere. The pressure in the cell of CO_2 in the optical path of the instrument is changed in a cyclical manner at a frequency of about 40 Hz. The spectral characteristics of the channel and, therefore, the height of the weighting function is then determined by the pressure in the cell during the period of integration. By using three cells filled with CO_2 at different pressures, weighting functions peaking at three different heights (within the stratosphere) can be obtained.

The purpose of the SSU is the determination of temperatures in the stratosphere (25–50 km above the Earth's surface) and thereby to contribute, in conjunction with the HIRS/2, to the determination of complete atmospheric profiles.

1.4.4   The MSU

The MSU, like the other instruments in the TOVS package, can be thought of as a very low spatial resolution scanner. It operates with four channels in the microwave part of the electromagnetic spectrum. Its characteristics are given in Table 1.8. Figure 1.18(b) shows the relation between the samples (fields of view) for the MSU and the HIRS/2. The size of the footprint on the ground for the MSU is about 109 km diameter at nadir. The swath width is only slightly narrower than the swath width of the AVHRR.

Taking all three instruments in the TOVS together this means that there are 27 channels of data and the principal characteristics of the various channels are identified in Table 1.9. One can use a selection of some of the data, from one or more of

Table 1.8 MSU instrument parameters (Schwalb 1978)

Characteristics	Value				Tolerance
	Channel 1	Channel 2	Channel 3	Channel 4	
Frequency (GHz)	50.3	53.74	54.96	57.05	± 20 MHz
RF band width (MHz)	220	220	220	220	Maximum
NEΔT (deg K)	0.3	0.3	0.3	0.3	Maximum
Antenna beam* efficiency	>90%	>90%	>90%	>90%	
Dynamic range (K)	0–350	0–350	0–350	0–350	
Calibration	Hot reference body and space background each scan cycle				
Cross-track scan angle	$\pm 47.35°$				
Scan time	25.6 s				
Number of steps	11				
Step angle	9.47°				
Step time	1.84 s				
Angular resolution	7.5° (3 db)				
Data rate	320 bps				

* >95% expected

Table 1.9 TIROS-N TOVS channel spectral response (Kidwell 1991)

Instrument channel	Number	Central wave number/ frequency/level of peak	Description
HIRS/2	1	668.00 cm^{-1}	15 μm CO_2 Band
	2	679.23 cm^{-1}	15 μm CO_2 Band
	3	691.12 cm^{-1}	15 μm CO_2 Band
	4	703.56 cm^{-1}	15 μm CO_2 Band
	5	716.05 cm^{-1}	15 μm CO_2 Band
	6	732.38 cm^{-1}	15 μm CO_2 Band
	7	748.27 cm^{-1}	15 μm CO_2 Band
	8	897.71 cm^{-1}	Window
	9	1027.87 cm^{-1}	Ozone-water vapour
	10	1217.10 cm^{-1}	Ozone-water vapour
	11	1363.69 cm^{-1}	Ozone-water vapour
	12	1484.35 cm^{-1}	Window
	13	2190.43 cm^{-1}	4.3 μm CO_2 Band
	14	2212.65 cm^{-1}	4.3 μm CO_2 Band
	15	2240.15 cm^{-1}	4.3 μm CO_2 Band
	16	2276.27 cm^{-1}	4.3 μm CO_2 Band
	17	2360.63 cm^{-1}	4.3 μm CO_2 Band
	18	2511.95 cm^{-1}	4.3 μm CO_2 Band
	19	2671.18 cm^{-1}	Window
	20	14 367.00 cm^{-1}	Visible window
MSU	1	1.6599 cm^{-1}	
	2	1.7734 cm^{-1}	
	3	1.6488 cm^{-1}	
	4	1.7385 cm^{-1}	
SSU	1	669.988 cm^{-1}/15 mb	
	2	669.628 cm^{-1}/5 mb	
	3	669.357 cm^{-1}/1.5 mb	

the three instruments, for studying some particular property, e.g. the temperature of the stratosphere. Alternatively, the data from all the channels of all three instruments can be used in an inversion scheme to determine atmosphere T–P (temperature–pressure) profiles; for further details see, for example, Olesen (1987, 1992).

1.5　THE FUTURE OF THE AVHRR

At any one time two satellites of the series are in use as operational satellites, one with a morning ascending node and one with an afternoon ascending node, see Figure 1.19 (in colour section). Two other earlier satellites in the series are under NOAA's Spacecraft Operations Control as back-up spacecraft. We shall describe the situation at the time of writing; it will be out of date by the time this book is published but it will be useful to give an indication of a typical situation. The status of the spacecraft launched before the time of writing has been described at the end of section 1.1.

It is not possible for Spacecraft Operations Control to interrogate and control more than four spacecraft at a time. At the time of writing this section (April 1993) the two operational spacecraft were

- NOAA-12 morning pass
- NOAA-11 afternoon pass

and the two back-up spacecraft were

- NOAA-10 morning pass
- NOAA-9 afternoon pass.

This means that data are routinely received and archived (see section 2.4) by NOAA from the two operational spacecraft. If one of the operational spacecraft fails then the back-up spacecraft can be used instead. However, in addition to the above we note the following.

- Data from the SBUV (Solar Backscatter UltraViolet Spectral Radiometer) was being routinely received from both NOAA-11 and NOAA-9. This instrument is designed to measure ozone concentrations and it has only been flown on these two of the satellites in the series (which are both afternoon-pass satellites); data from the SBUV are obviously of very great interest and importance in relation to the current environmental problem of the depletion of the ozone layer.

- Data from ERBE (Earth Radiation Budget Experiment) were being received from NOAA-10. The ERBE was only flown on two satellites in the series, NOAA-9 and NOAA-10. The ERBE on NOAA-9 failed in 1987. However, the ERBE consists of two instruments, a non-scanning radiometer and a scanner; of these, the scanner of the ERBE on NOAA-10 has failed and so it is only the non-scanner data from the ERBE which are being collected.

- The real-time data from the search and rescue system on the NOAA-9 satellite were still in use. The reason for this is simply to give extra coverage and better data for location (and tracking) of a platform associated with a disaster involving the need for a rescue. (The search and rescue system was not flown on NOAA-12.)

Table 1.10 Physical characteristics of NOAA-I

Main body: 4.18 m long, 1.88 m in diameter
Solar array: 2.37 by 4.91 m, 11.6 m^2
Weight: At liftoff, 1712 kg; in orbit, 1030 kg
Power: Orbit average end of life – 593 W for gamma angle = 0°,
533 W for gamma angle = 80°
Lifetime: Greater than two years

At the time of writing, NOAA-I (which would be expected to be known as NOAA-13 if it were to be launched successfully) was scheduled to be launched on 9 June 1993.

Obviously all the satellites in the system have slight differences from one another and there is nothing particularly special about NOAA-I. However, we give some substantial detail of NOAA-I because it is the one about which information was most easily available and this should give a general impression of how the AVHRR, which is the subject of this book, fits into the overall ATN (Advanced TIROS-N) programme. This book is not intended to be a comprehensive documented history of the whole TIROS-N and Advanced TIROS-N programme. The main physical characteristics of NOAA-I are indicated in Table 1.10. The spacecraft is actually longer (4.18 m) than in the original plans (3.71 m in Table 1.2(b)). The orbit into which the NOAA-I spacecraft will be launched was illustrated in Figure 1.14. The nominal orbit will be Sun-synchronous and precess eastward about the Earth's polar axis by 0.986° per day. It will cross the equator at about 1.40 p.m. northbound and 1.40 a.m. southbound, local solar time, i.e. it will be an afternoon satellite. The intention, given a successful launch, was for NOAA-11 to be relegated to being the back-up afternoon satellite, and for NOAA-9 to be taken out of service.[1]

The NOAA-I spacecraft was launched from the Air Force Western Space and Missile Center (WSMC) at Vandenberg Air Force Base, California, by an Atlas-E launch vehicle. The standard Atlas launch vehicle consists of an E-series Atlas ballistic missile that has been refurbished and modified to a standard configuration for use as a launch vehicle for orbital missions. It is capable of launching a spacecraft into a variety of low Earth orbits. The launch vehicle is manufactured and refurbished by GDSSD under contract to the USAF. The vehicle is 28.7 m tall and 3.05 m in diameter. The fairing is 7.42 m long and 2.13 m in diameter. At liftoff, it carries 70 kilolitres (kl) of liquid oxygen and 43 kl of RP-1 fuel, a highly refined kerosene, see Figure 1.20.

The NOAA-I spacecraft is a member of the Advanced TIROS-N programme (see Table 1.2(b)) which means that it is a larger spacecraft than the original TIROS-N

---

[1] In the event NOAA-I was actually launched on 9 August 1993 but it failed after only 12 days. This was journalistically described in the *New Scientist* of 1 October 1994 as 'A screw that cost less than 10 cents wrecked a $77 million weather satellite within two weeks of launch, a US Government investigation has concluded . . . The investigators concluded that a 30 mm long screw was a few mm too long and caused a short circuit. The screw, which fastened part of the satellite's battery charger to a heat sink, was at the same voltage as the solar panels. But it punctured a layer of electrical insulation and touched an earthed radiator, preventing the solar panels from producing any power. This forced the craft to run on its batteries until they ran down, leaving the satellite dead.' This illustrates rather dramatically the fact that, at least until we have an operational space station, one has no opportunity to make even minor repairs to a spacecraft after it has been launched.

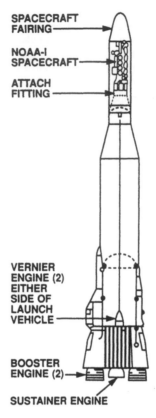

Figure 1.20   Atlas-E launch vehicle ('rocket').

spacecraft and those other earlier spacecraft which were almost exact direct copies of TIROS-N. In addition to the standard instruments carried on all the spacecraft of the series (i.e. the AVHRR, TOVS, SEM and the ARGOS data collection and platform location system), this particular spacecraft carries:

- SBUV/2, Solar Backscatter UltraViolet Spectral Radiometer, Mod 2
- SARR and SARM, search and rescue instruments
- EHIC, the Energetic Heavy Ion Composition Experiment
- MAXIE, the Magnetospheric Atmospheric X-ray Imaging Experiment.

The various members of the Advanced TIROS-N programme carry different combinations of instrumentation in addition to the standard instruments. For instance, we have noted above that NOAA-9 and NOAA-11 carried an SBUV/2 and that NOAA-9 and NOAA-10 carried an Earth Radiation Budget Experiment (ERBE). Of the additional instruments carried on NOAA-I the search and rescue and SBUV/2 are established instruments which have been flown successfully on previous spacecraft in the series. The other two instruments, the EHIC and the MAXIE, are experimental instruments. Table 1.11 shows the planned sets of instruments for the spacecraft of the Advanced TIROS-N series, as envisaged in 1982 (Schwalb 1982). This includes the SBUV/2, the ERBE and the search and rescue

Table 1.11 Projected payload complement for NOAA C–J satellites (Schwalb 1982)

Instrument	Spacecraft							
	C	D	E	F	G	H	I	J
AVHRR	*/2	*/1	/2	/1	/2	/2	/2	/2
HIRS/2	Yes	Yes	Yes	Yes	Yes	Yes	Yes	Yes
SSU	Yes	Dummy	Yes	Dummy	Yes	Dummy	Yes	Dummy
MSU	Yes	Yes	Yes	Yes	Yes	Yes	Yes	Yes
DCS	Yes	Yes	Yes	Yes	Yes	Yes	Yes	Yes
SAR	No	No	Yes	Yes	Yes	Yes	Yes	Yes
SBUV/2	No	No	Ballast	Ballast	Yes	Yes	Yes	Yes
ERBE Scanner	No	No	Ballast	Yes	Yes	Dummy	Dummy	Dummy
ERBE Non-scanner	No	No	Ballast	Yes	Yes	Dummy	Dummy	Dummy
SEM TED	Yes	Dummy	Yes	Dummy	Yes	Dummy	Yes	Dummy
MEPED	Yes	Dummy	Yes	Dummy	Yes	Dummy	Yes	Dummy
HEPAD	Yes	Dummy	No	No	No	No	No	No

NOTE: Ballast denotes addition to balance spacecraft centre of gravity. Dummy denotes a physical simulation model with proper weight, thermal characteristics and appropriate electrical terminations.
*/2 represents AVHRR/2, the five-channel version of the AVHRR
*/1 represents AVHRR/1, the four-channel version of the AVHRR

system but not the two experimental EHIC and MAXIE instruments. The SBUV/2, search and rescue system and ERBE will now be described briefly.

The Solar Backscatter UltraViolet Radiometer (SBUV/2) design is based on the technology developed for the SBUV/TOMS flown on the Nimbus-7 satellite; the NOAA instrument was built by the Aerospace Systems Division, Ball Corporation, Boulder, Colorado. The instrument was designed to provide data from which it is possible to compute the vertical distribution of ozone in the Earth's atmosphere. From these data, global maps of ozone concentration can be constructed; from continuous monitoring, long-term trends can then be estimated. To collect this information, two separate measurements in the 160–400 nm spectral range are made by the SBUV/2 instrument. These are (a) the spectral radiance of the solar ultraviolet radiation backscattered from the strong ozone absorption band of the Earth's atmosphere and (b) the direct solar spectral irradiance in the ultraviolet range.

The SBUV/2 instrument is a non-spatial-scanning nadir-viewing instrument which measures backscattered solar radiation in an 11.3° field-of-view in the nadir direction at 12 discrete, 1.1 nm wide, wavelength bands between 252.0 and 339.8 nm. The solar irradiance is determined at the same 12 wavelength bands by deploying, upon command, a diffuser which will reflect sunlight into the instrument's field-of-view. The atmospheric radiance measurement, relative to the solar irradiance, is the significant factor being determined. The SBUV/2 instrument can also measure the solar irradiance or the atmospheric radiance with a continuous spectral scan from 160 nm to 400 nm in increments nominally of 0.148 nm. These measurements provide data on photochemical processes in the atmosphere. A separate narrowband filter photometer channel, called the Cloud Cover Radiometer (CCR), continuously measures the Earth's surface brightness at 380 nm, i.e. outside the ozone absorption band. The Cloud Cover Radiometer is located in the same structure as the monochromator. The CCR field-of-view is the same size ($11.3° \times 11.3°$) as, and is co-aligned with, the monochromator's field-of-view. Further details of the SBUV/2 are given by Schwalb (1982).

The Search and Rescue (SAR) system is a random access system to acquire data from Emergency Locator Transmitters (ELTs) and Emergency Position Indicating Radio Beacons (EPIRBs) carried on general aviation aircraft and some classes of marine vessels. It is a development from the ARGOS data collection and platform location system and works on the same basic principles; it is, however, dedicated solely to search and rescue and its data are handled separately from all the rest of the data generated on board the spacecraft. The received data are transmitted (directly) to the ground where they are processed to locate the origin of the distress signal. These data are transmitted to the ground using a special radio frequency allocated for this purpose. SAR data are not included with the conventional meteorological data real-time transmission.

The Earth Radiation Budget Experiment (ERBE) instrument was built to gather data to permit a better understanding of climate and its potential predictability. It was planned to determine, for at least one year, the monthly average radiation budget on regional, zonal and global scales and to determine the equator-to-pole energy transport gradient. A second objective was to determine average diurnal variations in radiation budget. NASA Langley Research Center was responsible for the procurement of the ERBE instruments and for data management for the mission. There are two ERBE instrument packages which together provide the data required. One instrument package contains wide and medium field-of-view channels viewing in a fixed mode and is called the Non-Scanner. The second package consisting of narrow field-of-view channels is referred to as the Scanner.

The ERBE Non-Scanner instrument consists of five channels, four of which have, as a primary function, the viewing of the Earth. Of these channels, two are wide field-of-view (viewing the total Earth disc beneath the satellite) and two are medium field-of-view ($31.8°$ field-of-view, equivalent to a $10°$ Earth central angle beneath the satellite). These four channels are mounted on a single axis gimbal which, when activated by command (in the appropriate orbital location), allows these channels to view the Sun for periodic calibration. The fifth channel is not gimbaled but is so located as to view the Sun periodically and provide a measurement of the solar constant. The five sensors differ in the spatial and spectral scales of their measurements as shown in Table 1.12(a). The Non-Scanner instrument package is estimated to weigh 32 kg and require 20 W orbit average power. Data output is 160 bits per second.

The ERBE Scanner instrument is designed to scan an essentially continuous pattern from horizon to horizon in a direction normal to the orbit plane. The instantaneous field-of-view is $3° \times 4.5°$. The mechanical system is configured to provide the same capability for viewing the Earth and Sun as does the Non-Scanner instrument. As shown in Table 1.12(b), there are three wavelength ranges of operation for the scanning channels. The scanner assembly is a lightweight, compact, three-bearing, mechanical configuration which is scanned in sawtooth fashion across the ground track of the spacecraft. The scanner instrument package weighs about 29 kg and requires 25 W orbit average power. The data output is 960 bits per second.

Table 1.12 ERBE characteristics (Schwalb 1982)

(a) Non-scanner characteristics

Channel	Spectral interval (μm)	Field-of-view
1 (Wide FOV)	0.2–50+	Limb-to-limb
2 (Wide FOV)	0.2–5	Limb-to-limb
3 (Medium FOV)	0.2–50+	10° ECA*
4 (Medium FOV)	0.2–5	10° ECA
5 (Solar)	0.2–50+	18° conical

(b) Scanner characteristics

Channel	Spectral interval (μm)	Instantaneous field-of-view
6	0.2–5	3° × 4.5°
7	5–50	3° × 4.5°
8	0.2–50	3° × 4.5°

* Earth central angle

The EHIC (University of Chicago and the Canadian National Research Council HIA) is designed to measure the chemical and isotopic composition of energetic particles between hydrogen and nickel over the energy range 0.5–200 MeV/nucleon. The experiment will measure energetic solar flare particles in the polar regions where the Earth's magnetic field connects to the interplanetary field carried in the solar wind and will also measure trapped energetic particles in the magnetosphere. The primary scientific objective for the EHIC is to obtain elemental and isotopic composition data, which can be used to test models for solar flare ion acceleration, ion transport in interplanetary space, ion entry into the magnetosphere, and nucleo-synthetic processes leading to the elemental and isotopic mix found at the Sun. The EHIC will provide data on the fluxes and energy spectra of heavy ions, which can cause single-bit upsets and damage to integrated circuits, found in the Sun-synchronous orbits used by NOAA and other environmental satellites.

The MAXIE (Lockheed Missiles and Space Company's Palo Alto Research Laboratory with the assistance of the Aerospace Corporation and Norway's University of Bergen) maps the intensities and energy spectra of X-rays produced by electrons that precipitate into the atmosphere. With mechanical scanning, the MAXIE will obtain new high-resolution X-ray imaging data on auroral and sub-storm processes with a temporal resolution and repetition rate that has not previously been available. By repeated scans of the atmosphere, the MAXIE will provide the opportunity to study temporal variations on a fast timescale that has yet to be accomplished.

We have given information about the launch dates of those spacecraft in the TIROS-N, ATN programme which are already in space (see Table 1.1).[1] Projected launch dates for the next few spacecraft, NOAA-J, -K, -L and -M, are shown in Table 1.13. However, these projected launch dates are based on the assumption that

[1] This section was written in April 1993 and it has deliberately not been updated (except to add the previous footnote about the failure of NOAA-I) because I was trying to give a snapshot of the situation, at one (arbitrarily chosen) instant in a 20-year ongoing programme. If I had updated it at proof stage it would almost certainly still have been out of date by the time the book was published.

Table 1.13 Launch schedule for polar-orbiting environmental satellites (NOAA)

Satellite name	Projected launch date
NOAA-J (a.m.)	May 1994
NOAA-K (p.m.)	June 1996
NOAA-L (a.m.)	May 1997
NOAA-M (p.m.)	June 1999
NOAA-N (a.m./p.m.)	May 2000/June 2002
NOAA-N' (p.m.)	TBD (at three-year intervals)
NOAA-O (p.m.)	TBD
NOAA-P (p.m.)	TBD
NOAA-Q (p.m.)	TBD

Note: TBD = to be decided.

none of the spacecraft in operation suffers a major failure before the end of its nominal lifetime. Some spacecraft far outlive their design lifetime; for instance NOAA-6 which had a design lifetime of two years was only finally deactivated after about eight years of service. Some others do not. NOAA-B for example, never achieved a usable orbit as a result of a booster engine anomaly on the launcher and NOAA-E, though launched successfully into orbit, had some quite serious problems (see section 1.1). If there is a serious failure of one of the operational spacecraft then the next scheduled launch (see Table 1.13) is brought forward. There are three critical instruments for the purpose of deciding what constitutes a serious failure in this context. These are the AVHRR, the HIRS and the MSU. If any one of these instruments fails, or if some important channels, e.g. 4 and 5, of the AVHRR fail, then notice is given to the Astro Space Division of General Electric; under their contract General Electric are obliged to launch the next satellite within a prescribed period. For spacecraft up to NOAA-J this period is 120 days, for NOAA-K, -L and -M it is 200 days.

As far as the AVHRR is concerned we have already noted in section 1.2 that NOAA-N and -N' and 2000 mark the beginning of another phase in the development of the system. An agreement has been made with EUMETSAT for EUMETSAT to take over from NOAA the responsibility for the morning missions of the polar-orbiting spacecraft programme from 2000. EUMETSAT is the organisation which was initially formed to operate the present (second) generation of European geostationary meteorological satellites of the Meteosat series. NOAA-N is envisaged as the first afternoon spacecraft of this new series. The first morning spacecraft in the new series will be known as Metop-1. As part of the agreement for passing over the responsibility for the morning satellites to EUMETSAT, NOAA is currently planning to produce four sets of instruments, one set for NOAA-N, one set for NOAA-N', one set for Metop-1 and one spare set for EUMETSAT. After that, it will be EUMETSAT's responsibility to procure the instruments for subsequent morning spacecraft and it will be EUMETSAT's responsibility to supply the instruments for Metop-1 to the European Space Agency (ESA). The European Space Agency will be responsible for the construction and launch of the first and subsequent spacecraft of this new series of morning satellites for EUMETSAT. At the time of writing this section (April 1993) it was anticipated that ESA would undertake a phase-A study of the first of these spacecraft in 1994. The reasons for handing over part of the programme for the polar-orbiting meteorological satellites are several and they include (i) the wish to reduce the cost to the US taxpayers, (ii) the reflection of the much wider spread of space technology among industrialised countries of the world in the last 20 years or so and (iii) the wish to spread the responsibility for the gathering of meteorological data from space more widely among the international community. The reason for the choice of morning spacecraft for EUMETSAT and the afternoon spacecraft for NOAA, and not vice versa, was based on the quality of the sounding data for input to numerical weather forecast models. For this purpose in Europe the North Atlantic area is extremely important and the morning spacecraft provides better data than the afternoon spacecraft for this purpose. Similarly, for the same purpose in the USA the data from the Gulf of Mexico is very important and for this purpose the afternoon spacecraft provides better data than the morning spacecraft. A new generation of, or rather a successor to, the AVHRR is expected to be flown on these new satellites. Proposed as the Advanced Medium Resolution Imaging Radiometer (AMRIR), some of its proposed features are described by Needham (1988), Fischer (1989) and Sparkman (1989); it would involve better spatial resolution than the AVHRR (500 m) and more spectral channels. We note the change of name, from using 'Very High Resolution' to using 'Medium Resolution'. The name of the AVHRR was based on comparison with previous environmental satellites of NOAA, whereas the name of the AMRIR is balanced into the entire realm of remote sensing imaging systems from SPOT to passive microwave scanners.

CHAPTER TWO

The data

2.1 DATA STRUCTURE AND FORMAT

2.1.1 Introduction

The two most important documents for anyone who is seriously interested in using AVHRR data are

- the two main editions of the NOAA Technical Memorandum on *Data Extraction and Calibration of TIROS-N/NOAA Radiometers* (Lauritson *et al.* 1979, Planet 1988), which we shall refer to as NESS107, plus additional appendixes which are issued for each new satellite launched in the series
- the various editions of the *NOAA Polar Orbiter Data User's Guide* (Kidwell 1984, 1991).

Further information will be found in a variety of other documents, mostly from NOAA, which we shall cite at relevant points in this chapter. Various software packages for processing AVHRR data have been described in the literature (Van Camp and Schlittenhardt 1989, Baranski 1992).

The data from the AVHRR and the other instruments are archived in a raw form by NOAA after a minimum of pre-processing and are also used for the routine generation of a number of products such as sea surface temperatures and vegetation indexes; these activities and products will be described in more detail later in this chapter. However, as we have seen already in section 1.2, the AVHRR data are transmitted in real time by the spacecraft. These transmissions can be received at any ground station (direct readout station) in the world for that part of the spacecraft's orbit when it is visible above the horizon and there is no royalty fee charged by NOAA. There are now many such ground stations that have been set up (see section 2.5) and a very significant fraction (nobody really knows how much) of the usage of AVHRR data involves the use of data from direct readout stations rather than data obtained from the NOAA ground segment. Since what these direct readout ground stations receive is raw data, it is necessary for us to devote some attention to the format of the raw data. This is described in some considerable detail in NESS107.

The real-time data flow has already been illustrated in Figure 1.13. This applies to the original spacecraft in the series, TIROS-N, NOAA-6, etc. Various words and blocks of words were originally left spare in the data stream. Thus, as further instruments have been added in the Advanced TIROS-N series, it has been possible to slot in the data from the new instruments without disturbing the locations of words occupied by data from the original set of instruments. This is important because if compatibility from one spacecraft to the next had not been retained it would have led to enormous software problems for ground station operators and for data users.

The formats on which we shall concentrate in this section are the HRPT and APT formats which are the ones of interest to people handling data received by direct readout stations. The downlinking of LAC and GAC data to NOAA's own ground stations will be mentioned in section 2.4 but differences of the format details of LAC and GAC from those of HRPT and APT are of little interest to people outside NOAA.

The HRPT transmission is at S-band (at 1.6980 GHz/1.7070 GHz) and requires a steerable antenna with a parabolic reflector. This antenna system has to be pointed in exactly the right direction (i.e. towards the spacecraft). One therefore needs to know exactly when and where the spacecraft will appear above the horizon and one needs to be able to track the path of the spacecraft across the sky during its overpass; otherwise one will receive no signal. By contrast, the APT transmission is a VHF transmission and so it can be received with a simple omni-directional antenna (i.e. a piece of wire) and is very appropriate for low cost or amateur reception facilities with which only images, and not the digital data, are required (see section 2.5.2). It should, perhaps, be emphasised that while the APT transmission is of analogue data it is not derived directly from the detectors in the AVHRR. The analogue signals from the detectors are digitised, the resolution of the data is degraded digitally and these degraded digital values are then converted back to analogue data for the APT transmission (see section 2.1.3).

2.1.2   The HRPT format

The High Resolution Picture Transmission (HRPT) (see section 1.2.3) contains the AVHRR data with the full spatial resolution and in all five spectral bands/channels; it also contains the data from the other instruments as well. The main features of the HRPT transmission are indicated in Table 2.1. The HRPT provides a major frame and each major frame consists of three minor frames. The AVHRR data are updated at the minor frame rate while the TIP data are updated at the major frame rate, i.e. the three minor frames that make up a major frame contain the same TIP data. This has a useful consequence that errors can be reduced by using all three sets of TIP data. If there is an error in the transmission or reception of one word of data in the TIP data in one of the three minor frames it is almost certain that the same error will not occur in the same word in the TIP data in the other two minor frames within the same major frame. An error in the TIP data can be detected by selecting the three words from the same position in each of the three minor frames. The values of these words should all be the same and if this is so then there is no error present. If the values from two minor frames are the same but differ from that

Table 2.1 HRPT characteristics (Lauritson *et al.* 1979)

Line rate	360 lines/minute
Carrier modulation	Digital split phase, phase modulated
Transmit frequency	1698.0 MHz* or 1707.0 MHz
Transmit power	8 W nominal
EIRP (approximate)	39.0 dBm
Polarisation	Right-hand circular
Spectrum band width	3 dB band width of 2.4 MHz

* 1702.5 MHz left-hand circular polarisation available in the event of failure of the primary frequencies

in the third, then one assumes that the one value that is different is erroneous and that the other two identical values are correct. The chance of the same two words in two different frames having the same error is infinitesimally small. Finally, unless one is dealing with a very noisy signal at the ground or a very noisy receiver, the chance of having so many errors that one obtains three different values for the same word from the three minor-frame TIP data is very small indeed. If the data stream is that noisy, then it is not worth trying to use the data for any purpose whatsoever! The TIP data contain the data from all the instruments except the AVHRR (and, on later spacecraft, except the search and rescue package) and so the user of any of these data would be very unwise not to build in the use of a correction procedure for TIP data in any software designed for the use of these data. There is actually a further source of error detection (though not error correction) for the TIP which we have not mentioned so far. All the words in the HRPT format are 10-bit words but the TIP data are only transmitted as 8-bit data. This leaves two of the ten bits spare and one of these is used as a parity-check bit for that TIP data word. When it comes to the AVHRR, to sacrifice one of the available 10 bits as a parity bit would have reduced the resolution of the intensity from 1024 to 512; this would not have affected the image quality of the AVHRR but it would have seriously reduced the value of the AVHRR for quantitative radiometric work.

The error correction procedure that we have outlined above for the TIP data cannot be used for the AVHRR since the data from each scan line is only transmitted once. There are various reasons why there is no in-built error detection or error correction procedure for the AVHRR data. These include the fact that the AVHRR was planned as an imaging device and if one pixel in the image has an erroneous intensity value the effect on the overall appearance of the image (in which most features contain a very large number of pixels) is unimportant. Another reason is that the quantity of AVHRR data is very large and any scheme even to provide error detection, let alone error correction, would have added very considerably both to the quantity of data to be transmitted and stored and to the time that would be needed to process the data. While those may not be regarded as very serious problems now, it should be remembered that this system was designed and built in the mid-1970s when electronic systems and computing facilities were much less sophisticated than they are now.

Consider the data in Table 2.1 briefly. With 360 lines per minute, this means there are six lines per second and this, in turn, corresponds to six minor frames or two major frames per second. The format of a minor frame is shown in Figure 2.1. Bit 1 is the most significant bit and bit 10 is the least significant bit. Let us consider

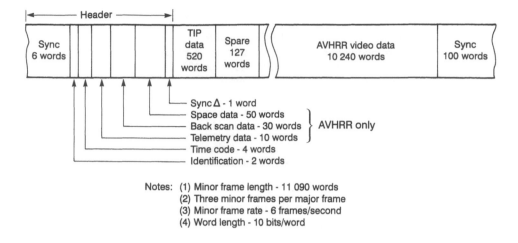

Notes: (1) Minor frame length - 11 090 words
(2) Three minor frames per major frame
(3) Minor frame rate - 6 frames/second
(4) Word length - 10 bits/word

TLM word allocations		ID word bit allocations	
		1st ID word	2nd ID word
1–5	Ramp calibration		(Spare)
6	Channel-3 target temp (5 pt subcom)	1 Sync ID	
7	Channel-4 target temp (5 pt subcom)	2–3 Frame ID 4–7 Spacecraft address	
8	Channel-5 target temp (5 pt subcom)	8 Resync marker	
9	Channel-3 patch temp	9 Data 0	
10	Spare	10 Data 1	

Figure 2.1   The format of an HRPT minor frame (Lauritson *et al.* 1979).

the general structure of this format. As the data are transmitted or received they will form a continuous stream of bits and the separation of this stream into frames is achieved by the use of sync words (synchronisation words). In the case of the minor HRPT frame this is a set of six 10-bit words, see Table 2.2. These six sync words form a particular string of 60 bits which will, almost certainly, not arise by chance within the data stream. The manner in which this particular bit pattern was generated is indicated at the right of the first entry of Table 2.2, but this in itself is not important. What is important is that this particular pattern of ones and zeros has been chosen and that it will be present at the start of every minor frame. Any ground station which receives HRPT data needs to be able, either with hardware or with software, to recognise this particular pattern of 60 bits and then to ensure that the following string of bits is stored in 10-bit words to constitute the entire minor frame. The first 102 words of the HRPT minor frame constitute the header. This includes the sync words, which form the first six words of the header (see Table 2.2), while the rest of the data in the header is concerned with some housekeeping for the spacecraft and with in-flight calibration data for the AVHRR. We shall have occasion to consider this inflight calibration data for the AVHRR in section 2.2.3.

After one word (no. 103) of Sync Δ there follow 520 words of TIP data (in positions 104–623), see Table 2.2, and this in turn consists of five minor frames of 104 words each. The data from the low data-rate instruments is distributed among the various positions in the minor frame as shown in Figure 2.2. The version in Figure

Table 2.2 HRPT minor frame format (Schwalb 1982)

Function	Number of words	Word position	Bit number 1 2 3 4 5 6 7 8 9 10	Plus word code and meaning
Frame sync	6	1	1 0 1 0 0 0 0 1 0 0	1st 60 bits from a 63-bit PN[1] generator started in the all 1's state. The generator polynominal is $X^6 + X^5 + X^2 + X + 1$
		2	0 1 0 1 0 1 1 1 1 1	
		3	1 1 0 1 0 1 1 1 0 0	
		4	0 1 1 0 0 1 1 1 0 1	
		5	1 0 0 0 0 0 1 1 1 1	
		6	0 0 1 0 0 1 0 1 0 1	
ID	2	7	Bit 1; 0 = internal sync; 1 = AVHRR sync	
			Bits 2 & 3; 00 = not used; 01 = minor frame 1; 10 = minor frame 2, 11 = minor frame 3	
			Bits 4–7; spacecraft address; bit 4 = MSB, bit 7 = LSB	
			Bit 8; 0 = frame stable; 1 = frame resync occurred	
			Bits 9–10; spare; bit 9 = 0, bit 10 = 1	
		8	Spare word; bit symbols undefined	
Time code	4	9	Bits 1–9; binary day count; bit 1 = MSB; bit 9 = LSB	
			Bit 10; 0; spare	
		10	Bits 1–3; all 0's; spare 1, 0, 1	
			Bits 4–10; part of binary msec of day count; bit 4 = MSB	
		11	Bits 1–10; part of binary msec of day count;	
		12	Bits 1–10; remainder of binary msec of day count; bit 10 = LSB	
Telemetry	10	13	Ramp calibration AVHRR channel 1	
		14	Ramp calibration AVHRR channel 2	
		15	Ramp calibration AVHRR channel 3	
		16	Ramp calibration AVHRR channel 4	
		17	Ramp calibration AVHRR channel 5	

Table 2.2 (Continued)

Function	Number of words	Word position	Bit number 1 2 3 4 5 6 7 8 9 10	Plus word code and meaning
		18	AVHRR channel 3 target temp.[2]	Each of these words is a 5-channel subcom — 4 words of IR data plus subcom sync (10 0's)
		19	AVHRR channel 4 target temp.	
		20	AVHRR channel 5 target temp.	
		21	Channel-3 patch temp.	
		22	0 0 0 0 0 0 0 0 1 spare	
Back scan	30	23 → 52	10 words of back scan data from each AVHRR channel 3, 4, and 5. These data are time multiplexed as Chan 3 (word 1), chan 4 (word 1), chan 5 (word 1), chan 3 (word 2), chan 4 (word 2), chan 5 (word 2), etc.	
Space data	50	53 → 102	10 words of space-scan data from each AVHRR channel 1, 2, 3, 4 and 5. These data are time multiplexed as chan 1 (word 1), chan 2 (word 1), chan 3 (word 1), chan 4 (word 1), chan 5 (word 1), chan 1 (word 2), chan 2 (word 2), chan 3 (word 2), chan 4 (word 2), chan 5 (word 2), etc.	
Sync Δ	1	103	Bit 1; 0 = AVHRR sync early; 1 = AVHRR sync late Bits 2–10; 9-bit binary count of 0.9984-MHz periods; bit 2 = MSB, bit 10 = LSB	
TIP data	520	104 → 623	The 520 words contain five frames of TIP data (104 TIP data words/frame) Bits 1–8: exact format as generated by TIP Bit 9: even parity check over bits 1–8 Bit 10: —bit 1	

Category	No. of words	Word	Bit pattern	Description
Spare words	127	624	1 0 1 0 0 0 0 1 1 1 0	Derived by inverting the output of a 1023-bit PN sequence provided by a feedback shift register generating the polynominal: $X^{10} + X^5 + X^2 + X + 1$ The generator is started in the all 1's state at the beginning of word 7 of each minor frame
		625	1 1 1 0 0 0 1 0 1 1 1	
		626	0 0 0 0 0 1 0 1 1 1 1	
		627	1 1 0 1 0 0 0 1 1 1 1	
		628	1 1 1 0 1 0 1 0 0 1 0	
		↓		
		748	1 0 0 1 0 1 1 0 1 0 0	
		749	1 1 0 0 1 0 0 0 1 0 0	
		750	1 0 0 0 0 0 0 0 0 0 0	
Earth data	10 240	751	Chan 1 – Sample 1	Each minor frame contains the data obtained during one Earth scan of the AVHRR sensor. The data from the five sensor channels of the AVHRR are time multiplexed as indicated
		752	Chan 2 – Sample 1	
		753	Chan 3 – Sample 1	
		754	Chan 4 – Sample 1	
		755	Chan 5 – Sample 1	
		756	Chan 1 – Sample 2	
		↓		
		10985	Chan 5 – Sample 2047	
		10986	Chan 1 – Sample 2048	
		10987	Chan 2 – Sample 2048	
		10988	Chan 3 – Sample 2048	
		10989	Chan 4 – Sample 2048	
		10990	Chan 5 – Sample 2048	
Auxiliary sync	100	10991	1 1 1 1 1 0 0 0 1 0	Derived from the non-inverted output of a 1023-bit PN sequence provided by a feedback shift register generating the polynominal: $X^{10} + X^5 + X^2 + X + 1$ The generator is started in the all 1's state at the beginning of word 10991
		10992	1 1 1 1 1 1 0 0 1 1	
		10993	0 1 1 0 1 0 1 0 1 1	
		10994	1 0 1 0 1 1 1 1 0 1	
		↓		
		11089	0 1 1 1 1 1 0 0 0 0	
		11090	1 1 1 1 0 1 0 1 0 0	

(1) PN = pseudo noise

(2) As measured by a platinum resistance thermometer embedded in the housing

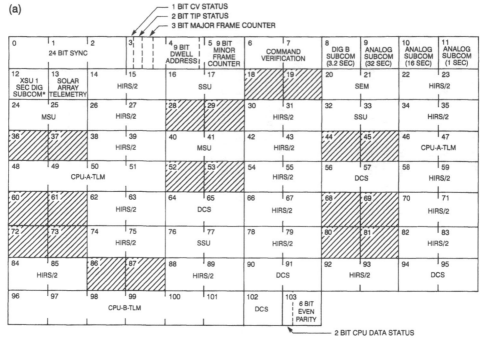

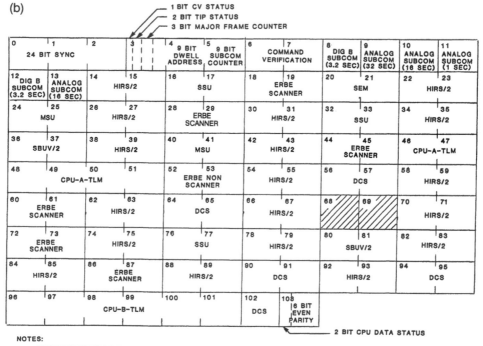

Figure 2.2 TIP (TIROS Information Processor) minor frame for (a) original spacecraft TIROS-N, etc., and (b) a later spacecraft of the Advanced TIROS-N series (Lauritson *et al.* 1979, Schwalb 1982).

2.2(a) is the original format for TIROS-N and the other early spacecraft of the first series and this included a number of spare words (shown shaded in the figure). Figure 2.2(b) shows how some of these spare words have been allocated to the ERBE and the SBUV/2 in a later spacecraft in the Advanced TIROS-N Series.

Following the TIP data there are 127 spare words (see Table 2.2) and this is followed by the AVHRR data, occupying positions 751–10 990. These data from one scan line of the AVHRR contain 2048 words (corresponding to 2048 samples or pixels) across the complete swath for each of the five bands, giving 10 240 words altogether. The data are transmitted as band interleaved data, see Table 2.2. The frame ends with an auxiliary set of 100 sync words.

2.1.3   The APT format

The APT transmission is important because it is a VHF transmission (at 137.60 MHz/137.62 MHz) and it can be received by using simple receiving equipment, especially without the need to have, and to be able to control, a steerable parabolic reflecting antenna system. The important characteristics of the APT transmission are given in Table 2.3. The video data for transmission on the APT transmission are a degraded version of the AVHRR data. Although they are analogue data, the APT data are not derived directly from the detectors of the AVHRR. The digital outputs from two selected AVHRR channels are processed in the MIRP (Manipulated Information Rate Processor) to reduce the ground resolution from 1.1 km to 4 km and to produce a linearised scan. The choice of AVHRR channels for the APT is any two of five and the selection is implemented by Spacecraft Operations Control. Channel A APT data are obtained from one spectral channel of the first AVHRR scan line and channel B from another spectral channel contained in the next AVHRR scan line. The third AVHRR scan line is omitted from the APT before the process is repeated. This processing results in the APT containing the data from one third of the AVHRR's 360 scan lines per minute. The resolution of the APT is, therefore, proportionally reduced and is received at the ground station at a rate of 120 lines per minute of video. During the APT formatting, the MIRP also inserts

Table 2.3   APT characteristics (Lauritson et al. 1979)

Line rate (lines per minute)	120
Data resolution	4 km nearly uniform
Carrier modulation	Analogue
Transmit frequency	137.50 MHz for VTX-1 or 137.62 MHz for VTX-2
Transmit power	6 W nominal
Transmit antenna polarisation	Right-hand circular
Subcarrier frequency	2.4 kHz
Carrier deviation	± 17 kHz
Ground station low-pass filter	1400 Hz seventh-order linear recommended
Synchronisation	7 pulses at 1040 pps. 50% duty cycle for channel A; 7 pulses at 832 pps, 60% duty cycle for channel B

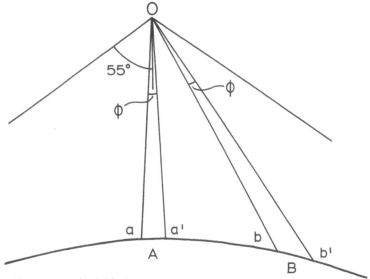

Figure 2.3   Variations in the field of view.

Figure 2.4   Example of a non-linearised AVHRR image (Dundee University).

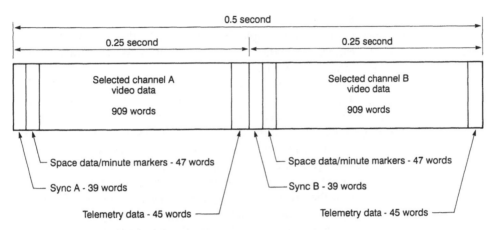

Notes: 1. Equivalent output digital data rate is 4160 words/second
2. Video line rate - 2 lines/second
3. APT frame size - 128 lines
4. Any two of the five AVHRR channels may be selected for use
5. Sync A is a 1040 Hz square wave - 7 cycles
6. Sync B is a 832 pps pulse train - 7 pulses
7. Each of 16 telemetry points are repeated on 8 successive lines
8. Minute markers are repeated on 4 successive lines, with 2 lines black and 2 lines white

Figure 2.5 APT video line prior to digital/analogue conversion (Lauritson *et al.* 1979).

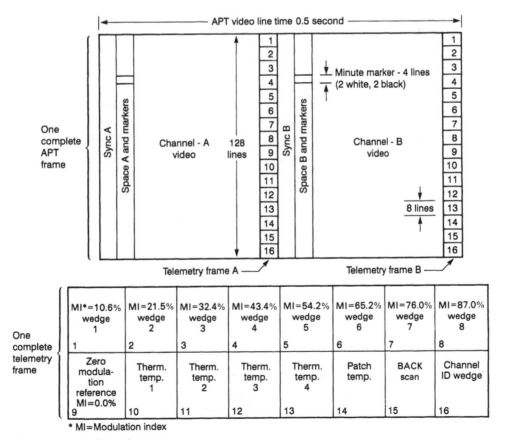

Figure 2.6 APT frame format.

appropriate calibration and telemetry data for each of the selected images being transmitted. This means that, in principle, it is possible to digitise the APT data again and to use the calibration data to determine brightness temperatures from thermal infrared APT data, see section 2.2.5.

We should explain what is meant by linearisation in this context. The AVHRR has a telescope with a fixed angular field of view and it scans with a mirror which rotates with a constant angular velocity and thus the field of view moves through the 110° of its Earth scan with a constant angular velocity. As the effective optic axis of the field of view scans through the angle of 110°, sampling is carried out at constant time intervals to generate 2048 output values (for each channel). The angular field of view of each of the 2048 samples is the same but the field of view, measured as a footprint on the ground, will vary across the scan line, see Figure 2.3. The distance bb' on the ground is larger than the distance aa' for several reasons: (a) the distance OB is greater than the distance OA, (b) the ground surface at B is being observed obliquely and (c) the Earth's surface is curved. If one simply printed the scan data without correcting for these factors the edges of the image would appear to be very distorted (squeezed along the scan direction, i.e. approximately E–W), see Figure 2.4. For HRPT data, no attempt is made to carry out linearisation on board the spacecraft; it has to be done on the ground after the reception of the data by the

Table 2.4   APT transmission parameters (Lauritson *et al.* (1979)

Type of transmitted signal	VHF, AM/FM
	2.4 kHz DSB-AM
	1.44 Hz video
System output	
Frequency, polarisation	137.50 MHz right circular polarisation
	or
	137.62 MHz right circular polarisation
EIRP at 63° from nadir	33.5 dBm worst case
	37.2 dBm nominal
Antenna	
Gain at 63° from nadir	−0.5 dBi, right circular polarisation
Ellipticity	4.0 dB, maximum
Circuit losses	2.4 dB
Transmitter	
Power	5.0 W minimum
Carrier modulation index	± 17, ± 0.85 kHz
Pre-modulation band width	0.1 to 4.8 kHz
± 0.5 dB	
Frequency stability	$+2 \times 10^{-5}$
Subcarrier modulator	
Subcarrier frequency	2400 ± 0.3 Hz
Subcarrier modulation index	$87 \pm 5\%$
Post-modulator filter, type	3-pole Butterworth
3 dB band width	6 kHz, minimum
Pre-modulator filter, type	3-pole Butterworth–Thompson
3 dB band width	2.4 kHz, minimum

Table 2.5   APT format parameters (Lauritson *et al.* 1979)

Frame	
Rate	1 frame per 64 seconds
Format	See Figure 2.6
Length	128 lines
Line	
Rate	2 lines/second
Number of words	2080
Number of sensor channels	Any 2 of the 5; selected by command
Number of words/sensor channel	909
Format	See Figure 2.5
Line sync format	See Figure 3-4 of Lauritson *et al.* (1979)
Word	
Rate	4160 per second
Analogue-to-digital	The 8 MSBs* of each 10-bit
Conversion accuracy	AVHRR word
Low-Pass Filter	
Type	3rd-order Butterworth–Thompson
3 dB band width	2400 Hz

* Most significant bits (MSBs)

ground receiving station. It is to provide an undistorted picture for a receiver of the APT transmission that linearisation is carried out on board the spacecraft.

After this (digital) processing the data are time multiplexed along with appropriate calibration and telemetry data, see Figure 2.5. The processor then converts the multiplexed digital data to an analogue signal, low-pass filters the output and modulates a 2.4 kHz subcarrier. The maximum subcarrier modulation is defined as the amplitude of grey scale wedge number eight (see Figure 2.6), providing a modulation index of $(87 \pm 5)\%$. Tables 2.4 and 2.5 give further information on the APT transmission.

2.2   CALIBRATION OF AVHRR DATA

There are various aspects of the calibration of the AVHRR instruments. These include:

- determination of the spectral response function of each spectral channel
- pre-launch radiometric calibration of all five spectral channels
- in-flight calibration of the thermal infrared channels, i.e. channels 3, 4 and 5 of HRPT/LAC data
- in-flight calibration of APT data
- post-launch calibration studies.

The first two of these are done on the ground before launch and give results which are provided by NOAA to the users of AVHRR data. The third and fourth (in-flight

calibration) occur on board and many users need to incorporate this in-flight calibration data into their own processing of the AVHRR data. The fifth is still an area
of active research and development work.

2.2.1   Spectral response functions

Nominal band widths of the various channels on the AVHRR instruments were
given in Table 1.4. It is important, however, to realise that the situation is not as
simple as Table 1.4 might seem to suggest. The radiation received by a detector in
the scanner has been passed through a filter, but the ideal filter which is perfectly
transparent for radiation with wavelength λ in the range $\lambda_1 \leq \lambda \leq \lambda_2$ and totally
opaque for $\lambda < \lambda_1$ and $\lambda > \lambda_2$ does not exist. Some radiation with $\lambda < \lambda_1$ or $\lambda > \lambda_2$
will be allowed through and there will be some blocking of radiation within the
range $\lambda_1 \leq \lambda \leq \lambda_2$. Figures 2.7 and 2.8 show the idealised and actual spectral
response functions for the AVHRR/2. While the spectral response for any given
channel of the AVHRR may be nominally the same for different instruments in the
series there will be differences in practice. This is illustrated in Figure 2.9 in which
the channel-1 response functions for the AVHRRs on NOAA-7, -8, -9, -10 and -11
are reproduced; except for NOAA-9 these are all five-channel (AVHRR/2) instruments. A full set of graphs of the spectral response functions for all the bands of
each AVHRR flown on every satellite in the series from TIROS-N to NOAA-12 is
given in the Users' Guide by Kidwell (1991). Tables of the spectral response functions for channels 3, 4 and 5 are given for each AVHRR instrument in the series in
Appendix B of NESS107. A new set of Appendix B tables is produced when each
new spacecraft is launched. The reason for providing tabulated values of the spectral

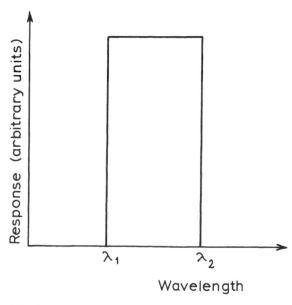

Figure 2.7   Sketch of idealised spectral response filter functions for nominal channel wavelengths
for NOAA-7.

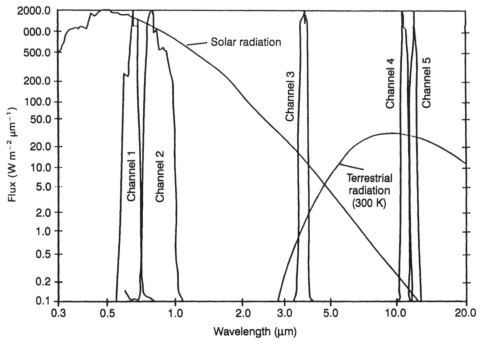

Figure 2.8 Normalised spectral response of AVHRR/2 (ITT).

response functions for the thermal infrared channels is because they are needed for the in-flight radiometric calibration of these channels, see section 2.2.3.

2.2.2 Pre-flight radiometric calibration[1]

The contract for building the AVHRR instruments requires that the builders, ITT (ITT Aerospace/Communications Division of ITT Defense & Electronics), carry out the calibration of each instrument once it has been built and before it is delivered to GE Aerospace for spacecraft integration and launch (ITT 1980). By radiometric calibration we mean establishing, for each individual AVHRR instrument, the relation between the output from the instrument and the intensity of the radiation incident on the instrument. The radiometric calibration involves exposing the radiometer to sources of radiation which have been calibrated against primary or secondary standards and thence establishing a relation between the output of the radiometer and the radiance (intensity of the incident radiation). For the three thermal infrared channels, 3, 4 and 5, there is also in-flight calibration of the data (see section 2.2.3) but there is no in-flight calibration of the visible and near-infrared channels, channels 1 and 2.

The pre-launch calibration of channels 1 and 2 of the AVHRR is described in detail by Rao (1987) who undertook the task of compiling relevant information obtained from diverse sources such as NOAA technical reports and memoranda, contractor reports, operating procedures developed under contract with NASA and NOAA and correspondence between scientists at NOAA, NASA and at ITT

[1] The understanding of section 2.2.2 is not a prerequisite for the understanding of sections 2.2.3–2.2.5.

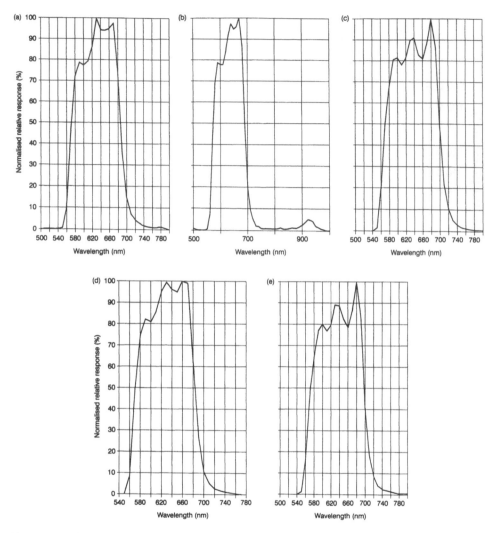

Figure 2.9 The spectral response function for channel 1 of the AVHRRs on (a) NOAA-7 (b) NOAA-8 (c) NOAA-9 (d) NOAA-10 and (e) NOAA-11 (Kidwell (1991).

Aerospace/Communications Division, Fort Wayne, Indiana, where the AVHRR instruments were built and calibrated. This involves a presentation of the underlying theory for establishing a simple linear regression relation between the reflectance factor of the Earth–atmosphere scene being viewed by the AVHRR and the AVHRR digital signals in terms of the reflectance factor of an integrating sphere source. The procedure adopted for the pre-launch laboratory calibration of channels 1 and 2 of the AVHRR is directed towards the determination of the response of the entire optical subsystem to changes in the ambient radiation field. This takes into account the wavelength-dependent nature of the absorption, reflection and transmission properties of the various components of the optical system and of the silicon detector.

Before discussing some of the details of the calibration procedure we describe the NASA 30-inch integrating sphere source which was used extensively by ITT

Aerospace/Optical Division, the AVHRR manufacturer, in the pre-launch calibration of channels 1 and 2 of the AVHRR. The inside of the sphere is coated with barium sulphate with a diffuse reflectivity very close to unity and can be illuminated with up to 12 quartz-halogen lamps matched, to the extent practicable, for spectral output and operating current. The radiance at the 12-inch exit aperture of the integrating sphere approximates very closely to that of an isotropic radiation field; it can be varied by changing the number of lamps used to illuminate the inside of the sphere. Available records indicate that the integrating sphere was calibrated in the laboratory for linearity of output and spectral radiance initially in 1974 and again in 1983. A slightly larger (40-inch diameter and 14-inch aperture) sphere is illustrated in Figure 2.10; the black honeycomb below it is used as a space reference during calibration.

Rao (1987) described the method of determination of spectral radiance and linearity of output of the integrating sphere using, in the main, data obtained during the calibration performed in 1983. An Optronic Laboratories Model 740A Optical Radiation Measurement System, essentially consisting of a single-grating monochromator with interchangeable gratings to cover the spectral region from 0.35–2.0 μm, and equipped with interchangeable ultraviolet-enhanced silicon and thermoelectrically cooled germanium detectors, and an infrared blocking filter, was used to compare the spectral output of the integrating sphere to that of an Optronic Laboratories Model 420 continuously variable integrating sphere source which in turn has been calibrated for spectral radiance over the 0.35–2.0 μm wavelength region relative to the National Bureau of Standards (NBS) radiance and spectral irradiance scale. The estimated uncertainty in the calibration of the Model 420 calibration source relative to the NBS standards is given in Table 2.6.

The radiometric quantities measured as part of the calibration of the integrating sphere are the near-normal spectral radiance of the 12-inch aperture and the relative radiances at selected wavelengths as the number of lamps illuminating the sphere is changed from 12 to 1 in steps of 1. The spectral radiance measurements have been made at 0.05 μm intervals over the spectral region of interest. It has been estimated that the uncertainty in the transfer calibration from the Optronic Laboratories Model 420 continuously variable integrating sphere source to the NASA integrating sphere is 1.5%.

Table 2.7 shows the measured values of the near-normal spectral radiance of the integrating sphere when it was illuminated with 12 lamps. Assuming that the integrating sphere is an isotropic source of radiation, the spectral irradiances are calculated as π times the spectral radiances. The discrepancies between the radiances measured at wavelengths common to the calibrations performed in 1974 and 1983 may in part be due to the fact that the integrating sphere was disassembled and repainted prior to the 1983 calibration and to the differences in the calibration procedures adopted.

The relative variations in the integrating sphere radiances at the wavelengths of 0.45 and 0.55 μm as the number of lamps was changed from 12 to 1 in steps of 1 have also been studied. It is observed that the mean and standard deviation of the ratio $(I_n/I_{12})/(n/12)$, where I_n and I_{12} are the integrating sphere radiances when illuminated with n ($n \leq 12$) and 12 lamps respectively, have values of 1.0026 and 0.0042 at 0.45 μm and 1.0011 and 0.0047 at 0.55 μm; these numbers can be considered a measure of the linearity of the integrating sphere source in the visible region of the spectrum.

Figure 2.10   This 40″ diameter calibration sphere with 14″ diameter opening is used to calibrate solar channels of space instruments (ITT). The black honeycomb in the space below it is used as a space reference during calibration.

Table 2.6   Uncertainty in the Model 420 calibration source (Rao 1987)

Spectral interval (μm)	Uncertainty (%)
0.35–0.40	3
0.40–0.90	2
0.90–1.30	3
1.30–2.00	4

Table 2.7 Near-normal spectral radiance of the NASA 30-inch integrating sphere source (Rao 1987)

Wavelength (μm)	Spectral radiance (W m^{-2} μm^{-1} sr^{-1})	
	a	b
0.500	178.2	180.0
0.550	—	264.0
0.600	349.7	351.0
0.650	—	429.0
0.700	487.8	505.0
0.750	—	568.0
0.800	618.8	606.0
0.850	—	610.0
0.900	655.5	613.0
0.950	—	606.0
1.000	—	602.0
1.050	569.8	589.0
1.100	—	560.0

Notes:
1. Values in column 'a' are based on the 1974 calibration of the integrating sphere.
2. Values in column 'b' are based on the 1983 calibration of the integrating sphere.
3. The spectral irradiance (W m^{-2} μm^{-1}) is given by π times the spectral radiance.

Figure 2.11 shows the spectral irradiances of the integrating sphere and of the Sun. The solar irradiance presented by three different sources is shown; these data were utilised to demonstrate the dependence of the reflectance factor of the integrating sphere on the solar spectral irradiances used. Tables of the values of the curves plotted in Figure 2.11 are given by Rao (1987). It is apparent from Figure 2.11 that the sphere and solar irradiances vary with wavelength in opposite senses, with the cross-over point lying around 0.69 μm. This wavelength dependence of the integrating sphere irradiance is such that the filtered sphere irradiance in channel 2 when the sphere is illuminated with 12 lamps is very nearly double the filtered solar irradiance in this channel; thus, in practice, since the reflectance factor $A \leq 1$, all calibrations of channel 2 are performed with the sphere illuminated with up to six lamps only.

The 'sphere ratio', which is assumed to be independent of wavelength, is a quantity which is used to calculate the reflectance factor of the integrating sphere at different levels of illumination, given the reflectance factor at the highest level of illumination (12 lamps for channel 1; 6 lamps for channel 2). It is defined as the normalised relative radiance of the integrating sphere at the wavelength of 0.55 μm (at the corresponding levels of illumination), the normalisation factor being the relative radiance at 0.55 μm at the highest level of illumination. Table 2.8 gives the

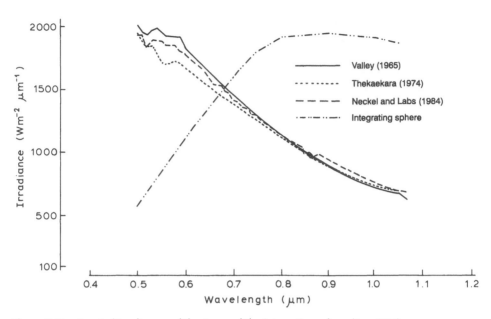

Figure 2.11   Spectral irradiances of the Sun and the integrating sphere (Rao 1987).

Table 2.8   Measured and calculated sphere ratios (Rao (1987)

	Sphere ratio					
	Observed				Calculated	
	Channel 1		Channel 2		Channel 1	Channel 2
Number of lamps (n)	a	b	a	b	($n/12$)	($n/6$)
12	1.0000	1.0000	—		1.0000	—
11	0.9154	0.9160	—		0.9167	—
10	0.8334	0.8340	—		0.8333	—
9	0.7504	0.7510	—		0.7500	—
8	0.6647	0.6690	—		0.6667	—
7	0.5801	0.5850	—		0.5833	—
6	0.4949	0.5040	1.0000	1.0000	0.5000	1.0000
5	0.4130	0.4200	0.8345	0.8330	0.4167	0.8333
4	0.3342	0.3360	0.6753	0.6667	0.3333	0.6667
3	0.2475	0.2510	0.5001	0.4980	0.2500	0.5000
2	0.1628	0.1660	0.3290	0.3294	0.1667	0.3333
1	0.0798	0.0834	0.1612	0.1655	0.0833	0.1667
0	0		0		0	0

Notes:

1. Values in column 'a' are based on the 1974 calibration of the integrating sphere.
2. Values in column 'b' are based on the 1983 calibration of the integrating sphere.

measured sphere ratios based on the 1974 and 1983 calibrations of the integrating sphere. Table 2.8 also includes the ratios $n/12$ and $n/6$, n being the number of lamps; these are referred to as the calculated sphere ratios.

It is apparent that there is very close correspondence, except at low levels of illumination ($n \leq 2$ for channel 1; $n = 1$ for channel 2), between the two sets of measured sphere ratios, and between the measured and calculated values. Regression of the measured ratios on the calculated sphere ratios yields slopes very close to unity and coefficients of determination in excess of 0.99; similar values for the regression parameters are obtained when the 1983 sphere ratios are regressed on the 1974 values. It is thus reasonable to assume that the integrating sphere is a linear source of illumination; however, we cannot comment upon the stability of spectral radiance of the integrating sphere as a function of time because of the sparsity of data and of the fact that the sphere was disassembled and repainted between the two calibrations performed in 1974 and 1983.

The laboratory calibration exercise consists in establishing how the upwelling radiances from the Earth–atmosphere system in the passbands of the AVHRR channels 1 and 2 can be estimated using the simple linear regression relation established in the laboratory between the variable radiances of an integrating sphere source and the corresponding AVHRR digital signals.

It has been the usual practice to express the radiances measured in channels 1 and 2 of the AVHRR in terms of an 'albedo' which relates the measured radiance to the filtered extraterrestrial solar irradiance in the passband of a given channel. However, in the light of accepted meteorological usage, Rao proposed that the term 'albedo' should be used exclusively to denote the ratio of the upward and downward fluxes of radiation at a surface and preferred to use the term 'reflectance factor' in connection with the calibration of channels 1 and 2.

Let the filtered radiance R_i (W m^{-2} sr^{-1}), when the AVHRR is viewing the integrating sphere source being used for calibration in the laboratory, be given by

$$R_i = \int_{\lambda_1}^{\lambda_2} S_\lambda \tau_\lambda \, d\lambda \tag{2.2.1}$$

where S_λ is the radiance of the integrating sphere source at the wavelength λ, τ_λ is the normalised response function of the given channel at the wavelength λ and λ_1, λ_2 are the cut-on and cut-off wavelengths of the passband of the given channel. We shall define the filtered extraterrestrial solar irradiance (W m^{-2}) as

$$F = \int_{\lambda_1}^{\lambda_2} F_{0\lambda} \tau_\lambda \, d\lambda \tag{2.2.2}$$

where $F_{0\lambda}$ is the extraterrestrial solar spectral irradiance at the wavelength λ.

The reflectance factor A of the integrating sphere source is defined as the ratio

$$A = \frac{\pi}{F} R_i . \tag{2.2.3}$$

The reflectance factor can thus be interpreted as the ratio of the irradiance of an isotropic radiation field with radiance R_i (the measured filtered radiance) to the filtered solar irradiance.

In a manner analogous to equation (2.2.3), we can define the reflectance factor A_e

of the Earth–atmosphere scene being viewed by the AVHRR in orbit as

$$A_e = \pi \int_{\lambda_1}^{\lambda_2} R_\lambda \tau_\lambda \, d\lambda / F = \frac{\pi R_e}{F} \qquad (2.2.4)$$

where R_λ is radiance of the Earth–atmosphere scene at the wavelength λ. It has been the practice to estimate A_e, expressed as a percentage, and subsequently the corresponding radiance R_e, using the transmitted AVHRR signals in the relation given in equation (2.2.4). It is thus tacitly assumed that the radiometer response is sensitive only to the total amount of radiant energy that is sensed by the instrument within the passband of the channel and not to the spectral energy distribution of the scene radiance over the passband.

The filtered radiance R_e (W m^{-2} sr^{-1}) of the Earth–atmosphere scene is given by

$$R_e = (FA_e)/100\pi \qquad (2.2.5)$$

and the corresponding spectral radiance R (W m^{-2} μm^{-1} sr^{-1}) by

$$R = R_e/w = FA_e/100\pi w . \qquad (2.2.6)$$

It is important that the F values used in equations (2.2.5) and (2.2.6) are based on the spectral solar irradiance data that were used in equation (2.2.2).

As mentioned earlier, the basic objective of the pre-launch calibration has been to establish a simple linear regression relation between the reflectance factor A_e, expressed as a percentage, of the integrating sphere source corresponding to different levels of illumination and the AVHRR digital signal X, in counts; thus

$$A_e = MX + I . \qquad (2.2.7)$$

What is finally tabulated in Appendix B of NESS107, for each spacecraft, is the values of M and I for each of channels 1 and 2 of the AVHRR on that spacecraft. In NESS107 A_e is actually expressed as a percentage instead of as a fraction. X (also described as counts, C, or digital number DN) is the pixel value from the raw data and is a 10-bit number; its value is thus an integer in the range 0–1023.

A_e, the reflectance (or target albedo as it is described in NESS107) (as a percentage) for channel 1 or channel 2 can thus be converted into the spectral radiance R (in W m^{-2} μm^{-1} sr^{-1}) by using equation (2.2.6). F is the integrated solar radiance weighted by the spectral response function of the channel (1 or 2 as appropriate) (see equation (2.2.2)) and w is the equivalent width of the spectral response function of the channel. We use the symbol R for the radiance; however, it should be noted that in NESS107 it is denoted by I. But NESS 107 also uses the symbol I for the intercept on the right-hand side of equation (2.2.7) and so we have made this change to avoid any possible confusion. The symbol R is used by Kidwell (1991).

One could substitute the value of A_e from equation (2.2.7) into equation (2.2.6) to give a relation between the radiance and the counts X, i.e.

$$R = (F/100\pi w)(MX + 1) \qquad (2.2.8)$$

which can be rewritten as

$$R = S(X - X_0) . \qquad (2.2.9)$$

X_0 is then described as an offset value. From these equations this means that

$$S = (FM)/(100\pi w) \qquad (2.2.10)$$

where S is in units of W m^{-2} μm^{-1} sr^{-1} count^{-1} and

$$X_0 = -I/M .\qquad(2.2.11)$$

S and X_0 are not given in Appendix B of NESS107 but their values can easily be calculated from the values of M and I which are given there.

We see from the above that the laboratory calibration of channels 1 and 2 of the AVHRR is essentially directed towards the estimation of the radiance of the Earth–atmosphere scene in terms of the radiance of the integrating sphere source. We shall now describe the various stages and procedures of establishing the aforementioned regression relation.

The experimental set-up for the determination of the normalised spectral response – also referred to as normalised response – of channels 1 and 2 is seen in Figure 2.12; it consists of a stable source (Nernst glower) of illumination, a 1/4 metre Jarrell–Ash Model 82-410 grating monochromator (resolving power: better than 3×10^{-4} μm at the wavelength of 0.313 μm (second-order spectrum) with 150 μm slits; dispersion: 3.3×10^{-3} μm/mm with the 1180 grooves/mm grating), a beam-folding mirror and collimator assembly, a pyroelectric detector and the AVHRR. The radiation emerging from the exit slit of the monochromator is directed towards the pyroelectric detector when the 'In-Out' mirror is in the light path; it is directed towards the AVHRR when the 'In-Out' mirror is removed from the light path.

The relative spectral radiance, in the form of voltages, of the Nernst glower, a standard temperature-controlled ceramic (zirconium oxide) rod which is heated to

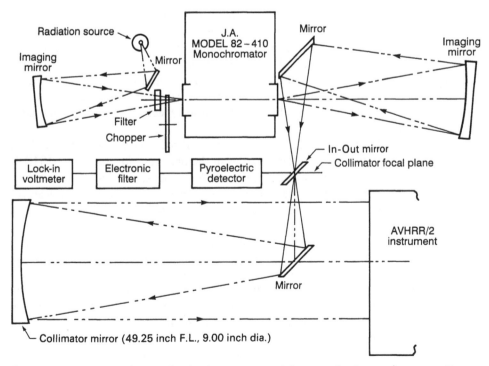

Figure 2.12 Experimental set-up for the determination of the normalised spectral response (Rao 1987).

incandescence (1200–1700°C) by passing an electric current through it, is measured at pre-determined wavelengths falling within the passband of each channel with the Jarell–Ash monochromator and pyroelectric detector. The radiation emerging from the exit slit of the monochromator is then directed towards the AVHRR by removing the 'In-Out' mirror from the light path and the AVHRR signals are recorded at the same wavelengths.

The relative spectral response of the AVHRR at the wavelength λ is then given by

$$\tau_i = V_i/(V_r R_i) \tag{2.2.12}$$

where V_i is the AVHRR signal at the wavelength λ, V_r is the pyroelectric detector or reference signal at the wavelength λ, and R_i is the reflectance of the collimator mirror at the wavelength λ. Similarly, the relative response of the AVHRR at the wavelength of peak or maximum response, λ_p, is given by

$$\tau_p = V_{ip}/(V_{rp} R_{ip}) \tag{2.2.13}$$

where V_{ip} is the AVHRR signal at the wavelength of peak response, V_{rp} is the pyroelectric detector or reference signal at the wavelength of peak response and R_{ip} is the reflectance of the collimator mirror at the wavelength of peak response. The normalised spectral response of the AVHRR at the wavelength λ is then given by

$$\tau_\lambda = \tau_i/\tau_p . \tag{2.2.14}$$

We shall illustrate this method of calculating the normalised spectral response of the AVHRR by referring to the entries in Table 2.9 where we have shown data obtained for channel 1 of the AVHRR on board NOAA-9. Let us consider the entries for the wavelength of 0.57 μm. We notice that at this wavelength $V_i = 990$ mV; $V_r = 5.034$ mV; and $R_i = 0.792$. Also, the AVHRR is observed to have peak response at the wavelength of 0.680 μm. The corresponding values of V_{ip}, V_{rp} and R_{ip} are respectively 1976 mV, 5.271 mV and 0.764. Substitution of these numerical values in equations (2.2.12) and (2.2.13) will yield values of 248.3115 and 490.6825 for τ_i and τ_p respectively. Using these relative spectral response values in equation (2.2.14), we obtain a value of 0.5061 or 50.61% for the normalised spectral response of channel 1 at the wavelength of 0.57 μm.

It should be observed that the maximum normalised spectral response of 1 (or 100%) can occur at different wavelengths within the passband for different AVHRRs because of differences in the wavelength-dependent characteristics of the various components of the optical train and of the responsiveness of the detector. Indeed the general shape of the normalised response curves could also be different as was illustrated in Figure 2.9. It is Figure 2.9(c) which shows the curve corresponding to the NOAA-9 channel-1 data in Table 2.9. The small but finite off-band response, at wavelengths longer than 0.8 μm, in channel 1 of the AVHRR on board NOAA-8 (Figure 2.9(b)), contributes about 2% of the various filtered radiances and irradiances, and affects the equivalent width of the channel to the same extent; the mean wavelength is affected to a smaller degree. Similar, off-band, responses in channel 1 of some of the other AVHRRs have smaller effects.

The experimental set-up of the determination of the reflectance factor of the integrating sphere in the laboratory is seen in Figure 2.13. It consists of the integrating sphere, the AVHRR, the space clamp target, the AVHRR Bench Control Unit (BCU) computer and associated electronics. A very detailed procedure has been

Table 2.9 Laboratory determination of the normalised spectral response of channel 1 of the AVHRR(FM 202) on board NOAA-9 (ITT 1980)

Wavelength (μm)	Signal amplitude		Channel 1 Reference	Mirror reflectance	Relative spectral response	Normal response
	Reference (mV)	Channel 1				
0.500	2.610	0	0	0.810	0	0.0000
0.510	3.055	0	0	0.806	0	0.0000
0.520	3.386	0	0	0.802	0	0.0000
0.530	3.874	0	0	0.800	0	0.0000
0.540	4.202	1	0.2380	0.797	0.2986	0.0006
0.550	4.508	25	5.5457	0.796	6.9670	0.0142
0.560	4.772	383	80.2598	0.794	101.0829	0.2060
0.570	5.034	990	196.6627	0.792	248.3115	0.5061
0.580	5.184	1370	264.2747	0.790	334.5249	0.6818
0.590	5.359	1669	311.4387	0.788	395.2268	0.8055
0.600	5.545	1743	314.3372	0.786	399.9202	0.8150
0.610	5.603	1681	300.0178	0.784	382.6758	0.7800
0.620	5.646	1758	311.3709	0.781	398.6823	0.8125
0.630	5.692	1950	342.5861	0.779	439.7767	0.8963
0.640	5.651	1963	347.3721	0.777	447.0684	0.9111
0.650	5.715	1794	313.9108	0.774	405.5695	0.8265
0.660	5.616	1721	306.4459	0.772	396.9506	0.8090
0.670	5.495	1830	333.0300	0.768	433.6328	0.8837
0.680	5.271	1976	374.8814	0.764	490.6825	1.0000
0.690	5.160	1681	325.7752	0.762	427.5265	0.8713
0.700	5.171	907	175.4013	0.759	231.0952	0.4710
0.710	5.080	403	79.3307	0.756	104.9348	0.2139
0.720	4.978	180	36.1591	0.753	48.0201	0.0979
0.730	4.781	89	18.6154	0.748	24.8868	0.0507
0.740	4.579	49	10.7010	0.743	14.4025	0.0294
0.750	4.394	29	6.5999	0.738	8.9430	0.0182
0.760	10.40	50	4.8077	0.734	6.5500	0.0133
0.770	10.55	37	3.5071	0.732	4.7911	0.0098
0.780	10.60	28	2.6415	0.731	3.6136	0.0074
0.790	10.75	24	2.2326	0.726	3.0751	0.0063
0.800	10.80	23	2.1296	0.722	2.9496	0.0060

established by ITT for the operation of the integrating sphere, with special reference to the manner and order in which the twelve matched quartz-halogen lamps should be switched on and off. The space clamp target – a cubic (~ 1 m on the side) honeycomb structure blackened on the inside – is used to simulate space-view conditions for the AVHRR.

The experiment is started with the illumination of the sphere at its highest, i.e. with 12 lamps on for channel 1 and, for reasons mentioned earlier, with 6 lamps on for channel 2. The AVHRR is directed to view the 12-inch aperture of the integrating sphere and 3600 measurements are made of the AVHRR signal, in volts, by the BCU computer; the mean and standard deviation of these measurements are then calculated and recorded as the uncorrected AVHRR signal and noise, respectively; the corresponding counts on a 10-bit scale are calculated by multiplying these AVHRR signals, expressed in millivolts, by 0.160 and rounding off the product

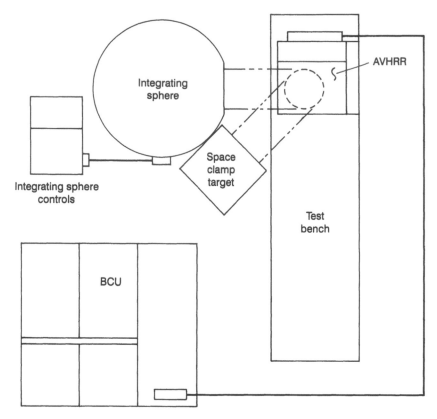

Figure 2.13 Experimental set-up for the radiance calibration of AVHRR channels 1 and 2 (Rao 1987).

to the nearest integer. The AVHRR is then directed towards the space clamp target and the corresponding space clamp signal and noise are recorded. This procedure is repeated, switching off one lamp at a time in a pre-determined order – thereby varying the integrating sphere radiance in a known, step-wise manner – until all the lamps have been switched off. It is observed that the AVHRR signal when it views the integrating sphere with all the lamps switched off is very close to the space clamp signal, as it should be. This zero-level signal is then subtracted from the signals measured at various levels of illumination of the integrating sphere and the result recorded as the corrected signal. The ratio of the corrected AVHRR signal when the radiometer is viewing the integrating sphere illuminated with n lamps (integer $n \leq 12$ for Channel 1; ≤ 6 for Channel 2) to the corrected AVHRR signal when the radiometer is viewing the integrating sphere illuminated with 12 or 6 lamps as the case may be is calculated; the degree of correspondence between this ratio and the sphere ratio (see above) is a measure of the linearity of response of the AVHRR over its dynamic range, assuming that the integrating sphere radiance varies linearly, over the spectral region of interest, with the number of lamps that have been switched on and that its spectral distribution is not affected by the number of lamps that are on. Experimental evidence indicates that the response of the AVHRR may be considered linear for all practical purposes. However, the purpose of the pre-launch calibration of the AVHRRs on board NOAA-6 to

Table 2.10 Reflectance factor of channel 1 (AVHRR FM 202; NOAA-9) (Rao 1987)

Spectral interval (μm)	Relative response	Sphere irradiance (W m^{-2} μm^{-1})	Solar irradiance (W m^{-2} μm^{-1})	Sphere output (W m^{-2})	Solar output (W m^{-2})
0.55–0.60	0.505	966.1	1702.2	24.39	42.98
0.60–0.65	0.840	1225.2	1588.0	51.46	66.70
0.65–0.68	0.879	1419.4	1470.0	37.43	38.76
0.68–0.70	0.781	1538.7	1399.3	24.03	21.86
0.70–0.75	0.147	1685.5	1302.0	12.39	9.57
0.75–0.80	0.010	1844.1	1172.2	0.92	0.59
			Sum:	150.62	180.46

Reflectance factor or albedo: $\dfrac{\text{Total sphere output}}{\text{Total solar output}}$

$: 150.62/180.46 = 0.8346$ or 83.46%

Notes:
Solar output: product of the width of the spectral interval, the relative response and the solar irradiance.
Sphere output: product of the width of the spectral interval, the relative response and the sphere irradiance.

NOAA-10 has been to relate the uncorrected AVHRR signals, expressed in counts, to the reflectance factor of the integrating sphere as the number of lamps is changed from 12 to 0 for channel 1 and from 6 to 0 for channel 2 in steps of 1.

We show in Tables 2.10 and 2.11 the procedure adopted in the laboratory to calculate the reflectance factor of the integrating sphere when all the 12 lamps are on, using data for channels 1 and 2 of the AVHRR (SN 202) on board NOAA-9; the passbands of the two channels have been divided into six intervals; the average

Table 2.11 Reflectance factor of channel 2 (AVHRR SN 202; NOAA-9) (Rao 1987)

Spectral interval (μm)	Relative response	Sphere irradiance (W m^{-2} μm^{-1})	Solar irradiance (W m^{-2} μm^{-1})	Sphere output (W m^{-2})	Solar output (W m^{-2})
0.69–0.72	0.224	1604.3	1357.3	10.78	9.12
0.72–0.78	0.912	1770.9	1236.3	96.90	67.65
0.78–0.83	0.904	1894.4	1097.2	85.62	49.59
0.83–0.88	0.881	1917.0	980.0	84.44	43.17
0.88–0.94	0.823	1918.7	882.7	94.74	43.59
0.94–1.07	0.292	1876.6	742.7	71.23	28.19
			Sum:	443.71	241.31

The sphere output corresponds to illumination with 12 lamps; however, only six lamps are on when channel 2 is calibrated; thus, the sphere output should be multiplied by the appropriate sphere ratio (Table 2.8) which, in this case, is 0.5040.

Reflectance factor or albedo: $\dfrac{0.5040 * \text{Total sphere output}}{\text{Total solar output}}$

$: 223.63/241.31 = 0.9267$ or 92.67%

Note: Solar output and sphere output defined as in Table 2.10.

Table 2.12 Data for the development of the regression relationship between the reflectance factor (albedo) of the integrating sphere and the AVHRR signals (SN 202; NOAA-9) (ITT 1980)

Number of lamps	Channel 1						Channel 2					
	Albedo	Signal (mV)	Counts	Std. dev. (mV)	Space signal (mV)	Space std. dev. (mV)	Albedo	Signal (mV)	Counts	Std. dev. (mV)	Space signal (mV)	Space std. dev. (mV)
12	82.43	5075.98	812	2.402	244.05	1.939						
11	75.46	4656.08	745	1.284	243.93	2.079						
10	68.70	4268.78	683	1.687	244.08	2.308						
9	61.86	3859.47	618	2.953	243.93	2.070						
8	54.79	3443.06	551	1.905	243.99	2.091						
7	47.82	3030.73	485	1.802	243.87	2.096						
6	40.79	2630.89	421	1.600	243.85	1.980	96.72	5862.49	940	2.106	250.25	1.940
5	34.04	2234.16	357	2.920	243.70	2.112	80.71	4930.74	789	1.749	250.14	1.636
4	27.55	1837.61	294	1.311	243.74	2.033	65.31	3998.90	640	2.800	250.55	2.145
3	20.40	1417.18	227	2.555	243.83	2.175	48.32	3024.98	484	1.973	250.15	1.714
2	13.42	1009.12	161	2.838	243.84	2.088	31.82	2074.18	332	1.871	250.33	1.948
1	6.58	604.19	97	2.783	243.76	2.073	15.60	1117.46	179	2.623	250.31	1.922
0	0	244.97	39	2.545	243.74	2.117	0	250.01	40	1.735	250.30	2.019

values of the normalised response, the sphere irradiance and the solar irradiance (Thekaekara *et al.* 1969, Thekaekara 1974) were then calculated over these intervals and used in the computation of the integrating sphere and solar output; the reflectance factor (or albedo) is then given by the ratio of the total sphere output to the total solar output; however, in the case of channel 2, since we have computed the total sphere output corresponding to illumination of the integrating sphere with 12 lamps, we have to multiply it by the sphere ratio corresponding to 6 lamps – 0.5040 (Table 2.8) for the 1983 calibration – before calculating the reflectance factor for reasons mentioned earlier.

It should be noted that the choice of the number of spectral intervals into which the passband of a channel is divided in the calculation of the reflectance factor of the integrating sphere is purely arbitrary; Rao (1987) showed that the calculated reflectance factors vary by about one percent as the number of spectral intervals is changed from 6 to 20 or more, provided the average values of the normalised response function of the channel and of the various irradiances have been properly computed.

We shall now illustrate the regression of the integrating sphere reflectance factor on the AVHRR digital signal, using the data in Table 2.12 where we have shown the measurements obtained with channels 1 and 2 of the AVHRR (SN 202) on board NOAA-9. The entries in the 'albedo' column are obtained by multiplying the reflectance factor of the integrating sphere at full illumination (12 lamps for channel 1; 6 lamps for channel 2) by the appropriate sphere ratios listed in Table 2.8; for example, the reflectance factor of 47.82% in channel 1 when the sphere is illuminated with 7 lamps is obtained by multiplying the 'albedo' of 82.43% at the level of highest illumination by the sphere ratio of 0.5801. The corresponding AVHRR digital signals, entered in the 'counts' column, are obtained, as was mentioned earlier, by multiplying the AVHRR signals in mV listed in the 'Signal (mV)' column by 0.160 and rounding off the product to the nearest integer. There are 13 pairs of 'albedo–counts' data, including measurements made when the integrating sphere is not illuminated (number of lamps: 0), for channel 1 and 7 for channel 2. Simple regression of 'albedo' on the counts yields the regression coefficients which are the required slope, M, and intercept, I, of equation (2.2.7).

The values of the slope, M, and the intercept, I, for channels 1 and 2 are given, for each of the AVHRRs in the series, in Appendix B of NESS107. These values are collected in Table 2.13. The dependence of the reflectance (or 'albedo') on counts for

Table 2.13 Pre-launch slopes and intercepts for AVHRR channels 1 and 2 (Kidwell 1991)

Satellite	M_1	I_1	M_2	I_2
TIROS-N	0.1071	−3.9	0.1051	−3.5
NOAA-6	0.1071	−4.1	0.1058	−3.5
NOAA-7	0.1068	−3.4	0.1069	−3.5
NOAA-8	0.1060	−4.2	0.1060	−4.2
NOAA-9	0.1063	−3.8	0.1075	−3.9
NOAA-10	0.1059	−3.7	0.1058	−3.6
NOAA-11	0.0950	−3.8	0.1061	−3.6
NOAA-12	0.1042	−4.4	0.1014	−4.0

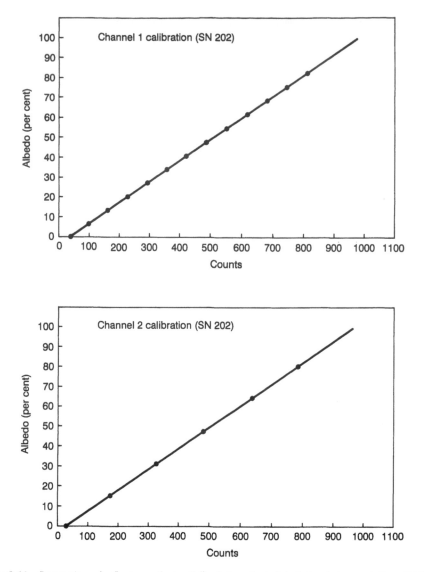

Figure 2.14   Regression of reflectance factor ('albedo') against digital signals (counts) (Rao 1987).

channels 1 and 2 is shown in Figure 2.14. Instead of the linear relation between the reflectance and counts, it is also useful to use the linear relation between radiance, R, and counts, X, given in equation (2.2.9) and that the constants S and X_0 in this equation can be obtained from M and I via equations (2.2.10) and (2.2.11), respectively. This requires a knowledge of the values of F and w. The values of F and w are also collected in Table 2.14.

An important point which is discussed in some detail by Rao (1987) is the question of determining the errors on the values of the calibration coefficients (slopes and intercepts) which are given in Table 2.13 and which are used in practice in the operational calibration of AVHRR data from the NOAA series of satellites. In Figure 2.11 we illustrated three slightly different curves that represent the spectral irradiance of the Sun; these three slightly different curves give rise to three slightly

Table 2.14 Values of w and F for AVHRR channels 1 and 2 (Kidwell 1991)

Satellite	w_1	F_1	w_2	F_2
TIROS-N	0.325	443.3	0.303	313.5
NOAA-6	0.109	179.0	0.223	233.7
NOAA-7	0.108	177.5	0.249	261.9
NOAA-8	0.113	183.4	0.230	242.8
NOAA-9	0.117	191.3	0.239	251.8
NOAA-10	0.108	178.8	0.222	231.5
NOAA-11	0.113	184.1	0.229	241.1
NOAA-12	0.124	200.1	0.219	229.9

different values of F for each channel of each AVHRR. Uncertainties in the extraterrestrial solar irradiance lead to uncertainties, i.e. errors, in the values of the calibration coefficients in Table 2.13. This will not be considered in detail here and for details one will need to refer to the original report by Rao (1987).

We have given considerable details of the pre-launch radiometric calibration of the visible and near-infrared channels; it was fortunate that this information had been collected together by Rao (1987). From what we have said about channels 1 and 2 it should be apparent that the pre-launch radiometric calibration is a complicated task which must be undertaken with very great care. We turn now to the thermal infrared channels, channels 3, 4 and 5, of the AVHRR. The pre-launch calibration of the thermal infrared channels of the AVHRR is described briefly by Weinreb et al. (1990) and Rao (1993a); otherwise a more detailed description can only be found in the manuals or contract reports from the original contractors (ITT Aerospace/Optical Division) which are cited by these authors. The AVHRR, operating as in its orbit, is situated in a thermal vacuum test chamber, see Figure 2.15, where it views a calibrated blackbody representing the Earth scene, a second blackbody which is cooled to about 77 K to simulate the in-orbit space view and the internal calibration target (ICT) on the baseplate of the AVHRR (see Figure 1.9). The Earth and space calibration simulation targets are the top and right, respectively, large cylindrical objects shown inside the chamber in Figure 2.15; these two targets are lined with black honeycomb for high emissivity. The AVHRR is upside down in the chamber, i.e. the Earth scan view is upwards. Below the right-hand cylinder is a target cooled with liquid helium to simulate space for the cooler (which is beneath the Earth shield). The rectangular box at the left of the chamber is a visibility stability target; this is a feedback-controlled very stable light source and it is used to ensure that the solar channel gain is not dependent on the temperature of the instrument. Finally, one can see that the AVHRR is mounted on a slide for easy installation and access. The scan mirror of the AVHRR rotates at six revolutions per second (6 Hz) and reflects radiation, in turn, from the Earth scene blackbody (top), from the cooled blackbody simulating the space view (right) and from the internal calibration target (ICT) on the baseplate on to the detectors of the AVHRR. The output from the detectors is amplified, digitised and recorded. Data are collected as the temperature of the laboratory blackbody is cycled between 175 K and 315 K, in 10 degK steps between 175 K and 290 K and in 5 degK steps above 290 K. This procedure is carried out for three values of the temperature of the ICT of

AVHRR thermal vacuum chamber is used to simulate space conditions in the laboratory

"Earth" and "space" infrared calibration targets are lined with black honeycomb for high emissivity. These targets give precise temperature measurements which are used to calibrate the infrared channels

Radiant cooler target is cooled with liquid He to simulate space for the cooler

Visible stability target is a feedback controlled light source which is very stable. It is used to ensure that the solar channel gain is not dependent on instrument temperature

AVHRR instrument is mounted on a slide for easy installation and access

Figure 2.15   An AVHRR and the thermal vacuum chamber used for the pre-launch calibration of the thermal infrared channels (ITT).

the AVHRR, nominally 10°C, 15°C and 20°C. (Note that the terms 'ICT temperature', 'baseplate temperature' and 'operating temperature' have been used interchangeably in the literature.)

The calibration of the blackbody sources is traceable back to NBS (the (US) National Bureau of Standards). The calibration coefficients for channels 3, 4 and 5 of each of the AVHRRs which has been launched are given in Appendix B of NESS107. As each new spacecraft is launched, so also a supplement to this Appendix B is published by NOAA. The part of Appendix B which refers to the AVHRR for NOAA-9 is reproduced in Table 2.15, as an example. The meaning of the various parts of Table 2.15 will become clear when we describe the in-flight calibration in section 2.2.3.

From our discussions of the NOAA polar-orbiting meteorological spacecraft and of the AVHRR in chapter 1 it should have become clear that for any given AVHRR instrument there are several stages in its life:

(i)   construction and pre-launch calibration by ITT Aerospace

(ii) delivery of AVHRR for storage, spacecraft integration and launch

(iii) launch

(iv) operation in space

(v)   decommissioning.

It should also have become apparent that, particularly because of the development of the five-channel AVHRR/2, for some instruments there was a long time lapse between the original calibration and launch. As a result of statements in the report by Rao (1987) and of studies by Weinreb *et al.* (1990), pre-launch calibration practices have been revised. Therefore it was argued that the data published by NOAA should be obtained from a pre-launch calibration carried out as close as possible to the launch date itself (and not years earlier) and so AVHRRs are now recalibrated before launch. Thus, for example, the AVHRR on NOAA-11 was originally calibrated but channels 1 and 2 were recalibrated in 1988 shortly before launch and fresh calibration coefficients were published by Abel (1990).

2.2.3 In-flight calibration of HRPT data

There is no in-flight calibration facility for channels 1 and 2 of the AVHRR but in-flight calibration is available for all three AVHRR channels 3, 4 and 5. These channels are calibrated by a two-point in-flight calibration using two sources of radiation, namely part of the housing of the internal calibration target, which is designed as a blackbody (see Figure 1.9), and deep space. The idea is that one obtains the digitised output from the detectors for these two sources. The temperature of the on-board target is measured in-flight, making use of the values of necessary constants that were determined from the pre-launch calibration, and the temperature of deep space is known. Thus for two points we have the values of the temperature (or the radiance) and the corresponding values of the digitised scanner output. A straight line can then be drawn through these two points. This straight line could then be used to calibrate the digital data from the scan line of the surface of the Earth; thus in using an equation such as (2.2.9) we do not have to rely on

Table 2.15   AVHRR part of Appendix B of Planet (1988) for NOAA-9

NOAA/F-9 coefficients*				
PRT	a_0	a_1	a_2	a_3	a_4
1	277.018	0.051 28	0.0	0.0	0.0
2	276.750	0.051 28	0.0	0.0	0.0
3	276.862	0.051 28	0.0	0.0	0.0
4	276.546	0.051 28	0.0	0.0	0.0

b_i – PRT Weighting Factors

b_1	b_2	b_3	b_4
0.25	0.25	0.25	0.25

N_{SP} = Radiance of space including non-linearity correction

Channel	N_{SP} (mW sr^{-1} m^{-2} cm)
3	0.0
4	-3.384
5	-2.313

Visible coefficients

	G	I
1	0.106 340 036 6	$-3.846 375 952$
2	0.107 491 067	$-3.877 005 621$

Normalised Response Functions

AVHRR Channel 3
Starting wave number   2469.135 50
Delta wave number   7.768 49
                                        60

0.0	0.75765E-05	0.14659E-04	0.23167E-04	0.43949E-04
0.91160E-04	0.18353E-03	0.35407E-03	0.68830E-03	0.12443E-02
0.18939E-02	0.24721E-02	0.29108E-02	0.31837E-02	0.33195E-02
0.33728E-02	0.33888E-02	0.34053E-02	0.34316E-02	0.34478E-02
0.34401E-02	0.34075E-02	0.33552E-02	0.33039E-02	0.32757E-02
0.32835E-02	0.33344E-02	0.34060E-02	0.34511E-02	0.34509E-02
0.34395E-02	0.34470E-02	0.34542E-02	0.34294E-02	0.33826E-02
0.33584E-02	0.33815E-02	0.34250E-02	0.34563E-02	0.34623E-02
0.34406E-02	0.33943E-02	0.33365E-02	0.32740E-02	0.31516E-02
0.28842E-02	0.24409E-02	0.19012E-02	0.13559E-02	0.87464E-03
0.51259E-03	0.30171E-03	0.19524E-03	0.13052E-03	0.74406E-04
0.31412E-04	0.75269E-05	0.0	0.0	0.0

AVHRR Channel 4
Starting wave number   862.068 85
Delta wave number   2.378 12
                                        60

0.0	0.30603E-04	0.64563E-04	0.10523E-03	0.17057E-03
0.37139E-03	0.85488E-03	0.17526E-02	0.29947E-02	0.43718E-02
0.56739E-02	0.67844E-02	0.77153E-02	0.84851E-02	0.91222E-02
0.96298E-02	0.10022E-01	0.10310E-01	0.10525E-01	0.10708E-01

Table 2.15   (Continued)

0.10903E-01	0.11130E-01	0.11370E-01	0.11596E-01	0.11786E-01
0.11949E-01	0.12111E-01	0.12299E-01	0.12523E-01	0.12746E-01
0.12926E-01	0.13022E-01	0.13039E-01	0.13030E-01	0.13047E-01
0.13135E-01	0.13274E-01	0.13419E-01	0.13522E-01	0.13518E-01
0.13274E-01	0.12640E-01	0.11466E-01	0.97239E-02	0.76698E-02
0.56031E-02	0.38225E-02	0.25039E-02	0.15835E-02	0.97002E-03
0.57192E-03	0.31626E-03	0.16604E-03	0.88422E-04	0.50625E-04
0.27594E-04	0.13455E-04	0.52455E-05	0.53119E-09	0.0

AVHRR Channel 5
Starting wave number   793.650 63
Delta wave number   1.710 45
                                              60

0.0	0.0	0.0	0.15207E-04	0.49409E-03
0.13229E-02	0.24498E-02	0.38133E-02	0.53498E-02	0.69507E-02
0.84644E-02	0.97377E-02	0.10632E-01	0.11173E-01	0.11486E-01
0.11700E-01	0.11932E-01	0.12210E-01	0.12526E-01	0.12868E-01
0.13226E-01	0.13583E-01	0.13923E-01	0.14227E-01	0.14479E-01
0.14678E-01	0.14826E-01	0.14928E-01	0.14989E-01	0.15030E-01
0.15082E-01	0.15175E-01	0.15339E-01	0.15557E-01	0.15773E-01
0.15930E-01	0.15971E-01	0.15888E-01	0.15756E-01	0.15658E-01
0.15675E-01	0.15847E-01	0.16041E-01	0.16079E-01	0.15785E-01
0.14993E-01	0.13702E-01	0.12032E-01	0.10104E-01	0.80408E-02
0.59652E-02	0.40025E-02	0.22783E-02	0.91823E-03	0.38213E-04
0.0	0.0	0.0	0.0	0.0

AVHRR non-linearity errors

		Error (degK)
Target temperatures (K)	10.3 μm	11.5 μm
315	+1.8	+0.9
305	+0.9	+0.5
295	+0.8	+0.1
285	−0.4	−0.1
275	−0.9	−0.5
255	−1.4	−0.8
245	−1.7	−1.0
225	−1.5	−1.3
205	−1.1	−1.4

The error is the difference between the best quadratic fit to the
actual target temperature and that temperature derived from a two-point
linear calibration using the artificial radiance of space.

When used by the inverse Planck function these central wave numbers
give the minimum error for the specified temperature bands.

Temp. (K)	3.7 μm	10.8 μm	11.8 μm
180–225	2670.93	928.50	844.41
225–275	2674.81	929.02	844.80
275–320	2678.11	929.46	845.19

* = a_{ij} of equation (2.2.15).

pre-launch values of the slope and intercept but can determine in-flight values of these quantities. However, to use such a straight line is not very good and it should first be modified to take account of the non-linearity of the relation between the digitised scanner output and the radiance input to the scanner. The non-linearity correction will be discussed in section 2.2.4.

It is only because of the pre-flight calibration work, which was carried out for each AVHRR instrument in the series and which has been described in some detail in section 2.2.2, that it is possible to determine satellite received radiances in absolute units. The in-flight calibration procedure for channels 3, 4 and 5 is described in detail in NESS107 and we shall follow that account quite closely. There is a separate version of Appendix B to NESS107 for each of the satellites in the NOAA series. The appropriate version of Appendix B contains the coefficients obtained from the pre-flight calibration measurements performed on the instruments on that satellite. It goes without saying that it is necessary, for any given AVHRR data, to identify which satellite in the NOAA series it has come from to make sure that the coefficients from the correct version of Appendix B of NESS107 are used. Although we are only concerned here with the AVHRR, it is worth noting in passing that the data from the other instruments on board the NOAA series of satellites can be calibrated in a similar manner.

The information required for producing AVHRR infrared channel calibration coefficients is located in the 103-word HRPT header (see Table 2.2). The onboard blackbody is maintained at a temperature which is approximately the operating temperature of the radiometer, namely 15°C or 288 K. This temperature is measured with four platinum resistance thermometers (PRTs) and the readings they produce are fed into the data stream; they are to be found among the 10 telemetry words (words 13–22). Three of these contain the Pt resistance thermometer readings; they are

    word   18   channel 3 blackbody temperature
    word   19   channel 4 blackbody temperature
    word   20   channel 5 blackbody temperature.

Any one of these words, when extracted from five consecutive HRPT minor frames, produces a reference (REF) value and one sample temperature value for each of the four PRTs. The pattern is as follows:

HRPT minor frame number	Parameter sample (word 18, 19 or 20).
.	.
.	
.	.
n	REF
$n + 1$	PRT1
$n + 2$	PRT2
$n + 3$	PRT3
$n + 4$	PRT4
$n + 5$	REF
⋮	⋮

where the symbols PRT1, PRT2, PRT3 and PRT4 refer to the data from the Pt resistance thermometers numbers 1, 2, 3 and 4, respectively, and REF refers to the reference value. If one starts with an arbitrary frame, as one has to, then there is no way of telling from that frame alone whether the data correspond to thermometer number 1, 2, 3 or 4. However, the reference value REF is a digital number with values less than 10 and such low values do not occur as thermometer readings for typical values of the on-board target temperature. Thus, looking at data from a set of successive minor frames, as above, it is quite obvious which frames contain the reference value. Then it is possible to identify which frames contain PRT1 values, which frames contain PRT2 values and so on. It is wise to average over several successive values for each of the PRTs to give a mean value for the PRT count value for conversion to temperature units.

To calculate the internal blackbody radiance, it is first necessary to compute the target temperature. The conversion of PRT mean counts to temperature uses the following:

$$T_i = \sum_{j=0}^{4} a_{ij}(\bar{X}_i)^j \qquad (2.2.15)$$

where T_i is in K, $\bar{X}_i$ is the mean value of PRTi (from header words 18, 19 or 20) and $i = 1, 2, 3, 4$ labels the four Pt resistance thermometers. a_{ij} are the coefficients of the conversion algorithm and T_i is the temperature of the internal blackbody calculated from PRTi. For example, the conversion of PRT1 values $(\bar{X}_1)$ into temperature (K) is

$$T_1 = a_{10} + a_{11}\bar{X}_1 + a_{12}(\bar{X}_1)^2 + a_{13}(\bar{X}_1)^3 + a_{14}(\bar{X}_1)^4 . \qquad (2.2.16)$$

$\bar{X}_1$ denotes not just the value PRTi from one minor frame, but an average of values of PRTi over several frames of values for that thermometer, i.e. say frames $n, n + 5$, $n + 10, n + 15$, etc., giving an average $\bar{X}_1$ over, for example, 10 or 20 frames.

The values of the coefficients a_{ij} have been determined for each of the AVHRR instruments individually in the pre-flight calibration process. The values of these coefficients have been published by NOAA (as an addition to Appendix B of NESS107) subsequent to the launch of each spacecraft. We have included one example, the data for NOAA-9, in Table 2.15. For different spacecraft in the series the values of these coefficients will be (slightly) different; it is important to use the correct set of coefficients for any given AVHRR dataset. The four values of T_i obtained from the four PRTs are then combined to form a weighted mean $\bar{T}$ where

$$\bar{T} = \sum_{i=1}^{4} b_i T_i \qquad (2.2.17)$$

to give the best estimate of the temperature to be used for the temperature of the on-board blackbody calibration target. The values of the coefficients b_i ($i = 1, 2, 3,$ 4) are also published by NOAA along with the values of a_{ij}, see Table 2.15 for the NOAA-9 values of b_i. In practice, the four thermometers are assigned equal weights so that $b_i = 0.25$ (see Table 2.15), i.e. $\bar{T}$ is just a simple average of the four values T_1, T_2, T_3 and T_4. However, the use of a formal published set of b_i values for each satellite means that, if one thermometer on one AVHRR instrument is less accurate than the other instruments, or if it fails completely, then the values obtained from it can be given a reduced weighting or even be totally excluded (i.e. with a value of $b_i = 0$).

It should now be appreciated that there are separate sets of data, in words 18, 19 and 20 for channels 3, 4 and 5 so that in practice one obtains separate values of $\bar{T}$, and therefore separate calibration curves, for each of channels 3, 4 and 5 of the AVHRR data.

In converting the data obtained from the scan of the Earth's surface into radiance we assume that there is a linear relation between the instrument's output, X (which it will be recalled is an integer in the range 0–1023) and the radiance, N, as we did for channels 1 and 2 in equation (2.2.7). We therefore need the radiance, not the temperature, for each of the two calibration points. The response of the detector will depend on the response function for the channel in question (see section 2.2.1) so that, as far as the data from a given channel are concerned, the size of the signal will depend on the integral of the product of the energy distribution function (the Planck function) and the spectral response function $\phi(v)$ over the range of the frequency v, i.e.

$$N(T_B) = \frac{\int_{v_1}^{v_2} B(v, T_B)\phi(v) \, dv}{\int_{v_1}^{v_2} \phi(v) \, dv} \, , \tag{2.2.18}$$

where v is wave number (cm^{-1}), $\phi(v)$ is the spectral response function, and v_1 and v_2 are its upper and lower limits. The Planck function $B(v, T_B)$ is given by

$$B(v, T_B) = c_1 v^3 / (\exp(c_2 v / T_B) - 1) \, . \tag{2.2.19}$$

The constants c_1 and c_2 are $1.191\,065\,9 \times 10^{-5}$ mW m^{-2} sr^{-1} cm^4 and $1.438\,833$ K cm, respectively.

During the pre-flight investigations the normalised response function $\phi(v)$ for a given spectral channel was determined at 60 discrete wave numbers, v, at equal wave number intervals Δv, i.e. $n = 60$. The spectral response function $\phi(v)$ has to be determined individually for each AVHRR instrument and for each channel (see section 2.2.1) and the numerical values of the function are tabulated and published by NOAA for each spacecraft in the series in Appendix B of NESS107. The values of $\phi(v)$ for the case of NOAA-9 are given in Table 2.15. Actually, to be quite correct the graphs of $\phi(v)$ are presented as functions of wavelength, λ, rather than of frequency, v, see Figure 2.9, while the tabulated values are given as functions of the wave number ($= 1/\lambda$). For each thermal infrared channel of the AVHRR the expression in equation (2.2.18) is evaluated numerically by

$$N(T_B) = \frac{\sum_{i=1}^{n} B(v_i, T_B)\phi(v_i)\Delta v}{\sum_{i=1}^{n} \phi(v_i)\Delta v} \, . \tag{2.2.20}$$

In practice, because equation (2.2.20) has to be used over and over again in calibration work, it is customary to evaluate this expression to generate look-up tables relating temperature to radiance. There will need to be one table for each of the three thermal infrared channels and, of course, there will need to be a separate set of tables for each spacecraft. At NOAA NESDIS (National Environmental Satellite Data and Information Service) the tables used specify the radiance at every tenth of a degree between 180 and 320 K. Thereafter, the tables are used whenever one needs to convert temperature to radiance or vice versa.

Returning to the header format (see Figure 2.1) we recall that in the header we have

	words	
Back scan data	30	23–52
Space data	50	53–102.

The back scan data contain 10 values of the detector output, X (digital number, DN, or counts, C) from the observation of the internal calibration target (the on-board blackbody) multiplexed as

channel 3 (word 1), channel 4 (word 1), channel 5 (word 1), channel 3 (word 2), channel 4 (word 2), . . ., channel 5 (word 10).

The space data contain the same, but for all five channels and looking at deep space, i.e.

channel 1 (word 1), channel 2 (word 1), channel 3 (word 1), channel 4 (word 1), channel 5 (word 1), channel 1 (word 2), channel 2 (word 2), channel 3 (word 2), . . . channel 5 (word 10).

Rather than use data from just one frame, one takes an average of the count values (e.g. 50 are used by NESDIS) to produce mean count values for the blackbody target and for space.

We have already mentioned the assumption that the count output, X, of each channel is a linear function of the observed radiance, R, so that

$$R = MX + I \qquad (2.2.21)$$

where M is termed the channel slope and I is termed the channel intercept. The quantity M (in units of radiance/count) is calculated for each channel from the equation

$$M = (\bar{N}_T - N_{SP})/(\bar{X}_T - \bar{X}_{SP}) \qquad (2.2.22)$$

where N_{SP} is the radiance of deep space, $\bar{N}_T$ is the radiance when the instrument views its internal radiance calibration target; the $\bar{X}_{SP}$ and $\bar{X}_T$ are the mean counts associated with several observations of space and the internal target, respectively. We have just described the calculation of $\bar{N}_T$. One would expect N_{SP} to be zero. However, for reasons associated with the non-linearity problem, non-zero values of N_{SP} were used for channels 4 and 5 of the AVHRRs on the early spacecraft of the series. For the moment we simply say that for any given spacecraft one should use the value of N_{SP} given in NESS107, see for example Table 2.15. This value has been supplied by NOAA in an attempt to overcome the non-linearity problem by an artifice; this will be explained in section 2.2.4. For $\bar{N}_T$, $\bar{X}_T$ and $\bar{X}_{SP}$ the number of observations in each case is sufficient effectively to eliminate the residual variances in $\bar{N}_T$, $\bar{X}_T$ and $\bar{X}_{SP}$ as contributors to the uncertainty in the derived value of M. The intercept, I, is calculated for each channel from the equation

$$I = N_{SP} - M\bar{X}_{SP} . \qquad (2.2.23)$$

We stress that one has a different set of values of M, the slope, and I, the intercept, for the three different channels 3, 4 and 5 and that these quantities have to be determined separately for each channel. For any given scene, once the values of M and I have been determined for a given channel (3, 4, or 5) by the procedure outlined above, then all the scan data in that channel can be converted from output count value, X, to the corresponding radiance, N (or R)[1], for each pixel in each scan line.

We converted temperatures into radiances for the purposes of having a linear, or nearly linear, calibration relation. However, having converted the raw digital data into radiances we then usually wish to convert these radiances into what are described as brightness temperatures. That is, we assume that the radiation arriving at the satellite is blackbody radiation of some temperature; this temperature is then called the brightness temperature, T_B. If there were no atmospheric effects at all this would be the same as the blackbody temperature corresponding to the radiation leaving the sea surface. To calculate the brightness temperature then, essentially, one has to invert equation (2.2.20). However, equation (2.2.20) cannot simply be inverted to give T_B as an explicit function of $N(T_B)$.

To find T_B for a given value of $N(T_B)$ it is then necessary to use trial values of T_B to calculate $N(T_B)$ using equation (2.2.20) and thus to determine, by a search operation, the value of T_B which reproduces the given value of $N(T_B)$. This would need to be repeated for each pixel and would be very time consuming. One can simplify this by generating a look-up table relating T_B and $N(T_B)$. However, there is an approximation which is sometimes used and which may be accurate enough for certain purposes. This involves assuming that for a given temperature the Planck distribution function (regarded as a function of wavelength, λ, or wave number, v) is approximately constant over the width of any one of the thermal infrared bands. In this situation, equation (2.2.18) simplifies so that

$$N(T_B) = \frac{B(v,\ T_B)\int_{v_1}^{v_2}\phi(v)\ \mathrm{d}v}{\int_{v_1}^{v_2}\phi(v)\ \mathrm{d}v} \tag{2.2.24}$$

i.e.

$$N(T_B) = B(v,\ T_B)\ . \tag{2.2.25}$$

Thus, using equation (2.2.19) we have

$$N(T_B) = c_1 v^3/(\exp(c_2\, v/T_B) - 1) \tag{2.2.26}$$

which can be rearranged to give

$$T_B = \frac{c_2\, v}{\ln\left(1 + \dfrac{c_1 v^3}{N(T_B)}\right)}\ . \tag{2.2.27}$$

The frequency v to be used in this approximation for T_B is also given, for each channel and for each spacecraft in the series, as the 'central wave numbers' in section 1.4 of Kidwell (1991). Since this is an approximation it is found that different values for these central wave numbers need to be used for different temperature ranges, see Table 2.16.

Over small ranges of temperature, such as 270–300 K, which covers most sea surface temperatures, it is convenient to establish an empirical relation from which T_B can easily be calculated directly. One can establish a direct and successful empirical relation of the form

$$\ln (N(T_B)) = a_1 + \frac{b_1}{T_B}\ . \tag{2.2.28}$$

---

[1] Both symbols R and N are used in the NOAA literature; we do the same in order to preserve with that literature.

Table 2.16 TIROS-N central wave numbers for AVHRR IR channels (Kidwell 1991)

Temperature range (K)	Channel 3 (cm^{-1})	Channel 4 (cm^{-1})
180–225	2631.81	911.13
225–275	2635.15	911.54
275–320	2638.05	912.01

This equation can be inverted to give

$$T_{\mathrm{B}} = \frac{b_1}{\ln (N(T_{\mathrm{B}})) - a_1} . \tag{2.2.29}$$

The values of a_1 and b_1 can be determined once and for all for each channel of each instrument (i.e. for each $\phi(v)$) so that equation (2.2.29) can be applied to the whole field of Earth-scan data to determine a brightness temperature, T_{B}, for each channel on a pixel-by-pixel basis. The values of a_1 and b_1 are given for the various spectral channels of the AVHRRs on the early spacecraft of the TIROS-N/NOAA series in Table 1 of Singh (1984).

We have used the term brightness temperature, T_{B}, the temperature which is obtained by inverting equation (2.2.20) using the satellite-received radiance. What this means is that we regard the radiation reaching the spacecraft as blackbody radiation and T_{B} is the temperature of a blackbody (i.e. a perfect emitter) that would give rise to that particular energy distribution of radiation. The brightness temperature is very unlikely to be the same as the temperature of the ground (or of the cloud) in the field of view of the scanner. There are two reasons for this; one is that the atmosphere affects the upwelling radiation passing through it and the second is that the emissivity of the ground or cloud also needs to be taken into account. The problem of atmospheric effects will be addressed in a later chapter (chapter 3).

This section has been concerned with the in-flight calibration of HRPT data. It is also possible to use the in-flight calibration data with the APT transmission to obtain brightness temperatures and this will be considered in section 2.2.5. However, before that we turn to the question of the non-linearity correction.

2.2.4 Non-linearity correction

The in-flight calibration that we have described in section 2.2.3 using two calibration points assumes that there is a linear relation between the AVHRR's output, in digital counts, and the radiance of the scene. For channel 3, which uses an InSb detector, the relation is indeed highly linear. However, as channels 4 and 5 use HgCdTe detectors, their calibrations are slightly non-linear. For channels 4 and 5 neglecting the non-linearity correction may cause errors of a degree Celsius (or kelvin) or more in a derived Earth surface temperature. Consequently NOAA NESDIS has provided information to enable users of AVHRR data to apply a non-linearity correction; this information is given in Appendix B of NESS107.

The approach originally adopted by NOAA and described in the first version of NESS107 (Lauritson *et al.* 1979) was the following; this procedure was adopted for TIROS-N and for the other earlier spacecraft in the series, up to NOAA-9 in fact. (a) Instead of taking the radiance of deep space as zero (its true value) an artifice of using a 'corrected' non-zero value of the radiance of deep space was estimated and given in Appendix B of NESS107. This is given under the heading 'N_{SP} = Radiance of space including non-linearity correction'. (b) A simple correction was published that was independent of the temperature of the on-board internal calibration target (blackbody), see under the heading 'non-linearity errors' in Table 2.15.

More recently this approach has been replaced. The idea of using an artificial non-zero value for N_{SP} was dropped and a table was published which simply gave the non-linearity correction. However, this correction varies according to the temperature of the internal calibration target (baseplate) of the AVHRR (Brown *et al.* 1985), and so different corrections were given for three different values of the temperature (10°C, 15°C and 20°C) of the ICT (baseplate), see Figure 2.16 and Table 2.17. To determine the appropriate correction the user must interpolate in that table to find the correction appropriate to the actual temperature of the ICT. This was

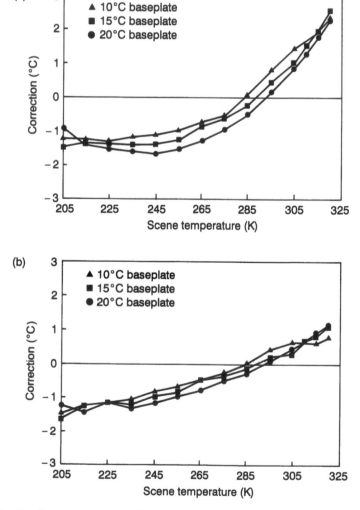

Figure 2.16 Non-linearity corrections for (a) channel 4 and (b) channel 5 of the AVHRR on NOAA-9 (Weinreb *et al.* 1990).

introduced at the time of the launch of NOAA-10 and so revised data, in this format, were produced for NOAA-9 which was then already in orbit and a revised version of Appendix B of NESS107 was produced for NOAA-9; both versions for NOAA-9 are given by Planet (1988). The new approach has been applied to the subsequent spacecraft and an extensive discussion of this change has been given by Brown *et al.* (1985), Dalu and Viola (1987), Weinreb *et al.* (1990), Steyn-Ross *et al.* (1992) and Rao (1993a). Tables for NOAA-9, -10 and -11 are given by Weinreb *et al.* (1990) and it is smoothed versions of these values that are now published in the updates for Appendix B of NESS107 (Planet 1988). A much more detailed study of the non-linearity correction has been published recently by Brown *et al.* (1993) and there are systematic differences from the results of Weinreb *et al.* (1990). The work of Brown *et al.* (1993) was motivated by the fact that the temperature of the internal target (the on-board blackbody) was observed to go outside the range (10°C to 20°C) for which the thermal vacuum test results and the published AVHRR calibration data of Weinreb *et al.* (1990) were given. Evidence of the temperature going outside this range (with temperatures as high as 24°C) is shown in Figure 2.17 (Gorman and McGregor 1994). Revised sets of calibration constants were given for the AVHRRs on NOAA-6, -7, -8, -9, -10, -11 and -12 by Brown *et al.* (1993). However, instead of presenting tables of the form of Table 2.17 of the non-linearity correction as a function of scene temperature and internal target temperature the non-linearity correction T_i^{nl} (for channel i) was given in terms of an empirical algorithm

$$T_i^{nl} = a_0 + a_1 T_i^* + a_2 T_i^{*2} \tag{2.2.30}$$

and the values of a_0, a_1 and a_2 were tabulated by Brown *et al.* (1993) for each channel of the AVHRR on each of NOAA-6 to NOAA-12 inclusive. T_i^* was defined as

$$T_i^* = T_i^{lin} - \bar{T}_i \tag{2.2.31}$$

where T_i^{lin} is the temperature obtained from the linear in-flight calibration and $\bar{T}_i$ is the mean internal target temperature.

The NOAA-NASA AVHRR Pathfinder Program, which has been entrusted with the task of the reprocessing and rehabilitation of the AVHRR records for the period 1981–1990 (which has been termed the Pathfinder period) for the purpose of the production of long-term records for climate studies, set up the AVHRR Pathfinder Calibration Working Group in March 1991 (see also section 7.9.6). The Working Group was charged with the tasks of (a) assessment of the in-orbit degradation of channel 1 and channel 2 of the AVHRR instruments on board NOAA-7, -9 and -11 and (b) development of a consistent set of in-flight calibration algorithms for the thermal infrared channels (3, 4 and 5) of the AVHRR. (a) will be considered in section 2.2.6. Under task (b) it was further understood that the development of appropriate correction procedures for the non-linear response of the detectors in channels 4 and 5 would be addressed first. The report on the non-linearity problem (Rao 1993a) recommended that a radiance-based non-linearity correction procedure based on applying a quadratic correction formula for the radiances should be used instead of, as previously, applying a correction to the brightness temperatures. This involves first of all a return to the original NOAA NESDIS approach of using a non-zero value of the deep-space radiance, N_{SP}, in the original linear in-flight calibration procedure (see equations (2.2.21)–(2.2.23)) and values of N_{SP} for the

AVHRRs on NOAA-7 and NOAA-9 are given in Table 3 of Rao (1993a). The non-linearity is then applied to the calculated radiances as follows. A pseudo-linear radiance N''_{LIN} is calculated from the (raw) counts X using

$$N''_{\text{LIN}} = M''X + I'' \tag{2.2.32}$$

where

$$M'' = M' \frac{\bar{N}_{\text{T}} - N''_{\text{S}}}{\bar{N}_{\text{T}} - N''_{\text{S}}} \tag{2.2.33}$$

and

$$I'' = N''_{\text{S}} - \frac{M''}{M'} N'_{\text{S}} + \frac{M''}{M'} I' . \tag{2.2.34}$$

Table 2.17   Non-linearity errors for NOAA-9 (Planet 1988)

The non-linearity errors are calculated in two ways. In the first table the error is the difference between the best quadratic fit to the actual target temperature and that temperature derived from a two-point linear calibration using a radiance of space of zero. The error in the second table is the difference between the quadratic fit and the temperature derived from a two-point fit using the corrected radiance of space.

Zero radiance of space:

| | | Error (K) |
Target temperatures (K)	10.8 μm	11.8 μm
315	+1.8	+1.0
305	+0.9	+0.6
295	+0.2	+0.2
285	−0.4	−0.1
275	−0.9	−0.5
255	−1.4	−0.8
245	−1.6	−1.0
225	−1.5	−1.3
205	−1.0	−1.4

Corrected radiance of space:

| | | Error (K) |
Target temperatures (K)	10.8 μm	11.8 μm
315	+1.2	+0.6
305	+0.5	+0.3
295	+0.1	+0.1
285	−0.2	0
275	−0.3	−0.1
255	+0.1	+0.1
245	+0.4	+0.2
225	+2.2	+0.8
205	+5.6	+2.1

The artifice of using a 'corrected' non-zero radiance of space was eliminated. The corrections were calculated for three temperatures of the internal blackbody, 10, 15, and

Table 2.17 *(Continued)*

20°C. To determine the appropriate correction the user must interpolate in the following tables on the actual blackbody temperature in orbit.

Target temperatures (K)	Correction at blackbody temperature (°C)		
	10	15	20
Channel 4 non-linearity correction table			
320	+2.3	+2.3	+2.3
315	+1.8	+1.9	+1.8
310		+1.4	+1.3
305	+1.3	+1.0	+0.9
295	+0.7	+0.4	+0.2
285	0.0		−0.5
275	−0.5	−0.7	−0.9
265	−0.8	−1.1	−1.2
255	−1.0	−1.3	−1.6
245	−1.1	−1.3	−1.7
235	−1.2	−1.4	
225	−1.3	−1.3	−1.5
215	−1.2	−1.5	−1.4
205	−1.6	−1.5	−0.7
Channel 5 non-linearity correction table			
320	+0.8	+1.0	+1.2
315	+0.6	+0.9	+0.9
310		+0.7	+0.7
305	+1.1	+0.4	+0.5
295	+0.4	+0.2	+0.1
285	0.0		−0.2
275	−0.3	−0.3	−0.5
265	−0.5	−0.6	−0.7
255	−0.7	−0.8	−1.0
245	−0.8	−0.8	−1.2
235	−1.1	−1.2	
225	−1.2	−1.0	−1.1
215	−1.2	−1.4	−1.4
205	−1.7	−1.6	−1.1

Central wave numbers

When used by the inverse Planck function these central wave numbers give the minimum error for the specified temperature bands. The fourth band may be useful for sea surface temperatures.

Temperature	3.7 μm	10.8 μm	11.8 μm
180–225	2670.93	928.50	844.41
225–275	2674.81	929.02	844.80
275–320	2678.11	929.46	845.19
270–310	2677.67	929.39	845.12

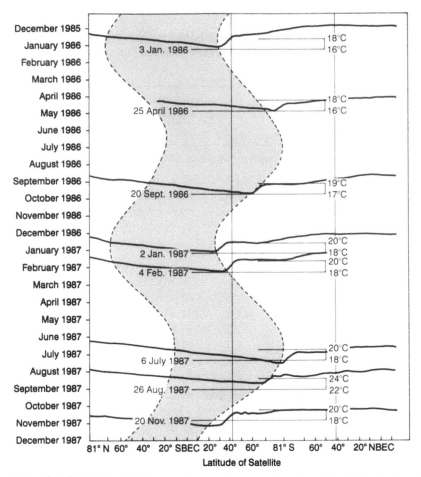

Figure 2.17 Plot of AVHRR orbital temperature profiles at various dates. The shaded region is where the spacecraft is in the Earth's shadow, the solid vertical lines are when the satellite passes into the study area (New Zealand) (Gorman and McGregor 1994).

In these equations M' and I' are the slope and intercept determined from the in-flight calibration data as described above using equations (2.2.21)–(2.2.23). The corrected radiance N is calculated from N''_{LIN} using a quadratic formula

$$N = AN''_{\text{LIN}} + B(N''_{\text{LIN}})^2 + C \; . \tag{2.2.35}$$

The values of N''_S, N'_S A, B and C are given for channel 4 and channel 5 of the AVHRRs on NOAA-7 and NOAA-9 in Tables 2 and 3 of Rao (1993a). Finally the corrected radiance N is converted back into a temperature by inverting equation (2.2.24) as before.

2.2.5   Calibration of APT data

The APT is an analogue transmission, rather than a digital transmission, and it was planned primarily as a low-resolution transmission to provide images which could be received with very simple equipment (see section 2.5.1). Nevertheless it is possible, if one is not too concerned about achieving high accuracy, to determine brightness temperatures from thermal infrared APT data with an accuracy of circa $\pm 2°C$. For example, one may find oneself able to obtain APT data easily but in a situation where it would be difficult to obtain HRPT data, e.g. on board an oceanographic cruise ship where the construction of a stabilised mounting for a steerable antenna would be difficult and expensive. If one wished to determine sea surface temperatures in near-real time the digitisation of APT data is a feasible proposition (see e.g. Wannamaker 1984). The best description of the calibration of APT data is given by Summers (1989), based on information from the appropriate NOAA manuals; although the process is never likely to be widely used, it is nevertheless worth describing it. The calibration technique is best understood with reference to the calibration described in section 2.2.3 for the high-resolution HRPT or LAC data.

As mentioned already in chapter 1, the APT signal is generated from the AVHRR digital data by subsampling and digital-to-analogue conversion. The transmission contains 120 lines of video per minute and the format is shown in Figure 2.6. The first stage in the calibration of the APT thermal infrared data involves an analogue-to-digital conversion of the APT data. This can be done relatively easily with a PC-based image display system. Such a system will demodulate the signal to remove the 2400 Hz subcarrier, digitise the demodulated signal with an analogue-to-digital converter and generate an array of (usually) 8-bit data in a computer file. Basically this process reverses the original processing which was done on board the spacecraft and reproduces the original 8-bit data used to establish the amplitude modulation of the 2400 Hz carrier. Ideally the ground station would recover the exact original digital values generated by the AVHRR. In practice, the transmission and reception processes will introduce some errors in the recovered values.

The array containing the digitised APT data from an infrared channel will then contain the image data along with the space data and telemetry data indicated in Figure 2.6. The temperature calibration of APT infrared channel data then involves the extraction of the calibration data from the space and telemetry frames. Table 2.18 shows the telemetry frame format for the APT. One complete frame contains 16 individual wedges, each of which is composed of eight successive video lines (one frame = 16 wedges × 8 lines = 128 lines per frame). These frames are continuously repeated during the satellite orbit but it is not necessary to repeat the extraction of the calibration coefficients for each separate frame. It should be noted that within a telemetry frame the first 14 wedges are identical in both the telemetry frames A and B; the only wedges which will be different will be 15 and 16 which are specific to the actual channel being used.

The voltage signal (for which sample values are given in Table 2.18) in each of the wedges 1–9 was generated on board the spacecraft by the digital signal indicated on the same line in the right-hand column of the table. Thus, if one imagines these digital values plotted against the digitised values of the voltages (or 'Modulation Index' MI) in the wedges, one obtains a calibration curve which can then be used for the remaining data in the telemetry frame and for the pixels in the image data to regenerate the digital data that originally existed on the spacecraft. At this stage, therefore, the whole APT data can be considered to have been converted back to 8-bit digital data. Thus, with the remaining wedges redigitised, wedges 10–13 contain the values of PRT1, PRT2, PRT3 and PRT4 which are the readings of the four Pt resistance thermometers which we have already described in section 2.2.3.

Table 2.18 Telemetry frame format used in TIROS-N series satellite APT (Summers 1989)

Wedges 1–8: APT	Analogue voltage	Digital value
1	0.757 V	
	MI = 10.6%	31
2	1.538 V	
	MI = 21.5%	63
3	2.319 V	
	MI = 32.4%	95
4	3.101 V	
	MI = 43.4%	127
5	3.881 V	
	MI = 54.2%	159
6	4.663 V	
	MI = 65.2%	191
7	5.444 V	
	MI = 76.0%	223
8	6.225 V	
	MI = 87.0%	255
9	ZERO	
	MODULATION	0
10	THERM TEMP	
	PRT #1	
11	THERM TEMP	
	PRT #2	
12	THERM TEMP	
	PRT #3	
13	THERM TEMP	
	PRT #4	
14	PATCH TEMP	
15	BACK SCAN	
16	CHANNEL	
	IDENT	

Taking the mean of these four (digitised) values as $\bar{X}_i$ as before, then the temperature $\bar{T}$ of the on-board blackbody is given by (see Summers 1989)

$$\bar{T} = 0.206\bar{X}_i + 276.943 .\tag{2.2.36}$$

One can then determine the slope, M, and the intercept, I, in the calibration curve given by equation (2.2.21). This involves the use of the temperature $\bar{T}$ of the on-board blackbody and the temperature of deep space, or rather the corresponding radiances. The value of $\bar{X}_T$, the digital value of the output from the scanner when it observes the on-board blackbody, is given by the value of wedge 15 and the value of $\bar{X}_{SP}$, the value when the scanner observes deep space, is given by the space data which is immediately following the synchronisation (sync) pulse, see Figure 2.6. The calibration of the APT data from this point, therefore, proceeds in the manner already described in section 2.2.3 for the full-resolution HRPT data, except that the digitised APT data will (usually) be 8-bit data rather than 10-bit data.

2.2.6 Post-launch calibration

Since there is no in-flight calibration of channels 1 and 2 of the AVHRR, the question arises as to the validity of the pre-launch calibration coefficients, both in the early days after launch and, perhaps more seriously, after the AVHRR has been in space for a long time. There is clear evidence in several environmental products, such as the normalised difference vegetation index (see sections 5.1, 5.2 and 5.4), global cloud morphology (see section 7.9.4) and Earth radiation budget (see section 4.4), that are generated from channel-1 and channel-2 AVHRR data to indicate that the performance of the instrument in these two channels has deteriorated after launch. The need to correct for this in-orbit degradation has been keenly felt recently since it is now being proposed to use the long-term records of AVHRR-derived environmental products in climate and global change studies (Price 1987a, b, Bocoum 1991, Brest and Rossow 1992, Rao and Chen 1993, 1994, Tarpley 1993) and the degradation of the instrument with time is clearly illustrated by the results shown in Figures 2.18 and 2.19. In Figure 2.18 the albedo calculated using the pre-launch calibration coefficients (thin lines) shows an apparent steady decrease with advancing time whereas in fact there is no real change in the target surface over this period; rather it is that the sensitivity of the detectors of the AVHRR is decreasing. The albedo calculated using post-launch calibration coefficients shows no apparent trend (thick lines) in Figure 2.18. With the increasing interest in the quantitative use of data from channels 1 and 2 of the AVHRR, the importance of checking on the calibration of these channels in this way is now quite important. A general discussion of the importance of the calibration of satellite-flown radiometers on a variety of satellites (Landsat, NOAA, SPOT and the geostationary/geosynchronous satellites) is given by Price (1987a); indeed this was one paper in a special issue of *Remote Sensing of Environment* (volume 22, number 1, June 1987) devoted to papers on the calibration of space-borne radiometers. Until recently there was no routine or systematic attempt to conduct any post-launch operation to attempt a recalibration of channels 1 and 2 to account for the degradation of the instruments. However, as mentioned at the end of section 2.2.4, the NOAA-NASA AVHRR Pathfinder Calibration Working Group was asked in the Spring of 1991 to assess the degradation of channels 1 and 2 on the NOAA-7, -9, and -11 spacecraft (referred to as the afternoon satellites since they cross the equator travelling northwards in the afternoon) and to make recommendations for implementing appropriate corrections for the observed degradation. The report of this Working Group has now been published (Rao 1993b). The question of calibration is also very important in the context of the International Satellite Cloud Climatology Project (ISCCP) (Brest and Rossow 1992), see section 7.9.4.

Teillet *et al.* (1990) identified three broad categories of methods available for the post-launch calibration of satellite radiometers with no on-board calibration devices. These are (i) methods based on simultaneous aircraft and satellite observations of a given area of the ground, (ii) using a combination of model simulations and satellite measurements and (iii) using statistical procedures on large bodies of data to determine the trends in the calibration of the radiometer. The first method involves using a calibrated airborne radiometer, making allowance for the different atmospheric paths and effectively transferring the calibration from the airborne instrument to the satellite-flown instruments. In the second method the general idea is to use a ground site that is homogeneous over a large area, in relation to the size

Southeastern Libyan Desert

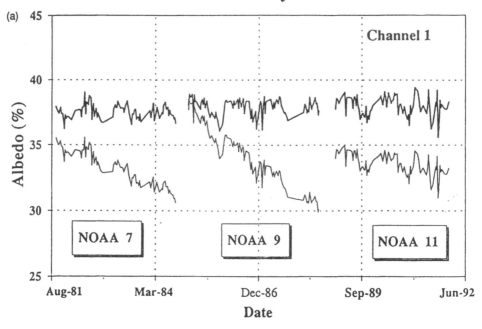

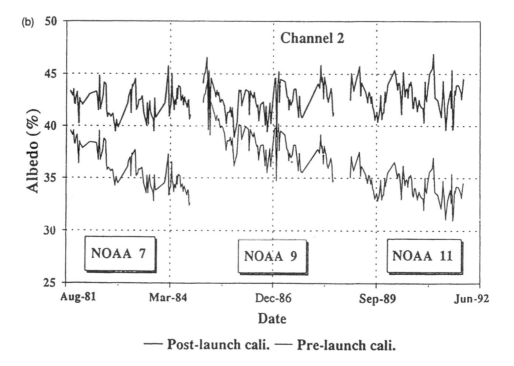

— Post-launch cali. — Pre-launch cali.

Figure 2.18 Isotropic albedo in (a) channel 1 and (b) channel 2 of the southeastern Libyan desert (Rao and Chen 1994).

Degradation of AVHRR

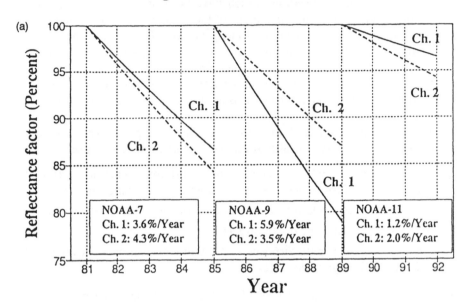

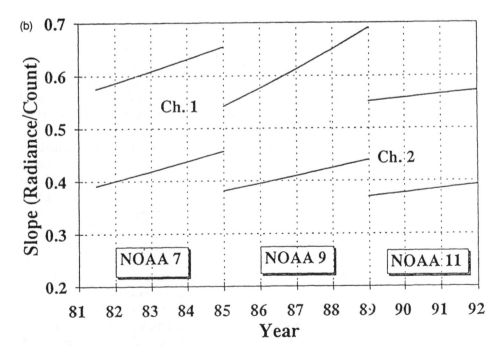

Figure 2.19   (a) Relative degradation rates and (b) time-dependent slopes of channels 1 and 2 of the AVHRRs on NOAA-7, -9 and -11 (Rao and Chen 1994).

of an AVHRR IFOV, and with a homogeneous surface for which the value of the albedo is known; suitable locations which have been used include large salt flats, smooth uniform areas of desert sand, areas of the ocean surface and clouds. Such an area is then regarded as a source of (reflected) radiation from which one attempts to calculate the satellite-received radiance and to use this to calibrate the instrument (see Frouin and Gautier 1987 for example). In the third method one also uses a ground area that is homogeneous over an area that is large with respect to the AVHRR IFOV and for which, further, it is assumed that its reflective properties do not vary with time.

In the Pathfinder Calibration Working Group report (Rao 1993b) four approaches for the post-launch calibration of channels 1 and 2 were used:

(a)   the determination of relative trends in the calibration of channels 1 and 2 using the southeastern Libyan desert as a time-invariant calibration target

(b)   the determination of relative trends based on the global ISCCP data sets

(c)   absolute calibration using ocean targets

(d)   the use of congruent path airborne and satellite measurements over the White Sands Missile Range in New Mexico.

These various approaches are described by Rao (1993b) in some detail and with many references cited.

A survey of work on post-launch calibration of the three AVHRRs on NOAA-7, NOAA-9 and NOAA-11 is given by Che and Price (1992) who analysed the results from the published work which had recently been done using various combinations of the above methods; they also give references to the work involved. Che and Price (1992) summarised the results of the published material that was available at that time and presented those results in terms of values of the gain values for each of the three AVHRR instruments on NOAA-7, -9 and -11 at various times. The gain value is the reciprocal of the slope S in equation (2.2.9). From an analysis of the measured variation of the gain values with time, it was found that (a) there was a significant reduction in the gain almost immediately after launch and (b) there was a steady degeneration of the instrument thereafter. The work of Kaufman and Holben (1993) is representative of AVHRR calibration research. They showed that, for NOAA-7 and NOAA-9, sensor gain decreased by approximately 10% between pre-launch and post-launch and following launch continues to decrease at a rate of approximately 3% per annum. NOAA-11 data show a larger apparent decrease in gain of 22% and 32% for the visible and near-infrared sensors, respectively (Holben et al. 1990). Contributions to the loss in sensitivity soon after launch can be expected to have arisen both from changes during the long period of storage before launch and as a result of outgassing in the first few days or weeks in orbit. As already mentioned at the end of section 2.2.2, it is now accepted that each AVHRR should be recalibrated just before it is launched so that the first of these problems can be eliminated. An extensive analysis was used to obtain formulae to represent the gain value as a function of the time (in months) that had elapsed since launch, see Table 2.19. Although, for studying the degeneration of the instrument in physical terms it is easier to think in terms of the gain value, the user of AVHRR data is more likely to want to use the slope and so the corresponding formulae for the variation of S as a function of time are given in Table 2.20. The offset, X_0, also varies with time, although its value generally lies in the range 35–40. The same studies that were used to study the

Table 2.19 Formulae for gain values (count (W m^{-2} sr^{-1} μm^{-1})) (Che and Price 1992)

Channel	Name of Sensor	Launch Date	Pre-launch Gain	Formulae for Gain
1	NOAA-7	23 Jun. 1981	1.88	$1.707 - 6.425 \times 10^{-3} \times$ months
	NOAA-9	12 Dec. 1984	1.95	$1.752 - 6.425 \times 10^{-3} \times$ months
	NOAA-11	24 Sep. 1988	2.04	$1.864 - 6.425 \times 10^{-3} \times$ months
2	NOAA-7	23 Jun. 1981	2.88	$2.411 - 2.167 \times 10^{-3} \times$ months
	NOAA-9	12 Dec. 1984	2.86	$2.411 - 2.167 \times 10^{-3} \times$ months
	NOAA-11	24 Sep. 1988	3.32	$2.725 - 8.250 \times 10^{-3} \times$ months

Table 2.20 Formulae for calibration coefficients (W m^{-2} sr^{-1} μm^{-1} /count) (Che and Price 1992)

Channel	Sensors	Pre-launch Calib. Coeff.	Formulae
1	NOAA-7	0.532	$0.591 + 2.23 \times 10^{-3} \times$ months
	NOAA-9	0.513	$0.576 + 2.23 \times 10^{-3} \times$ months
	NOAA-11	0.490	$0.534 + 2.23 \times 10^{-3} \times$ months
2	NOAA-7	0.347	$0.420 + 2.33 \times 10^{-3} \times$ months
	NOAA-9	0.350	$0.420 + 2.33 \times 10^{-3} \times$ months
	NOAA-11	0.301	$0.369 + 1.20 \times 10^{-3} \times$ months

variation of the slope, and that were reviewed by Che and Price (1992), also studied the variation of the offset. This was usually done by taking the radiance values obtained when the scanner is observing deep space to provide the offset values. This means that the offset values can be taken from the in-flight data (the space data) in the AVHRR data that one is actually using and one does not need to rely on formulae obtained from post-launch calibration exercises. A comparison between pre-launch values and some in-flight values of the offsets for channels 1 and 2 of the AVHRRs on NOAA-7, -9 and -11 are given in Table 2.21 (Holben *et al.* 1990).

Table 2.21 Pre-flight and deep space offsets for the AVHRR on board NOAA-7–NOAA-11 (Holben *et al.* 1990)

Satellite	Year	Pre-flight offset band 1	band 2	Deep space band 1	band 2
NOAA-7	1981	32.2	32.6	36.0	38.0
NOAA-7	1982	32.2	32.6	36.0	37.7
NOAA-7	1983	32.2	32.6	35.8	37.4
NOAA-7	1984	32.2	32.6	35.4	37.2
NOAA-9	1985	36.2	36.1	38.0	39.9
NOAA-9	1986	36.2	36.1	37.9	39.3
NOAA-9	1987	36.2	36.1	37.8	39.1
NOAA-9	1988	36.2	36.1	37.8	39.0
NOAA-11	1989	38.0	41.0	40.0	40.0

We have not described all the individual work on post-launch calibration that was reviewed by Che and Price (1992) but have just quoted the main conclusions of that review. However, we shall just mention post-launch calibration work which has been published since that review. Rao and Chen (1993), for example, have used the statistical approach with data from the southeastern part of the Libyan desert; this was chosen as having long-term stability, a high reflectance combined with low to moderate solar zenith angles, surface uniformity over a few hundred square kilometres and low cloudiness and precipitation. Rao and Chen (1993, 1994) used NOAA-7, NOAA-9 and NOAA-11 data and performed regression on the reflectance factor against time, see Figure 2.20. The use of the relative degraduation rates illustrated in Figure 2.20 in the determination of the rate of variation of the slope – the reciprocal gain – in units of $W\ m^{-2}\ sr^{-1}\ \mu m^{-1}\ counts^{-1}$ that could be used in the calculation of the upwelling radiances from the quantities $(X - X_0)$ was illustrated for the NOAA-9 data, where X is the digital number (counts) on the 10-bit scale (between 0 and 1023) and X_0 is the offset value. This was done by using simultaneous satellite and airborne measurements over the White Sands area of New Mexico obtained over a two-week period in October/November 1986. For channel 1 of the NOAA-9 data they concluded that the calibration of the data should be as follows:

$$S_d = 0.5465 \times \exp[1.66 \times 10^{-4} \times (d - 65)] \tag{2.2.37}$$

where S_d is the value of the slope S (in equation (2.2.9)) to be used for data acquired on day d after launch. The upwelling radiance is then given in units of $W\ m^{-2}\ sr^{-1}$ μm^{-1} when S_d is multiplied by $(X - X_0)$. The values of X_0 recommended by Rao and Chen (1993) to be used for channels 1 and 2 were 37 and 39.6, respectively. Similar equations for channel 2 of NOAA-9 and for both channel 1 and channel 2 of NOAA-7 and NOAA-11 are given by Rao and Chen (1994, 1995), see Table 2.22.

Kaufman and Holben (1993) also studied the NOAA-7, -9 and -11 AVHRR data for the period 1981–1990. This was independent of ground support and relied on

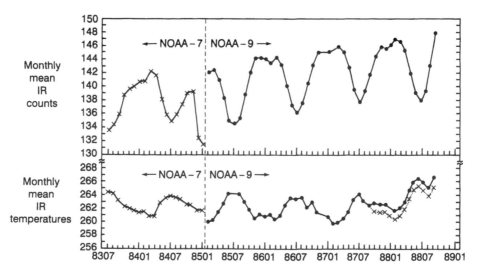

Figure 2.20 Demonstration of the effectiveness of on-board calibration of the thermal infrared channels of the AVHRR. Dates are given as 8307 = July 1983 etc. (Brest and Rossow 1992).

Table 2.22 Formulae for the calculation of (a) calibrated radiances and (b) calibrated AVHRR albedos (Rao and Chen 1994)

(a)

Spacecraft	Radiance ($W\ m^{-2}\ sr^{-1}\ \mu m^{-1}$)
NOAA-7	
Channel 1	$0.5753 \times \exp(1.01 \times 10^{-4} \times d) \times (X - 36)$
Channel 2	$0.3914 \times \exp(1.20 \times 10^{-4} \times d) \times (X - 37)$
NOAA-9 (Set A)	
Channel 1	$0.5465 \times \exp[1.66 \times 10^{-4} \times (d - 65)] \times (X - 37)$
Channel 2	$0.3832 \times \exp[0.98 \times 10^{-4} \times (d - 65)] \times (X - 39.6)$
NOAA-9 (Set B)	
Channel 1	$0.5406 \times \exp(1.66 \times 10^{-4} \times d) \times (X - 37)$
Channel 2	$0.3808 \times \exp(0.98 \times 10^{-4} \times d) \times (X - 39.6)$
NOAA-11	
Channel 1	$0.5496 \times \exp(0.33 \times 10^{-4} \times d) \times (X - 40)$
Channel 2	$0.3680 \times \exp(0.55 \times 10^{-4} \times d) \times (X - 40)$

(b)

Spacecraft	Albedo (per cent)
NOAA-7	
Channel 1	$0.1100 \times \exp(1.01 \times 10^{-4} \times d) \times (X - 36)$
Channel 2	$0.1169 \times \exp(1.20 \times 10^{-4} \times d) \times (X - 37)$
NOAA-9 (Set A)	
Channel 1	$0.1050 \times \exp[1.66 \times 10^{-4} \times (d - 65)] \times (X - 37)$
Channel 2	$0.1143 \times \exp[0.98 \times 10^{-4} \times (d - 65)] \times (X - 39.6)$
NOAA-9 (Set B)	
Channel 1	$0.1039 \times \exp(1.66 \times 10^{-4} \times d) \times (X - 37)$
Channel 2	$0.1136 \times \exp(0.98 \times 10^{-4} \times d) \times (X - 39.6)$
NOAA-11	
Channel 1	$0.1060 \times \exp(0.33 \times 10^{-4} \times d) \times (X - 40)$
Channel 2	$0.1098 \times \exp(0.55 \times 10^{-4} \times d) \times (X - 40)$

Note: The two sets of formulae given for NOAA-9 yield the *same* radiances/albedos; the quantity $\exp(-65k)$ occurring in Set A has been incorporated into the numerical coefficient appearing at the beginning of the formulae in Set B to render their format the same as that of the formulae for the AVHRRs on NOAA-7 and NOAA-11.

three distinct Earth–atmosphere phenomena, namely (i) molecular scattering over the ocean for absolute visible-channel calibration, (ii) ocean glint (sunglint) to transfer the calibration from the channel 1 (visible wavelengths) to channel 2 (near-infrared wavelengths) and (iii) desert reflectance to monitor the stability of both channels 1 and 2. For NOAA-9 both the ocean and desert calibrations yielded very similar values for the degradation of both channels 1 and 2. However, for NOAA-7, while a similar degradation was obtained using the desert calibration an opposite behaviour was obtained using the ocean calibration method. For NOAA-11 only the desert calibration was used. Kaufman and Holben (1993) discuss the reasons for the possible discrepancy between the results from the ocean and desert calibrations for NOAA-7; their conclusion was that further work was needed on studying the accuracy of the calibration methods but that, at present, the indications were that the desert method should be regarded as the more reliable (for a detailed discussion see section 4 of the paper by Kaufman and Holben 1993). Another set of formulae for the degradation of the gain values for channels 1 and 2 of NOAA-7, -9 and -11 is given by Kaufman and Holben; these are in terms of r_1 and r_2, which are the ratios of the post-launch gain value to the pre-launch gain value for the channels 1 and 2 respectively:

for NOAA-7:

$$r_1 = 0.916 - 0.049(Y - 1981.5) + 0.0050(Y - 1981.5)^2 \tag{2.2.38}$$

$$r_2 = 0.882 - 0.080(Y - 1981.5) + 0.0125(Y - 1981.5)^2 \tag{2.2.39}$$

for NOAA-9:

$$r_1 = 0.953 - 0.051(Y - 1985) \tag{2.2.40}$$

$$r_2 = 0.866 - 0.026(Y - 1985) \tag{2.2.41}$$

and for NOAA-11:

$$r_1 = 0.797 - 0.010(Y - 1989) \tag{2.2.42}$$

$$r_2 = 0.683 - 0.020(Y - 1989) \tag{2.2.43}$$

where Y, the year, is given on a continuous scale, so that for example 1 April 1985 will be 1985.25. The work of Kaufman and Holben was motivated by various applications of long-term AVHRR data sets, principally in the field of vegetation studies.

In their work in relation to calibration for the ISCCP Brest and Rossow (1992) concentrated on the intercalibration of the AVHRRs on NOAA-7, NOAA-8 and NOAA-9. This was done by normalising the NOAA-8 and NOAA-9 data to NOAA-7 data to allow the ISCCP calibration to be maintained. Corrections were also given to allow for the steady degradation of the AVHRR on NOAA-9 in orbit, see Table 5 of Brest and Rossow (1992). The Pathfinder Program Calibration Working Group (Rao 1993b) concentrated on producing graphs for channels 1 and 2 of the AVHRR on NOAA-9, and using all the data available at that time, to show the increasing value of the slope as a function of time since launch. The results are a slightly refined version of the NOAA-9 part of Figure 2.20. Vermote and Kaufman (1995) have developed a method for post-launch absolute calibration of channels 1 and 2 of the AVHRR using ocean and cloud views and have applied it to NOAA-7, NOAA-9 and NOAA-11. The approach includes two steps. The first step is intercalibration between channels 1 and 2 using high-altitude (12 km and above) bright clouds as 'white' targets. This cloud intercalibration is compared with intercalibration using sunglint on the ocean. The second step is an absolute calibration of channel 1 employing an ocean off-nadir view (40–70°) in channel 1 and channel 2 and correction for the aerosol effect. In this process the satellite measurements in channel 2, corrected for water vapour absorption, are used to correct channel 1 for the aerosol effect. The net signal in channel 1 composed from the predictable Rayleigh scattering component is used to calibrate this channel. The result is an absolute calibration of the two AVHRR channels. NOAA-9 channels 1 and 2 show a degradation rate of 8.8% and 6%, respectively, during 1985–1988 and no further degradation during 1988–1989. NOAA-11 shows no degradation during the 1989 to mid-1991 period. Although this trend is similar to the trend obtained using desert site observations, the absolute calibration found in this work for both sensors is

lower by 17% to 20% (suggesting higher degradation) than the absolute calibration
of Abel *et al.* (1993) which used aircraft measurements. Furthermore, it was shown
by Vermote and Kaufman (1995) that the application of both their calibration and
that of Abel *et al.* for remote sensing of aerosols over Tasmania, Australia, failed to
predict correctly the aerosol optical thickness measured there. The only way to
reconcile all these differences was found to be by allowing a shift of 17 nm towards
longer wavelengths of the AVHRR channel 1 effective wavelength. It was shown
that, with this shift, one obtained agreement between the two absolute calibration
techniques ($\pm 3\%$) and that both of them do predict correctly the optical thickness
in the two channels (± 0.02).

The post-launch calibration work described above has all been concerned with
channel 1 and channel 2 data and has concentrated on NOAA-7, NOAA-9 and
NOAA-11 over the period 1981–1990. Channels 3, 4 and 5 have in-flight calibration
available and therefore post-launch studies of the calibration is a less serious
problem than for channels 1 and 2. Nevertheless it is important to consider the
post-launch behaviour of channels 3, 4 and 5. Gorman and McGregor (1994), who
were interested in using AVHRR data for climatological studies, considered data
from channels 3, 4 and 5 for a seven-year period from five satellites NOAA-5,
NOAA-7, NOAA-8, NOAA-9 and NOAA-10. Their study of the calibration data
showed that a number of changes in channels 3, 4 and 5 occurred during the lifetime
of NOAA satellites. These can be categorised into three main groups (i) noise and
sudden short-duration excursions, (ii) sudden large-scale changes and (iii) seasonal
variations.

Studies of noise in the infrared channels 3, 4 and 5 of the AVHRR have been
carried out by Dudhia (1989). Noise is, at certain times during the satellites' lifetime,
a particular problem for channel 3 data, but channels 4 and 5 show little evidence of
serious noise problems. The noise in channel 3 makes it difficult to use data from
this channel in climatological studies. The removal of the herringbone noise pattern
that is commonly found in channel 3 AVHRR data has been considered by Warren
(1989) and Simpson and Yhann (1994) who used Fourier transform techniques. This
can be done by taking the Fourier transform of a channel-3 image, examining the
Fourier transform to identify the noise components, removing these components
and taking an inverse Fourier transform to produce a less noisy image. This
approach achieved some success. The short-duration excursions observed in channels 4 and 5 are sufficiently infrequent to be of minor importance in climatological
studies. Sudden large-scale changes and seasonal variations in the calibration constants have been shown to be associated with changes in the environment of the
AVHRR and, in particular, are related to changes in temperature on board the
spacecraft. Sudden changes are often related to changes in power demand on board
the satellite but seasonal changes are related to the satellite's orbit and to the time
spent in or out of the Earth's shadow of the Sun, see Figure 2.17. A rise in temperature of about 2°C occurs regularly shortly after the spacecraft leaves the Earth's
shadow. However, it can be concluded (Brest and Rossow 1992, Gorman and
McGregor 1994) that, from the point of view of climatological studies, the evidence
seemed to support the case that these changes in the operating temperature of the
AVHRR are adequately compensated for by the on-board calibration procedures.
This is illustrated by Figure 2.20 which shows the history of channel 4 calibration
for NOAA-7 and NOAA-9 as the global monthly mean value of counts from the
AVHRR brightness temperature and the inferred on-board calibration. Let us

assume that the global annual mean temperature of the Earth is also constant over time periods of 5–10 years. The average count values suggest that the NOAA-9 channel sensitivity changed slowly with time and differed significantly from that of the NOAA-7 channel; however, the calibration procedure has eliminated these variations and the evolution of global monthly mean temperatures shown in Figure 2.20 exhibits a seasonal variation, but no trend, over 4–4½ years, although there is clearly some problem with the data for the last 12 months of the period studied.

2.3 SCHEDULING

We have already mentioned that the AVHRR generates

- HRPT, High Resolution Picture Transmission, data which are a direct readout transmission
- APT, Automatic Picture Transmission, data which are analogue data at degraded resolution and are a direct readout transmission
- GAC, Global Area Coverage, data which are spatially and spectrally degraded data which are also on-board recorded data
- LAC, Local Area Coverage, data which are on-board recorded data.

As the satellite orbits the Earth the HRPT and APT data are transmitted continuously, whether or not there is actually any ground station within range to receive the data. The selection of the two channels to be included in the APT is made by the Control Center at Suitland, Maryland. The LAC and GAC data are recorded on board for later playback and transmission to Earth. NOAA NESDIS (National Environmental Satellite Data and Information Service) operates two Command and Data Acquisition (CDA) stations, one in Wallops Island, Virginia, and one in Fairbanks, Alaska (Gilmore Creek before 1984) to receive both tape-recorded (GAC and LAC) and direct readout (HRPT) data from the spacecraft and send the data to the NOAA NESDIS Satellite Operations Control Center (SOCC) at Suitland, Maryland. The spacecraft, however, remains out of contact with these stations for three (or sometimes four) successive orbits. In order to save time in the recovery of data, particularly TOVS data which are urgently needed in near-real-time for meteorological purposes, from such orbits, arrangements have been made for downlinking of tape-recorded data at the station operated by the Centre National d'Etudes Spatiales (CNES) at Lannion in France for onward transmission to Suitland, Maryland.

Each spacecraft in the TIROS-N series carries five digital tape recorders, each with a single electronic module and dual tape transport, to record data for subsequent transmission to the Command and Data Acquisition stations and thence to the data processing facility (see Kidwell 1991). Each tape transport has the capacity to record one of the following:

(i)   110 minutes (slightly more than a full orbit) of GAC data with embedded TIP data

(ii) 10 minutes of HRPT data (which is called LAC data when recorded on-board)

or

(iii) 250 minutes of TIP data only (called stored TIP).

It is clear from this that the recorded LAC data constitute only a small fraction (ca. 10%) of the total global high resolution data gathered by the AVHRR. The archive of Global Area Coverage (GAC) data, maintained by NOAA/NESDIS, is vital for many studies, although the reduced resolution of 4 km is a limit to its use, as is the time delay in obtaining the data from the USA. For most research applications digital AVHRR data are required in order to have access to all five imagery channels, to obtain the full 1 km horizontal resolution data and because of the improved radiometric calibration.

Users may request scheduling of Local Area Coverage (LAC) data which are recorded outside the direct readout range of Wallops Island, Virginia or Fairbanks, Alaska. Because tape-recorder space and transmission time must be shared by many requesters, requests must be received at least one month prior to the data acquisition period. Requests are considered on a first-come, first-served basis, and according to the following priority considerations:

1   national emergencies, as specified in the various national emergency plans

2   situations where human life is in immediate danger (i.e. search and rescue operations)

3   US strategic requirements

4   commercial requirements

5   scientific investigations and studies

6   other miscellaneous activities.

Requests must also be accompanied by the following information:

(a)   brief description of application

(b)   geographical area (i.e. East Greenland, Korea Straits, etc.)

(c)   latitude and longitude coordinates bounding the area of interest

(d)   desired frequency of coverage (i.e. once weekly, etc.)

(e)   spectral channels required for image processing; range of expected brightness values of temperatures for image enhancement purposes for the thermal infrared channels

(f)   type of data – digital data available on CCT, and/or analogue data, available as photographic prints

(g)   beginning and ending dates of the study period

(h)   satellite preference: NOAA-12 – daylight descending, night-time ascending; NOAA11 – daylight ascending, night-time descending

(i)   name, address, and telephone number of requester.

Failure to provide this information at the time of the request may cause a delay in scheduling of the LAC data. Requests for AVHRR LAC data may be phoned in, but must be followed by written documentation. They should be submitted to:

Chief, Interactive Processing Branch (IPB),
NESDIS,
Room 510, World Weather Building,
Washington, DC 20233,
USA
telephone (. . . 1) 301-763-8142.

The staff of the Interactive Processing Branch produce a monthly schedule which is submitted to Satellite Control and a confirmation letter is sent to the requester to say that the request has been scheduled. Every effort is made to accommodate each request, for example, by combining requests of overlapping areas. However, because the number of requests for LAC coverage almost always surpasses scheduling resources, NESDIS does not guarantee complete or even partial fulfilment of LAC requirements. When lack of scheduling resources severely limits the acquisition of LAC coverage, requesters will be notified by Image Processing Branch LAC scheduling personnel. There is no charge to the user for the scheduling of a request.

The Image Processing Branch receives a report, once or twice a month, on the acquisition of data requested. After data have been recorded they are processed and archived in the usual way. The user needs to appreciate that a request for LAC scheduling is not an implicit request for data. Users must also contact the Satellite Data Services Division (SDSD) and meet all pre-payment requirements before the actual processing of a data request can begin (see section 2.4 for more details on ordering products from SDSD).

2.4   NOAA'S ARCHIVAL AND DISTRIBUTION OF DATA

2.4.1   Reception, pre-processing and archival of AVHRR data

In this section we shall be concerned with the handling of AVHRR data by NOAA. That is, we are concerned with (i) HRPT received by NOAA's own ground stations, (ii) LAC data recorded on board the spacecraft for about 10% of each orbit and downlinked to a NOAA ground station and (iii) GAC data (global coverage) recorded on board and downlinked to a NOAA ground station. We shall not attempt to describe the archiving and distribution activities at other direct readout ground stations which are now to be found all over the world (see section 2.5). The stages that we shall need to consider in this section include:

 (i)   ingestion and pre-processing of the data (to level 1b)

 (ii)   archive of data at NOAA and distribution to users

(iii)   generation of NOAA NESDIS operational products.

Following reception at one of the CDA stations (Wallops Island or Fairbanks) the data are transmitted to NOAA NESDIS at Federal Building 03, Suitland, Maryland, where they undergo processing to produce level 1b products, see Figure 2.21. By level 1b we mean (following FGGE (First Global GARP Experiment) terminology) raw data that have been quality controlled, assembled into discrete data sets and to which Earth location and calibration information have been appended (but not applied). The data are present on this database as a collection of data sets. Each data set contains data of one type for a discrete time period. Thus there are separate data sets for HRPT, LAC and GAC and for each of the various other instruments on board the spacecraft (HIRS, MSU, SSU). Time periods are arbitrary subsets of orbits for HRPT and LAC data. Generally GAC, HIRS, MSU and SSU data sets (which are gathered for complete orbits) will be available for corresponding time periods and usually have a three-to-five minute overlap between consecutive data sets. The storage of the level 1b data in the archive is depicted in Figure 2.22; a duplicate copy of the archive is now held at the US National Climatic

Data Center in Asheville (North Carolina) for security, i.e. in case of loss or destruction of the archived data at NOAA NESDIS in Maryland.

Between October 1978 (when TIROS-N was launched) and 11 April 1985 HRPT and recorded GAC, LAC and TIP data were ingested by NESDIS computers and stored temporarily on staging disks used as work space and for interfacing these computers with the NESDIS Terabit Memory (TBM) mass storage system on 9-track 1600 b.p.i. computer compatible tapes (CCTs). The ingested data were then retrieved from disk storage on a time-available basis, processed to level 1b format and returned to the disks for subsequent processing and the Satellite Data Services Division (SDSD) archive. On 11 April 1985 NESDIS abandoned the old TBM system as a means of storing ingested polar orbiter data. The data are currently stored on IBM 3380 disks attached to several IBM 4381 computers. From April 1985 until June 1986 all level 1b data were archived on 9-track 6250 bpi CCTs.

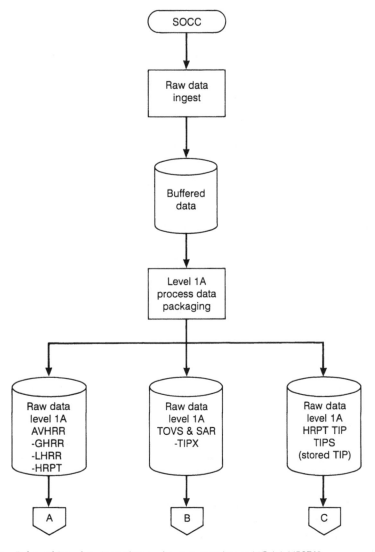

Figure 2.21   Polar orbiter data ingestion and pre-processing at NOAA NESDIS.

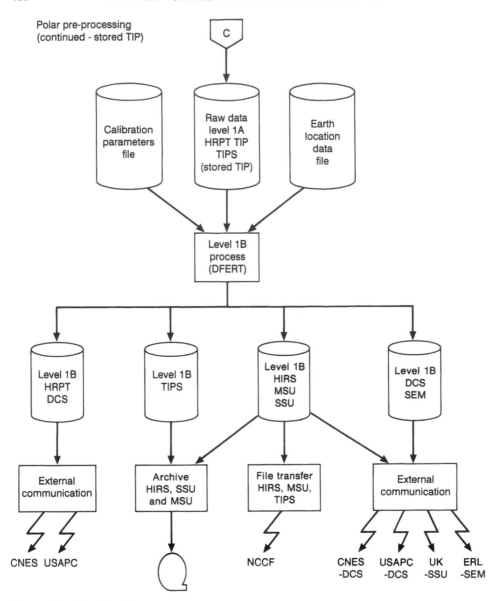

Figure 2.21 *(Continued)*

Beginning in June 1986 the data have been archived on IBM 3480 (magnetic tape) cartridges. Over the years there has been an extensive programme to recover the old (pre-June 1986) data from CCTs and to transfer them to IBM 3480 cartridges, making two copies of each tape – one for the working archive and the other one for back-up storage in Asheville. However some of the pre-1985 data appear to have been lost. The changes in archive medium are intended to be transparent to the user.

For anyone who needs a description of the level 1b database, tape formats and data record formats, this information will all be found in sections 2 and 3 of Kidwell (1991). It was decided not to reproduce it all here. One point, however, is worth making here. The AVHRR data, which consist of 10-bit words, are generally packed

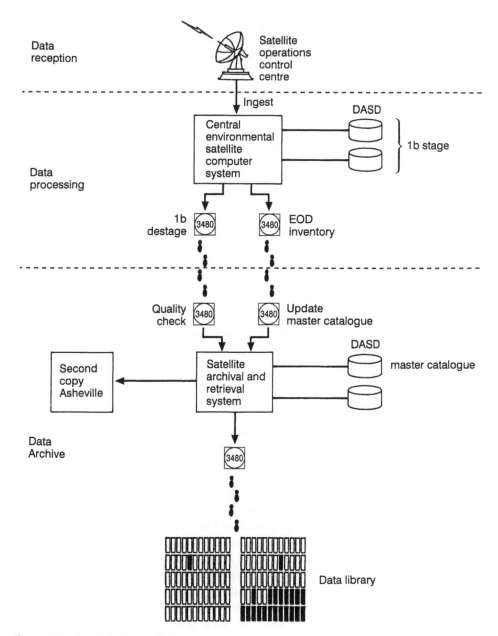

Figure 2.22   Level 1b data path for NOAA NESDIS.

```
Bytes   |<--------4---->|<------3------>|<------2------>|<-----1------->|
Bits    .3.3.2.2.2.2.2.2.2.2.2.2.1.1.1.1.1.1.1.1.1.1.9.8.7.6.5.4.3.2.1.0.
        .1.0.9.8.7.6.5.4.3.2.1.0.9.8.7.6.5.4.3.2.1.0. . . . . . . . . . .
                  Scan Point 1            Scan Point 1          Scan Point 1
        |0 0|<----Channel 1---->|<-----Channel 2--->|<---Channel 3----->|
                   (10 bits)              (10 bits)            (10 bits)
```

Figure 2.23   Packed AVHRR data format, NOAA NESDIS level 1b (Kidwell 1991).

to save space on a magnetic tape. Three 10-bit samples are packed in four bytes, right justified, see Figure 2.23. Appendix B of Kidwell (1991) gives useful guidance for analysts and programmers who need to unpack the data. It should be noted that non-NOAA ground stations (see section 2.5) often pack the data in some other way; this is because in the past there was little pressure on individual independent ground stations to standardise their tape formats.

2.4.2 Operational products

As we have already noted, level 1b data are essentially raw data with some Earth location and calibration data attached to them. In addition to level 1b products, NOAA NESDIS also generates and archives a number of operational products. These include atmospheric soundings (from TOVS data) which we shall not consider here (for details see Kidwell (1991) sections 5.0 and 5.1),

- mapped/gridded AVHRR data
- sea surface temperatures
- heat budgets.

This list of products may seem rather restricted and there are two reasons for this. First, other operational products are now produced by other organisations, based on the use of AVHRR data, although the initial development work in the past was done by NOAA NESDIS. For instance in the USA snow cover mapping was handed over to the National Weather Service in 1982 and sea ice charts are now produced by the US Navy/NOAA Joint Ice Center (also at Suitland, Maryland). Various other organisations in different countries produce their own snow maps and sea ice charts from satellite data. Secondly, some products, such as aerosol concentration maps for example, are still regarded as experimental products rather than operational products.

The task of making atmospheric corrections to visible band data from any satellite will be discussed in section 3.3. The calculation of some contributions is routine, though tedious. However, the one contribution that is much more difficult to calculate reliably is the aerosol contribution; the aerosol optical path length is commonly written as $\tau = A\lambda^B$ and the difficulty which arises is that the values of the parameters A and B in this expression vary widely, both spatially and temporally. Aerosols arise from various sources, for example wind-blown dust from desert areas, smoke from industrial or domestic fires, forest fires or volcanoes, sulphuric acid arising from fossil fuel burning or from dimethyl sulphide produced in the oceans, sea salt produced from ocean spray, etc. In recent years NOAA has been producing an experimental product giving aerosol concentrations over the oceans (see, for example, Hastings *et al.* 1989a, Rao *et al.* 1989). The experimental product is a contour map of atmospheric optical thickness. The aerosol amount is expressed as an optical thickness, scaled by 100. Thus, a value of 20 can be interpreted to mean that 82% of a vertically incident beam of solar radiation at a wavelength of 0.5 μm would be transmitted to the surface, i.e. solar transmittance = 100 times exp(0.2) = 82%. Observations are composited over a seven-day period before being analysed and contour-mapped at a resolution of 100 km.

We show portions of the weekly aerosol optical thickness charts in Figure 2.24. Enhanced values of atmospheric turbidity are seen to the west and north-west of

MONITORING GLOBAL AEROSOL MOVEMENTS

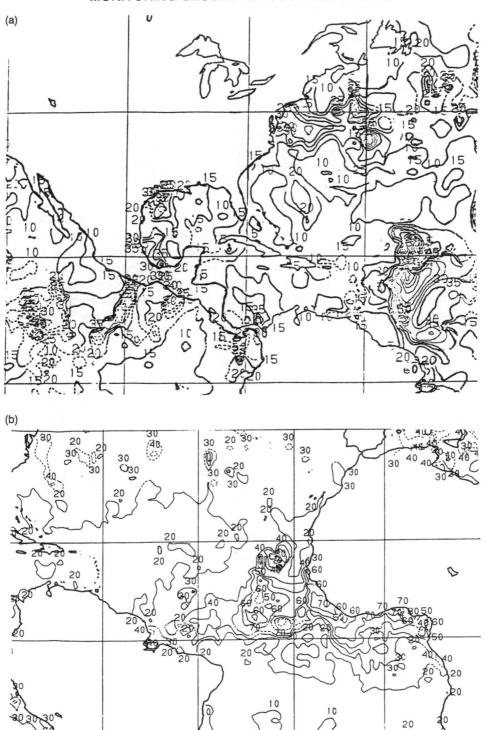

Figure 2.24 Examples of NOAA's experimental aerosol products for periods ending (a) 4 June 1987 and (b) 1 January 1988 (Hastings *et al.* 1989b).

Africa in Figure 2.24(a). These constitute part of the Saharan dust cloud that normally would be driven along the equatorial Atlantic Ocean by the prevailing easterlies. In this particular instance, however, there was a breakdown of the easterlies and the flow was directed northwards. Available surface reports of sky conditions and surface and upper-level meteorological data, and analysis of high-resolution (~ 1 km) satellite imagery, support this conclusion. We see in Figure 2.24(b) large values for atmospheric turbidity in the western part of the Mediterranean close to the coast of Africa. It is likely that the source of these enhanced turbidities is the northern Saharan desert.

Further information about the experimental aerosol product can be obtained by contacting Dr. L. L. Stowe, NOAA/NESDIS/Office of Research and Applications, Satellite Research Laboratory (E/RA11), World Weather Building, 7th Floor, Washington, DC 20233, tel. (001) 301 763-8102. The discussion of the theory behind the method used to generate this product will be discussed in section 7.5.

The most important use for the mapped/gridded GAC data is in connection with vegetation indexes. The detailed discussion of vegetation indexes will be postponed until sections 5.1 and 5.2; we simply note for present purposes that the various vegetation indices can be derived from a ratio involving the channel 1 and channel 2 AVHRR data. The detailed discussion of sea surface temperatures will be postponed until section 4.2. For the present we note that the history of the development of the sea surface temperature and heat budget products is outlined in Table 2.23 and the steps involved in the processing of the AVHRR data to generate sea surface temperature products are illustrated in Figure 2.25.

The gridded AVHRR products consist of mosaics of orbital passes of unmapped data and the mapped AVHRR products consist of Mercator and polar stereographic projections, see Table 2.24. The mapping mosaics are of day-time visible and infrared (channels 1 and 4) and night-time infrared (channel 4). The stereographic polar mosaics are available on CCT and the Mercator mapped GAC mosaics are available on CCT from 31 March 1985. The full details of the formats of the CCTs are given in section 5.3 of Kidwell (1991).

NOAA NESDIS has been generating vegetation index products since May 1982. The first generation product was an experimental product and this was followed

Table 2.23 Oceanographic products system – major events (NOAA)

1972–1979	ITOS Series Satellites (Single Window) (SR)
	Global Operational Sea Surface Temperature Computation (GOSSTCOMP)
28 February 1979	TIROS-N (GOSTCOMP III) (AVHRR/HIRS)
25 January 1980	NOAA-6
19 August 1981	NOAA-7
17 November 1981	Multichannel Sea Surface Temperature (MCSST) Operational Technique
05 February 1985	NOAA-9
01 July 1985	NOAA-8 Radiation Budget in addition to NOAA-9
28 September 1985	Terminated processing of NOAA-8 data (oscillator malfunctioned)
10 December 1986	NOAA-10 Radiation Budget in addition to NOAA-9
10 September 1987	Experimental Aerosol Optical Thickness
08 November 1988	NOAA-11

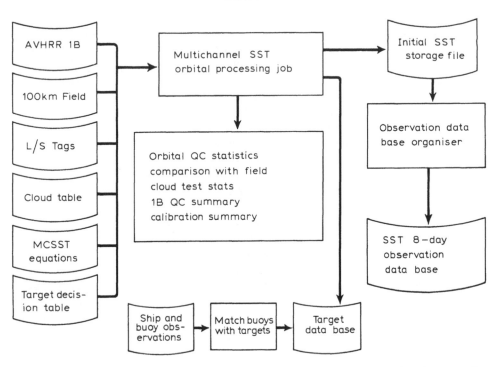

Figure 2.25 Multichannel sea surface temperature (MCSST) orbital processing and quality control (NOAA NESDIS).

from April 1985 by the operational second generation product. This included the normalised difference vegetation index (see section 5.1) calculated with the digital numbers from channels 1 and 2. The product is described in the *Global vegetation index user's guide* (Kidwell 1990). The process by which GVI products are produced takes place in two phases. The first phase involves the production of a daily GVI map by processing AVHRR data orbit by orbit and mapping it on to a standard base projection, the Plate Carrée projection, using only data from alternate scan lines of the GAC data so that the spatial resolution is about 8 km. A single day's map consists of a mosaic of the daytime portions of 14 orbital swaths of the AVHRR. The data are sampled in the mapping process so that the GVI in an array location consists of a value computed from a single AVHRR pixel; there is no averaging. The second phase consists of the production of a set of final weekly composite products. On any single day, about half of the Earth's surface is obscured by clouds. To remove clouds, seven-day maximum vegetation index composites are produced from the daily arrays. For each composite period of seven days, the pixel from the daily data having the largest (channel 2 − channel 1) difference (computed from raw data or counts, i.e. uncalibrated data) is retained at each array location (i.e. the 'greenest' of the seven daily values for each array location is retained in the composite). This eliminates clouds from the composite except for any areas that were cloudy at the time of the satellite overpass on all seven days. In addition to the NDVI value, six further pieces of data are also given in the weekly composite product for each array location and taken from the same day as the selected NDVI

Table 2.24   Mapped/gridded AVHRR product (Kidwell 1991)

Product description	Accuracy goals	Coverage/spatial resolution	Processing format	Processing schedule
Hemisphere mapped GAC polar mosaics IR and VIS mosaics	Nominal ± 5 km for polar and Mercator mapping location	Northern/ southern hemispheres 1024 × 1024 14.8 km equator 29.6 km poles	Mapped imagery CCT	Daily
Mercator mapped GAC mosaics IR/VIS	See above	360° longitude 40°N–40°S 9.8 km equator increasing poleward	Mapped imagery CCT (begin 31 May 1985)	See above
Polar mapped GAC composites IR/VIS (minimum brightness/ maximum temps)	See above	North/south polar regions 1024 × 1024	Mapped	Conforms to compositing period (7 days)
Pass-by-pass gridded GAC imagery VIS/IR (one satellite)	Nominal ± 5 km grid placement	Global 4 km	Gridded imagery	Orbit by orbit
Imagery from Local Area Coverage (LAC) data: both recorded and direct readout (ungridded)	N/A	Recorded Data: Selectable; two 11.5 min segment/orbit. Direct readout continental US	Imagery	Recorded: Variable two 11.5 min segments/ orbit. Direct readout all continental US

value; these are:

• the values of channel 1 and channel 2 raw data (truncated to 8-bit precision)
• channel 4 and channel 5 data converted into 'GOES counts' which are, essentially, temperatures which can be identified in absolute units from a standard look-up table given in Appendix B of Kidwell (1990)
• the solar zenith and scan angles.

The weekly Plate Carrée projection composite is used to generate weekly Mercator and polar stereographic projections as well. From July 1990 onwards a calibrated vegetation index was added to the product making eight items that are specified for each location in the map projection. This is a vegetation index which is calculated using the albedoes derived from channel 1 and channel 2 data, rather than the raw channel 1 and channel 2 data (see section 5.2). Details of the map projections and the format of the GVI product are given by Kidwell (1990).

For the sea surface temperature product the GAC data are processed into a basic set of sea surface temperature (SST) observations at 8 km resolution (50 km resolution prior to 17 November 1981) over the global oceans. All observations are values which have been integrated over an 8 km diameter spot; however they have variable spacing, ranging from 8 km (contiguous) in the coastal waters of the USA to 25 km in the open ocean. This database is further processed to generate gridded analyses at the global, regional and local scales, see Table 2.25. The global-scale analysis covers the global oceans at 100 km resolution (1° latitude/longitude grid). The regional-scale analyses (50 km resolution) are over five selected regions on a 0.5° latitude/longitude grid. The local-scale analyses are at 14 km resolution (0.125° latitude/longitude grid) over eight selected regions. The geographical locations of the regions involved in the regional- and local-scale products are identified in section 5.2.1 of Kidwell (1991). The nature of the product (CCT, image or chart) and the frequency are indicated in Table 2.26. Full details of the formats of the CCTs of the sea surface temperature products are given in section 5.2 of Kidwell (1991).

Beginning in 1974 estimates of the planetary albedo and the outgoing longwave radiation (OLR) (i.e. thermal infrared radiation) were made from scanners on NOAA spacecraft, starting with the Very High Resolution Radiometer (VHRR) which was the predecessor of the AVHRR (Gruber 1977, Gruber and Winston 1978, Ohring and Gruber 1983, Henderson-Sellers 1984, Gutman, 1988, Jacobowitz 1991, Arino *et al.* 1992 and many other references given in chapter IX-1.1 of Rao *et al.*

Table 2.25 Oceanographic products system – analysed fields

Global MCSST analysis
    Daily analysis covers the world from 70°S to 70°N
    using a 1° lat/long grid
Regional MCSST analyses
    Twice weekly composite field analyses generated on 1/2° lat/long grid
        Atlantic Coat
        Pacific Coast
        Hawaii/Alaska
        EPOCS (Eastern Tropical Pacific)
        Pacific Islands
Local MCSST analyses
    Twice weekly high-resolution automated analyses
    produced on a grid spacing of 1/8°
        Northeast Atlantic Coast
        Southeast Atlantic Coast
        Gulf of Mexico
        Southwest Pacific Coast
        Northwest Pacific Coast
        Gulf of Alaska
        Bering Sea
        Gulf of California

Table 2.26　Sea surface temperature products (Kidwell 1991)

Product description	Accuracy goals	Spatial resolution/ geographical coverage	Format/ schedule
Sea surface temperature observations	±1.5°C Abs ±1.5°C Rel	8 km (nominal/ global)	CCT weekly
Sea surface temperature regional-scale analysis	See above	0.5° lat/long grid (50 km) 5 regions	Image weekly; CCT monthly
Sea surface temperature global-scale	See above	1° lat/long grid (100 km)	Contour chart weekly; CCT bimonthly; Image daily
Sea surface temperature local-scale analysis	See above	0.125° lat/ long grid (14 km) 8 regions	2 CCTs monthly; Image daily
Sea surface temperature monthly mean	See above	Global 2.5° lat/long grid (250 km)	CCT yearly; Contour chart monthly

1990). Good estimates of seasonal averages of albedo, outgoing longwave radiation and net radiation can be made and are very useful as input to climatological models; they are discussed in some detail in section IX-1.3 of Rao *et al.* (1990). They are, however, essentially long-term (seasonal) averages of rather low spatial resolution. The heat budget archive consists of four types of similar data. One type is the monthly heat budget consisting of daily values of day flux, night flux, absorbed solar energy and available solar energy for a month. These data are available in polar stereographic and Mercator projections. Another type of heat budget product is the seasonal heat budget which has the same daily radiation fields as the monthly heat budget product, but, in addition to polar stereographic and Mercator projections, there is a smaller (45 × 45) polar chip included and the data range over a three-month season. A subset of the seasonal heat budget product is now available for 10 years in the Mercator format, and is called the 10-year Mercator heat budget product. The fourth type of heat budget product is the monthly mean heat budget which contains monthly averages for the above-mentioned radiation fields for the period from June 1974 to February 1978. These data are also available in polar stereographic and Mercator projections. Some further details are given in Table 2.27. Full details of the formats of the CCTs of the heat budget products are given in section 5.4 of Kidwell (1991).

Table 2.27   Heat budget products (Kidwell 1991)

Product description	Accuracy goals	Coverage/ spatial resolution	Format/ schedule
Monthly heat budget parameters	±7 W m⁻² reflected and outgoing energy	Global	CCT
Day-time longwave flux		Observations 50 km	
Night-time longwave flux			a. 2.5° × 2.5° Mercator map array
Absorbed solar radiation			b. 125 × 125 polar map array
Available solar energy (calculated field to be included in output form)			2 times/day IR flux; 1 time/day reflected energy; monthly (time average)
Seasonal heat budget parameters	±7 W m⁻²	Global	CCT a. 125 × 125 polar-stereo map array
Day-time outgoing longwave			b. 45 × 45 polar-stereo (chip) map array
Absorbed solar energy			c. 144 × 72 (2.5° × 2.5°) Mercator map array
Available solar energy			Daily data for a three-month season
Night-time outgoing longwave			
10-year Mercator heat budget parameters	±7 W m⁻²	Global	CCT a. 144 × 72 (2.5° × 2.5°) Mercator map array
Day-time outgoing longwave			Daily data for 10 years (June 1974–March 1978, Jan. 1979–Feb. 1986)
Absorbed solar energy			

Table 2.27   (*Continued*)

Product description	Accuracy goals	Coverage/ spatial resolution	Format/ schedule
Available solar energy			
Night-time outgoing longwave			
Monthly mean heat budget parameters	± 7 W m^{-2}	Global	CCT
Day-time IR flux			a. 144 × 72 (2.5° × 2.5°) Mercator map array
Night-time IR flux			b. 125 × 125 polar-stereo map array
Absorbed solar energy			c. Monthly mean for 45 months (June 1974–Feb. 1978)
Available solar energy			

2.4.3   Data distribution to users

We have seen already that AVHRR data and derived products are available as hardcopy images or as digital data and sometimes as both. At the time of writing (April 1993) the archive of NOAA NESDIS contains more than 135 000 CCTs and more than 5 000 000 images. Of the CCTs about 77 000 are of AVHRR data and they contain about 7.6 Tbyte of data from 1978 until 1993. Also at the time of writing orders are received and processed at about 3500–4000 per annum and the fulfilment of these orders involves shipping about 9000 CCTs and 4000 hardcopy images per annum.

All requests for digital tapes or hardcopy prints, etc., should be addressed to:

National Climatic Data Center
Satellite Data Services Division
Room 100, Princeton Executive Square
Washington, DC 20233
telephone: (... 1) 301 763-8400 or FTS 763-8400
telemail: SDSDDSB/NESDIS
OMNET: K. METCALF
fax: (... 1) 301 763-8443 or FTS 763-8443

Standing orders for certain specified products to be produced on a regular basis are accepted. Users may obtain three-to-five day turnaround on requests for the current operational polar orbiter level 1b data sets provided that the order is placed either before the time of collection of the data or within 24 hours after its collection. Standing orders will be accepted. Users may request routine delivery of data sets with coverage of a specified area. Users may place a standing order and arrange to confirm delivery of each day's pass with the Satellite Data Services Division (SDSD) within 24 hours of the time of data collection. (Note that the foregoing applies to data received by SDSD. In the case of AVHRR LAC data, SDSD receives only those data which have been scheduled for collection by the Interactive Processing Branch (IPB), see section 2.3). It should be noted that 'one-to-two-day delivery' means that the requested data will be put in the mail (or provided to Federal Express, etc.) within one to two days. Users wishing further information about the 'near-real-time' service should contact the Data Services Branch of SDSD at the above telephone number.

Retrospective orders for complete or selective tape copies or hardcopy imagery which are not too expensive can generally be fulfilled within ten days, depending upon the correct number of orders being processed. Digital data are available from SDSD on either 9-track 1600 bpi CCTs or 6250 bpi CCTs. SDSD guarantees the contents of each CCT for 60 days.

When ordering data from SDSD the following parameters should be specified to ensure the fulfilment of your data needs:

1   type of data (also specify whether level lb or operational products)
2   dates and times of data
3   channels
4   satellite name
5   day, night, or both
6   area (latitude/longitude box)
7   orbit numbers if known
8   cartridge/CCT (density)
9   packed or unpacked (level 1b only)
10 capability of handling split data sets.

Some of these parameters are redundant but may assist SDSD in providing the user with the exact data desired.

SDSD charges the user according to the source of the input data. The charges for both imagery and digital satellite data are subject to change; they are contained in a separate document, the NOAA *Retrospective satellite data price list and ordering procedures*, which is available from SDSD upon request. All orders or requests for satellite data or information must be prepaid and details of the acceptable methods are available from the SDSD. Because of the custom nature of most requests for satellite data and information, it is not always possible for SDSD to establish a firm price quotation for its data and information. One-time-only or intermittent users are encouraged to contact the SDSD for a price estimate (valid for 60 days) and then send their payment.

It became obvious to SDSD personnel that it was necessary to devise an efficient method of accessing the data inventories to fulfil customer requests. Since the most common customer requirement for these data was to cover a specific geographical location, it was often necessary to cross-reference each data set manually with an ephemeris (equator crossing information) data set. This process was very tedious and very time-consuming, resulting in many delays in responding to customer requests. So in 1985, the Electronic Catalog System (ECS) was designed and developed for SDSD. It is also available to users and potential users of NOAA satellite data (see Hastings *et al.* 1988b).

The ECS includes inventories for all the level lb HRPT and LAC data as well as all of the level lb TOVS data available from SDSD since 7 April 1985. In addition, it includes inventories for all the GAC data available since October 1978. The system is stored on a commercial time-sharing system, and is available through direct dial-up; it is completely menu driven. SDSD is in the process of converting ECS (renamed OSCAR system for On-line Satellite Catalog Access and Request system) to a Microvax 3300, which is planned to be accessible through a commercial network.

SDSD provides complete instructions for accessing the system, as well as an account and a password. All that is required of the user is a terminal with a telephone line. An SDSD representative may be contacted at (...1) 301 763-8400 for information on how to access the system at no charge (other than the cost of the telephone call).

The search process begins with the selection of the appropriate data type. The system is then directed to specific data files, depending upon the selection. Next, a data range must be entered. This range must be from 1 to 180 days, and all data sets within the applicable range are then extracted. This constraint is made to keep searches from being too slow. Longer periods of time can be searched by breaking up the time period into increments of 180 days or less and running the search on each increment.

An array of all the data sets that meet the requested criteria is then generated, broken down by satellite and data type, and from this the user must decide which satellite to use (data for TIROS-N to NOAA-12 are included, depending upon the dates of interest). Once all the preliminary limiting has been completed, the geographical search can begin. The user enters either a point of interest, or an area of up to 90° of longitude. The system then uses either the single point or the corner points of the rectangle to calculate an acceptable equator crossing range based upon the rotation equations of the Earth, and the speed of the satellite, which determine the boundaries of the suborbital track. The ephemeris database is then accessed, and the record for each previously selected data set is compared to the calculated time and node range. Those data sets that fall within the proper range are then written to a temporary file to be displayed to the user.

From this point the user can: (1) choose to print (or, using his or her communications software, to download the file of data sets for review or future reference, (2) view each data set individually, and specify a keep/delete option, (3) specify a minimum time duration and delete all scenes that are shorter than this specified minimum, (4) perform any combination of the above functions, or (5) submit an order to the SDSD Master Work Order File if prepayment has been made. Once this option is exercised, SDSD personnel will retrieve the work order and begin processing the request with no further contact with the user, unless special instructions have been indicated.

Since the system was made available to the public, several enhancements have been initiated. OMNET Electronics Mail Service has made the service available to their subscribers by building a gateway, so that a simple 'GO TO ECS' command will take the user from his or her mailbox to the ECS, and 'GO TO MAIL' will take him or her back again.

ECS does not yet include information such as cloud cover. Users in the land sciences who are more concerned with cloud-free conditions than with imagery of a specific date and time may still wish to have SDSD perform a search of archival imagery to confirm the cloud cover or other easy-to-define characteristic of the imagery before purchase. Such searches may be arranged by contacting SDSD (with the understanding that major searches take time and effort, and may need to be paid for). But even in such cases, ECS can be used to narrow down the range of images that need searching if the user knows the most promising general times for acquisition of appropriate imagery.

SDSD now has ILABS (Image Library And Browse System) software (De Cotiis *et al.* 1991) which can process level 1b GAC, LAC, and HRPT data sets into image browse files, see Figure 2.26. These image files are stored on optical disk and can be viewed using the ILABS PC workstation located at SDSD's Data Services Branch in Camp Springs, Maryland.

SDSD currently archives ILABS browse images for NOAA-11 GAC data sets (from 18 April 1991 to the present) and LAC data sets (from 11 October 1990 to the present). Image files for HRPT data sets and for NOAA-10 level 1b data sets are not currently being archived by SDSD. Current plans are to create browse images for the NOAA-12 GAC data. Image files for these data sets and for retrospective level 1b data sets are created on a customer request basis. Customer requests should be placed with SDSD's Data Services Branch.

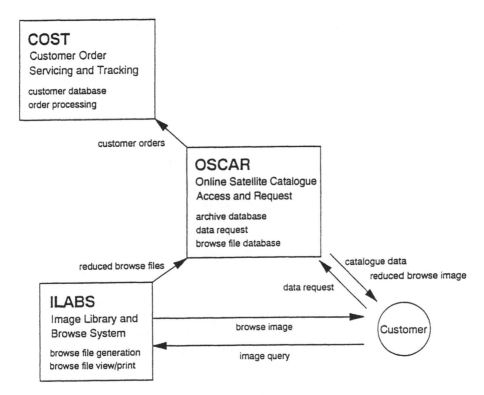

Figure 2.26 Customer data services at NOAA NESDIS.

2.5 DIRECT READOUT STATIONS

2.5.1 Introduction

In this section we shall be concerned with the real-time or direct-readout transmissions of AVHRR data. These real-time transmissions can only be received on the ground at points above the horizon, as viewed from the instantaneous position of the satellite carrying the AVHRR. Since the NOAA TIROS series and Russian Meteor series satellites have planned orbital altitudes between 833 and 900 km, a ground station can expect to receive APT signals if these satellites pass through a circular area with a radius of about 3100 km with the ground station at the centre. This area will vary somewhat with the exact altitude of a given satellite but can be used for routine work. More exact calculations can be made using the *TIROS-N series direct readout services user's guide* (Barnes and Smallwood 1982). The reception of these data at a receiving station on the ground provides a complementary source of data to the data set received and archived by NOAA. In the previous section we have given an extensive account of the archiving and distribution of AVHRR data, and of AVHRR-derived products, by NOAA. However, NOAA is not able to obtain all the data generated by the AVHRR for reasons that should, by now, be apparent. The low spatial resolution GAC (global area coverage) data are tape-recorded on board from the complete orbit and downlinked to a North American ground station. A proportion of the LAC (local area coverage – high resolution) data from each orbit (about 10%) is also tape-recorded on board and downlinked over North America. Thus large amounts (ca. 90%) of the full-resolution AVHRR data are not collected by NOAA itself and can only be recovered by a local ground-based receiving station within direct line-of-sight of the satellite.

From Figure 1.13 it can be seen that there are three real-time transmissions of data from the satellites of the TIROS-N, NOAA-6, etc. series. These are:

(i)   the VHF beacon

(ii) the analogue APT (automatic picture transmission) at frequencies of between 137 and 138 MHz and

(iii) the HRPT digital data (= real-time version of the LAC data) which are transmitted at 1698.0/1707.0 MHz.

Of these, (i) is of little interest, (ii) is technically easier than (iii) to receive. If one reads of high schools which receive data or images from meteorological satellites, it is almost certainly either data from a geostationary satellite or the analogue APT data from the NOAA series of polar-orbiting satellites that is involved. The APT signal can be received with a fixed antenna and relatively simple receiving equipment. To receive the digital HRPT data is a considerably more complicated operation. First, it is necessary to have a steerable antenna with a large reflecting dish behind it in order to be able to receive a good detectable signal (i.e. to have a high signal-to-noise ratio). Moreover, it is necessary to know the direction in which to point the antenna for the satellite to rise above the horizon of the ground station. This requires a rather detailed knowledge of the parameters of the orbit of the spacecraft. It is then also necessary to be able to change the orientation of the antenna so that it continues to point directly at the satellite as it travels across the sky and eventually drops below the horizon again. Handling a steerable antenna

Table 2.28 Breakdown of APT and HRPT stations known to NOAA (Hughes 1993)

	Government	Military	University	School
APT	105	83	91	122
HRPT	96	43	39	9
	Commercial/Business	Amateur	Other	Total
APT	60	602	106	1169
HRPT	21	28	43	279

and being able to control it so that it tracks the position of the moving satellite for HRPT data is much more sophisticated than operating with a fixed antenna for APT data. Also, the fact that the HRPT data are digital, whereas the APT data are analogue, means that the task of HRPT data reception is more sophisticated than that of the APT data. The cost of purchasing and running an HRPT ground receiving station is very much higher than for an APT station. We shall consider the two cases of APT and HRPT stations in turn.

It should be noted that NOAA extracts no fees for establishing and operating an APT or HRPT direct readout ground station. Indeed, it does not even require station operators to make themselves known to NOAA. In mid-1993 there were worldwide over 1000 APT stations and nearly 300 HRPT stations known to NOAA; the breakdown among different types of operators is given in Table 2.28 (Hastings and Emery 1992, Hughes 1993). Doubtless other stations exist unknown to NOAA. A list of direct readout stations given by the International Geosphere–Biosphere Program (1992) is reproduced in the Appendix. NOAA recommends, however, that operators be on NOAA's mailing list and make use of its on-line bulletin board, so that they can keep themselves informed with news of current and planned satellite operations. NOAA has several references (most notably Barnes and Smallwood 1982 and Summers 1989) available to potential operators of HRPT or APT ground stations. It maintains an office to support such stations:

Coordinator, Direct Readout Services,
NOAA NESDIS,
World Weather Building,
Washington, DC 20233,
USA

2.5.2 APT data reception

The first APT system was pioneered on TIROS-VIII (Television InfraRed Observational Satellite) which was launched in December 1963. This satellite was one of the early polar-orbiting weather satellites and was a predecessor of TIROS-N. Several US weather offices were equipped to receive transmissions from this satellite and plans for building relatively simple low-cost ground receiving stations were widely distributed to national meteorological services outside the United States. By 1965, radio amateurs (hams) were designing stations for home reception and publishing design information in popular electronics magazines. Activity and interest in receiving direct readout transmission by members of the academic community also developed. The current series of NOAA satellites still continue to transmit images of the Earth via APT and these have been joined by the Russian Meteor satellites with transmission systems similar to the APT. This is fortunate because it means that a ground station capable of receiving US polar-orbiting satellite APT data can also receive Meteor images.

APT transmissions from the polar-orbiting satellites have traditionally been on radio frequencies between 137 and 138 MHz FM. The NOAA series currently use 137.50 and 137.62 MHz (see Figure 1.13), while the frequencies used by the Russian polar-orbiting satellites vary but have used 137.30 and 137.85 MHz on a regular basis. The FM signal from the satellites contains a subcarrier, the video image itself, as a 2400 Hz tone which is amplitude modulated (AM) to correspond to the light and dark areas of the Earth as seen by the detecting instrument (in our case the AVHRR) on the satellite. The louder portion of this tone represents the lighter portions of the image while the lower volumes represent the darkest areas of the image. Intermediate volumes form the shades of the grey scale needed to produce the complete image. As explained earlier (see section 2.1.3) this analogue transmission is obtained by degrading the higher resolution data generated by a selection of channels from the AVHRR.

An excellent guide to APT data reception is the NOAA Technical Report NESDIS 44 by Summers (1989). It starts with a general description of the NOAA system and of the AVHRR and gives information that has already been given in the book. Section IV of Summers (1989) describes a basic ground station for receiving both APT data and data from the geostationary meteorological satellites, see Figure 2.27. Summers describes two alternative antenna systems for the reception of the APT transmission. Having said above that the APT transmission can be received with an omni-directional antenna, one will nevertheless obtain better reception, i.e. coverage of a larger geographical area, if one uses a steerable antenna. Thus Summers (1989) describes two alternative antenna systems for the reception of APT data. One is a crossed yagi directional antenna and this requires tracking of the spacecraft, while the other is an omni-directional antenna which is simpler to make but gives a slightly reduced reception range. Both antenna systems are described in considerable detail in section V of Summers (1989).

The question of locating and tracking polar-orbiting satellites is discussed in section VII of Summers (1989) and it is not necessary to repeat that material in detail here; we simply note a few salient points. Although all polar-orbiting satellites have basic orbital characteristics in common, each spacecraft is unique in its orbital parameters and needs to be tracked individually. The data necessary to locate and track the meteorological satellites are generally not difficult to obtain. The future orbits of a given satellite can easily be calculated and, if a directional antenna is used, determining the azimuth and elevation of the satellite as it passes over the ground station is not difficult after the basic orbital patterns are understood. Since all of the TIROS series satellites are inserted into Sun-synchronous orbits, which place the spacecraft in a relatively constant relation to the Sun, the ascending node (northbound equator crossing) will remain at a constant solar time, see Figure 2.28. This permits images and other meteorological data to be received by direct broadcast at about the same local time each day.

The time required to complete one orbit is referred to as the nodal period of that satellite. For polar-orbiting satellites this is measured from the time it crosses the

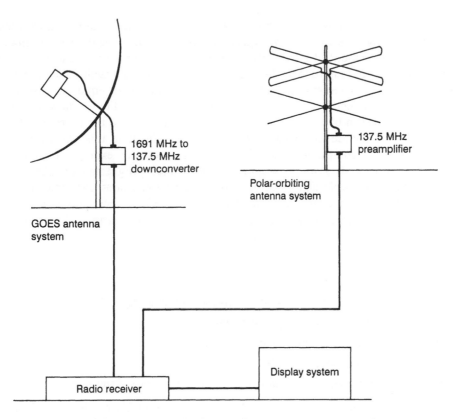

Figure 2.27 Generalised components of a direct readout station (Summers 1989).

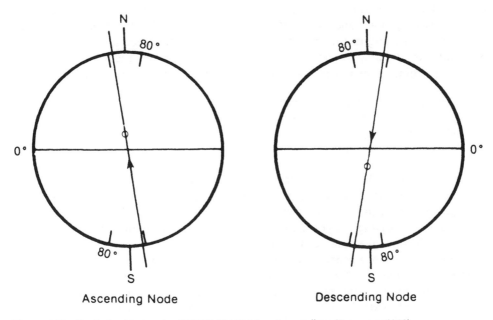

Ascending Node

Descending Node

Figure 2.28 Typical orbital path of TIROS-N NOAA series satellites (Summers 1989).

equator (0° latitude) moving northward (ascending node) until the next northbound equator crossing. The southbound equator crossing is called the descending node of that orbit. The amount of Earth rotation between two successive equator crossings, given in degrees of longitude at the equator, is known as the satellite increment. If a satellite's period, increment and the time and longitude of an equator crossing are known, it is not difficult to predict future orbits for that satellite for days or months in advance. This can be done by simply adding increments and the times of orbits to get the next longitude of an equator crossing and the time this will occur. It is natural to use a computer to do this calculation and an example of a BASIC program to do this is given by Summers (1989). Such programs can take various approaches from simple listings of equator crossing longitudes and times to more complex programs that give local station times, orbital numbers, antenna tracking data for azimuth and elevation and a variety of other information. The orbit input data required for such a program include:

1   month

2   day

3   longitude of the northbound equator crossing (ascending node)

4   hour of the reference orbit equator crossing

5   the minute and decimal minute of the equator crossing

6   the orbital increment of the satellite

7   the time, in minutes, of the orbital period (nodal period) of the satellite.

These data are available from the NOAA Direct Readout Users Electronic Bulletin Board (EBB). Details on this information service can be found in the appendix of Summers (1989). First-time users of the system should contact the Help Center at (...1) 800 638-8742 for start-up information on the EBB and for telephone numbers and access codes for the user's location. An example of orbital data from NOAA is given in Table 2.29. The output of the computer program will provide future dates, equator crossing times and longitudes until, say, the end of the current month. The last equator crossing in the output can be used to continue for a longer period of time, but it is advisable to update this data monthly for greater accuracy.

The recording of the image data in the following ways is described by Summers:

(i)   CRT (cathode ray tube) monitors
(ii)   photographic drum recorders
(iii) electrostatic recorders and
(iv)   computer display systems.

It is a measure of the relentless progress of computer technology that since 1989 low-cost very powerful computer systems have become so much more widely available and one is far more likely now than in 1989 to adopt the fourth of the above methods for displaying images. To do this does, of course, involve first of all digitising the APT analogue data, see Figure 2.29.

Figure 2.29 shows a generalised view of the hardware components that are found in most computer graphic APT (and WEFAX) display systems. At the ground station radio receiver, the satellite transmissions are detected as a 2400 Hz amplitude modulated (AM) signal transmitted at either 120 or 240 lines per minute from the TIROS (or GOES) satellites. At this point the image exists as an analogue repre-

Table 2.29 NOAA orbital predictions for 1 May 1988 (Summers 1989)

	NOAA-9	NOAA-10
ORBIT NUMBER	17430	8408
EQ. CROSSING	0010.71Z	0055.08Z
LONG. ASC. NODE	125.7W	79.43W
NODAL PERIOD	102.0710 min	101.2855 min
FREQUENCY	137.62 MHz	137.50 MHz
INC. BET. ORBITS	25.52 deg	25.32 deg

Note: EQ. CROSSING is given in Greenwich Mean Time (GMT) in hours, minutes and decimals of minutes. 0055.08Z = 0 hours (24 hour clock) 55.08 minutes.
LONGITUDE OF ASCENDING NODE is the longitude where the satellite will cross the equator (northbound at 0 degrees latitude) during the reference orbit.
NODAL PERIOD is the time in minutes for one complete orbit from the northbound equator crossing until the next northbound crossing. (101.2855 = 1 hour 41.2855 minutes.)
FREQUENCY is the FM radio frequency on which the APT images are being transmitted by this satellite in megahertz.
INCREMENT BETWEEN ORBITS is given in degrees of Earth rotation during one orbit of the satellite. (NODAL PERIOD × 0.25 degrees/minute of Earth rotation = INCREMENT BETWEEN ORBITS in degrees).

sentation of the original image created by the satellite's imaging instrumentation. The varying amplitude can be measured as a varying voltage having a discrete voltage range. The 2400 Hz tone, referred to as the video subcarrier, carries the image as a function of its amplitude. Two electronic processes must be accomplished

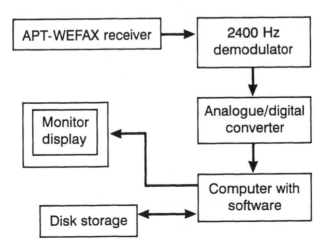

Figure 2.29 Generalised components for computer display of analogue APT data (Summers 1989).

before this analogue image can be managed within a computer system:

1   The 2400 Hz subcarrier must be removed and only the amplitude variations of this carrier, which is the actual image, allowed to pass. This process is known as demodulation and is necessary so that the 2400 Hz, which in itself contains no information, does not become a part of the finished image.

2   The demodulated video, in the form of a varying voltage, must be changed into relative digital values so that these data can be handled in the digital domain of the computer. This step in the process can be accomplished by an analogue-to-digital converter (A/D) which is built to detect a voltage at a given instant and represent that reading as a positive integer number. In an 8-bit computer system this will be a number between 0 and 255. This value can then be stored in computer memory and the next A/D conversion made. Each of these digital values then becomes a discrete element of the image and is referred to as a pixel or picture element. It is important to note that the speed or frequency of the sampling process will influence resolution of the image and the relative width of each scan line but is limited by the resolution of the original data.

Two additional steps are needed in order to display these digital pixels as a coherent image on the computer video monitor. Both of these require software programs written specifically for the computer and graphic display hardware that are available.

1   Each digital picture element must be assigned a specific intensity or brightness proportional to the original amplitude of the image. In black and white displays this can be used to form a linear grey scale or, in instances where enhancement of a certain portion of the image is desirable, other intensities can be used. Alternatively a colour-coded density sliced image can be generated by assigning specific colours to ranges of digital values.

2   The picture segments, or scan lines, must be precisely aligned to form a final coherent image. This requires that the beginning of each scan line can be recognised by the software and positioned in the proper location on the monitor screen. Hard copy of the images can be made by photographing the monitor screen or by special graphic printing programs. There are two approaches to obtaining a satellite computer display system. If a computer with graphics capabilities is already available, the only additional hardware necessary is a 2400 Hz demodulator and an analogue to digital (A/D) converter. Software development will require time and effort; alternatively there are numerous software packages available now which provide a wide range of image processing facilities. The alternative is to purchase a commercial satellite computer display system. If a computer is already available the cost of the additional hardware and software may not be very high.

2.5.3   HRPT data reception

It is possible to buy off-the-shelf a satellite data ground receiving station for HRPT data from any one of a variety of suppliers. Useful accounts of the general features of such a station are given by Baylis (1981) and Baylis *et al.* (1989), see Figure 2.30. After a brief summary of some important aspects of signal analysis, Baylis gives a

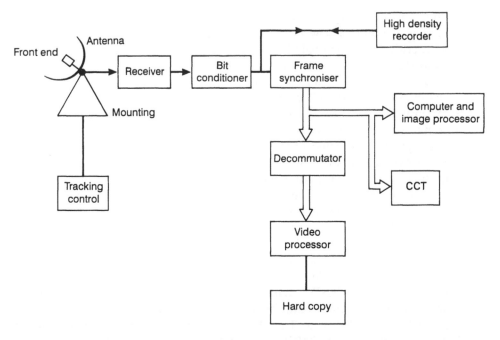

Figure 2.30 Block diagram of HRPT ground station (Baylis 1981).

link calculation to determine the size of the antenna and the quality of the receiver front end that are needed. Some aspects of the design of the antenna and front end of the antenna mounting are given. In order to achieve hemispherical coverage of the sky, the antenna must be mounted on two axes of rotation at right angles to each other. The most common arrangement has the first rotation axis vertical and the second axis horizontal; this is called elevation on azimuth and is illustrated in Figure 2.31. If an elevation on azimuth mount is turned on its side so that the first axis of rotation is horizontal and the second axis is vertical, the mount is called an *X-Y* mount. With such an arrangement the horizontal first axis is usually aligned east–west. Whatever scheme is used, there is always a problem in tracking a satellite

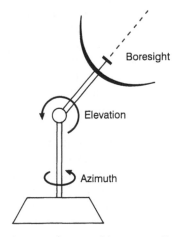

Figure 2.31 Elevation on azimuth mount for steerable antenna (Baylis 1981).

across that part of the sky towards which the first axis of rotation is pointing. This is because of the infinite slew rate called for, about the first axis, as the satellite crosses the critical point (gimbal flip). At first sight this suggests that the X-Y mount (which was mentioned above in connection with an APT receiving station) is better than the elevation on azimuth for tracking overhead satellite passes. However, while they may be preferred for APT reception, X-Y mounts for dish antennae are much more expensive than elevation on azimuth ones and they are not commonly used for this purpose. Despite the tracking problem at zenith, elevation on azimuth mounts are preferred, because of their operational convenience, less complex construction, lower weight, and lower cost. There are two types of tracking control, program track and autotrack, for azimuth on elevation mounts described by Baylis (1981) and there are also brief outlines of the principles of the electronics between the receiver on the one hand and the various display and storage systems and computers on the other hand.

As we have already mentioned, the data from the NOAA polar-orbiting satellites are received at two US Command and Data Acquisition (CDA) ground stations in Fairbanks, Alaska, and Wallops Island, Virginia; they are relayed from these two stations to the NOAA NESDIS Satellite Operations Control Center at Suitland, Maryland. These data include both the real-time (direct readout) data as well as the downlinked tape-recorded data. The Dundee receiving station was constructed in 1975 and daily archiving of VHRR (Very High Resolution Radiometer) data commenced in 1976 and continued until February 1979. AVHRR data have been received and archived from the launch of TIROS-N in October 1978 and this continues until the present time. Currently this station is funded by the UK Natural Environment Research Council (Baylis 1983). At the Berlin Meteorological Institute the first APT reception system was installed in 1966, followed by a VHRR (Very High Resolution Radiometer) receiving system in 1974. AVHRR data have been received operationally there since late 1981 (Eckhardt 1989, Eckhardt et al. 1992). As another example we mention the Tromsø station in the north of Norway (see Ellingsen 1989, Hamnes et al. 1991, Seljelv and Enoksen 1993). The Tromsø Telemetry Station (at 69° 39′ N, 18° 56′ E) was established in 1967 by the Space Activity Division of the Royal Norwegian Council for Scientific and Industrial Research. From 1968 to 1974 the station worked on contract with the European Space Research Organisation (ESRO), the predecessor of the European Space Agency (ESA). Since 1974 the station has been working with remote sensing satellites, in particular meteorological satellites of the TIROS-NOAA series. The station produces AVHRR quicklook images; on the basis of the quicklook images users can order high-quality processed images or digital data. Pre-processed and processed AVHRR data are now distributed. The fact that the Tromsø station is at such a high latitude means that it can receive data from 11 of the 14 passes each day for each NOAA satellite that is in operation. With two satellites it is then possible to receive about 210 minutes of AVHRR data per day there. We have already seen in Table 2.28 that nearly 300 HRPT ground stations are known to NOAA. Most of these stations collect imagery primarily for meteorological forecasting.

We have mentioned the Tromsø receiving station at nearly 70°N which, being so far north, receives about 210 minutes of AVHRR data per day. Although the presence of cloud means that only a fraction of these data is useful for Earth-surface monitoring, nevertheless the station is very well placed to establish a near-real-time environmental surveillance and monitoring system for the Arctic, particularly in connection with sea ice monitoring to provide information to ice-breakers and to shipping (Øvergård 1989).

While the most common way of using digital HRPT data for near-real-time support services during ship cruises for oceanographic and sea-ice studies is the transmission of analysed data products by remote receiving stations on land, the disadvantage of this method is the time lag between the reception of the data and the availability to the user on board the ship. Additionally it is necessary to reduce the information content because of data transmission limitations, so the full satellite information cannot be made available on board ships in this way. While there is a good network of stations to provide these data for the Arctic, a second problem in acquiring the high resolution data arises for many other parts of the world because of the absence of direct readout stations and the very small amount of the AVHRR data which can be stored on the tape recorders on board the satellites. Thus for considerable areas, as for example the Weddell Sea, no continuous supply of AVHRR data from ground stations is available. Thus the question of the development of shipborne reception facilities arises. Wannamaker (1984) described a shipborne APT reception system using a stationary mast antenna. The disadvantages of the APT data are their reduced spatial and radiometric resolution and the analogue character of the signal. However, for the study of features with short time scales and small spatial scales, such as the velocity fields of ice floes in shear zones of strong oceanic surface currents or the development of mesoscale thermal features at frontal zones, the APT data are not sufficient and the HRPT digital data need to be used. For the reception of the HRPT data a tracking system for the satellite is necessary. An obvious problem that arises if one considers putting an HRPT receiving station on board a ship is the problem of compensating for the ship's pitch and roll in tracking the satellite's position. Viehoff (1990) has described the mounting of an HRPT station on the research ice-breaker RV *Polarstern* and its operation in the eastern Greenland Sea in 1988. The steerable antenna which comprised a 2 m disk parabolic antenna with an x-y pedestal was mounted on a 20 ft container palette (for easy handling). The x-y pedestal has the advantage that its singularity in continuous tracking occurs only at $0°$ elevation in the lower axis directions. The antenna beam width was $5°$. Neglecting any ship motion this means that the position of the ship has to be known to an accuracy of about 20 nautical miles in order to enable undisturbed tracking of the satellites. Even if most of the antenna beam width is required to take care of the ship's motion it follows that for a mean ship speed of 10 knots a satisfactory receiving operation while underway is still possible without continuous correction of the ship position.

The RV *Polarstern* with a weight of 11 300 tonnes, a length of 118 m, a breadth of 25 m and a draught of 11 m is a sufficiently stable platform, especially when operating in sea ice, that no automatic compensation of ship motion was required. Due to the short period of operation Viehoff (1990) was not able to make any general statement about the critical weather conditions under which data reception would no longer be possible without automatic compensation for the ship's motion. However it was observed that in open water the long wave swell is the most important hindrance. Even at wind speeds greater than 15 m s^{-1} no frame-loss occurred. The critical values of wind speed and wave height for undisturbed data reception are very much dependent on the size of the vessel, especially the draught. Therefore an icebreaker has the best characteristics for a non-ship-motion-compensated HRPT station on board a ship.

Most ground stations do not archive data; others save a few select scenes based on institutional interests. The length of archive available for different parts of the globe will depend on the period of operation of stations offering cover, and this is highly variable. The most complete archive for western Europe is provided by the ground receiving station at the University of Dundee in Scotland, which has been archiving AVHRR data since the launch of TIROS-N in 1978. The Maspalomas station (Canary Islands) has been archiving daily HRPT data for west Africa since 1986. Cover of west and central Africa has recently been extended with the opening of the Niamey station (Niger) in 1988; however this station's archive of daily five-channel data only exists from 1990 onwards, although decadal composites of the Normalised Difference Vegetation Index (NDVI) are available dating back to 1988. Cover of eastern and southern Africa provided by the Nairobi (Kenya), Harare (Zimbabwe) and Réunion Island (off the east of Madagascar) stations is even more recent. However, as interest in regional and global environmental studies increases, efforts are being made to develop internationally cooperative ventures to save data from several HRPT stations, supplementing these data with LAC coverage to obtain periodic global full resolution AVHRR data, see section 7.9.6. For example, the European Space Agency has anticipated increased interest in AVHRR data for Europe and Africa by developing a coordinate archive and dissemination system with on-line catalogue.

The current network of AVHRR direct readout stations developed in an *ad hoc* manner so that, for instance, there was no standard adopted for tape format for CCTs (computer compatible tapes) and there was no common catalogue of available scenes. Some attempts have now begun to be made to address these problems, for instance on a European scale by Earthnet, the operational Earth observation branch of the European Space Agency (ESA) (Fusco 1989, Fusco *et al.* 1989, Pittella and Bamford 1989, Tobiss and Muirhead 1989, Marelli 1992). In 1986 the Earthnet Programme Office commissioned a study to assess the requirements of Europe's AVHRR remote sensing users (excluding meteorology) and to survey the existing HRPT facilities. The study showed that at that time

1 there was a significant interest in Europe to apply AVHRR data to remote sensing problems

2 the geographical areas of interest encompassed the entire reception capability of all European HRPT stations and beyond

3 there were at least 11 HRPT stations in Europe which regularly disseminated data to external users and

4 acquisition and archiving practices and data dissemination formats (computer compatible tape and quicklook) were different for every station.

In summary, the report found that a huge redundancy existed across Europe in terms of AVHRR data acquisition but potential users would experience considerable problems in accessing or even determining the existence of specific data sets. It was apparent that Europe's remote sensing scientists would benefit from a single point of access to AVHRR data and that, for maximum efficiency, the enquiry facility should be on-line and remotely accessible. Earthnet took the decision to establish such a scheme with the following components:

1 Central User Service including on-line catalogue accessible worldwide

2 Common Archive feeding selectively from a number of strategic stations

3 Common Tape Format – for the convenience of users, adoption of the standard format is to be encouraged for all AVHRR data suppliers and not only the stations participating in Earthnet's scheme and

4 High Level AVHRR products.

This coordinated network should have no adverse effect on existing facilities which would continue to serve their national requirements, i.e. for the most part, meteorological applications. More details of the system set up by Earthnet to provide these services are described by Hastings *et al.* (1989b) and Fusco *et al.* (1989). Terms such as the **SHARK** (Station HRPT Archiving and Reprocessing Kernel) workstation and the **SHARP** (Standard-family HRPT Archive Request Product) format are

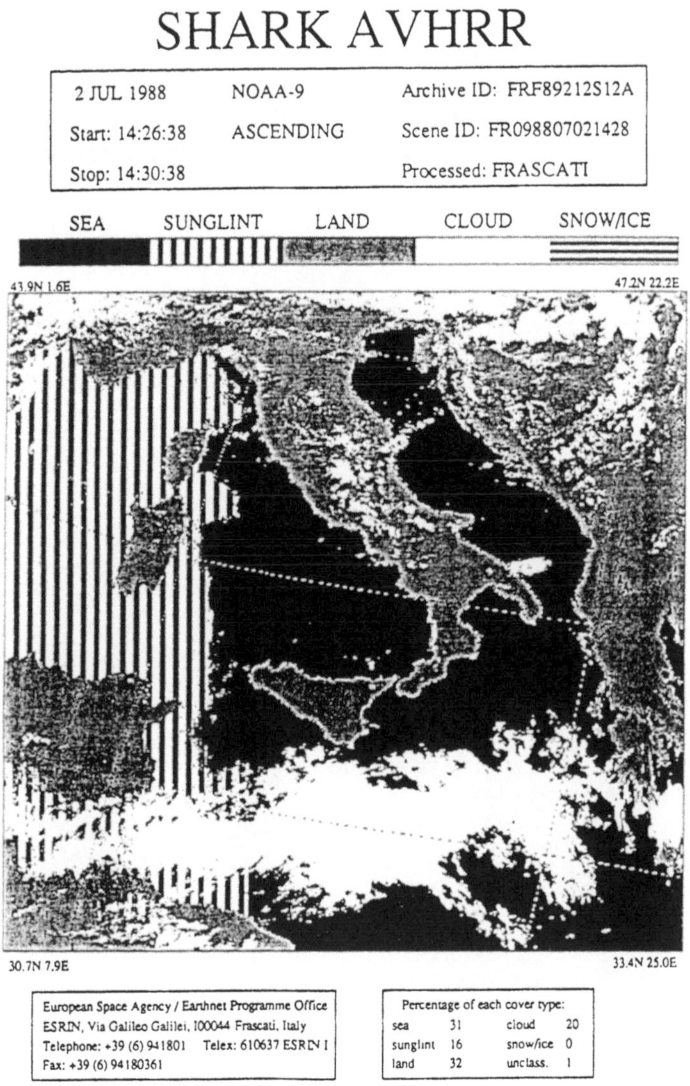

Figure 2.32 Standard ESA classified AVHRR quicklook (Muirhead and Malkawi 1989).

explained in that paper. ESA has developed a tool called 'CD-Browser' for digital quicklook consultation and product selection from CD-ROMs (Melinotte and Arino 1993). The AVHRR CD-Browser Ionia is a demonstration CD-ROM for AVHRR data. The AVHRR Work Plan of the CEOS (Committee on Earth Observation Satellites) Working Group on Data is described by Fusco (1991). This Working Group is actively coordinating international efforts in the area of Earth Observation Data Management.

To assist in the classification of AVHRR (quicklook) images at ESA Earthnet Muirhead and Malkawi (1989) developed a simple algorithm to classify daylight AVHRR data on a pixel-by-pixel basis into land, cloud, sea, snow or ice and sunglint. This was developed to provide an automatic alternative to a visual and subjective assessment of the cloud cover on quicklook images of data in the archive. The classified quicklook is presented in appropriate colours (green: land, blue: sea, white: cloud, etc.); it is also output in black-and-white to a laser printer to become the standard browse quicklook for the ESA network stations, see Figure 2.32. The developments at Earthnet have been paralleled by developments at individual ground stations to make their catalogues available on line (see, for example, Soukeras and Hayes 1991). Information about the acquisition and archiving programmes of various ground stations is given in the table in the Appendix.

CHAPTER THREE

Pre-processing

In this chapter we shall consider a number of pre-processing operations that are very commonly performed on AVHRR data sets before they can be used in any of the various applications we shall be considering in later chapters. The methods and techniques that are to be discussed in this chapter are widely used in many different applications of AVHRR data.

3.1 GEOREFERENCING/GEOLOCATION/GEOMETRICAL RECTIFICATION

3.1.1 Introduction

Raw data collected from the AVHRR do not have geographical coordinates attached to them. Data are essentially obtained as a rectangular array of numbers and any pixel (picture element) in the image is located by its row, or scan line, number S and within the scan line by its column, or pixel, number P. Thus the row corresponds to a scan line and the column corresponds to the position within the scan line. The raw data can be used to produce an image. For many applications of AVHRR data it is necessary to be able to determine the geographical coordinates of the pixels in the scheme. It is also of course, in many cases, necessary to remove the various distortions that are present in a raw image. Three possible treatments are to be considered:

- for a single image one can generate a distorted geographical grid to superimpose on the original satellite image
- in a set of satellite images one of them can be regarded as a master image and all the others can be distorted or warped to match the master image
- an image or a set of images can be corrected and resampled to fit a desired geographical projection.

The centre point of a pixel corresponds to a point on the surface of the Earth which we suppose has geographical coordinates of E (longitude) and N (latitude). The identification of the values of E and N for each pixel, i.e. for each pair of values of S and P in the image, is variously referred to as image referencing (e.g. by Ho and Asem 1986) or georeferencing or geolocation. This corresponds to distorting the geographical grid to match the AVHRR image. This process is sometimes carried out electronically, for instance the quicklook images produced by the satellite data receiving station in Dundee University are generated as rectangular images from the raw data and unconventional distorted grids are produced and superimposed on them, see Figure 3.1. Before producing an image such as that shown in Figure 3.1 the data are linearised; this means that corrections are applied (electronically) to allow for the obliquity of viewing and for the increase in the size of the IFOV that occurs near the ends of the scan lines. However, apart from the production of quick-looks, the number of situations in which one is prepared to use an image with an unconventional distorted geographical grid is rather small. Time-lapse sequences of images treated in this way cannot easily be intercompared.

For very many applications of AVHRR data one needs to make comparisons among several different images received at different times for a common geographi-cal area or to make comparisons between the images and maps or charts. If it is not necessary to relate the information in the image to geographical coordinates, but just to compare different images obtained at different times, then one can take one of the images as a master and all the others can be distorted or warped to match the master image. This approach is sometimes, but not frequently, adopted; for example Gupta (1992a) used this approach in some work on vegetation indexes. Time sequences of images corrected in this way can be animated to show the temporal evolution of some geophysical feature (see section 4.3).

If one seeks to fit a satellite image to a geographical projection one has to remove the various distortions that are present in the image; these distortions are due to Earth shape, Earth rotation, variations in satellite orbit and satellite attitude. The variations in satellite attitude, however, are quite small (Krasnopolsky and Breaker 1994), see Figure 1.5. One then needs to transform the image to the geo-graphical grid and to resample it to fit a desired geographical map projection. The transformation of an array of image data to a standard geographical projection is described as inverse image referencing by Ho and Asem (1986) but it is more com-monly described as geometrical rectification or as image navigation.

There are two approaches that one can take to image referencing or to inverse image referencing. The first is to attempt to determine the exact position and atti-tude (orientation) of the satellite at the time that the radiation was received from the surface of the Earth. This involves knowing the parameters of the orbit of the satel-lite and also involves using the quite complicated spherical trigonometry of the system (see below). The alternative approach, which is very widely used, is a more empirical approach; it involves assuming a form for a mathematical transformation relation between S and P (the coordinates of a pixel in the original raw data array) and E and N (the geographical coordinates). The coefficients in the transformation are determined by using a number of control points (or ground control points, GCPs) (see below). A GCP is a point or fixed feature that can be recognised both on a map and in the image that one is trying to rectify. In the case of AVHRR data a GCP is almost always a sharp feature on a land–water boundary.

The use of ground control points is quite time consuming because the operator must interact with the image and a map (unless pre-selected points are always used) to locate the satellite view of the selected ground features used as reference. It is, of course, possible to construct libraries of ground control points and to develop auto-

Figure 3.1   A linearised unrectified AVHRR image with a geographical grid superimposed on it
(NOAA-11, orbit 9367, 20 July 1990, 1330 GMT), see also Figure 2.4 (Dundee University).

matic pattern recognition techniques to identify the ground control points in an image, see, for example, Cracknell and Paithoonwattanakij (1989a). The method also assumes that the features selected as ground control points are clearly visible in the image which, in the presence of cloud cover, may not always be the case for optical and infrared wavelength images. Also, a ground control point approach only works when land is present in the image; there are very few, if any, ground control points in the open ocean. It is not satisfactory to use the bare minimum number of ground control points that would be necessary for the determination of the values of the coefficients in the mathematical transformation chosen to relate S and P to E and N. Although this would give an exact fit at the GCPs themselves it would give very poor results for the rest of the image. It is necessary to use an over-determined set of equations, i.e. to use a larger number of ground control points than would be needed for an exact determination of the coefficients in the chosen transformation, and then to perform a least squares fit to determine the values of the coefficients in the transformation (see section 3.1.3). Finally, having found a set of coefficients for a transformation using the ground control point approach, the coefficients only apply to that particular scene. The whole process has to be carried out afresh for each new scene that one chooses to study. A useful set of references on the navigation of data from instruments on polar-orbiting satellites that were flown before the first AVHRR was flown on TIROS-N is cited by Krasnopolsky and Breaker (1994).

We shall outline both the approach of using the orbital parameters of the satellite and that of using a set of GCPs. Sometimes both of these methods are used together, i.e. the orbital parameters are used to provide an approximate rectification which is accurate to the nearest few pixels. A refinement to this can then be achieved very quickly with a rather small number of ground control points, see section 3.1.4. Reviews of the methods used in AVHRR image navigation have been given by Emery *et al.* (1989) and Krasnopolsky and Breaker (1994).

One or two other general points should be made before we pursue the details of navigation or geometrical rectification methods. The first concerns the limit to the accuracy of geometrical rectification. In general one can expect to carry out a geometrical rectification to an accuracy of the order of the edge of the instantaneous field of view (IFOV) (or the scale translation of that on to the image). However, by using special techniques (Cracknell and Paithoonwattanakij 1989b) one can improve the accuracy and reduce the error to about only 20–30% of the length of the edge of the IFOV. This is described in a little more detail in section 3.1.6. The second point concerns the possibility of misregistration among data from different channels in one AVHRR scene. Almost without exception, people have assumed that in a raw image the five channels of the AVHRR are exactly coregistered. In other words, one assumes that the geographical coordinates of the centre of the pixel (P, S) are exactly the same for all five channels. However, Allam (1986) demonstrated a misregistration, in the direction along the scan lines, of about one quarter of a pixel edge between channel 3 and channel 4 in one particular night-time scene from NOAA-7. The question of the impact of misregistration on the detection of changes in land cover has been studied by Townshend *et al.* (1992) using Landsat MSS data.

3.1.2 The use of ephemeris data

A dictionary definition of the word ephemeris is 'an astronomical almanac or table of the predicted positions of celestial bodies'. In this case the celestial bodies in question are the NOAA satellites. In this section we use the notation of Emery and Ikeda (1984) as modified in Emery *et al.* (1989):

Notation

a = semi-major orbital axis

α_s = angle formed by the arc from the pixel to the ascending node and the equatorial plane (rad)

β = angle from ascending node to the pixel (rad)

δ = off-nadir viewing angle (rad)

δ_i = scan step angle (rad)

D' = position at the end of the scan line

e_0 = orbital eccentricity

γ = satellite azimuth angle relative to the orbit plane (rad)

h = satellite altitude (km)

i_s = orbital inclination angle (degrees)

i,j = counters in image arrays

j_2 = Earth oblateness coefficient (0.001 082 63)

λ_0 = longitude of equator crossing computed from T_0, e_0 and i_s

λ_D = geographical Earth longitude (degrees)

$\lambda_{D'}$ = static Earth longitude (rad)

λ_e = nodal longitude or longitude of equator crossing (degrees)

M_0 = mean anomaly

M_1 = mean orbital motion

M_1' = equivalent mean orbital motion

M_2 = orbital decay rate

μ = Earth gravitational constant ($3.986\,03 \times 10^{14}$ m^3 s^{-2})

ω = argument of perigee

ω_0 = argument of perigee at T_0

$\Delta\omega$ = daily average orbit-to-orbit fluctuation in ω

ϕ_D = geographical Earth latitude (degrees)

$\phi_{D'}$ = static Earth latitude (rad)

ψ = arc between the viewing pixel and the subsatellite point (rad)

R_e = Earth radius (6 378 160 m)

S' = subsatellite point

T_0 = epoch (midnight GMT)

t = time from the ascending node (equator crossing, s)

$t_{1/2}$ = half scan line time (s)

t_f = time of first scan line (s)

t_0 = nodal time or time of equator crossing (s)

t' = time from equatorial passage (t_0) to the scan line of interest t

t^* = time of image centre (s)

θ = geocentric angle from the ascending node to the subsatellite point (rad)

θ_1^* = latitude of image centre (rad)

θ_1 = geocentric latitude of nadir on the scan line of interest (degrees)

V_e = Earth angular velocity (rad s^{-1})

V_s = satellite angular velocity (m s^{-1})

In the ephemeris navigation procedure it is assumed that the Earth is an oblate spheroid. In some treatments an assumption is made about the satellite orbits being circular. However, since the circular orbit assumption is a simplification of the elliptical orbit treatment it does not really need to be treated separately.

Figure 3.2 gives a flow chart of the various causes of image distortion as a key to a discussion of their magnitudes and correction procedures. Earth shape (oblateness) contributes to image distortion through changes in the force of gravitational attraction and in the geometrical distance to the satellite (altitude). The former results in both long-term (orbit-to-orbit) and short-term (within one orbit) changes in orbital shape and spacecraft velocity. Geometrical distance variations contribute directly to changes in altitude, a fundamental satellite orbital parameter. Other effects include the small difference between the elliptical satellite orbit and the assumed locally circular orbit, along with the long-term changes in orbit due to drag on the spacecraft.

Of these mechanisms only those due to variations in orbital shape (numbers 1 and 2 in Figure 3.2) result in changes large enough to be considered for image navigation. Orbital eccentricity is so small that the orbit can be considered circular within each image and the aerodynamical drag is important only if one is addressing the long-term degradation of the orbit rather than single-image navigation. Most of the short-term changes are also ignored except for variations in the semi-major axis which are assumed to be almost completely compensated for by corresponding changes in spacecraft altitude (number 3 in Figure 3.2). Such compensating variations in semi-major axis and satellite height can individually lead to deviations of up to 20 km in altitude with a maximum at the equator and a minimum at the poles. Suborbital changes are also accounted for by using an averaged value for the longitude of the ascending node.

The changes in orbital character most important to image navigation are those long-term intra-orbital variations indicated as number 1 in Figure 3.2. These include variations in: the semi-major orbital axis (a), orbital eccentricity (e_0), orbital inclination (α_s, relative to the Earth's axis), the longitude of equatorial passage (λ_0) and the argument of perigee (ω). Variations in a, e_0 and in inclination angle (α_s) are small and affect orbital parameters only to higher order. Thus only fluctuations in equatorial passage longitude ($\Delta\lambda_0$) and the argument of perigee ($\Delta\omega$) need to be considered to first order. These variations ($\Delta\lambda_0$ and $\Delta\omega$) are calculated from the ephemeris data taken on the day of image acquisition and the day immediately following.

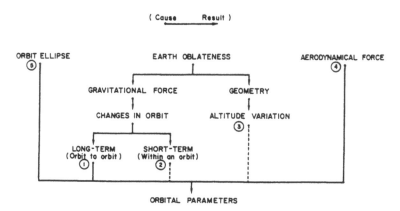

Figure 3.2   Flow chart of causes of image distortions (Emery and Ikeda 1984).

The orbital parameters, which are needed for image navigation, must be computed from the ephemeris data and they are: the longitude of equatorial passage for the particular orbit of interest (λ_0), the local value of satellite angular velocity (V_s), the local value of satellite altitude (h), the equivalent equatorial passage time (t_0) extrapolated using V_s, along with ephemeris values of the mean anomaly (M_0) at Epoch (T_0; midnight GMT) and the argument of perigee (ω_0) at T_0. In addition to these last two quantities the ephemeris data consist of values for: mean orbital motion (M_1), orbital decay rate (M_2), longitude of equatorial passage (λ_0) calculated for T_0, eccentricity (e_0) and inclination (i_s) (see Figure 3.3). These ephemeris data are used to calculate the orbital parameters as follows.

The orbit inclination of the orbit is given directly by i_s. The satellite's velocity V_s can be calculated using

$$V_s = M_1'(1 + e_0 \cos A)/(1 - e_0 \cos E) \tag{3.1.1}$$

where

$$A = M^* + O(e_0^2) \tag{3.1.2}$$

(for most cases $A = M^*$)

$$M^* = M_0 + M_1(t^* - T_0) + M_2(t^* - T_0)^2 \tag{3.1.3}$$

$$E = M^* + e_0 \sin M^* \tag{3.1.4}$$

$$\sin \theta_1^* = \sin(\omega + A) \sin i_s \tag{3.1.5}$$

$$\omega = \omega_0 + \Delta\omega \, N \tag{3.1.6}$$

$$N = M_1(t^* - t_0)/2\pi \tag{3.1.7}$$

$$M_1' = M_1(1 + \Delta\omega/2\pi) \tag{3.1.8}$$

$$t_0 = t^* - (\omega + A)/V_s. \tag{3.1.9}$$

Note that M_1 is measured relative to perigee while M_1' is measured relative to the equator. The time t^* is the time at the centre of the image of interest. Equations (3.1.1)–(3.1.7) are iterated with increasing t^* until θ_1^* is close to the latitude of the centre of the image of interest.

The parameter E, as solved for above, is also used to compute the satellite altitude as

$$h = a(1 - e_0 \cos E) - R_e \tag{3.1.10}$$

where

$$a = (\mu/M_1^2)^{1/3}. \tag{3.1.11}$$

The equatorial crossing longitude (λ_0) is computed as

$$\lambda_0 = \lambda_0' + M_1 \, \Delta\lambda_e/2\pi - (t_0 - T_0) - V_e(t_0 - T_0). \tag{3.1.12}$$

Let us now consider the question of determining the (geocentric) latitude, θ_2, and longitude, ϕ_2, of a pixel in an AVHRR image. Using solid geometry we can derive expressions for ϕ_2 and θ_2 of the pixel

$$\sin \theta_1 = \sin(V_s t') \sin i_s \tag{3.1.13}$$

$$\sin(\delta + \psi) = (1 + h/R_e) \sin \delta \tag{3.1.14}$$

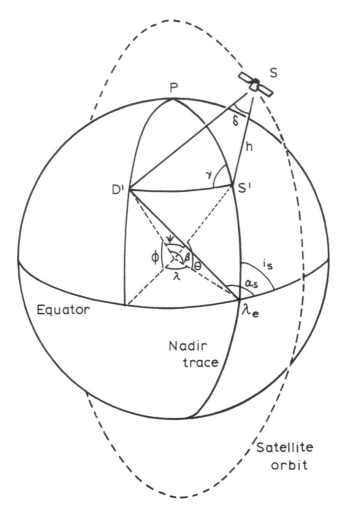

Figure 3.3   Satellite orbit geometry (Emery *et al.* 1989).

$$\sin \theta_2 = \cos \psi \sin \theta_1 - \sin \psi \cos i_s \qquad (3.1.15)$$

$$\cos \theta_2 \sin \phi'_2 = \cos \psi \sin(V_s t') \cos i_s + \sin \psi \sin i_s \qquad (3.1.16)$$

$$\phi_2 = \phi_0 - V_e t' + \phi'_2 \qquad (3.1.17)$$

where $0 < |\phi'_2| < \pi/2$ for northbound passes and $\pi/2 < |\phi'_2| < \pi$ for southbound passes. In these equations

θ_1 = geocentric latitude of nadir on the scan line
t' = time from equator passage (t_0) to the scan line (t) $(t' = t - t_0)$
δ = scan angle between the pixel and nadir
ψ = Earth centre angle between the pixel and nadir.

The relation between geocentric (θ) and geographic $(\bar{\theta})$ latitudes is given by

$$\bar{\theta} - \theta = 694'' \sin 2\bar{\theta} \qquad (3.1.18)$$

where a term $\sin 4\bar{\theta}$ has been neglected due to its small magnitude.

Using this procedure, digital image data for each pixel are transferred to a selected geographic map projection. Emery and Ikeda (1984) used a conical projection. The location of each pixel (ϕ_2, θ_2) on this map projection is found using

$$x = \{Y - y_0(\theta_2 - \theta_c)\} \sin\{(\phi_2 - \phi_c) \sin \theta_c\}$$

$$y = Y - \{Y - y_0(\theta_2 - \theta_c)\} \cos \{(\phi_2 - \phi_c) \sin \theta_c\}$$

(3.1.19)

where $Y = y_0/\tan \theta_c$, y_0 is a scale corresponding to 1 radian of latitude, and ϕ_c and θ_c are the longitude and geocentric latitude of the map centre. Again geocentric latitude is related to geographical latitude by equation (3.1.18). The fully elliptical calculation is also considered by Brush (1988), with particular reference to the specific characteristics of the facsimile machine used for image output, with grid lines included, at the Dundee University AVHRR receiving station. He demonstrated sufficient accuracy to allow the calculation and fitting of a grid including coastlines, such that a precision of better than 3 km is obtainable. Kloster (1989) was concerned with ice mapping in the vicinity of Svalbard and described an approach which is based on a model that takes as input the TBUS (TIROS Bulletin United States) mean orbital elements and an improved value of a parameter giving mean satellite speed. An accurate value of the speed of the satellite can be found if the satellite is close to the equator at subsequent TBUS times. The resultant gridding errors found by comparison with Svalbard coastlines are attributed to scan line timing and satellite attitude. Although these methods make no use of GCPs in the rectification process, if one wishes to test the accuracy of the rectification one does need to make use of a set of GCPs. When Kloster's method was tested with sets of GCPs for AVHRR scenes of the vicinity of Svalbard, using 14 scenes from the summer of 1984 and 13 scenes from the spring of 1987, Kloster (1989) found a mean along-track error of 3.9 km and a mean across-track error of 1.5 km.

Equations (3.1.19) can be used to calculate the geographical coordinates of any pixel in the image. However, the use of equations such as (3.1.19) for every pixel in an image is likely to be extremely expensive in terms of computing time. One could speed up the navigation computation by navigating only a limited portion of the pixels in the image and then using an efficient interpolation scheme to adjust the remainder of the pixels in the image, see for example Emery et al. (1989).

An alternative method to reduce the expense, in terms of computer time, involved in the geolocation of satellite-based Earth sensor data has been developed by Patt and Gregg (1994). This method is intended to produce fast, accurate, efficient geolocation which is generalised for use with both tilting and non-tilting sensors. The method involves vector and matrix algebra, so that the use of computationally expensive trigonometric functions is minimised, and it is completely general with respect to sensor position and orientation. Perhaps more importantly, the method is closed form and exact, depending only upon the assumption of an ellipsoidal Earth model. In addition to determining the geodetic coordinates (either geocentric vectors or latitude and longitude), the work of Patt and Gregg included the determination of algorithms for computing the sensor zenith, sensor azimuth, solar zenith and solar azimuth angles; these angles are required by the radiative transfer models used for remotely sensed ocean data. The method has been accepted for use in the Sea-viewing Wide Field-of-view Sensor (SeaWiFS) mission, a global ocean colour mission due for launch soon, and it has already been used in the global AVHRR

Pathfinder Land Project. For details of this method we refer the reader to the paper of Patt and Gregg (1994).

As we have indicated already, the use of ephemeris data alone is generally relatively fast and leads to an error of a few pixel edge lengths in the registration of AVHRR data. For instance Legeckis and Pritchard (1976) proposed a simple algorithm for correcting only the geometrical distortions due to the Earth curvature, Earth rotation and spacecraft roll. Other effects were ignored. In their procedure, for the Earth rotation correction, the altitude was assumed constant for the whole scene (about 1000 km). The satellite roll correction assumed a spherical Earth, but the asymmetry due to the Earth's oblateness and the satellite orbit was not taken into account. Their average error, representing the misalignment over a distance of 1000 km was 5 km. Clark and La Violette (1981) tested the accuracy of the geographical coordinates provided by NOAA NESS (National Environmental Satellite Service) at 51 equally spaced intervals along each scan line of TIROS-N data; they found that the positioning errors are from 2 to 4 pixels. An example of the errors obtained with a set of seven AVHRR images over the Persian Gulf, based on using navigation data from NOAA NESDIS by Krasnopolsky and Breaker (1994) is shown in Figure 3.4.

At this point we include a brief description of the NESDIS navigation procedures. The NESDIS navigation procedures are based on high precision orbital elements which are provided by the US Navy Space Command (USNSC). These elements include an inertial position, a velocity vector with orbit numbers, a ballistic coefficient, solar flux, average solar flux and planetary index. A numerical integrator is used to predict the velocity vector ahead for inclusion in a User Ephemeris File

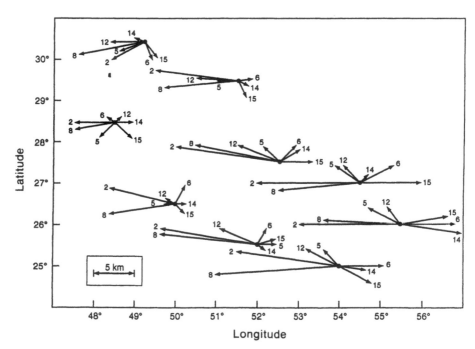

Figure 3.4 Space and time variation of Earth location errors for nine landmarks in a sequence of seven AVHRR images. The numbers on the arrowheads indicate the date in August 1991 (Krasnopolsky and Breaker 1994).

(UEF). The UEF contains the original USNSC velocity vector and data records at 60-second intervals for a 10-day period. The UEF is used to generate the Earth location data and the TBUS messages. The 10-day UEF is used to compute the Earth location and gridding parameters for the AVHRR data. A start time, taken from other instruments aboard the spacecraft, is used together with instrument scanning parameters and vectors from the UEF to produce the Earth location and gridding parameters. The Earth location and gridding data files are updated for a 29-hour period that starts on the following day. With respect to the orbit prediction model presently in use by NOAA/NESDIS, geographical coordinates are produced with an accuracy of approximately 0.05–0.1°. The availability of the orbital parameters on the NOAA Direct Readout Users Electronic Bulletin Board has already been described in section 2.5.2.

3.1.3 The use of ground control points

A set of ground control points can be used either with a raw image or with an image which has been partially rectified using ephemeris data. In either case the general idea is the same. We suppose (see above) that a pixel in the image file that is to be rectified is identified by its scan line or row number, S, and by its pixel or column number, P, within that scan line. The values of the intensity $I(P, S)$, denoting the intensity in any given channel of pixel number P in row S of the image, form a rectangular matrix or array. We suppose that the coordinates of the centre point of this pixel, referred to a chosen map projection, are x and y. The process of geometrical rectification using ground control points therefore proceeds as follows.

We assume that there is some mathematical transformation relating the geographical coordinates (x, y) of a pixel to its coordinates (P, S) in the unrectified image. If one assumes linear transformations they will be of the form

$$x = a_{11} + a_{12} P + a_{13} S$$
$$y = a_{21} + a_{22} P + a_{23} S$$

(3.1.20)

and

$$P = b_{11} + b_{12} x + b_{13} y$$
$$S = b_{21} + b_{22} x + b_{23} y.$$

(3.1.21)

A linear transformation is likely to be adequate if one is rectifying a relatively small extract from the data from a complete AVHRR overpass or if one is attempting to obtain a more accurate rectification of image data which has already been approximately rectified using ephemeris data as described in section 3.1.2. However, if one is trying to rectify directly to map coordinates a large area of a raw AVHRR image, it will probably be necessary to include second-order terms in x^2, xy and y^2 and possibly even higher-order terms (see, for example, Muirhead and Cracknell 1984a). We shall restrict the discussion here to linear transformations; it can be extended in a quite straightforward way to transformations that use higher-order polynomials instead of the linear transformations in equations (3.1.20) and (3.1.21).

The next stage is to select a number of ground control points, that is features which are identifiable in both the uncorrected image and on a map. Thus for each of these ground control points (GCPs), both pairs of coordinates (P, S) and (x, y) are known. In high resolution satellite data, such as that from the scanners on Landsat,

SPOT or IRS, there are many different types of recognisable features that can be used as GCPs. However, in AVHRR data many such features (intersections of runways on airports, major road intersections, river confluences, sharp bends in a river and so on) are not able to be identified and for AVHRR data the main features that can be identified for use as GCPs are sharp features on the coastline or on the shores of very large inland water bodies. In equation (3.1.20) there are six unknowns, $a_{11}, a_{12}, \ldots, a_{23}$ and so in principle one could use just three GCPs. Each GCP would give a pair of (P, S) values and the corresponding pair of (x, y) values thereby leading to a set of six equations in which the unknowns would be the six coefficients $a_{11}, a_{12}, \ldots, a_{23}$. A set of values of these six coefficients could then be obtained from an exact solution of these six equations. However, if one did this and then attempted to use the transformation in equation (3.1.20) to rectify the whole scene one would obtain very poor results. This arises, essentially, because of the errors that would necessarily be present in the P, S, x and y values used for the GCPs. In practice what one does, to reduce the effect of these errors, is to use a much larger set of GCPs, say 15–20 or so, to give a heavily over-determined set of equations and the best set of values of the coefficients is determined by a least squares fitting procedure. The coefficients $b_{11}, b_{12}, \ldots, b_{23}$ in equation (3.1.21) can be found in a similar way. Having found the coefficients in equation (3.1.20) this equation can be used with any pixel from the original unrectified image to determine the geographical coordinates of the centre of that pixel. The accuracy of the rectification of a scene using the transformation (in equation (3.1.20)) can be estimated from the standard deviation of the best fit results. Alternatively, to estimate the accuracy of the values of the coefficients, one can exclude some of the GCPs from the least squares fitting and then apply equation (3.1.20) to the (P, S) values for these GCPs and compare the values of x and y calculated using this equation with the known (x, y) values for these GCPs. One would normally aim to achieve an accuracy of better than the length of the edge of the IFOV for a rectification, provided the GCPs used were well distributed in the scene that was to be rectified. However, if the GCPs used are not well distributed in the scene the accuracy may be poorer than this.

Although the use of an extensive set of GCPs leads to an error of less than the length of the edge of the IFOV, there are several problems:

- it is very expensive in terms either of operator time, if the GCPs are located interactively, or of computer time, if an automatic pattern recognition procedure is attempted

- there may be problems in finding a large and well-distributed set of GCPs away from the coastline and without large inland water bodies

- many interesting AVHRR images are of ocean areas or polar regions with very few identifiable control points and possibly even none at all

- very large numbers of AVHRR images are generated every day and therefore, in practice, any operational rectification scheme must be quick and efficient and not involve enormous computational resources.

What we have described so far, involving the use of ground control points to refine the accuracy of geolocation obtained from ephemeris data, is only applicable where a suitable number of GCPs can be found within the scene. However, as we have just noted, there may not always be an adequate number of GCPs available for an accurate geolocation to be achieved. This arises, for instance, in the case of the study of oceanic features based on their thermal manifestations at the surface. As the oceanographic applications for AVHRR data have become more sophisticated, the need for greater Earth-location accuracy has likewise increased. For example, by measuring the displacements experienced by a selected thermal feature between successive images, it is often possible to infer the apparent adjective motion that transports these features (see section 4.3). Such feature-tracking methods place stringent requirements on Earth-location accuracy since the associated displacements may not be large compared to the uncertainties in the navigation. Although careful coregistration of successive images in coastal regions usually produces close alignment near coastlines which are often used as a common reference in the coregistration process, there is no guarantee that close alignment will occur between the images for oceanic locations far removed from land. The question of the accuracy of geolocation of AVHRR data in regions that are distant from an adequate number of identifiable GCPs has been addressed by Krasnopolsky and Breaker (1994). This involved the development of a correction technique based on using correction functions $F(u,v)$ which relate the approximate geographical coordinates, u and v, of a pixel to the true coordinates, x and y; thus

$$x = u + F(u, v)$$
$$y = v + G(u, v).$$

(3.1.22)

The method of Krasnopolsky and Breaker (1994) then involves modelling the correction functions and determining the coefficients in the model from those regions where GCPs are available. The correction functions are then used to extrapolate the corrections for navigational errors to regions that are remote from areas containing GCPs. For details see the paper by Krasnopolsky and Breaker (1994) where two examples of AVHRR data for the Gulf of Mexico and for the Persian Gulf are used to test the method.

The question of achieving accurate geometrical rectification of AVHRR data that is to be used for the calculation of NDVIs was considered by Gupta (1992a). Use of third-order polynomial transformation techniques in the geometric correction of 185 km swath width data sets from Earth Resources Satellites could give satisfactory results for the geometrical correction of very wide swath width AVHRR data sets when applied to only a part of the AVHRR image. The distortion dominance for AVHRR images is in the across-track direction. Thus, having a GCP distribution as narrow as feasible surrounding the area of interest, especially in the across-track direction, ensures higher relative geometrical accuracies in the temporal analysis associated with district-level agrometeorological/agriculture monitoring studies. Geometric corrections need to be applied to raw visible and near-IR band data sets, and thereafter vegetation index images need to be generated from these geometrically corrected primary images. This would ensure the representativeness of the vegetation features (as seen in the primary data) in the geometrically corrected vegetation index images.

3.1.4 Combined ephemeris data and ground control points

There is one important aspect of geometrical rectification which arises with AVHRR data and which is much less important with high resolution data such as Landsat or

SPOT data. This arises from the need to be able to rectify accurately AVHRR data from the open sea or a polar region where ground control points are not available. This can, for instance, be done by using GCPs in another part of the orbit where the AVHRR obtains an image containing a land area in order to improve the ephemeris data so that one can carry out an accurate rectification later (or earlier) in the same orbit when the image gathered is of an area where no GCPs are available.

What is needed in practice is therefore a method of geometrical rectification that is accurate and rapid and this can be achieved by using a combination of ephemeris data and a rather small number of GCPs. This combined approach is described, for instance by Emery and Ikeda (1984), Bachmann and Bendix (1991) and Brunel et al. (1991). Emery and Ikeda described two different versions of this approach. One method assumes a circular spacecraft orbit based on the initial ephemeris data provided by NOAA. After initial corrections for rotation and curvature were made (assuming a circular orbit) further corrections were carried out using a linear correction to seven selected GCPs to improve the image-to-map registration to an accuracy of ± 1.5 km. A second method using high quality US Navy ephemeris data, including Earth oblateness, for the initial correction leads to sufficiently accurate rectification that only a single GCP was needed to achieve a similar image accuracy. Ho and Asem (1986) used an orbital model and only one GCP; their procedure assumed a spherical Earth and circular orbit and took into account the effects due to the Earth's rotation and oblateness, and the scan skew. Inputs to the procedure were the ascending nodal longitude and time, the time of the first scan line and the coordinates of one GCP. The effects of an ellipsoidal Earth and an elliptical orbit are corrected by using the GCP to adjust the spacecraft altitude and inclination angle. No detailed ephemeris data are required. The method was tested on a set of ten AVHRR images, using an independent set of well-distributed ground control points for each image, and the results obtained are given in Table 3.1. The errors in displacements are calculated from the geographical shifts of the GCPs. It should be noted that these errors are different from those often discussed in other geometrical correction procedures, where the errors are often calculated from the same GCPs that are used to determine the coefficients in the transformation equations. The

Table 3.1 Accuracy of combined orbital and GCP rectification method due to Ho and Asem (1986).

			Root mean square error		
Number	Date	Orbit	Line	Pixel	Distance (km)
1	3 November 1982	7025	1.98	2.64	3.91
2	18 February 1983	8543	1.89	3.22	3.45
3	3 March 1983	8726	1.26	2.19	3.51
4	8 March 1983	8797	2.45	2.43	2.45
5	15 April 1983	9334	2.09	1.73	3.17
6	14 July 1983	10605	2.32	2.45	3.67
7	21 July 1983	10704	1.90	2.17	3.82
8	4 December 1983	12624	1.09	2.17	2.33
9	5 December 1983	12638	0.95	1.21	1.85
10	27 April 1984	14672	1.64	2.75	3.25
Average r.m.s. error			1.76	2.30	3.15

Table 3.2 Accuracy of geometrical rectification achieved by Bachmann and Bendix (1992)

Number	Date	Satellite	Offset	GCPs	Mean error Line	Mean error Pixel
1	4 January 1989	NOAA-11	210	3	0.47	1.0
2	10 January 1990	NOAA-11	874	3	1.43	1.33
3	19 January 1989	NOAA-11	1300	2	1.15	1.00
4	15 August 1989	NOAA-10	550	10	0.69	2.30
5	1 February 1989	NOAA-10	985	2	0.95	4.50
6	3 February 1990	NOAA-10	1300	2	0.65	1.00
			Average		0.89	1.80

GCPs used in the test referred to in Table 3.1 were totally independent of the GCP used in the original rectification of the images. Bachmann and Bendix (1992) were able to modify the method of Ho and Asem and to achieve a substantial reduction in time (by a factor of about 6) for the processing. The method was tested with six AVHRR scenes and mean errors of 0.89 of a pixel edge along-track and 1.80 across-track were obtained, see Table 3.2; these represent 50% and 20% improvements, respectively, on the results of Ho and Asem shown in Table 3.1.

There have been various subsequent pieces of work related to this problem, i.e. attempting to achieve improved accuracy without the use of excessive computational resources, often without making any use of GCPs in the rectification process. Thus Brunel and Marsouin (1987) used the ARGOS data collection and platform location system as a means of improving the ephemeris data and thereby improving the accuracy of geometrical rectification of AVHRR data, without the need to identify any ground control points in the AVHRR image. When the method was tested by Brunel and Marsouin (1987) on ten NOAA-9 AVHRR scenes (using ground control points) the positioning errors were found to have a mean value of 4.7 km. In some later work (Brunel and Marsouin 1989, Marsouin and Brunel 1991) they compared the results of their method using ARGOS orbital elements and TBUS orbital elements. Results were obtained using 16 NOAA-9 and NOAA-10 orbits from between 19 June 1988 and 26 September 1988; they found a mean error of 3.5 km and a standard deviation of this error of 1.2 km with ARGOS and a mean error of 2.3 km and a standard deviation of 0.6 km with TBUS. The positioning error consists of a pixel error which is systematic for each satellite (-1.6 pixels for NOAA-9 and 2.0 pixels for NOAA-10) and a variable line error which is mainly due to the positioning error of the satellite on its trajectory.

O'Brien and Turner (1992) described a method of rectification which involves using a non-re-entrant, or star-like, coastline and instead of using a set of GCPs one uses the digitised coastline to match with the AVHRR image. The coastline is represented in terms of polar coordinates ρ and θ, referred to some convenient origin, where

$$\rho = f(\theta) . \tag{3.1.23}$$

That is, the coastline is referred to the chosen origin and the digitisation involves determining the function $f(\theta)$. O'Brien and Turner (1992) illustrated the use of their method with an AVHRR image of Kangaroo Island off the coast of South Australia. Bordes *et al.* (1992) applied template-matching between selected coastal landmarks in AVHRR images with corresponding landmarks taken from a digital coastline reference file. Accuracy of the mapped AVHRR data using this technique was on the order of one pixel. Maitre and Wu (1987) applied dynamic programming to the registration of AVHRR images of the Normandy coastline.

3.1.5   Resampling

Equation (3.1.19) in section 3.1.2 or equation (3.1.20) in section 3.1.3 can be used to calculate (x, y), the geographical coordinates of any pixel in an AVHRR image. However, there is another problem. In printing an image, or in seeking to transfer features from an image to a map, or in attempting to compare different members of a set of images, it is common to use a grid set out in a regular network of geographical coordinates (see the solid grid in Figure 3.5(a)) and not the grid of points that comes about accidentally by rectification of the pixel coordinates in an image (see the dashed grid in Figure 3.5(a)). Data from different AVHRR images would lead to a quite different grid (see the dashed grid in Figure 3.5(b)). The generation of image intensity values at a regular grid of geographical coordinates, i.e. the solid grid, from the navigated grid, i.e. one of the dashed grids, is an interpolation procedure which is usually described as resampling, and this is a very important procedure.

To consider the process of resampling we need to be careful about the notation. x and y are the geographical coordinates of a transformed pixel from the image. They correspond to the grid points of the dashed grid in Figure 3.5(a); we are not free to choose their values. We also introduce another set of coordinates, which we shall denote by e, for the easting, and n, for the northing. These correspond to the grid points of the solid grid in Figure 3.5(a) or (b) and we are free to choose some convenient set of these points. Suppose that one is trying to generate an image for comparison with, or superposition on, a given portion of a map. Then what one requires, in order to be able to print the image on a grey scale hardcopy device, is to be able to generate the image intensities $I'(e, n)$, in a given spectral channel, over the rectangular grid with each grid point specified by a pair of values of e and n, i.e. the chosen points of the solid lattice in Figure 3.5. Thus one needs to be able to work systematically through the ranges of values of e and n and for each pair of (e, n) values to generate $I'(e, n)$. For a given pair of values (e, n) we can apply equation (3.1.21) to generate the corresponding pair of values (P, S) to locate the corresponding pixel in the original unrectified image. This point is denoted by Q in Figure 3.6 where it is shown in relation to a portion of the unrectified image. If P and S happen to be integers, then Q will coincide with the centre of one of the pixels in the unrectified image, i.e. one of the points $FGJK$ in Figure 3.6(a). In that (unlikely) case the pixel intensity $I(P, S)$ with those values of P and S is taken as the pixel intensity $I'(e, n)$ in that spectral channel in the rectified image. However, it is most unlikely that the values of P and S calculated in this way will be integers and in this situation it is necessary to use an interpolation method. There are several interpolation methods which can be used to estimate $I'(e, n)$, the pixel intensity at the point (e, n) in the rectified image; these include

- the nearest neighbour method
- the bilinear interpolation method
- the bicubic interpolation method.

(a)

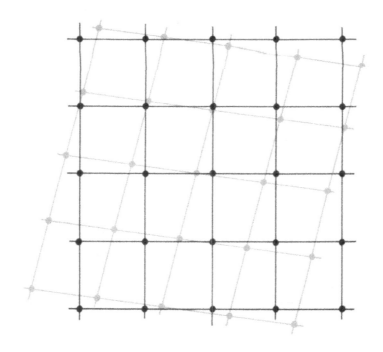

(b)

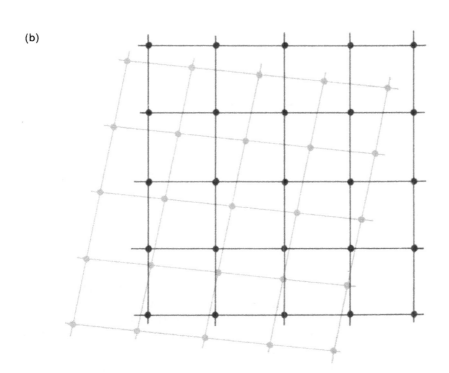

Figure 3.5 Diagram to illustrate resampling.

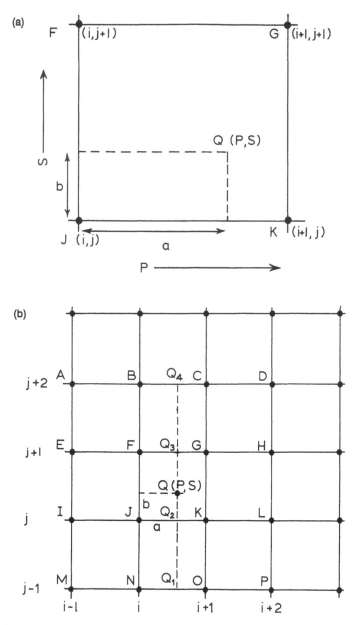

Figure 3.6 Grid for the definition of an interpolation (resampling) formula. In (b) the coordinates of Q_1, Q_2, Q_3 and Q_4 are Q_1 $(P, j - 1)$, Q_2 (P, j), Q_3 $(P, j + 1)$ and Q_4 $(P, j + 2)$.

In the nearest neighbour method one just takes the intensity $I(P, S)$ for the point nearest to Q; in Figure 3.6(a) this would be the point K. This method is fast to use and does not lead to loss of information by smoothing. However, spurious effects may appear as a result of the repetition or omission of pixels, particularly if there is a large difference between the spacings of the two grids.

The bilinear interpolation method is based on the assumption that a surface of intensity, as a function of P and S, would be a plane over the region $FGKJ$ in

Figure 3.6(a). The intensity $I'(e, n)$ is then taken to be

$$I'(e, n) = (1 - a)(1 - b)I(i, j) + a(1 - b)I(i + 1, j)$$
$$+ (1 - a)bI(i, j + 1) + abI(i + 1, j + 1) \tag{3.1.24}$$

where $a = P - i$ and $b = S - j$. Bilinear interpolation thus has a smoothing effect and sharp boundaries may, therefore, become blurred. It also involves more computer time than the nearest neighbour method.

The bicubic interpolation method involves fitting a two-dimensional third-degree polynomial surface in the region surrounding the point $Q(P, S)$. The 16 nearest pixels in the unrectified image are used, see Figure 3.6(b). First of all one interpolates along a row m ($m = j - 1, j, j + 1, j + 2$) to find the interpolated intensities at Q_1, Q_2, Q_3 and Q_4 at points $(P, j - 1)$, (P, j), $(P, j + 1)$ and $(P, j + 2)$, respectively. We find

$$I'(P, m) = -(\tfrac{1}{6})a(1 - a)(2 - a)I(i - 1, m) + (\tfrac{1}{2})(1 + a)(1 - a)(2 - a)I(i, m)$$
$$+ (\tfrac{1}{2})(1 + a)a(2 - a)I(i + 1, m) - (\tfrac{1}{6})(1 + a)a(1 - a)I(i + 2, m). \tag{3.1.25}$$

That this is cubic is obvious from inspection. That it reduces to the four values $I(i - 1, m)$, $I(i, m)$, $I(i + 1, m)$ and $I(i + 2, m)$ is apparent by substituting the appropriate values of a:

if $a = -1$, $P = i - 1$ and $I'(P, m) = -(1/6)(-1)(2)(3)I(i - 1, m) = I(i - 1, m)$;
if $a = 0$, $P = i$ and $I'(P, m) = (1/2)(1)(1)(2)I(i, m) = I(i, m)$;
if $a = 1$, $P = i + 1$ and $I'(P, m) = (1/2)(2)(1)(1)I(i + 1, m) = I(i + 1, m)$; and
if $a = 2$, $P = i + 2$ and $I'(P, m) = (1/6)(3)(2)(-1)I(i + 2, m) = I(i + 2, m)$.

Having found the four interpolated values at Q_1 Q_2, Q_3 and Q_4, these are themselves interpolated to find $I'(P, S)$ which is given by

$$I'(P, S) = -(\tfrac{1}{6})b(1 - b)(2 - b)I'(P, j - 1) + (\tfrac{1}{2})(1 + b)(1 - b)(2 - b)I'(P, j)$$
$$+ (\tfrac{1}{2})(1 + b)b(2 - b)I'(P, j + 1) - (\tfrac{1}{6})(1 + b)b(1 - b)I'(P, j + 2) , \tag{3.1.26}$$

where the values of $I'(P, m)$ ($m = j - 1, j, j + 1, j + 2$) are given by equation (3.1.25). This can again be seen to be cubic in b by inspection and it can be shown that it reduces to the values $I'(P, j - 1)$, $I'(P, j)$, $I'(P, j + 1)$ and $I'(P, j + 2)$ for $b = -1, 0, 1$ and 2, respectively.

The bicubic interpolation method is both more complicated to program and also more demanding in terms of computer time than the previous two methods; however, it is a quite popular method. It avoids the oversimplistic nearest neighbour method which can lead to blockiness and it avoids the excessive smoothing produced by the bilinear method. However, it does lead to some loss of high-frequency information.

The choice of which of these three methods to use will depend on two factors, the use to which the data will be put and the computer facilities that are available. If the image is to be subjected to classification then the replacement of the raw data by the interpolated data may well have some adverse effect on the final classification because, as we have already noted, the interpolation involves some smoothing of the data. It may, therefore, be decided that it would be better to perform the classification on the raw data and to perform the geometrical rectification subsequently. A restoration technique, based on modelling known sensor-specific parameters and developed at the Environmental Research Institute of Michigan (ERIM), is described by Chiesa and Tyler (1994). This involves using the system's point spread function (PSF) which will be discussed later in this section. The system's PSF, which is a function of sensor noise, detector size, optical blur circle and electronic filter characteristics, can be modelled and used to synthesise a new (output) PSF. This new PSF is then used to derive the best estimation of the radiometric value of the pixel to be resampled. By incorporating noise and signal statistics, input PSF, and desired output PSF, restoration compensates for the blurring that has already occurred and boosts those frequencies that have already been suppressed during the imaging process. Restoration should not be confused with edge enhancement or other heuristic spatial filtering techniques which are designed to enhance visual discrimination of features. Restoration also enhances edges and other features, but it does so in a physically meaningful and controllable way which results in less radiometric error. Restored images are visually distinguished from interpolated or convolved images by their sharp, crisp appearance.

3.1.6   Subpixel accuracy

We have seen in section 3.1.3 that if one uses a well-distributed set of GCPs one can expect to achieve an accuracy of geometrical rectification with an error (mean standard deviation) of rather less than the length of the edge of the instantaneous field of view (IFOV). A standard deviation of, say, 80–90% of the edge of the IFOV is commonly regarded as a reasonable achievement. In some circumstances it may be highly desirable to achieve a considerably better accuracy. Let us consider just one example, briefly. In section 4.3. we shall consider ocean circulation and, in particular, the determination of current vectors (i.e. current speed and direction) by feature tracking using sequences of AVHRR thermal infrared images. With rectification of individual scenes to ±0.8 km by conventional rectification with GCPs one can determine current vectors with an error of ±1.8 cm s^{-1}. Typical magnitudes of the current itself may be within a range of about 2–50 cm s^{-1} so that ±1.8 cm s^{-1} may correspond to quite a large percentage error in the calculated current velocity. If the accuracy of the geometrical rectification of each individual AVHRR scene can be reduced to about ±0.2 km (i.e. to about $\pm20\%$ of the edge of the IFOV) then the percentage error in the derived current vectors could be reduced very significantly.

A method for achieving this level of accuracy with AVHRR data was developed by Cracknell and Paithoonwattanakij (1989b). This method for AVHRR data was an adaptation of a method developed previously by Torlegård (1986) for very accurate registration of Landsat MSS images. Torlegård had demonstrated that it was possible, using digitised small extracts (or chips) from air photos, to achieve geometrical rectification of a Landsat MSS image with an error of only 20% of the edge of the IFOV.

The procedure adopted in Torlegård's method for Landsat MSS data was as follows. First, select a GCP from a digitised air photograph which is a known ground coordinate and called a control point chip (CPC). Secondly, find the image coordinates of the resampled CPC. Thirdly, find the image coordinates in the weighted resampled CPC by using a weight matrix that was obtained from the point spread function of the Landsat MSS. Finally, use this CPC to achieve registration in the Landsat MSS data. For more details see Torlegård (1986).

In the adaptation of this method for rectifying AVHRR data (Cracknell and Pai-thoonwattanakij 1989b) one needs to establish a set of CPCs of Landsat MSS data for the GCPs that are used for the AVHRR scene that is to be rectified. Landsat MSS band 4 (0.8–1.1 μm), which is in the near-infrared, is the best suited for land and sea discrimination. Cracknell and Paithoonwattanakij made a collection of GCPs all over the UK, with about 380 points in all, and obtained a CPC of Landsat MSS data for each of these GCPs. Most of the points were along the coastline, since land and sea are separated quite distinctly, which is suitable for the evaluation of shape similarity. The reason for using such a large number of points is because, for any given scene, part of the area studied is almost certain to be under cloud. For convenience the Landsat MSS data were resampled to pixels correspond-ing to 50 m by 50 m.

First of all it is necessary to perform an approximate rectification and this was done with an orbital model. Thus, for each GCP a search window can be defined from this approximate rectification. In an automatic rectification using the Landsat CPCs, the CPC is moved by only one-tenth of an AVHRR pixel edge between matching attempts. For each trial position of the CPC in the window the MSS data are degraded to AVHRR resolution, using the AVHRR point spread function, and the correlation with the AVHRR image evaluated. The best correlation between the CPC and the window in the unrectified data can therefore be identified to an accu-racy of one-tenth of a pixel edge rather than to the nearest pixel. With a set of GCPs located in the image to this accuracy it is then possible to achieve a rectification of the whole image to subpixel accuracy. The method was tested with only seven GCPs and even with such a small number of GCPs it was possible to achieve a rectification with a standard deviation of only 26% of the pixel spacing.

3.1.7 The point spread function

To attempt to understand the concept of the point spread function which we have now mentioned in the two previous sections we need to devote a little more atten-tion to the idea of the instantaneous field of view (IFOV) of the AVHRR. The concept is not trivial. It is convenient, and common practice, to construct an image from square or rectangular picture elements (or pixels). The simple-minded idea that one often finds presented in elementary textbooks, when describing the operation of a scanner, is that the instrument receives all the radiation from a certain area on the ground (the (instantaneous) field of view, IFOV) and generates a response that is proportional to the quantity of radiation received. For convenience that field of view is commonly regarded as a rectangular (or square) piece of the Earth's surface so that, from the array of scanner data, an image can be constructed where the pixels fill a two-dimensional plane surface as a scale reduction representing the ground. There are several reasons why this simple description is very far from the truth. For instance, the simple description above would mean that in collecting the radiation from the surface the scanner would pause and collect the radiation from within the IFOV, then jump to view the next adjacent IFOV and pause there too; however, in practice the scanner's mirror is rotating continuously and the output from the detectors is integrated over a time interval as the mirror rotates. It would also mean that the scanner would respond uniformly to radiation from all points within the IFOV and would give no response to radiation from outside the IFOV,

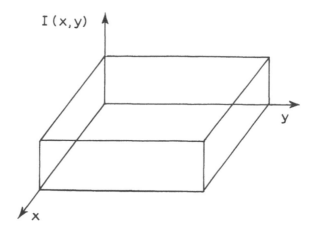

Figure 3.7 Idealised response from nominal IFOV.

see Figure 3.7; this also is not what really happens and there are, in practice, contributions from outside the nominal IFOV. What also makes the situation complicated is that (a) radiation originates with different intensities from different locations in the field of view and (b) the response of the sensor to a source of radiation with a given intensity will vary according to the location of the source within the field of view (in addition to the fact already mentioned that the response to sources just outside the nominal field of view will not be zero). Looked at in another way, point (b) means that the signal we attribute to any given pixel actually arises as a result of contributions not only from the field of view corresponding to that pixel but includes contributions that properly belong to neighbouring pixels, see Figure 3.8. In other words the pixel intensities are not independent but there is autocorrelation among them. There has been a discussion of this complicated situation published for the Landsat MSS and TM by Markham (1985) and by a few other authors, while Breaker (1990) has attempted to remove autocorrelation from sea surface temperatures determined from thermal infrared AVHRR data. We shall outline what is involved in adapting Markham's argument to the AVHRR.

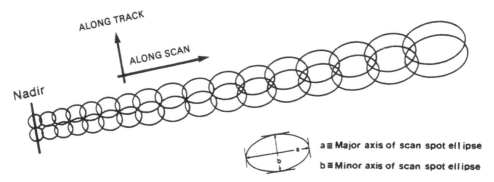

Figure 3.8 Sketch of pixel geometry for the AVHRR for adjacent scan lines showing the scan spot (IFOV) overlap in the along-track and cross-track directions (Breaker 1990).

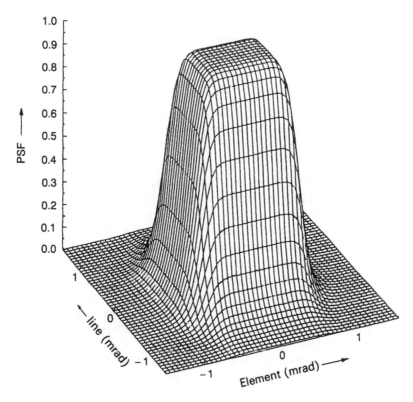

Figure 3.9 Sketch of point spread function (Mannstein and Gesell 1991).

The spacecraft is so far away from the Earth that the object distance can be considered to be infinite and the image is formed in the focal plane. The optical point spread function, $PSF(x - u, y - v)$, describes the intensity as a function of position (x, y) in the focal plane arising from an object which is a point source with its geometrical image at the point (u, v) in the focal plane. The sketch in Figure 3.9 shows a point spread function. For theoretical considerations it is convenient to separate the point spread function or each of the line spread functions into three separate stages, namely

- the optical aspects
- the detectors
- the electronics

see Figure 3.10. The final response of the system to a signal received from the surface of the Earth is the product of three separate responses corresponding to these three stages. Let us consider these three stages in turn. The first stage is primarily concerned with diffraction effects. For an ideal distortion-free imaging system, namely the pinhole camera, the image of a point source would be a point and the location, (u, v), of this image would be determined by simple geometry, i.e. by a straight line connecting the source to the pinhole and projected to the focal plane. If the pinhole is replaced by a lens system the image will become a circular diffraction pattern centred at the point located by a ray of light passing through the

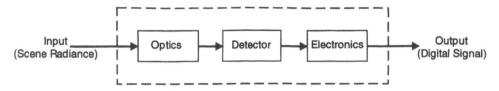

Figure 3.10 Schematic diagram of a typical spaceborne radiometer depicting the basic functions that occur from the input scene radiance to the digital signal at the output (Breaker 1990).

nodal point, i.e. the 'centre', of the lens. This diffraction pattern contributes to the optical factor in the point spread function. The passage of the light through further optical components will introduce further diffraction effects. In practice, however, one is not observing a point source. The object plane is a two-dimensional array of sources of various intensities. Each of these sources gives rise to a diffraction pattern that is (a) centred at the geometrical image position in the focal plane and (b) of height (or maximum intensity) that is proportional to the intensity of the original source (neglecting atmospheric corrections). The total intensity at any point in the image plane arises as the sum of contributions from a large number of diffraction patterns which have their centres in the vicinity of the point in question. The second stage is concerned with the detectors, their shape, their location in the final image plane of the optics, and their physical construction. The third stage is concerned with the electronic response of the detectors, in terms of the distribution of the radiation across the surface of the detectors, the electronics of the amplification, digitisation, etc., stages. To calculate the effects of all these stages and obtain a meaningful value for the PSF from these calculations is not feasible.

The approach used by Markham (1985) for Landsat and by Breaker (1990) for AVHRR is to work in the spatial frequency (or Fourier transform) domain and thus to write

$$G_o(v_x, v_y) = \text{OTF}(v_x, v_y)G_i(v_x, v_y) \tag{3.1.27}$$

where $G_o(v_x, v_y)$ and $G_i(v_x, v_y)$ are the two-dimensional spatial spectra (i.e. Fourier transforms) of the output and input scenes of the AVHRR, respectively, and $\text{OTF}(v_x, v_y)$, the optical transfer function, is the Fourier transform of the point spread function. In practice it is not feasible to measure the two-dimensional point spread function directly because of the problems of producing point sources of significant intensity. Consequently the point spread function is expressed as a product of two line spread fucntions $\text{LSF}(x)$ and $\text{LSF}(y)$ in the x and y directions, i.e.

$$\text{PSF}(x, y) = \text{LSF}(x)\text{LSF}(y) . \tag{3.1.28}$$

A line spread function is the response of the system to an infinitesimally narrow line source and is the integral of the PSF in the direction of the line source. The determination of the line spread functions in the along-scan (cross-track) direction and the along-track direction for the AVHRR was carried out by Breaker (1990) and some of the effects of autocorrelation on satellite-derived sea surface temperatures were discussed.

With the simple response function of Figure 3.7 all sources within the geometrically defined field of view, which is a square of side $2a$, would contribute equally to the signal. In practice, points near to the nodal point N, the intersection of the

1 km

1	5	8	11	17	21	25	28	30	30	30	30	28	25	21	17	11	8	5	1
5	9	13	20	26	31	35	38	40	41	41	40	38	35	31	26	20	13	9	5
8	13	20	28	34	40	44	48	50	50	50	48	44	40	34	28	20	13	8	
11	20	28	36	43	49	54	57	59	60	60	59	57	54	49	43	36	28	20	11
17	26	34	43	50	56	62	66	68	69	69	68	66	62	56	50	43	34	26	17
21	31	40	49	56	63	69	73	76	77	77	76	73	69	63	56	49	40	31	21
25	35	44	54	62	69	75	79	82	83	83	82	79	75	69	62	54	44	35	25
28	38	48	57	66	73	79	84	87	88	88	87	84	79	73	66	57	48	38	28
30	40	50	59	68	76	82	87	90	92	92	90	87	82	76	68	59	50	40	30
30	41	50	60	69	77	83	88	92	94	94	92	88	83	77	69	60	50	41	30
30	41	50	60	69	77	83	88	92	94	94	92	88	83	77	69	60	50	41	30
30	40	50	59	68	76	82	87	90	92	92	90	87	82	76	68	59	50	40	30
28	38	48	57	66	73	79	84	87	88	88	87	84	79	73	66	57	48	38	28
25	35	44	54	62	69	75	79	82	83	83	82	79	75	69	62	54	44	35	25
21	31	40	49	56	63	69	73	76	77	77	76	73	69	63	56	49	40	31	21
17	26	34	43	50	56	62	66	68	69	69	68	66	62	56	50	43	34	26	17
11	20	28	36	43	49	54	57	59	60	60	59	57	54	49	43	36	28	20	11
8	13	20	28	34	40	44	48	50	50	50	50	48	44	40	34	28	20	13	8
5	9	13	20	26	31	35	38	40	41	41	40	38	35	31	26	20	13	9	5
1	5	8	11	17	21	25	28	30	30	30	30	28	25	21	17	11	8	5	1

1 km

Figure 3.11   AVHRR point spread function (Paithoonwattanakij 1989).

optical axis with the object plane, contribute most strongly, points 0.25 km away contribute less strongly, points at the nominal boundary contribute even less strongly but points beyond the nominal boundary make some contribution. The values of this response function are given in Figure 3.11. These values have not, apparently, been published; they were obtained from NOAA or from the manufacturers of the AVHRR (ITT Aerospace) by K. Paithoonwattanakij a few years ago and included in his PhD thesis (Paithoonwattanakij 1989). The signal output by the AVHRR can therefore be thought of as the convolution of this function with the function representing the source intensity in the object plane. This function is of practical importance in a number of situations, particularly when considering objects that are very small in relation to the IFOV, when considering mixed pixels and when considering off-nadir viewing of the ground; we shall have occasion to refer to this function, which can be considered as a discretised form of the point spread function, at several later stages in this book. The function represented in Figure 3.11 was derived from laboratory measurements made before launch. There is a further complication in flight that the AVHRR's scan mirror is rotating and (less importantly) the spacecraft is advancing along its orbit. The signal output by the instrument results from an integration over a period of time during which there will have been some movement of the field of view of the instrument relative to the surface of the Earth.

3.2   CLOUD DETECTION TECHNIQUES

Textbook or review article examples using AVHRR data to show sea surface temperatures retrieved from the thermal infrared channels, or vegetation indices derived from the visible and near-infrared channels, are often taken from cloud-free areas. They are, in that case, very untypical of the vast majority of AVHRR data that are obtained in practice. Cloud cover in AVHRR data is of interest to various scientists. For the climatologist the quantitative determination of cloud cover is important because small changes in cloud cover may lead to major changes in climate (Coakley and Bretherton 1982). For oceanographers, the accuracy of satellite observations of sea surface temperature (SST) is critically dependent upon the ability of satellite radiometers to view the sea surface unobstructed by cloud. We have already mentioned in section 2.4 the need to eliminate cloudy pixels from data that are used to determine operational vegetation index products. If one is concerned with studying the surface of the Earth with AVHRR data it is important first to identify areas where there is cloud that obscures the surface of the Earth and secondly to eliminate those areas of cloud from the area for which results are presented. Large areas of thick cloud can be identified very easily by visual inspection by an operator using an image display system; the area in question can be delineated with a cursor and that area can then be masked out from the subsequent processing of the AVHRR data for the generation of Earth-surface parameters. Alternatively a quite simple classification algorithm for the classification of pixels as cloud or noncloud can be implemented digitally. What is much more difficult to handle are areas of very thin cloud or haze or small clouds that are smaller than the size of the IFOV. Thin clouds include high cirrus and very low stratus clouds. In particular, cirrus clouds are very much colder than the sea surface; even a few small cirrus clouds can add large errors to estimates of sea surface temperature (Stewart 1985). Subpixel clouds are usually cumulus or thin scattered clouds. Typically, trade wind cumulus clouds are less than 1 km in diameter. Thin cloud or haze is difficult to identify, whether this is done visually or automatically with some classification scheme based on thresholding in one or more channels. Sub-IFOV-size clouds, i.e. small and relatively isolated (cumulus) clouds which are smaller than about 1 km in size, are also difficult to detect in an AVHRR image. In some applications, particularly the determination of vegetation indices over large areas on a regular and frequent basis, no attempt is made to identify and mask out areas of haze and thin cloud or sub-IFOV-size clouds; instead such areas are eliminated at the end by multitemporal compositing.

We have already attempted to make it clear in the Preface that this book is not concerned with the study of clouds *per se* using AVHRR data, since that is very well covered in other books (Scorer 1986, Rao *et al.* 1990); see also recent articles by Reiff (1988), Allam *et al.* (1989), Kriebel (1989), Winiger *et al.* (1989), Saunders and Seguin (1992) and Yamanouchi and Kawaguchi (1992). We should perhaps also mention the pioneering book by Barrett (1974) although at the time that book was published it was only VHRR, not AVHRR, data that were available. We are only concerned in the present book to be able to identify a pixel as cloud-free or (partially or totally) cloud-covered. We have no need to be able to distinguish between pixels corresponding to partially and totally cloud-covered areas of the surface of the Earth. In either case the pixel would be rejected and no attempt would be made to recover the value of any Earth surface geophysical parameter from that pixel of the AVHRR data. We also have no interest in being able to classify cloudy pixels according to the type of cloud, except in so far as that may be involved in the formulation of algorithms for the detection of clouds. The main purpose of our discussion of cloud study with AVHRR data is to identify all cloud-contaminated pixels and to eliminate them from subsequent processing.

Before considering cloud detection and elimination, we shall just mention two examples of the use of AVHRR data in studying phenomena which are related to clouds but which are anthropogenic in origin and which will not be discussed elsewhere in this book. These are (i) aircraft contrails and (ii) ship trails. Aircraft contrails are familiar to us from ground observations in everyday experience; they are commonly visible in satellite imagery and they have been studied with AVHRR data by Engelstad *et al.* (1992) and Forkert *et al.* (1993). Ship trails are much less familiar (Scorer 1986, 1987, Hastings *et al.* 1989b). Indeed, their existence was probably unknown before the images from meteorological satellites became available since their scale makes them very difficult to study except from satellite data; an example is shown in Figure 3.12 (which is actually a CZCS image, not an AVHRR image). When the ship trails appear they are very obvious, they widen slowly and persist for a long time (typically 8–30 hours). But their appearance is rare; from an analysis of satellite data for a six-year period from 1980 to 1985 Scorer identified only 47 occasions on which these ship trails appeared in the northeast Atlantic. There are ships in the same shipping lanes every day and the cloud pattern may be quite similar in appearance to that in Figure 3.12 but yet trails are only seen, on average, about once in six weeks in this area which is one of the busiest shipping areas of the world. They have never been observed in the Mediterranean. There has been little incentive to study them since no significant consequences result from them.

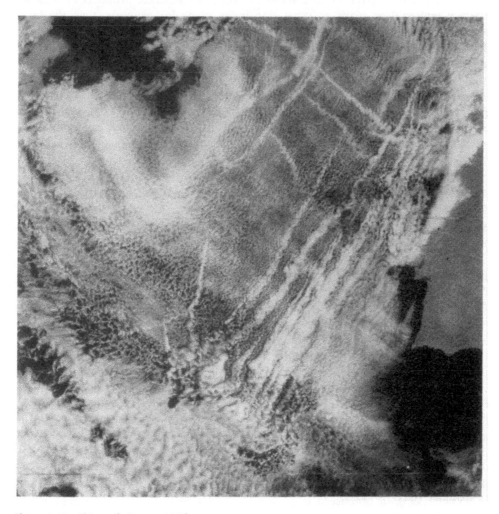

Figure 3.12   Ship trails (Scorer 1987).

However, their rarity and very irregular occurrence demands an explanation. The explanation offered by Scorer (1987) is that the ship trails only occur in very clean air of recent Arctic origin which has not been mixed with air from a continental region and has not been subject to strong convection from the sea; it is therefore very short of condensation nuclei. When there is a dearth of such nuclei the condensed water is in the form of a small number of large cloud droplets. The abundance of hygroscopic nuclei from the ship replaces these by a large number of smaller droplets whose cross-section, i.e. optical opacity, is much greater and there is much more multiple scattering. Thus the polluted air contains a much whiter cloud.

If one is handling a modest amount of data in a research mode then the quickest and safest way to identify areas of cloud is by visual inspection of the image. Areas of thin haze and sub-IFOV-size clouds may, however, be difficult to detect. If they cannot be detected visually in an image, even after varying the contrast in the image by contrast stretching, then it may be difficult to detect them from an automated study of the digital data either. However, if one is concerned with handling all the data received by a given ground station, in order to generate products such as sea surface temperature maps or vegetation index maps for example, then the large volume of data involved makes it highly desirable to develop an automated process for the elimination of cloudy areas. The analogue, in digital image processing terms, of the visual inspection of an image by a trained meteorologist would be a method that studied a region of an image in terms of its context or used some pattern recognition process or neural network approach for groups of pixels in the image (Garand and Weinman 1986, Ebert 1987, 1989, Garand 1988, Gu et al. 1989, Welch et al. 1989, Galladet and Simpson 1991). Such methods, however, are likely to be complicated to develop and validate and are also likely to be expensive in terms of the consumption of computer time.

For speed of processing it is desirable to have a simple test that can be applied to an individual pixel on its own, with no reference to surrounding pixels, so that the whole scene can be processed on a pixel-by-pixel basis. There are various levels of sophistication in single-pixel cloud detection algorithms or methods. Over the years a number of algorithms have been proposed for pixel-by-pixel processing, using varying numbers of tests. Coakley and Bretherton (1982), the World Climate Programme (1984), Kelly (1985), Olesen and Grassl (1985), Rossow et al. (1985), Saunders (1986), Kaufman (1987), Saunders and Kriebel (1988), Le Gléau et al. (1989), Galladet and Simpson (1991), Wald et al. (1991), Thiermann and Ruprecht (1992) and others have developed cloud masking and cloud classification algorithms which are commonly based on thresholds obtained from visible and infrared satellite data. One of the simplest can be used quite successfully over the surface of the sea. This is based on the idea that clouds are significantly colder than the surface of the sea and that one could therefore use a threshold value of the brightness temperature in channel 4 or channel 5 to distinguish between pixels from cloudy areas and pixels from clear areas. For the determination of sea surface temperature charts this approach can be quite successful, although in polar regions with sea ice present it may be more difficult. However, there is no reason to suppose that the best results will be obtained with a common value of the threshold temperature for all times of day, for all seasons and for all geographical areas (Eck and Kalb 1991, França and Cracknell 1995). In practice it is quite a good idea to utilise the histogram of the whole sea area that is being studied. The histogram can be expected to be bimodal

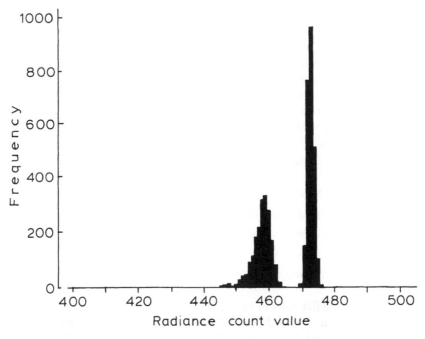

Figure 3.13 Example of bimodal histogram of brightness temperatures.

with, hopefully, a clear separation between the digital numbers (DNs, counts) for the colder clouds and the warmer sea surface, see Figure 3.13. The threshold can then be chosen to be between the two separated regions of the histogram. This method works satisfactorily provided the histogram does actually take a bimodal form, i.e. provided the clouds are dense and cold enough and the water is warm. It will not work in general for clouds smaller than the pixel size, like trade wind cumulus or thin scattered clouds, because the pixel signal is then an average of cloud and sea temperatures and may easily fall above the threshold. It will also not work in regions with ice cover. It also cannot be expected to work satisfactorily over land areas because the spatial and diurnal variations of the land surface temperature are much greater than those of the sea. For general use, therefore, any cloud detection scheme for AVHRR data should involve the use of several, or all, of the channels. One development, therefore, is to involve visible channel data as well as the thermal infrared channel data. This involves trying to find pixels with either low radiances in the visible channel of the AVHRR (Bernstein 1982) or a combination of low radiances in the visible and high radiances in the thermal infrared (Maul 1981). Bernstein (1982) developed a procedure for identifying cloud-free pixels by comparing measured solar radiances at 1 μm (the centre wavelength for AVHRR channel 2) backscattered from the atmosphere with those expected from a simple linear Rayleigh scattering model. The first part of this procedure incorporated a criterion which accepted only pixels with an AVHRR channel-2 albedo of less than 2%. Then pixels that satisfied this criterion were required to satisfy a mean empirical linear relation between percent albedo and the simple Rayleigh scattering model. This mean relation (shown in Figure 7 of Bernstein 1982) was derived from a limited set of data subjectively determined from a series of AVHRR images. This method has three serious disadvantages: (1) it includes the limitations (discussed below) of the

static threshold method, (2) it implicitly assumes that the statistically derived relation in Figure 7 of Bernstein (1982) can be applied universally to any AVHRR
image and (3) in most cases it imposes a total set of criteria which rejects more
pixels than necessary to detect clouds (hence valuable data are lost).

Perhaps the best approach to the description of cloud-detection algorithms is to
follow the method of Saunders and Kriebel (1988). This incorporates various tests,
several of which have been used in various combinations by various other previous
workers. The scheme presented by Saunders and Kriebel is based on five day-time
or five night-time tests which are applied to each individual pixel to determine
whether that pixel is cloud-free, partly cloudy or cloud-filled. These are:

- thermal infrared threshold test
- local uniformity or spatial coherence test
- dynamic reflectance threshold test
- channel 2/channel 1 ratio test
- channel 4 — channel 5 brightness temperatures test.

We shall describe these five tests in turn.

The test that we have already described, based on the brightness temperature
from channel 4 or channel 5 is the first of these five tests. The other four tests will be
described shortly. A pixel is only identified as cloud-free if it passes all five tests. The
method was not only described but it was also tested by Saunders and Kriebel
(1988) using some AVHRR data captured over the UK. An alternative method for
the study of day-time AVHRR data, based on similar ideas to those of Saunders and
Kriebel (1988) has also been described by Gutman (1992). A multispectral colour-
composite presentation, using three 8-bit images superimposed, where each image
results from some combination of AVHRR channels and is assigned to one of the
colours, blue, green or red, has been developed (Derrien et al. 1989, Bellec and Le
Gléau 1992, Derrien and Le Gléau 1993). It is routinely used with AVHRR data at
the Centre de Météorologie Spatiale, at Lannion in France, for the detection of
clouds and of snow-covered ground. The 'AVHRR Processing scheme Over cLouds,
Land and Ocean' (APOLLO) (Saunders and Kriebel 1988), as developed in its
present form by DLR, has been combined with the 'Station HRPT Archiving and
Reprocessing Kernel' (SHARK), developed by ESA/ESRIN, to realise an easily
accessible tool for quantitative AVHRR data processing (Gesell et al. 1993). We
shall now give a few more details of the method described by Saunders and Kriebel
(1988).

The overall philosophy of the cloud detection scheme of Saunders and Kriebel
(1988) is to apply five tests to detect cloud and then to identify a pixel as cloud-free
only if all the tests prove negative, see Figures 3.14 and 3.15. There are differences
between the tests used for day-time data and for night-time data and there are also
differences in the tests depending on whether the underlying ground area is sea, land
or coast (i.e. mixed land and sea). Using five tests in this way does lead to the
possibility that some tests will incorrectly identify some cloud-free pixels as cloud-
contaminated, but this is the safest way to ensure that no cloud-contaminated pixels
escape detection. The scheme uses data from channels 1, 2, 4 and 5 for day-time
data and uses channel 3 as well for night-time data. For data from those AVHRR
instruments without a separate channel 5 the tests involving channel-4 — channel-5

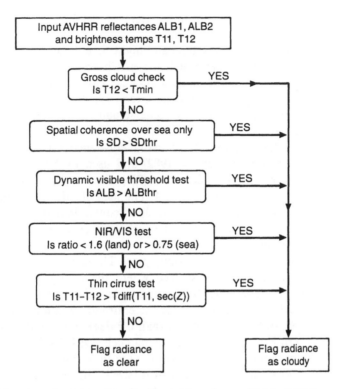

Figure 3.14 Day-time cloud detection algorithm of Saunders and Kriebel (1988).

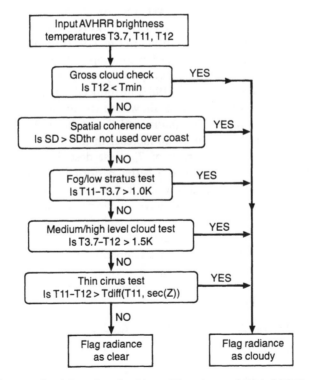

Figure 3.15 Night-time cloud detection algorithm of Saunders and Kriebel (1988).

differences are omitted and elsewhere in the test channel-4 data are substituted for channel-5 data.

The first test applied both to day-time and night-time data is the threshold technique which has already been described above. Channel-5 data are used (where available) in preference to channel-4 data because clouds have a greater optical depth at channel-5 wavelengths (Olesen and Grassl 1985). The determination of the threshold on a dynamic basis, i.e. for the scene actually being studied, is important, particularly over land areas.

The second test is a local uniformity or spatial coherence test. This was originally developed by Coakley and Bretherton (1982) and used by Saunders (1986). This technique is based on the standard deviation (SD) of a 3×3 pixel array composed of brightness temperature, T_{B4}, of channel 4. The idea of this technique is that the variability of brightness temperature over a cloud-contaminated pixel array should be higher than for a clear pixel array. The technique criterion flags a pixel as a cloudy pixel if the standard deviation of this array of nine values of T_{B4} is found within a certain empirical range for a cloudy pixel. The spatial coherence technique has been mostly applied over the surface where the background brightness is assumed to be uniform. Therefore the standard deviation thresholds are usually empirically determined and set constant for the entire observational area. Saunders (1986) has established an SD value of less than 0.2 degK for cloud-free pixels over the sea in northeastern Europe. França and Cracknell (1994, 1995) found SD values less than 0.4 degK for cloud-free pixels over the sea in northeastern Brazil. A value of 1 degK was used for data around the British Isles in the tests in Figures 3.14 and 3.15. The main problem of this technique is its bad performance in detecting cloudy pixels over coastal areas, land and regions of strong gradient of sea surface temperature (see Saunders 1986). It is, however, very useful for the delineation of limits between land and sea. A few investigators have, however, applied the spatial coherence method over land; for example Gutman et al. (1987) used a spatial coherence method for cloud screening of a reduced resolution data set over the US Great Plains. Since the land surface is much less spatially uniform than the sea and varies with the season as well, Gutman et al. (1987) introduced standard deviation thresholds as functions of space and time. This allows the technique to work in areas with high surface albedo variability. Thiermann and Ruprecht (1992) have suggested a new coherence technique that includes an increase in the weight of the central pixel of the 3×3 standard deviation array of T_{B4} just described. According to their results the new technique has produced a better performance in comparison to the classical standard deviation, especially when it was applied over a low cloud area. Besides, the new technique was found to have an enormous increase in its sensitivity in relation to the classical standard deviation technique when only one or a few pixels of the array are cloud contaminated. However, the detection of low clouds (e.g. stratus and stratocumulus), which have temperatures that are normally close to the sea surface temperature, or of large flat clouds having a low temperature variance, is still a problem for the technique. The technique is based on four empirical thresholds given as follows:

$$P_1 = (\,|\,T_{B4}(1, 2) - T_{B4}(2, 2)\,| + |\,T_{B4}(3, 2) - T_{B4}(2, 2)\,|\,)/2 \qquad (3.2.1)$$

$$P_2 = (\,|\,T_{B4}(2, 1) - T_{B4}(2, 2)\,| + |\,T_{B4}(2, 3) - T_{B4}(2, 2)\,|\,)/2 \qquad (3.2.2)$$

$$P_3 = (\,|\,T_{B4}(1, 1) - T_{B4}(2, 2)\,| + |\,T_{B4}(3, 3) - T_{B4}(2, 2)\,|\,)/2 \qquad (3.2.3)$$

$$P_4 = (\,|\,T_{B4}(1, 3) - T_{B4}(2, 2)\,| + |\,T_{B4}(3, 1) - T_{B4}(2, 2)\,|\,)/2 \qquad (3.2.4)$$

where $T_{B4}(i, j)$ is the brightness temperature of channel 4 and i and j are the line and column of the local array respectively. The pixel will be identified as a cloudy pixel if one of the four quantities P_1, P_2, P_3 and P_4 exceeds a specified empirical threshold. Thiermann and Ruprecht (1992) have established the critical value of P as being equal to 0.25 degK for their study area in Europe. For data from northeastern Brazil França and Cracknell (1995) obtained a threshold equal to 0.7 degK for cloud-free pixels.

The third test proposed by Saunders and Kriebel (1988) for day-time data is the dynamic visible threshold test. Most cloud types have a high reflectance at visible and near-infrared wavelengths when compared with the low reflectance of the sea surface away from areas of specular reflection. This allows a visible threshold to be applied where all reflected radiances, normalised by solar zenith angle, which are above a certain level are identified as cloud contaminated. Over the land the reflectivity in channel 1 tends to be lower than in channel 2. Therefore over land the contrast between land and cloud will be greater in channel 1 than in channel 2. Over the sea the opposite is true, the reflectivity is lower in the near-infrared (channel 2) than in the visible region of the spectrum (channel 1). Therefore over land channel-1 (visible) data are best used and over the sea channel-2 (near-infrared) data are best used for the visible threshold test. For both channels the threshold is determined dynamically from the histogram of all the visible radiances normalised by solar zenith angle over the region of interest. In Figure 3.13 a typical visible radiance histogram with a cloud-free peak is shown, together with the various parameters used to compute the threshold value. First the number of radiances that make up the histogram peak value and the corresponding radiance I_{pk} are determined and if the peak is significant (i.e. contains more than 0.5% of the total population) the process continues. The lower I_1 and upper I_2 limits of the histogram are then determined. If the low radiance end I_1 of the histogram is within a reflectance m of the histogram peak and the peak radiance is less than that over a typical cloud scene I_{max}, then the peak is assumed cloud-free and the visible threshold is set at a reflectance n above I_{pk}. Values from m, n and I_{max} over land and sea were determined empirically for AVHRR data from western Europe by Saunders (1986) by studying a number of histograms from different images at different times of the year.

If a cloud-free peak is not found in the histogram then all pixels are flagged as cloudy. The visible histogram can have two cloud-free peaks over coastal regions. In this case the dynamical threshold technique does not work successfully. Either the peak due to the low radiances from the sea surface is selected as the cloud-free peak or the peak due to the cloud-free land surface is detected. In the former case the higher radiances from the cloud-free land will be incorrectly identified as cloud, and in the latter case the value of m will be too great for the peak to be identified as cloud-free. To overcome this problem a straightforward constant threshold method is used in coastal areas which flags all pixels with a reflectance greater than I_{max} as cloud-contaminated. This does mean, however, that cloud over coastal regions can more easily escape detection. A more complex method such as that described by England and Hunt (1985) could be used when over coastal areas. This is similar to the method described above as the thresholds are determined dynamically but it is computationally more expensive. A major disadvantage of the visible threshold technique is that it can only be used during the day. In addition it cannot be used over sea in areas of sunglint.

The fourth test used on day-time data makes use of the ratio between the reflectance in the near-infrared and visible bands which corresponds to $Q = R_2/R_1$, where R_1 and R_2 represent the reflectivity of channels 1 and 2, respectively. The Q values for cloudy pixels are close to unity due to quite similar scattering effects (Mie scattering) of the reflectance for both channels. Over land (vegetated areas), the Q values are higher than unity due to the higher reflectivity in the near-infrared, R_2, than in the visible channel, R_1. Over the sea R_1 is much greater than R_2. Two thresholds are then chosen. Over land, if $Q < 1.6$ it is assumed that the pixel is taken to be cloudy, while over the sea if $Q > 0.75$ the pixel is taken to be cloudy. This method does not work in an area of sunglint because the value of Q from an area of sunglint is close to 1, i.e. is similar to the values obtained from cloud.

The third and fourth tests which we have just described involve reflected solar radiation (i.e. in channels 1 and 2) and therefore are only relevant to day-time data. On the other hand, during the day-time, channel 3, the mid-infrared channel, contains both reflected solar radiation and emitted infrared radiation. Channel-3 data are not used in any of the day-time tests of Saunders and Kriebel (1988). Saunders (1986) pointed out that channel 3 could be used instead of channel 1 and channel 2 in the second and third tests on day-time data if channel 3 were less noisy than it is. Channel-3 data are a useful discriminator between different cloud types (see chapter 4 of Scorer 1986) and thus they can be useful in cloud-detection tests too. They are used in the third and fourth tests on night-time data, see Figure 3.15. The third night-time test uses the difference between the brightness temperatures calculated from channel 4 and channel 3, i.e. $(T_{B4} - T_{B3})$. If this difference is larger than 1 degK this is taken to indicate fog, the difference arising from the different values of the emissivity of fog at the channel-3 and channel-4 wavelengths. The fourth test involves $(T_{B3} - T_{B5})$; if the value of this exceeds 1.5 degK the pixel is flagged as cloud-contaminated. This will detect pixels partially filled with cloud, semi-transparent (thin cirrus) cloud and most medium- and high-level cloud.

The fifth test, which is applied to both day-time and night-time data, uses the difference between the channel-4 and channel-5 brightness temperatures. This test can only be applied to data from those AVHRR instruments that generate channel-5 data. The use of a suitable threshold for this test (see Table 1 of Saunders and Kriebel 1988) enables one to detect most types of cloud, apart from uniform low cloud.

As presented by Saunders and Kriebel (1988) the method is based on using raw data and threshold values that are fixed, at least for any one given AVHRR instrument in the series. If one considers data from AVHRR instruments on different satellites in the series or if one considers data from other parts of the world, such as tropical regions (França and Cracknell 1995) then the actual values of the thresholds may need to be altered. Sub-IFOV clouds and cloud-IFOV misalignment (when the field of view falls on the edge of the cloud and therefore the IFOV is only partially covered by the cloud) can lead to considerable errors in cloud cover obtained from threshold methods. The way in which these problems are usually handled is to be over-cautious in setting the values of the thresholds used. Consequently some cloud-free pixels are likely to be rejected as cloudy; this is generally considered to be preferable to the alternative of failing to reject some (partially) cloudy pixels and thus leading to significant errors in sea surface temperatures. Gallegos et al. (1993) described a cloud masking scheme based not just on algorithms of the type described above which are, essentially, applied on a pixel-by-pixel basis, but in which cloud edges were detected and thus cloudy areas identified.

Simpson and Humphrey (1990) developed a new cloud removal method for day-time AVHRR data. This operates on a pixel-by-pixel basis. However, it does not use either a fixed set of thresholds with the raw digital numbers, as done by Saunders and Kriebel (1988) for example, or an empirically and subjectively determined criterion, as done by Bernstein (1982). Instead it involves determining the albedo for each pixel and thereby eliminating variations in the raw data arising from variations in solar illumination, view angle, size of IFOV, etc. This method uses calibrated information in channels 2 and 4 of the AVHRR. Channel 2 was selected in preference to channel 1 because the channel-1 spectral response is unable to detect cirrus clouds. Channel 4 was selected because it provides the best infrared data used to compute SST and is available on all AVHRR instruments. Knowledge of cloud type within the image is not needed for this procedure. The channel-2 data are used to calculate the channel-2 albedo. This is the ratio of the effective radiance as seen by channel 2 to the effective radiance of the radiometer viewing reflected sunlight. An empirical model of albedo as a function of solar zenith angle θ_s is computed. The details are described in the appendix of the paper by Simpson and Humphrey (1990). This method requires that the pertinent Sun-IFOV-satellite angles be computed for each pixel in the image. A subset of albedo values is then made by sampling AVHRR channel 2 along the image diagonals and at equally spaced grid locations throughout the image. This grid is proportional to image size. The population of data values subsampled along the image diagonals and at the grid points is called the initially subsampled data set. The initially subsampled data set provides a large number of data values which includes the range of Sun-IFOV-satellite angles associated with that image. This minimises errors which might result from an incomplete representation of the range of Sun-IFOV-satellite angles in the image. Then an iterative procedure is used to develop a model of cloud-free albedo as a function of solar zenith angle ($\alpha = F_1(\cos \theta_s)$ where $\theta_s = $ solar zenith angle). This procedure requires an initialisation constant α_0 which is objectively and automatically determined. The solar zenith angle (or its complement, the solar altitude) was chosen as the independent variable for this empirical model because observations have shown that albedo can be parametrised well by solar altitude for a given value of atmospheric transmittance.

Dynamic local rejection of cloud-contaminated pixels is now done using this empirical model. A localised albedo threshold, α_1, is computed from the empirical model on a pixel-by-pixel basis. Thus each pixel in the image has its own rejection criterion defined by its particular cross-track scan geometry within the image. Then a non-linear statistical model of cloud-free albedo, α_2, as a function of the Rayleigh scattering cross-section, R, ($\alpha_2 = F_2(R)$) is developed.

Pixel rejection based upon albedo from channel 2 uses a combination of these empirical and theoretical criteria. The empirical criteria require that each pixel in the image (channel 2) must satisfy the condition $0 < \alpha \le \alpha_1$. The theoretical criteria require that each pixel in the image (channel 2) must satisfy the condition $\alpha_2 - \sigma \le \alpha \le \alpha_2 + \sigma$, where σ is the standard deviation of the cloud-free data set obtained from the F_1 criterion. A pixel is retained as valid (uncontaminated) if its channel-2 albedo satisfies either of these constraints.

Channel-4 data are used as well because clouds in the shadow of other clouds may have low radiances similar to sea surface values; thus the use of channel 2 alone to mask clouds may be insufficient for some images. These shadow areas, however, can be identified with channel-4 data. All channel-4 pixels whose corresponding channel-2 albedos satisfy the criteria mentioned above are then high-pass filtered to exclude pixels whose temperatures are below $-2.0°C$. This minimum value of temperature is consistent with the minimum global sea surface temperature, which may of course be lower than $0°C$ because of the lowering of the freezing point of water by dissolved salt. Optimisation of the final cloud-masked sea surface temperature is achieved by filtering this data set with a band-pass filter whose width is ± 4 standard deviations about the mean of the high-passed SST image. In practice these channel-4 criteria were found to reject only a small number of additional pixels. Simpson and Humphrey (1990) presented the results of the application of this method to a set of 10 AVHRR images from NOAA-6, -7, -8 and -9 and from a variety of different geographical locations. Their results were evaluated (a) by comparison with results of other methods and (b) by studying the histograms of channel 2 and of the surface temperature for the original data and for the cloud-masked data.

Accurate cloud detection in AVHRR data over land is a difficult task because of the complications arising from spatially and temporally varying land surface reflectances and emissivities. An AVHRR Split-and-Merge-Clustering (ASMC) algorithm for cloud detection in AVHRR designed scenes over land has been developed by Simpson and Gobat (1996) and it provides a computationally efficient, scene-specific, objective way to circumvent these difficulties.

In our discussion so far it has been assumed that our purpose in cloud detection is to identify totally cloudy pixels and partially cloudy pixels, to flag them and then to eliminate them from all further consideration. This is fine if one is concerned with processing large amounts of AVHRR data on a routine and regular basis, e.g. for producing sea surface temperature charts or ice-cover maps for example. However, there are two other aspects that we should, perhaps, consider a little. The first concerns partially cloudy pixels. In determining sea surface temperature charts, ice charts or vegetation index maps, one is receiving data on a regular basis and if one particular area is partially or totally covered by cloud that is no serious problem. The chances are that within a few days an AVHRR scene will be obtained in which this area will be clear of cloud. The phenomena being studied change relatively slowly and over a period of several days a set of data giving complete ground coverage can be built up. However, one may be in a situation in which one is particularly interested in one scene and wishes to obtain the maximum possible information from that scene, perhaps because some short-lived event had occurred. In this case one may not be content simply to identify partially cloud-covered pixels and to reject the data in those pixels. One may wish to determine the percentage cloud cover in a partially cloud-covered pixel and use that to try to estimate or recover the signal originating from the ground in the part of the IFOV that was free of cloud. The paper by Kaufman (1987) and section 4 of the paper by Saunders and Kriebel (1988) include a discussion of the estimation of the effective cloud cover; for details the reader should consult those papers. The question of identifying semi-transparent cloud in the infrared channels has been studied by Lin and Coakley (1993).

In this section we have been addressing the question of identifying whether a pixel is cloudy or not and this was simply in order that the pixel in question could be masked so that the data from that pixel would not be used in the algorithm being used to determine whatever geophysical parameter was under investigation. In other words we were only concerned to identify whether or not there was cloud present.

We were not interested to know what type of cloud was present. It is, however, possible to use the data in the five channels of the AVHRR in a multispectral classification scheme to identify the various types of cloud present in an image. This is analogous to the studies, which are very common, of land areas using Landsat MSS, Landsat TM or SPOT data. The question of cloud classification has been studied by a number of people (Parikh 1977, Coakley and Baldwin 1984, Eyre *et al.* 1984, Olesen and Grassl 1985, Pairman and Kittler 1986, Liljas 1987, 1989, Garand 1988, Berger 1989, Derrien *et al.* 1989, Karlsson 1989a, Lee *et al.* 1990, Masuda and Takashima 1990a,b, Khazenie and Richardson 1991, Barker and Davies 1992, Ebert 1992). Much of the earlier work was concerned with using a small number of different classes, but more recently classification schemes involving 20 or more classes have been developed (Berger 1989, Liljas 1989, 1991) and software systems have been developed, particularly at the Swedish Meteorological and Hydrological Institute, for carrying out this classification on an operational basis. Comparison of the AVHRR cloud classification scheme of the SMHI to surface observations has been carried out by Karlsson (1993). However, as indicated at the beginning of this section we do not propose to consider the details of this work in this book.

3.3 ATMOSPHERIC CORRECTIONS

3.3.1 Introduction

The output from the instruments on board an Earth-observing satellite depends on the intensity and spectral distribution of the energy that is received at the satellite. This is not the same as the intensity and spectral distribution of the energy that left the surface of the Earth. The useful information about the target area of the land, sea or clouds is contained in the physical properties of the radiation leaving that target area, whereas what is measured by a remote sensing instrument are the properties of the radiation that arrives at the instrument. This radiation has travelled some distance through the atmosphere and accordingly has suffered both attenuation and augmentation in the course of that journey. The problem that faces the user of remote sensing data is that of being able to regenerate the details of the properties of the radiation that left the target area from the data generated by the remote sensing instrument. An attempt to set up the radiative transfer equation to describe all the various processes that corrupt the signal that leaves the target area on the land, sea or cloud from first principles is a nice exercise in theoretical atmospheric physics and, of course, it is a necessary starting point for any soundly based attempt to apply atmospheric corrections to satellite data. However, in a real situation the problem soon arises that suitable values of various atmospheric parameters have to be inserted into the radiative transfer equation in order to arrive at a solution; however, accurate values of these parameters are often not available.

3.3.2 Radiative transfer theory

To try to make quantitative calculations of the difference between the satellite-received radiance and the Earth-leaving radiance, one has to use what is commonly known as radiative transfer theory. In essence, this consists of studying radiation travelling in a certain direction, specified by the angle ϕ between that direction and the vertical axis z, and setting up a differential equation for a small horizontal element of the transmitting medium (the atmosphere) with thickness dz. It is necessary to consider

- the radiation entering the element dz from below
- the attenuation suffered by that radiation within the element dz
- the additional radiation that is either generated within the element dz or scattered into the direction ϕ within the element dz

and thence to determine an expression for the intensity of the radiation leaving the element dz in the direction ϕ.

The resulting differential equation is called the radiative transfer equation. Although it is not particularly difficult to formulate this equation, it is not commonly used. In practice, the details of the formulation will be simplified to include only the important effects. The equation will therefore be different for different wavelengths of electromagnetic radiation because of the different relative importance of different physical processes at different wavelengths. Suitable versions of the radiative transfer equation for optical and near-infrared wavelengths and for thermal infrared wavelengths will be presented in this section.

If the values of the various atmospheric parameters that appear in the radiative transfer equation are known, this differential equation can be solved to determine the relation between the satellite-received radiance and the Earth-leaving radiance. However, the greatest difficulty in making atmospheric corrections to remotely sensed data lies in the fact that it is usually impossible to obtain accurate values for the various atmospheric parameters that appear in the radiative transfer equation. The atmosphere is a highly dynamic physical system and the various atmospheric parameters will, in general, be functions of the three space variables, x, y and z and of the time variable, t. Because of the paucity of the data it is common to assume a horizontally stratified atmosphere, in other words the atmospheric parameters are assumed to be functions of the height z but not of the x and y coordinates in a horizontal plane. The situation may be simplified further by assuming that the atmospheric parameters are given by some model atmosphere based only on the geographical location and the time of the year. However, this is not a very realistic approach because the actual conditions at any given time will differ quite considerably from such a model. It is clearly much better to try and use values of the atmospheric parameters that apply at the time that the remotely sensed data are collected.

It is important, however, to realise that there is a fundamental difficulty, namely that the problem of solving the radiative transfer equation in the situations described is an example of an unconstrained inversion problem. That is, there are many unknowns (the atmospheric parameters for a given atmospheric path) and a very small number of measurements (the intensities received in the various spectral channels from the given instantaneous field of view of the instrument). The solution will, inevitably, take some information from the mathematical and physical assumptions that have been built into the method of solution adopted.

A general formalism for atmospheric absorption and transmission is required. Consider a beam of radiation with wavelength λ and wavevector of magnitude $\kappa(= 2\pi/\lambda)$ travelling in a direction θ to the normal to the Earth's surface. After the radiation has travelled a distance l, the flux (radiance) of the wavelength λ, $\phi_\lambda(l)$, is related to its initial value $\phi_\lambda(0)$ by

$$\phi_\lambda(l) = \phi_\lambda(0) \exp\left\{ - \sec \theta \int_0^l K_\lambda(z) \, dz \right\} \tag{3.3.1}$$

where $z = l \cos \theta$ and $K_\lambda(z)$ is the attenuation coefficient. Notice that the attenuation coefficient is a function of height as well as of wavelength. These quantities can be expressed in terms of κ instead of λ giving

$$\phi_\kappa(l) = \phi_\kappa(0) \exp\left\{ - \sec \theta \int_0^l K_\kappa(z) \, dz \right\}. \tag{3.3.2}$$

The dimensionless quantity $\int_0^z K_\kappa(z) \, dz$ is called the optical thickness and is commonly denoted by $\tau_\kappa(z)$ and the quantity

$$T_\kappa(z) = \exp\left(- \int_0^z K_\kappa(z) \, dz \right) \tag{3.3.3}$$

is called the beam transmittance.

3.3.3 Physical processes involved in atmospheric correction

In atmospheric correction processes, the first distinction to be made is whether the radiation leaving the surface of the land, sea or clouds is radiation emitted by that surface or whether it is reflected solar radiation. The relative proportions of reflected and emitted radiation will vary according to the wavelength of the radiation, the time and the place of observation. At optical and very-near-infrared wavelengths the emitted radiation is negligible compared with the reflected radiation, while at thermal infrared and microwave wavelengths it is the emitted radiation which is important, with the reflected radiation being of negligible intensity. Within the wavelength range of channel 3 of the AVHRR, the emitted and reflected radiation are both important. The problem is to relate data usually consisting of, or derived from, the output from a passive scanning instrument, to the properties of the land, sea or clouds under investigation.

The approach adopted to the question of the contribution of the intervening atmosphere to remotely sensed data is governed both by the characteristics of the remote sensing system in use and by the nature of the environmental problem to which the data are to be applied. In most cases the user of remote sensing data is interested in knowing how important the various atmospheric effects are on the quality of image data or on the magnitudes of the derived physical or biological parameters; the user is not usually interested in the magnitudes of the corrections to the radiance values *per se*. However, to assess the relative importance of the various atmospheric effects it is necessary to devote some attention to

- the physical processes occurring in the atmosphere
- the magnitudes of the effects of these processes on the radiance reaching the satellite
- the consequences of these effects on images or on derived physical or biological parameters.

The cases of emitted radiation and reflected solar radiation will be considered separately, with consideration also being given to atmospheric transmittance.

Estimates of corrections to remotely sensed data are based, ultimately, on solving the radiative transfer equation although, as has been indicated in the previous section, accurate solutions are very hard to obtain and one is forced to adopt an appropriate level of approximation.

The importance of understanding the effect of the atmosphere on remote sensing data and of making corrections for atmospheric effects depends very much on the use that is to be made of the data. There are many meteorological and land-based applications of remote sensing for which in the past there has been no need to carry out any kind of atmospheric correction. This may be because the information that is being extracted is purely qualitative or because, though being quantitative, the information being extracted is statistical or geographical and not radiometric or, even if radiometric, because the remotely sensed data are calibrated by the use of *in situ* data within a training area. Nevertheless it is anticipated that in the future some of these studies will become more exact, particularly as more careful multitemporal studies are undertaken of environmental systems that exhibit change. This is likely to mean that it will become increasingly important to include atmospheric corrections for some of these applications in the future. In the case of oceanographic and coastal work the information for extraction consists of quantitative values of physical or biological parameters of the water, such as the surface temperature (see section 4.2) and concentrations of suspended sediment or chlorophyll (see section 7.3). The theory and techniques of the determination of atmospheric corrections for the thermal infrared channels (channels 3, 4 and 5) over sea areas, for the determination of sea surface temperatures, are very well developed and the atmospheric corrections are routinely applied to such data (see section 4.2). The application of atmospheric corrections to channel-1 and channel-2 AVHRR data is a much more difficult problem than for the thermal infrared channels. For the wavelengths of these channels the atmospheric effects are very much larger, relative to the signal (the satellite-received radiance), than in the thermal infrared case and the techniques are, in consequence, much less well develoed than in the thermal infrared case (see, for example, Koepke 1989). While it is interesting to consider the importance of atmospheric effects in terms of the magnitude of the attenuation relative to the magnitude of the signal from the target area, these effects should not be considered in isolation, but should rather be considered in conjunction with the use to which the data are to be applied.

3.3.4 Emitted radiation

At long wavelengths, i.e. for microwaves and for thermal infrared radiation, it is the emitted radiation and not the reflected solar radiation that is important. There are several contributions to the radiation received at the instrument, see Figure 3.16; these contributions, identified as T_1, T_2, T_3 and T_4 are described below. Each can be considered as a radiance $L(\kappa)$ where κ is the wave number, or as corresponding to an equivalent blackbody temperature.

Surface radiance: $L_1(\kappa)$, T_1 The surface radiance is the radiation which is generated thermally at the Earth's surface and which undergoes attenuation as it passes through the atmosphere before reaching the scanner; this radiance can be written as

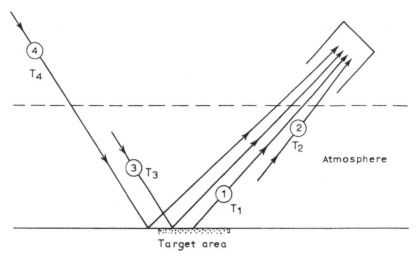

Figure 3.16 Contributions to satellite received radiation for emitted radiation.

$\varepsilon B(\kappa, T_S)$ where ε is the emissivity, $B(\kappa, T_S)$ is the Planck distribution function and T_S is the temperature of the surface. In general, the emissivity ε is a function of wave number and temperature. For example, the emissivity of gases varies very rapidly with wave number in the neighbourhood of the absorption (emission) lines. For sea water ε may be treated as constant with respect to κ and T_S, provided any material which is not part of sea water is ignored, e.g. oil pollution, industrial waste, etc. Let p_0 be the atmospheric pressure at the sea surface. By definition, the pressure at the top of the atmosphere is zero. Thus, the radiance reaching the detector from the view angle θ is

$$L_1(\kappa) = \varepsilon B(\kappa, T_S)\tau(\kappa, \theta; p_0, 0) \tag{3.3.4}$$

where $\tau(\kappa, \theta; p, p_1)$ is the atmospheric transmittance for wave number κ and direction θ between heights in the atmosphere where the pressures are p and p_1.

Upwelling atmospheric radiance: $L_2(\kappa)$, T_2 The atmosphere emits radiation at all altitudes. As this emitted radiation travels upwards to the scanner it undergoes attenuation in the overlying atmosphere. It is possible to show (see, for example, Singh and Warren 1983) that the radiance emitted by a horizontal slab of the atmosphere lying between heights z and $z + \delta z$, where the pressure is p and $p + \delta p$, respectively, and arriving in a direction θ at a height z_1 where the pressure is p_1, is given by

$$dL_2(\kappa) = B(\kappa, T(p))\, d\tau(\kappa, \theta, p, p_1) \tag{3.3.5}$$

or

$$dL_2(\kappa) = B(\kappa, T(p))\, \frac{d\tau}{dp}(\kappa, \theta; p, p_1)\, dp \ . \tag{3.3.6}$$

The upwelling emitted radiation received at the satellite can thus be written as

$$L_2(\kappa) = \int_{p_0}^{p} B(\kappa, T(p))\, \frac{d\tau}{dp}(\kappa, \theta; p, 0)\, dp \tag{3.3.7}$$

where p_0 is the atmospheric pressure at the sea surface and $T(p)$ is the temperature at the height at which the pressure is p. This expression is based on the assumption of local thermodynamic equilibrium and the use of Kirchhoff's law to relate the emissivity to the absorption coefficient.

Downwelling atmospheric radiance: $L_3(\kappa)$, T_3   In this case allowance is made for atmospheric emission downwards to the Earth's surface where the radiation undergoes reflection upwards to the scanner. Attenuation is undergone as the radiation passes through the atmosphere. The total downwelling radiation from the top of the atmosphere, where $p = 0$, to the sea surface, where the pressure is p_0, is given by

$$\int_0^{p_0} B(\kappa, T(p)) \frac{d\tau(\kappa, \theta; p, p_0)}{dp} \, dp \, . \tag{3.3.8}$$

A fraction $(1 - \varepsilon)$ of this is reflected at the sea surface and a fraction $\tau(\kappa, \theta; p_0, 0)$ of this passes through the atmosphere so that the radiance reaching the satellite is given by

$$L_3(\kappa) = (1 - \varepsilon)\tau(\kappa, \theta; p_0, 0) \int_0^{p_0} B(\kappa, \tau(p)) \frac{d\tau}{dp}(\kappa, \theta; p, p_0) \, dp \, . \tag{3.3.9}$$

Space component: $L_4(\kappa)$, T_4   Space has a background brightness temperature of about 3 K. The space component passes down through the atmosphere, is reflected at the surface, and passes up through the atmosphere again to reach the scanner.

The total radiance $L^*(\kappa)$ received at the satellite can be written as

$$L^*(\kappa) = L_1(\kappa) + L_2(\kappa) + L_3(\kappa) + L_4(\kappa) \, . \tag{3.3.10}$$

Alternatively the same relation can be expressed in terms of the brightness temperature T_B and the equivalent temperatures for each of the contributions already mentioned, i.e.

$$T_B = T_1 + T_2 + T_3 + T_4 \, . \tag{3.3.11}$$

Sea surface temperatures are studied quite extensively using both infrared and microwave passive instruments. In both cases the problem is to estimate or eliminate T_2, T_3 and T_4 so that T_1 can be determined from the measured values of T_B. There is a further complication in the case of microwave radiation as, for certain parts of the Earth's surface at least, a significant contribution also arises from microwaves generated artificially for telecommunications purposes. It is simplest, from the point of view of the above scheme, to include this contribution in T_2.

Apart from information about the equivalent blackbody temperature of the surface of the land, sea or cloud, the brightness temperature measured by the sensor contains information on a number of atmospheric parameters such as water vapour content, cloud liquid water content and rainfall rate. By using multichannel data it may be possible to eliminate T_2, T_3 and T_4 and hence to calculate T_1 and T_B.

3.3.5   Reflected radiation

The reflected radiation case concerns radiation that originates from the Sun and eventually reaches a remote sensing instrument on an aircraft or spacecraft, the energy of the radiation that arrives at the instrument being measured by the sensor. Hopefully the bulk of this radiation will come from the instantaneous field of view, IFOV, on the target area of land, sea or cloud that is the observed object of the remote sensing activity. However, in addition to radiation that has travelled directly over the Sun–IFOV–sensor path and which may contain information about the area that is seen in the IFOV, there will also be some radiation which reaches the sensor by other routes. This radiation will clearly not contain information about the IFOV. Accordingly various paths between the Sun and the sensor are considered for reflected radiation reaching the sensor (see Figure 3.17).

$L_1(\kappa)$ denotes radiation that follows a direct path from the Sun to the target area and thence to the sensor. As illustrated in Figure 3.17 the radiation is shown reflected at the surface. If the surface is a water surface there will be some penetration of the light below the water surface and the reflectivity of the water will be affected by processes that occur below the surface. This enables information to be obtained about the properties and contents of the water near the surface (see section 7.3).

$L_2(\kappa)$ denotes radiation from the Sun that is scattered into the field of view of the sensor, either by single or multiple scattering in the atmosphere, without the radiation ever reaching the target area at all.

$L_3(\kappa)$ denotes radiation which does not come directly from the Sun but which has first undergone some scattering event before reaching the target area; this radiation then passes to the sensor directly.

$L_4(\kappa)$ denotes radiation which has been reflected by other target areas of the land, sea or clouds and is then scattered by the atmosphere into the field of view of the sensor.

These four processes may be regarded, to some extent, as analogues, for reflected radiation, of the four processes outlined in section 3.3.4 for emitted radiation.

It is $L_1(\kappa)$ that contains the useful information. $L_2(\kappa)$ and $L_4(\kappa)$ do not contain useful information about the target area. While, in principle, $L_3(\kappa)$ does contain some information about the target area, it may be misleading information if the radiation is mistakenly regarded as having travelled directly from the Sun to the target area.

It cannot be assumed that the spectral distribution of the radiation reaching the outer regions of the Earth's atmosphere, or its intensity integrated over all wave-

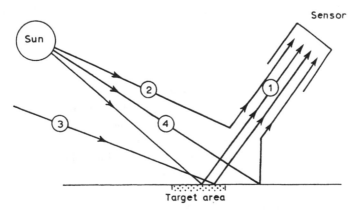

Figure 3.17   Contributions to satellite-received radiation for reflected solar radiation.

lengths, is constant. The extraterrestrial solar spectral irradiance and its integral over wavelength, which is called the solar constant, have been studied experimentally over the last 50 years or more. The technique which is used is due originally to Langley and involves the extrapolation of ground-based irradiance measurements to outside the Earth's atmosphere. A review of such measurements, together with recommendations of standard values was given by Labs and Neckel (1967, 1968, 1970), see also section 2.2.2 and Figure 2.11. While various experimenters acknowledge errors in the region of $\pm 3\%$ following the calibration of their instruments to radiation standards, the sets of results differ from one another by considerably more than this; in some parts of the spectrum they differ by as much as 10%. It will be recalled from section 2.2.2 that this variation led NOAA to tabulate three different possible values of F in Appendix B to NESS107. Some of the discrepancy between the results of different sets of measurements is explained by the fact that the radiation from the Sun itself varies. Annual fluctuations in the radiance received at the Earth's atmosphere associated with the variation of the distance from the Sun to the Earth can be taken into account mathematically. The eccentricity of the ellipse describing the orbit of the Earth is 0.0167 and the minimum and maximum distances from the Sun to the Earth occur on 3 January and 2 July, respectively. The extraterrestrial solar irradiance for Julian day D is given by the following expression:

$$E_0(D) = \bar{E}_0[1 + 0.0167 \cos\{(2\pi/365)(D - 3)\}]^2. \qquad (3.3.12)$$

3.3.6   Atmospheric transmission

3.3.6.1   Introduction

The possible origins of the radiation that finally reaches a remote sensing instrument, and the possible routes that the radiation may take in travelling from its source to the sensor, were considered in sections 3.3.4 and 3.3.5. It is also necessary to consider the scattering mechanisms involved, both in the atmosphere and at the target area on the surface of the Earth or the clouds. While the reflection or scattering at the target area is relevant to the use of all remotely sensed data, the details of the interaction of the radiation with the target area are not considered here, where attention is focused on the scattering and absorption of the radiation which occurs during the passage of radiation through the atmosphere.

Three types of scattering are distinguished depending on the relation between a, the diameter of the scattering particle, and λ, the wavelength of the radiation. If $a \ll \lambda$ Rayleigh scattering is dominant. For Rayleigh scattering the scattering cross-section is proportional to $1/\lambda^4$; for visible radiation this applies to scattering by gas molecules. Other cases correspond to scattering by aerosol particles. If $a \approx \lambda$ Mie scattering is dominant. Mie scattering involves water vapour and dust particles. If $a \gg \lambda$ non-selective scattering is dominant. This scattering is independent of wavelength; for the visible range this involves water droplets with radii of the order of 5–100 µm.

The mechanisms involved in scattering or absorption of radiation as it passes through the atmosphere can be conveniently considered as follows. The attenuation coefficient $K_\kappa(z)$ mentioned in section 3.3.2 can be separated into two parts:

$$K_\kappa(z) = K_\kappa^M(z) + K_\kappa^A(z) \tag{3.3.13}$$

where $K_\kappa^M(z)$ and $K_\kappa^A(z)$ refer to molecular and aerosol attenuation coefficients, respectively. Each of these absorption coefficients can be written as the product of $N^M(z)$ or $N^A(z)$, the number of particles per unit volume at height z, and a quantity σ_κ^M or σ_κ^A, known as the effective cross-section, i.e.

$$K_\kappa(z) = N^M(z)\sigma_\kappa^M + N^A(z)\sigma_\kappa^A \ . \tag{3.3.14}$$

The quantities

$$\tau_\kappa^M(z) = \sigma_\lambda^M \int_0^z N^M(z) \, dz \tag{3.3.15}$$

and

$$\tau_\kappa^A(z) = \sigma_\lambda^A \int_0^z N^A(z) \, dz \tag{3.3.16}$$

are called the molecular optical thickness and the aerosol optical thickness, respectively. It is convenient to separate the molecular optical thickness into a sum of two components

$$\tau_\kappa^M(z) = \tau_{\kappa_s}^M(z) + \tau_{\kappa_a}^M(z) \tag{3.3.17}$$

where $\tau_{\kappa_s}^M(z)$ corresponds to scattering and $\tau_{\kappa_a}^M(z)$ corresponds to absorption. Thus the total optical thickness can be written as

$$\tau_\kappa(z) = \tau_{\kappa_s}^M(z) + \tau_{\kappa_a}^M(z) + \tau_\kappa^A(z) \ . \tag{3.3.18}$$

These three contributions will be considered briefly in turn.

3.3.6.2   Scattering by air molecules

At optical wavelengths this involves Rayleigh scattering. The Rayleigh scattering cross-section is given by a well-known formula

$$\sigma_{\lambda_s}^M = \frac{8\pi^3(n^2 - 1)^2}{(3N^2\lambda^4)} \tag{3.3.19}$$

where n is the refractive index, N is the number of air molecules per unit volume, and λ is the wavelength. This contribution to the scattering of the radiation can be calculated in a relatively straightforward manner. The λ^{-4} behaviour of the Rayleigh scattering (molecular scattering) means that this mechanism is very important at short wavelengths but becomes unimportant at long wavelengths. The blue colour of the sky and the red colour of sunrises and sunsets is attributable to the difference between this scattering for blue light and red light. This mechanism becomes negligible for near-infrared wavelengths (see Figure 3.18) and is of no importance for thermal infrared wavelengths.

3.3.6.3   Absorption by gases

In remote sensing work it is usual to use radiation of wavelengths that are not within the absorption bands of the major constituents of the atmosphere. The gases

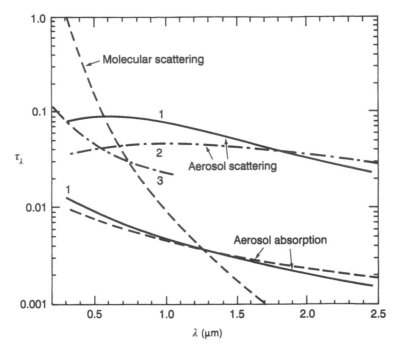

Figure 3.18   Normal optical thickness as a function of wavelength (Sturm 1981).

to be considered are O_2 and N_2, the main constituents of the atmosphere, and carbon dioxide, ozone and water vapour. At optical wavelengths the absorption by O_2, N_2 and CO_2 is negligible. Water vapour has a rather weak absorption band for wavelengths from about 0.7 to 0.74 μm. The only significant contribution to atmospheric absorption by molecules is by ozone. This contribution can be calculated and, although it is small in relation to the Rayleigh and aerosol contribution, it should be included in any calculations of atmospheric corrections to optical scanner data. For scanners operating in the thermal infrared and microwave regions of the electromagnetic spectrum absorption by gases constitutes the major absorption mechanism. The attenuation experienced by the radiation can be calculated using the absorption spectra of the gases involved, carbon dioxide, ozone and water vapour. The relative importance of the contributions from these three gases will depend on the wavelength range under consideration. As indicated, it is only the ozone absorption that is significant at visible wavelengths.

3.3.6.4   Scattering by aerosol particles

The aerosol scattering also decreases with increasing wavelength. It is common to write the aerosol optical thickness as

$$\tau_\lambda^A = A\lambda^{-B} \tag{3.3.20}$$

where B is referred to as the Ångström exponent. However, the values of the parameters A and B do vary quite considerably according to the nature of the aerosol particles. Quoted values of B vary from 0.8 to 1.5 or even higher. At optical wavelengths the aerosol scattering is comparable in magnitude with the Rayleigh scattering; see Figure 3.18. In practice, however, it is more difficult to calculate because of the great variability in the nature and concentration of aerosol particles in the atmosphere. Indeed, when dealing with data from optical scanners it is the calculation of the aerosol scattering which is the most troublesome part of the atmospheric correction calculations. While being of some importance in the near infrared, aerosol scattering can be ignored in the thermal infrared for clear air, i.e. in the absence of cloud, haze, fog or smoke.

3.3.7   Thermal infrared scanners and the radiative transfer equation

The discussion in this section will be expressed in terms of thermal infrared scanners but it would apply equally well to (passive) microwave scanners, though there are no microwave channels in the AVHRR. For a thermal infrared scanner we are concerned with emitted radiation and we consider the radiative transfer equation in the following form

$$\frac{dI_\nu(\theta, \phi)}{ds} = -\gamma_\nu I_\nu(\theta, \phi) + \psi_\nu(\theta, \phi) \tag{3.3.21}$$

where $I_\nu(\theta, \phi)$ is the intensity of electromagnetic radiation of frequency ν in the direction (θ, ϕ), s is measured in the direction (θ, ϕ) and γ_ν is an extinction coefficient. The first term on the right-hand side of this equation describes the attenuation of the radiation both by absorption and by scattering out of the direction (θ, ϕ). The second term describes the augmentation of the radiation, both by emission and by scattering of additional radiation into the direction (θ, ϕ); this term can be written in the form

$$\psi_\nu(\theta, \phi) = \psi_\nu^A(\theta, \phi) + \psi_\nu^S(\theta, \phi) \tag{3.3.22}$$

where $\psi_\nu^A(\theta, \phi)$ is the contribution corresponding to the emission and can, in turn, be written in the form

$$\psi_\nu^A(\theta, \phi) = \gamma_\nu^A B(\nu, T) \tag{3.3.23}$$

where γ_ν^A is an extinction coefficient and $B(\nu, T)$ is the Planck distribution function for blackbody radiation which is given in equation (2.2.19). $\psi_\nu^S(\theta, \phi)$ is the contribution to scattering into the direction (θ, ϕ) and can be written in the form

$$\psi_\nu^S(\theta, \phi) = \gamma_\nu^S J_\nu(\theta, \phi) \tag{3.3.24}$$

where $J_\nu(\theta, \phi)$ is a function that depends on the scattering characteristics of the medium. Accordingly, equation (3.3.21) can be rearranged to give

$$-\frac{1}{\gamma_\nu} \frac{dI_\nu}{ds} (\theta, \phi) = I_\nu(\theta, \phi) - \frac{\gamma_\nu^A}{\gamma_\nu} B(\nu, T) - \frac{\gamma_\nu^S}{\gamma_\nu} J_\nu(\theta, \phi) \tag{3.3.25}$$

or

$$\frac{dI_\nu}{d\tau} (\theta, \phi) = I_\nu(\theta, \phi) - (1 - \omega)B(\nu, T) - \omega J_\nu(\theta, \phi) \tag{3.3.26}$$

where $d\tau = -\gamma_\nu ds$, τ is the optical thickness, $\gamma = \gamma_\nu^A + \gamma_\nu^S$ and $\omega = \gamma_\nu^S/\gamma_\nu$. The differential equation is then expressed in terms of optical thickness rather than the geometrical path length s.

There are two slightly different ways in which the problem of solving the radiative transfer equation is likely to be approached. We have introduced it in terms of thinking about it as a means to correct satellite-received radiances, or aircraft-received radiances, to determine the Earth-surface-leaving radiance. In this case one must either have independent data on the physical parameters of the atmosphere or one must assume some values for these parameters. Alternatively, the radiative transfer equation may be used in connection with attempts to determine the atmospheric profile or conditions as a function of height. Atmospheric profiles have been determined for many years by radiosondes which are launched at regular intervals by weather stations. However, since radiosonde stations are relatively sparse, use may also be made of sounding instruments flown on various satellites for determining atmospheric profiles. Perhaps the best known of these sounding instruments are the TOVS (TIROS Operational Vertical Sounder) series flown on the TIROS-N, NOAA series of satellites themselves (see section 1.4). Until the split-channel version of the AVHRR was introduced on NOAA-7, data from these sounding instruments was used routinely by NOAA in the application of atmospheric corrections to AVHRR data for the production of sea surface temperature charts. A near-real-time atmospheric correction algorithm for channel-1 and channel-2 AVHRR data has been described by Singh (1991).

CHAPTER FOUR

Earth surface temperatures

Most of this chapter is devoted to the study of the temperature of the surface of the sea; only the last section (section 4.4) is devoted to the temperature of the surface of the land. A useful survey of sources of thermal infrared data from various past Earth-orbiting satellite missions is given by Lynn (1986).

4.1 SATELLITE OCEANOGRAPHY

Satellite remote sensing has revolutionised the gathering of oceanographic data over the last couple of decades (see Cracknell 1981, Robinson 1985, Stewart 1985, Victorov 1996, for example). This revolution concerns, first, the quantity of data that can be gathered. Traditionally oceanographic data have been gathered on an occasional basis by scientific oceanographic cruises and on a regular basis, in certain areas, from commercial and naval vessels. The data obtained have been for very restricted geographical areas and on an irregular temporal basis. Satellites provide regular repetitive coverage of the global oceans and provide large quantities of data that it would be impossible to obtain by conventional methods. There are problems, however, and because of these problems there was, for a while, a great reluctance on the part of the oceanographic community to accept the potential and importance of satellite data; happily this reluctance is now rapidly being overcome. The main problem is that, whereas instruments on a ship or buoy measure marine parameters rather directly, the data obtained from satellites are not usually direct measurements of the physical, chemical or biological parameters that are of interest to the oceanographer; there is a considerable amount of processing and interpretation necessary before one can actually obtain oceanographic parameters from the satellite data.

Oceanographic data from satellites can be obtained from instruments of various sorts (both active and passive) which operate at various wavelength ranges in the electromagnetic spectrum (visible, near-infrared, thermal infrared (all passive) and microwave (both passive and active)). The AVHRR, while being an important spaceborne instrument, is only one of several sources of oceanographic remote sensing data. High spatial resolution (but low temporal resolution) data for estuarine and coastal work can be obtained from systems such as Landsat, SPOT or Meteor, while passive and active microwave data can be obtained from other instruments and systems.

The most successful use of AVHRR data in oceanographic work is in the determination of sea surface temperature using data from the thermal infrared bands. Unlike many of the other successful applications of AVHRR data, which were not originally foreseen by the planners and designers of the system, this use was foreseen. It was really for this purpose that good provision was made for the calibration of the thermal infrared data (see section 2.2). The discussion of the determination of absolute values of sea surface temperatures (in section 4.2) illustrates very well the point made above that satellite-borne instruments do not give direct measurements of oceanographic parameters and that a substantial amount of work has to be done to extract the values of the parameters. However, for some purposes, such as the study of the location of shelf fronts or the use of the temperature as a tracer to study ocean circulation (see section 4.3) the absolute values of the sea surface temperature are not important and an uncalibrated thermal band image or digital data set is perfectly adequate.

4.2 SEA SURFACE TEMPERATURES

4.2.1 Introduction

The traditional method for collecting sea surface temperature data was extremely tedious. It involved scattered point measurements obtained on an irregular basis. The data might come from

(a) data buoys and weather ships

(b) engine-room intake temperatures from ships, either plying regular routes or on irregular trips

(c) systematic – but very infrequent – data obtained from scientific oceanographic cruises.

To build up a synoptic view of the sea surface temperature from such data involves a lot of extrapolation or interpolation and only gives a rough idea of the situation. It produces charts like the Meteorological Office five-day-mean chart shown in Figure 4.1. While this gives a good overall impression it does not reveal fine detail. It does not, for instance, reveal the fact that there may be very sharp fronts in the sea surface temperature, see section 4.3, or eddies on the side of the Gulf Stream, see Figure 4.2 (in colour section). This figure shows the Gulf Stream coming along the East Coast of the USA and one can see very clearly the rings which have broken off on both sides of the Gulf Stream. These are all revealed in very considerable detail in an image which it took the satellite-flown scanner system about 10 minutes to collect. However, the original oceanographic cruise data that had to be collected and put together to demonstrate the existence of these rings a few decades ago took several years of patient cruise work and interpretation of the data to analyse. Even then one could not study these rings and their temporal evolution in the detail that is now possible.

An overall view of the use of thermal infrared scanner data from a satellite for the determination of sea surface temperature (SST) is indicated in Figure 4.3. The left-

Figure 4.1   Meteorological Office five-day-mean sea surface temperature chart.

hand side of Figure 4.3 illustrates the physical processes involved. Radiation is emitted at the surface of the sea; its spectral distribution is governed by the Planck distribution function. To a very good approximation the sea surface can be regarded as a blackbody, i.e. a perfect emitter; if a very high accuracy is required in the retrieved sea surface temperature a more exact value (of 0.98) can be used for the emissivity. The radiation then travels through the entire atmospheric column before it reaches the satellite and during this passage its spectral distribution is significantly changed, see section 3.3. The radiation arrives at the satellite and enters the objective aperture of the AVHRR. A more extensive discussion of these physical

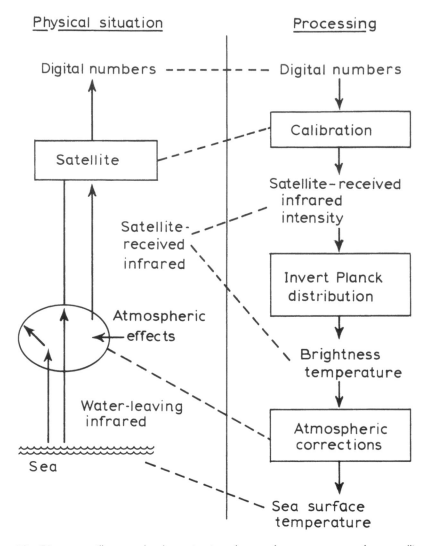

Figure 4.3 Diagram to illustrate the determination of sea surface temperatures from satellite thermal infrared data.

processes is given, for example, by Fiúza (1992). Eventually a signal is generated as 10-bit digital data which is broadcast directly or tape-recorded on board and subsequently transmitted to Earth (see section 2.1). The right-hand side of Figure 4.3 is intended as an outline of the processing operation which seeks to invert the physical processes just mentioned so that the digital data can be processed to retrieve the sea surface temperature on a pixel-by-pixel basis. The main steps involved are

- cloud detection and elimination
- geometrical rectification
- calibration of data (determination of brightness temperature)
- conversion of satellite-received radiance into water-leaving radiance
- conversion of water-leaving radiance into sea surface temperature.

Now the calculation of SSTs is carried out routinely at a number of ground receiving stations around the world to produce charts or maps of SSTs on a regular basis at various spatial scales. An example is shown in Figure 4.4. Alternatively, the data may be provided as digital data, through some computer network or on a suitable medium, for ingestion into numerical weather forecast models or even numerical climate models. An important additional aspect, of course, is that for any routine operational use of SSTs derived from satellite data it is necessary to have some appropriate method for the validation of the results, both in terms of the accuracy of any one set of results and in terms of the reliability of the results from one time to another. We shall see that it is reasonable to expect an accuracy of considerably better than 1 K for SSTs derived from AVHRR data, though weather forecast modellers and climate modellers have a tendency to ask for an accuracy of,

(a)

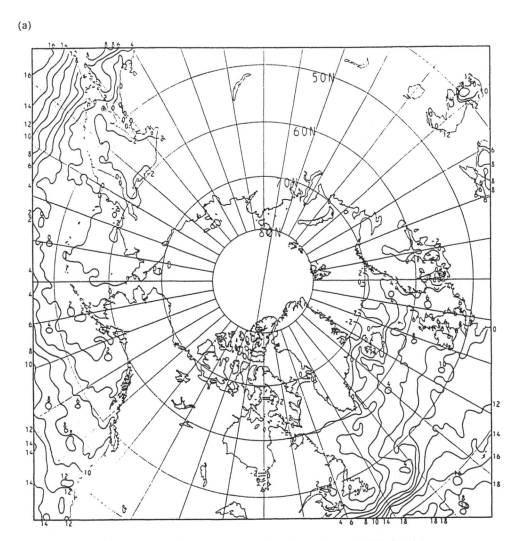

Figure 4.4 Example of a sea surface temperature chart derived from AVHRR data (a) GOSSTCOMP polarstereographic MCSST 1°C grid chart for 5 April 1976;

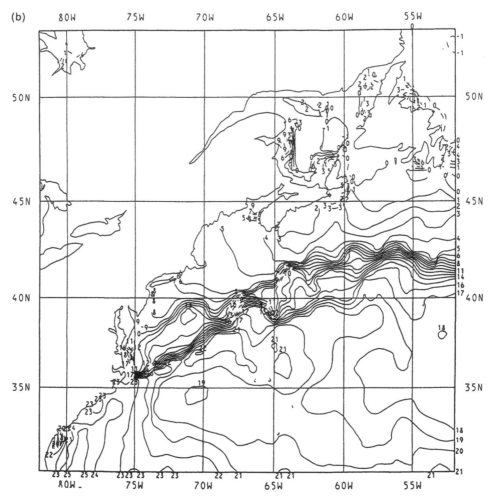

Figure 4.4 (*Continued*) (b) NESS 50 km SST analysis Atlantic Coast 0.5°C grid chart for 27 April 1982 and

say, 0.2 K or 0.3 K which is rather more stringent than one can reasonably expect to obtain from AVHRR data.

The AVHRR is not the only satellite-borne instrument that generates, or has generated, thermal infrared data. One could, however, argue that it is by far the best source of such data for the following reasons:

- The thermal bands of the AVHRR are less noisy than those of some other instruments (e.g. the CZCS (Coastal Zone Colour Scanner) on Nimbus-7).

- The AVHRR is an operational system whereas other systems (e.g. CZCS or HCMM (Heat Capacity Mapping Mission)) were experimental missions which are no longer operating so that their data are only of historic interest.

- The thermal bands on the scanners on the geostationary satellites have much poorer spatial resolution and much poorer calibration than those on the AVHRR. They do, however, have the advantage that they generate data much more frequently than the AVHRR, namely every 30 minutes.

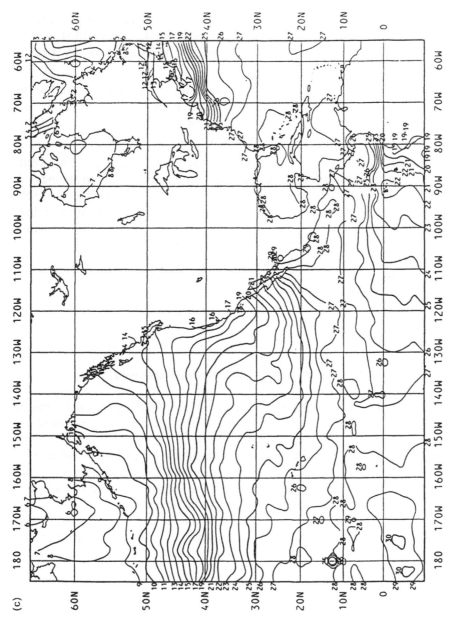

Figure 4.4 (Continued) (c) MCSST 2.5°C grid GOSSTCOMP monthly mean chart for September 1982 (NOAA NESDIS).

- The Landsat TM (Thematic Mapper) has a thermal infrared channel with a much finer spatial resolution than the AVHRR but the coverage is very infrequent and the data are very expensive, so that TM data are not widely used in the oceanographic context; it is, however, useful for the study – on an occasional basis – of some small-scale phenomena in lakes, wide rivers and estuaries and coastal waters.

- The ATSR (Along-Track Scanning Radiometer) flown on ERS-1 is a conical scanner that provides two-look data intended to give SST values of greater accuracy than can be obtained from the AVHRR. However, at least at present, ATSR data are not available so easily, frequently, rapidly or cheaply as AVHRR data and at present are not yet seriously challenging the position of the AVHRR for routine SST determination.

- The AVHRR data can be received in real time with a direct readout station (see section 2.5) and can be processed locally within a very short time to give sea surface temperature maps or charts in near-real time, i.e. within a matter of minutes, rather than hours, of the original generation of the data.

Before considering the details of the processing of AVHRR thermal infrared data to determine sea surface temperatures, there are two further points to be made at this stage. First, it should be recalled that during daylight hours data from channel 3 of the AVHRR contain both reflected solar radiation and infrared radiation emitted from the surface of the sea. At night, of course, there is only the emitted thermal radiation. Thus while channel 3 can be used in sea surface temperature determination at night it cannot be used for data collected during daylight hours. Secondly, sea surface temperatures can also be determined from satellite-flown passive microwave scanners which have been flown on various satellites. However, the spatial resolution of a passive microwave scanner is much poorer (i.e. the instantaneous field of view is much larger) than that of the AVHRR.

4.2.2   Calibration of AVHRR thermal infrared data

The in-flight calibration of channels 3, 4 and 5 of the AVHRR was described in section 2.2.3. The question of the non-linearity of the detectors was considered in section 2.2.4.

It should, perhaps, be stressed that the set of calculations that lead to the value of the brightness temperature, T_B, can be done – and should be done – separately for each of the three thermal infrared channels with the appropriate data in each case. Therefore, we ought to speak of T_{Bi}, the brightness temperature calculated from data from channel i. It also has to be realised that the calculations based on the use of the in-flight calibration data to calculate the slope and intercept have to be performed afresh for each new AVHRR scene that is processed.

4.2.3   Atmospheric corrections

Let us suppose that the data in the thermal infrared channels of the AVHRR have been processed, on a pixel-by-pixel basis, to generate brightness temperatures, T_{Bi}, for each channel of the AVHRR data. The question arises as to whether one should carry out atmospheric corrections to the data or not. The brightness temperatures may differ from the sea surface temperatures by as much as 5°C or more. While it would be naive to regard the correction as a simple fixed offset over the whole scene, it is, nevertheless, reasonable to assume that the value of the correction varies relatively slowly across a scene. Therefore for some purposes, such as for example the location of fronts or the study of the temporal evolution of eddies on the edge of the Gulf Stream, there is no need to carry out atmospheric corrections. In many other situations, however, the actual values of the sea surface temperature, not just relative values, are required and so it is necessary to make corrections to account for the atmospheric effects.

What we have described in section 2.2 has been concerned with the determination, in absolute units, of the energy received at the satellite and its equivalent blackbody temperature or brightness temperature, T_{Bi}. What we actually need is information about the radiation leaving the surface of the Earth; this involves radiative transfer theory, see section 3.3.

There are several different approaches that one can take to the question of applying atmospheric corrections to thermal infrared scanner data from Earth-observing satellites for the determination of sea surface temperatures, i.e. for the determination of the sea surface temperature, T_S, from the brightness temperatures, T_{Bi}. These include the following.

- Calibration with *in situ* measurements.

- The use of computer programs for numerical solution of the radiative transfer equation. Such programs require various parameters of the atmosphere to be specified. It is not adequate to use some seasonal and regional average of the atmospheric conditions; for accurate retrieval of sea surface temperatures it is necessary to make use of the values of the atmospheric parameters that correspond to the actual atmospheric path of the infrared radiation between the surface of the Earth and the satellite. In practice this method involves the acquisition of sophisticated computer packages, the acquisition of relevant atmospheric data and the consumption of very substantial amounts of computing resources for the processing of each set of data.

- One can attempt to eliminate the atmospheric effects by using two or more simultaneous, or almost simultaneous, measurements of the brightness temperature for a given sea surface area.

The selection of the appropriate option will be governed by considerations both of the sensor that is being used to gather the data and of the problem to which the data are being applied. We shall consider these options in turn.

(i) One can make empirical corrections to the data with the results of some simultaneous *in situ* sea surface temperature measurements. These *in situ* measurements may be obtained for a training area or at a number of isolated points in the scene. It is important that the measurements are actually made simultaneously with the gathering of the data by the satellite. This is not a good method for general use, although it is capable of yielding quite accurate results on a very local basis. In terms of processing data for the global oceans on a regular basis one cannot expect to obtain a regular supply of simultaneous *in situ* data at a large number of points for such calibration purposes. In addition to the problems associated with variations in the atmospheric conditions from day to day, there is also the quite serious problem that there are likely to be significant variations in the atmospheric conditions even within a given scene at any one time. To allow for the variations that may exist between atmospheric conditions in different parts of a given scene it would be necessary to have available *in situ* calibration data for a much finer network of closely packed points than would be at all feasible. While it is, of course, necessary to have some *in situ* data available for initial validation checks and for occasional monitoring thereafter on results derived from satellite data, to use a large network of *in situ* calibration data largely negates the value of using remote sensing data anyway, since one important objective of using remote sensing data was largely to eliminate costly field work.

(ii) One can use a model atmosphere; this involves calculating atmospheric corrections to the satellite data based on the physical principles that are known to be involved and using some kind of model for them based on solution of the radiative transfer equation, see section 3.3 (Agarwal and Ashajayanthi 1982, Price 1983, Li and McDonnell 1988, and Saunders 1989). Computer programs exist (LOWTRAN, etc.) to enable one to carry out these calculations. However, one needs data on atmospheric parameters, namely the pressure, relative humidity and temperature as functions of height. For such data one could use an atmospheric profile that was based on a model adjusted according to the geographical location and the time of year. However, atmospheric conditions vary quite rapidly, both spatially and temporally (see e.g. Callison and Cracknell 1984) with the result that any model calculation really needs to be carried out on a quite closely spaced network of points (if not on a pixel-by-pixel basis) and to make use of values of atmospheric parameters that are simultaneous and spatially coincident with the acquisition of the AVHRR thermal infrared data that one is using. The use of a model atmosphere based on the geographical location and the time of the year is more likely to be successful if one is dealing with an instrument with low spatial resolution that is gathering data over wide areas for an application that involves taking a global view of the surface of the Earth. In this situation the local spatial irregularities and rapid temporal variations in the atmosphere are likely to cancel out and fairly reliable results may be obtained. This approach is also likely to be relatively successful for situations in which the magnitude of the atmospheric correction is relatively small compared with the signal from the target area that is being observed. All these conditions are satisfied for passive microwave radiometry and so this approach is moderately successful for SMMR data; they are, however, unlikely to be satisfied for the thermal infrared wavelengths at which the AVHRR operates. As an improvement one could make use of such simultaneous meteorological data (atmospheric profiles) as may actually be available instead of using only assumed values based on geographical location and time of year. These simultaneous data may be obtained from radiosondes, or they could come from satellite-flown sounding instruments such as the TOVS. The advantage of using the TOVS data, when making atmospheric corrections to AVHRR data, is that since both instrument systems are flown on the same satellite, the atmospheric data are coincident both in space and in time with the AVHRR data that one is trying to correct. The available radiosonde data will have been collected at a different time from the AVHRR data and the spatial distribution of radiosonde data is very non-uniform over the Earth's surface.

For a period TOVS data was used by NOAA in its production of SST charts from AVHRR data. However, such calculations consume a large amount of computer time which was not justified by the results. Moreover, the results proved to be less accurate than those which can be obtained using a multichannel approach which was possible once some of the five-channel versions of the AVHRR were in operation.

(iii) Finally, rather than try to calculate the atmospheric correction with a model (such as LOWTRAN) plus some atmospheric data, one can try to eliminate the atmospheric effects. There are various ways of trying to do this. Basically they fall into two categories:

- the multiple-look or the two-look method
- the two-channel method or the multichannel method.

In the first of these methods one tries to eliminate the effects of the atmosphere by a multilook approach in which a given target area on the surface of the sea is viewed from two different directions. Attempts have been made to do this using data from two different satellites, one being a geostationary satellite, Meteosat, and the other being a polar-orbiting satellite of the TIROS-N/NOAA series (Chedin *et al.* 1982, Holyer 1984). There are, however, a number of reasons why this method, though interesting in concept, is unlikely to prove to be particularly accurate in practice. These include (a) problems of the registration of one data set to the other, given the differences of the geometries of the systems and differences in the spatial resolution, (b) the lack of in-flight calibration for the geostationary satellite data, (c) the non-simultaneity of the data acquisition by the geostationary and polar-orbiting satellites, (d) differences between the spectral characteristics of the two radiometers and (e) systematic errors in the individual radiometers. Holyer studied satellite data from a day on which *in situ* data were available and differences between the calculated sea surface temperature and the *in situ* sea surface temperature data varied from about $+0.9$ to -1.4 degK.

The multilook idea has now been developed on the Along Track Scanning Radiometer, ATSR, which is being flown on ERS-1 (Prata *et al.* 1989, Sobrino *et al.* 1993) This instrument was designed with a conical scanning system. This instrument, by looking both forward and vertically downwards, views a given target area on the sea surface twice in rapid succession through two different atmospheric paths. On the Earth's surface these two scans are both curved and are separated by about 900 km on the subsatellite track. The early results from this instrument are encouraging in terms of accuracy.

We shall now outline the approach to the two-look situation given by Singh (1984). It is supposed that the atmospheric transmittance, in the view direction θ, is $t(\kappa, \theta)$ where κ is the wave number characterising the spectral channel in question. The radiance reaching the satellite, $B(\kappa, T_{Bi})$, is then related to the radiance leaving the surface of the sea, $B(\kappa, T_S)$, by

$$B(\kappa, T_{Bi}) = t(\kappa, \theta)B(\kappa, T_S) . \qquad (4.2.1)$$

It is further supposed that the transmittance $t(\kappa, \theta)$ for a viewing angle θ can be written as

$$t(\kappa, \theta) = \exp(-\tau(\kappa)/\cos \theta) \qquad (4.2.2)$$

where $\tau(\kappa)$ is the effective optical path length for nadir viewing. If we write the Planck distribution function as $c_1\kappa^3/[\exp(c_2\kappa/T) - 1]$ and assume that $\exp(c_2\kappa/T) \gg 1$, which is valid for the values of κ with which we are concerned in the thermal infrared, then equation (4.2.1) becomes

$$\frac{c_2 \kappa}{T_{Bi}} = \frac{c_2 \kappa}{T_S} + \frac{\tau(\kappa)}{\cos \theta} . \tag{4.2.3}$$

Suppose that θ_1 and θ_2 are the two viewing angles used for viewing the same sea surface area, then T_S must be the same for these two directions and we have two equations

$$\frac{c_2 \kappa}{T_{Bi}(\theta_1)} = \frac{c_2 \kappa}{T_S} + \frac{\tau(\kappa)}{\cos \theta_1} \tag{4.2.4}$$

$$\frac{c_2 \kappa}{T_{Bi}(\theta_2)} = \frac{c_2 \kappa}{T_S} + \frac{\tau(\kappa)}{\cos \theta_2} \tag{4.2.5}$$

where $T_{Bi}(\theta_1)$ and $T_{Bi}(\theta_2)$ are the brightness temperatures obtained from the two sets of AVHRR data obtained for the two look directions θ_1 and θ_2. Equations (4.2.4) and (4.2.5) contain two unknowns, T_S and $\tau(\kappa)$.

The two-look approach described by Singh (1984) has been applied to the study of atmospheric corrections to satellite-derived sea surface temperatures by Al-Taee *et al.* (1993) using two AVHRR scenes from successive orbits of the NOAA-7 satellite. They used data from two scenes in the western Mediterranean Sea from 1227 GMT on 8 July 1982, orbit number 5364, and 1406 GMT on 8 July 1982, orbit number 5365. For a set of 10 000 pixels in the overlap area of the two scenes the in-flight calibration data were used to calculate the brightness temperatures $T_{Bi}(\theta_1)$ and $T_{Bi}(\theta_2)$ from the two scenes (for which the sea surface was observed at off-nadir angles θ_1 and θ_2, respectively, in the two successive scenes). These brightness temperatures were corrected for the emissivity of sea water being not exactly equal to unity to give corrected brightness temperatures $T_W(\theta_1)$ and $T_W(\theta_2)$ for each pixel. The two equations (4.2.4) and (4.2.5) were solved on a pixel-by-pixel basis, to determine a value of T_S for each pixel throughout the study area. If we consider the effect of the viewing angle on the atmospheric attenuation then we would expect the difference between T_S and T_{Bi} to increase with increasing angle and this was indeed observed (see also Antoine *et al.* 1992). However, the standard deviation in the values of T_S was larger than those obtained from multichannel calculations.

The multichannel approach, or split-window approach as it is sometimes called, originated in the mid-1970s, has been developed by various workers since then, and is at present widely used operationally for the determination of sea surface temperatures from AVHRR data, especially since AVHRRs have been flown with the split channel in the 10–12 µm range. One can regard the method as being simply an empirical method in which the sea surface temperature is expressed as a linear combination of the brightness temperatures in the individual channels of the AVHRR. However, the method rests on a fairly sound basis; it works because the effect of the atmosphere on infrared radiation is different for radiation in the different wavelength channels of the scanner. One can start with the radiative transfer equation and, making certain assumptions, one can derive a relation of the form

$$T_S = a_3 T_{B3} + a_4 T_{B4} + a_5 T_{B5} + a_6 \tag{4.2.6}$$

and obtain expressions for the coefficients a_3, a_4, a_5 and a_6 in terms of the parameters obtained from an atmospheric profile (Anding and Kauth 1972, Maul and Sidran 1972, McMillin 1975, Prabhakara *et al.* 1975, Deschamps and Phulpin 1980,

Sidran 1980, Becker 1982, 1987, Maul 1983, Price 1983, McMillin and Crosby 1984, Singh 1984, McClain *et al.* 1985, Eyre 1986, Ho *et al.* 1986, Singh *et al.* 1986, Malkevich and Gorodetsky 1988, Becker and Li 1990, Harris and Mason 1992). It is possible to obtain formal expressions for the coefficients in equation (4.2.6) in terms of the atmospheric parameters (see for example equation (20) of Becker 1987). However, it is not practical to use the values of the atmospheric parameters to calculate the values of these coefficients. First, the results are not particularly accurate and secondly the processing would be very costly in terms of the use of computing resources. It suffers from the problems we have already noted in relation to the use of atmospheric models. In practice what is done is to regard equation (4.2.6) as an empirical equation and determine the values of the coefficients a_3, a_4, a_5 and a_6 by calculating the brightness temperatures T_{B3}, T_{B4} and T_{B5} separately from the data in channels 3, 4 and 5, respectively, and fitting to some *in situ* data obtained from ships or buoys. For day-time data, although we speak of multichannel data from the AVHRR, we are only really concerned with two-channel data; this is because channel 3 of the AVHRR contains both emitted infrared radiation and reflected solar infrared radiation. At night channel 3 data can be used so that the multichannel approach can become a three-channel approach at night. There is an extensive literature on multichannel atmospheric correction methods for sea surface temperatures derived from AVHRR data (see for instance Robinson *et al.* 1984, McClain *et al.* 1985, McClain 1989 and references therein). We have seen in sections 2.4.2 and 4.2.1 that sea surface temperature maps are routinely generated from AVHRR data by NOAA using their multichannel sea surface temperature (MCSST) algorithm. The coefficients introduced by NOAA/NESS for use with data from NOAA-7 (the first of the five-channel AVHRRs) in 1982 were as follows, where T_S is the sea surface temperature (in °C):

Day-time (bands 4 and 5)

$$T_S = 1.0351 T_{B4} + 3.046(T_{B4} - T_{B5}) - 283.93 \tag{4.2.7}$$

Night-time

Split window (bands 4 and 5)

$$T_S = 1.0527 T_{B4} + 2.6272(T_{B4} - T_{B5}) - 288.23 \tag{4.2.8}$$

Triple window (bands 3, 4 and 5)

$$T_S = 1.0239 T_{B4} + 0.9936(T_{B3} - T_{B5}) - 278.46 \tag{4.2.9}$$

Dual window (band 3 and 4)

$$T_S = 1.0063 T_{B4} + 1.4544(T_{B3} - T_{B4}) - 278.47 \ . \tag{4.2.10}$$

Tests with 94 drifting-buoy temperatures have indicated accuracies as follows:

	Bias (degK)	Scatter (degK)	RMSD (degK)
Split window	0.06	0.60	0.61
Triple window	0.14	0.64	0.66
Dual window	0.16	0.72	0.73

A revised set of coefficients and much more comprehensive comparisons between MCSST and *in situ* data is given by McClain *et al.* (1985). Thus equations (4.2.7)–(4.2.10) become replaced by

$$T_S = 1.0346 T_{B4} + 2.5779(T_{B4} - T_{B5}) - 283.21 \qquad (4.2.11)$$

for day-time and

$$T_S = 3.6139 T_{B4} - 2.5789 T_{B5} - 283.18 \qquad (4.2.12)$$

$$T_S = 1.0170 T_{B4} + 0.9694(T_{B3} - T_{B5}) - 276.58 \qquad (4.2.13)$$

$$T_S = 1.5018 T_{B3} - 0.4930 T_{B4} - 273.34 \qquad (4.2.14)$$

respectively, for NOAA-7. Yet another different set of coefficients in these equations is given for NOAA-9 data by McClain *et al.* (1985).

The multichannel method has the advantage of being very economical of computing resources so that a pixel-by-pixel calculation can be carried out very easily and quickly. Sea surface temperature products are now routinely generated using it by NOAA NESDIS (see section 2.4) and by various other ground stations around the world. For example, the operational production by the Centre de Météorologie Spatiale at Lannion in France is described by Castagne *et al.* (1986) and Antoine *et al.* (1991) and a product from the National Remote Sensing Agency at Hyderabad in India is shown in Figure 4.5 (in colour section). Many other ground receiving stations all over the world now produce SST products for their own local areas. It would not be very easy or useful to produce a complete list of available products; a reader who is interested in SST products for a given geographical area should make contact with a ground station near to that area (see the list in the Appendix). A novel extension of the split-window method, based on the concept of spatially varying surface temperature beneath a constant atmosphere, has been described by Harris and Mason (1991) and tested in an initial study of the Great Lakes.

The details of pre-processing, cloud screening, atmospheric corrections and the procedures for producing the output products for NOAA's operationally produced multichannel sea surface temperatures (MCSSTs) are given by McClain *et al.* (1985). According to McClain (1989) the basic algorithms have been little changed since operational processing began in 1981. The coefficients of the atmospheric correction algorithms are tuned shortly after the launch of each satellite using a large and geographically diverse set of close match-ups (within 25 km and 6 h) between MCSSTs and measurements at 1 m depth from drifting buoys, and the coefficients are rarely changed thereafter until the next satellite. Although these buoys are not uniformly distributed across the world's oceans, in fact being mostly in the southern hemisphere, the drifting buoys are located in widely varying water masses and atmospheres and the basic idea of the split-window technique is to 'cancel out' the effect of the local atmosphere through its having different effects in the different spectral channels. Thus the coefficients determined in this way are then taken to be fixed for worldwide application. NOAA NESDIS has now accumulated several years of month-by-month match-up statistics over the globe. Figure 4.6 is a scatter plot of satellite buoy data for November 1987. Match-ups peaked at over 500 per month in 1986.

The retrieval of accurate values of sea surface temperatures from the AVHRR thermal infrared data can be quite adversely affected by high tropospheric aerosol concentrations arising from severe volcanic eruptions, desert dust storms or the

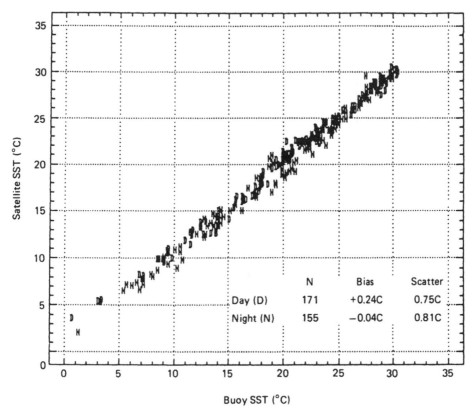

Figure 4.6 Global match-ups between NESDIS operational satellite MCSSTs and SSTs measured at depths of 1 m from drifting buoys in November 1987 (McClain 1989).

Kuwait oil fires that were a sequel to the Gulf War. For example, the eruptions of the El Chichón volcano in Mexico in April 1982 resulted in very large volumes of sulphuric acid droplets (aerosol) being injected into the stratosphere and distributed initially into a relatively narrow zonal band around the world. Eventually it spread in a much less dense aerosol cloud over much of the remainder of the globe. This volcanic aerosol layer had a significant effect on the retrieval of MCSSTs for several months and was detectable throughout the remainder of 1982 (Walton 1985, McClain 1989, Rao 1992), see Figure 4.7. The effects of some other recent volcanic eruptions are discussed in section 7.5.2.

Aerosol contamination greatly alters the normal atmospheric spectral absorption characteristics, which in turn increases the attenuation of the infrared signal reaching the satellite. The attenuation degrades the accuracy of MCSST retrievals and also interferes with cloud screening tests that depend on visible band measurements. Aerosol types differ in chemical composition, size distribution and normal atmospheric locale. Volcanic aerosols typically consist of sulphuric acid particles formed from sulphur gases that are injected into the stratosphere during large volcanic eruptions. Although this contamination is irregular, its effects on satellite-derived sea surface temperature accuracy can be quite widespread for several months. A night-time MCSST algorithm (Walton 1985) and a day-time correction algorithm

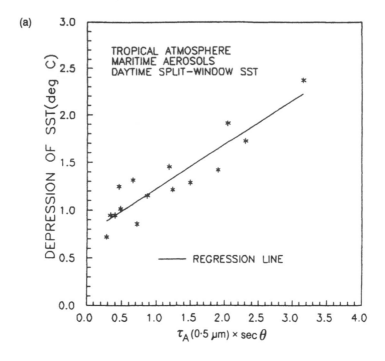

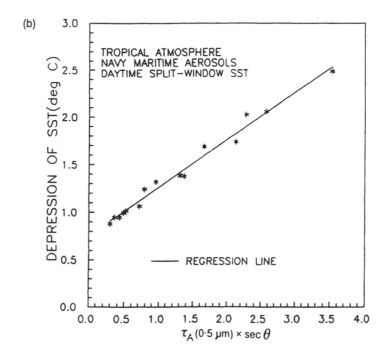

Figure 4.7   Variation of the day-time split-window sea surface temperature with the line-of-sight aerosol optical thickness with (a) maritime aerosols and (b) navy maritime aerosols (Rao 1992).

(Griggs 1985) have been developed for volcanic aerosols. Saharan dust aerosols generally contaminate the low-to-middle troposphere with silicate particles (Carlson and Prospero 1972, Prospero and Carlson 1972, Carlson 1979). This contamination is more regular than volcanic aerosol contamination, affecting satellite SST accuracy in ocean areas located downwind from large desert regions. Algorithms to perform corrections for the effects of Saharan dust aerosol contamination in the determination of sea surface temperatures from AVHRR thermal infrared data have been developed by May *et al.* (1992). The MCSST, multichannel sea surface temperature, technique which we have described above, has recently been improved by incorporating the cross-product SST (CPSST) technique (Walton 1988). This new algorithm no longer follows the MCSST assumption that attenuation due to atmospheric water vapour is a linear function of channel brightness temperature difference. Rather, the CPSST technique assumes the function to be non-linear. For NOAA-7 data this took the form

$$T_S^C = \frac{0.1761\,T_{B5} - 47.56}{0.1761\,T_{B5} - 0.117\,T_{B4} - 15.72} + (T_{B4} + 0.2 - T_{B5}) + T_{B5}\,. \tag{4.2.15}$$

A modified version of this was introduced by NOAA in 1990 for NOAA-11 data, namely

$$T_S^C = \frac{a_0\,T_{B5} - a_1}{a_2\,T_{B5} - a_3\,T_{B4} - a_4}\,(T_{B4} - T_{B5} + a_5)$$

$$+ a_6\,T_{B5} + a_7(T_{B4} - T_{B5})(\sec\theta - 1) - a_8 \tag{4.2.16}$$

where different sets of values of the coefficients $a_0, \ldots, a_8$ are used for day-time and night-time data, see Table 4.1 (Wick *et al.* 1992). The CPSST technique has shown improvement over the MCSST in warm, moist tropical atmospheres and in cold, dry polar atmospheres. In the method developed by May *et al.* (1992) the satellite-derived sea surface temperatures are matched with coincident satellite optical depth measurements and *in situ* sea surface temperature data from drifting buoys. The optical depth values, τ, are determined from the channel-1 AVHRR data (see section 7.5).

The regression relations presented by May *et al.* (1992) took two forms. First, a correction algorithm was derived that predicts the difference in temperature (DT)

Table 4.1   CPSST coefficients (Wick *et al.* 1992)

	Day	Night
a_0	0.190 69	0.195 96
a_1 (°C)	49.16	48.61
a_2	0.205 24	0.205 24
a_3	0.173 34	0.173 34
a_4 (°C)	6.78	6.11
a_5 (°C)	0.789	1.46
a_6	0.929 12	0.954 76
a_7	0.81	0.98
a_8 (°C)	254.18	263.84

between CPSSTs and buoy SSTs as a function of τ. The SST was then estimated by adding DT to the original CPSST retrieval. The second approach was to derive an SST algorithm that incorporates an optical depth term directly into the CPSST algorithm. This approach results in a retrieval that is corrected for both water vapour and aerosol contamination in one step. This algorithm is called AERSST to indicate aerosol SST. Each algorithm was then applied to the independent data set to validate its accuracy under Saharan dust conditions. The derived algorithms are presented in Table 4.2. The performance of these algorithms was analysed by May *et al.* (1992). While the performance was good, it should be noted that the study was restricted to Saharan dust over the North Atlantic Ocean. The algorithm may need modification for other sea areas affected by dust from other sources, for example the Arabian Sea.

Apart from the work by NOAA, comparisons between sea surface temperatures derived from AVHRR data and *in situ* sea surface temperature measurements, for various geographical areas, have been described by Legeckis *et al.* (1980), McConaghy (1980), Huh and DiRosa (1981), McKenzie and Nisbet (1982), Pathak (1982), Llewellyn-Jones *et al.* (1984), Ho *et al.* (1986), Gastellu-Etchegorry and Boely (1988), Minnett and Saunders (1989), Pearce *et al.* (1989), Robinson and Ward (1989), Viehoff (1989), Desjardins *et al.* (1990), Yokoyama and Tanba (1991) and França and Cracknell (1994). The use of *in situ* sea surface temperature measurements, both for determining the coefficients in a multichannel (split-window) algorithm and for the validation of satellite-derived sea surface temperatures is a widespread and important activity. In practice, the desirable constraints of exact spatial and temporal coincidence often have to be relaxed, and the limits defining acceptable near-coincidence appear in many cases to be arbitrary. With the current accuracy of AVHRR, and the improvement associated with the ATSR on ERS-1, the correct choice of these limits, or an estimate of the consequent errors, is increasingly important. Minnett (1989) made a study of this question. Data from the AVHRR, a free-drifting meteorological buoy and a research ship were examined to determine the acceptable limits of spatial displacement and elapsed time. While the acceptable time lapse proved to be difficult to quantify, displacement of ~ 20 km seemed likely to contribute no more than ± 0.2 K to the validation error budget.

It will be recalled that certain of the satellites in the TIROS-N/NOAA-6, -7, ..., etc. series carried the first version of the AVHRR which was only a four-channel version of the instrument. For data from the instruments on these satellites there is thus only one thermal infrared channel of data. The split-window, or multichannel, sea surface temperature algorithms thus cannot be used with day-time data from these satellites. (Night-time data can, of course, be used with channels 3 and 4.) A technique was, however, proposed by Singh *et al.* (1985). This method arose because

Table 4.2   Satellite SST and aerosol correction algorithms (May *et al.* 1992)

Algorithm	Formula
DT1	$5.1578\tau \sec \theta - 0.3003$
DT2	$2.9355(R_1/R_2)\tau \sec \theta - 0.4605$
AERSST1	$0.9391 T_{B4} + 2.27(T_{B4} - T_{B5}) + 4.7665\tau \sec \theta + 1.44(\sec \theta - 1) - 255.66$
AERSST2	$0.9565 T_{B4} + 2.25(T_{B4} - T_{B5}) + 2.6812(R_1/R_2)\tau \sec \theta + 1.41 (\sec \theta - 1) - 260.86$

Singh *et al.* noticed that with data from the five-channel version (AVHRR/2) the sea-surface brightness temperatures T_{B4} in channel 4 and T_{B5} in channel 5 are linearly related to a good degree of accuracy, i.e. $T_{B5} = \alpha + \beta T_{B4}$. Using AVHRR/2 data for various dates and from different parts of the world's oceans, the values of the parameters α and β were determined. It was therefore suggested that this linear relation could be used for simulating T_{B5} for those cases for which only T_{B4} is available (i.e. for the AVHRR on TIROS-N, NOAA-6, NOAA-9, and NOAA-12). It was then further suggested that the brightness temperature T_{B4} and this simulated brightness temperature T_{B5}, obtained as just described, could be used in a split-window algorithm for estimating atmospherically corrected sea surface temperatures from the channel-4 data alone. This is based on an assumption that the channel 4s of the various four-channel and five-channel AVHRR instruments are sufficiently close to being identical that the values of α and β determined from data from the five-channel AVHRR can actually be used with channel-4 data from the four-channel instrument.

This technique was used by Singh *et al.* (1985) with two split-window algorithms for correcting the data from channel 4 of the AVHRR instrument on the TIROS-N satellite obtained off southwestern Portugal. The resulting SSTs for twelve dates from 15 June 1979 to 14 June 1980 were compared with sea surface temperatures which were obtained with airborne radiometer data obtained on the same dates and good agreement was generally obtained. However, the method was demonstrated by Minnett (1986) to be less accurate than the use of real channel-5 data (where available). The basis of the method has been further discussed by Eyre (1986) and Singh *et al.* (1986). Essentially the conclusion was that the method of Singh *et al.* (1985) can be regarded as using the channel-4 data in a non-linear sea surface temperature algorithm. Though interesting, that discussion is largely academic now since the last remaining four-channel AVHRR is at present only on a back-up spacecraft and is likely to disappear from that role very soon. The use of four-channel AVHRR data for sea surface temperature studies is thus only of interest now in certain cases of using historical data.

Although MCSST bias and scatter statistics relative to the data from drifting buoys and moored buoys have tended to be reasonably stable for the overall period and these measures of accuracy have been good enough to encourage the use of MCSSTs for many purposes, the question of regional representativeness of the global statistics has been raised. In particular, there are problems in tropical regions where the atmospheric humidity is often very high; some recent work in waters off the northeastern coast of Brazil is described by França and Cracknell (1994).

La Violette and Holyer (1988) observed that if the output from a multichannel sea surface temperature algorithm is represented as an image it is of poorer quality, in terms of increased noise levels and reduced sea surface temperature gradients, than images obtained simply from the original channel-4 or channel-5 data; possible reasons for this have been discussed by Barton (1989).

In all our discussion of the split-window and two-look approaches we have been concerned with trying to eliminate the effect of the atmosphere, and primarily the effect of the water vapour content of the atmosphere. However, another possible use of the data is to determine the water vapour content of the atmospheric column from the thermal infrared AVHRR data. This has been considered by Dalu (1986) and using both AVHRR and HIRS data by Schluessel (1989) (see also equation (7.9.12)).

4.2.4   Accuracy

It is important to consider the question of the accuracy of satellite-derived SSTs. There are various possible sources of error. What has been described so far is the calculation of the equivalent blackbody radiation temperature for the Earth-leaving radiance. We ought also to take into account various other things:

- the emissivity of sea water
- point measurements versus pixel averages
- the skin effect and the diurnal thermocline
- cloud cover
- sunglint.

We shall consider these in a little more detail in turn.

(i)   *Emissivity*   If one is determining the values of the coefficients in an empirical split-window formula by using *in situ* temperature data then the emissivity will be accounted for in the values of the coefficients obtained. However, if one has applied the radiative transfer equation and essentially calculated the water-leaving radiance, or the brightness temperature of the water, then one needs to take account of the true value of the emissivity of sea water. In some work on the determination of sea surface temperatures from thermal infrared data acquired by the AVHRR the value of the emissivity of the surface of the sea is taken to be 1. For many purposes this is adequate. However, this is not quite correct and the emissivity of sea water is close to 0.98 rather than being exactly equal to 1. To use a value of $\varepsilon = 1$ is to introduce an unnecessary error into the conversion of the water-leaving radiance into a water surface temperature. The correction of this error has been considered by Singh (1984) and Dalu (1985). The error may vary from about 0.5 to 1.4 degK. The dependence of the emissivity on the angle of observation and on the near-surface windspeed has been studied by Masuda *et al.* (1988b) and the effect of this in terms of corrections to satellite-derived SST values has been studied by Harris *et al.* (1994). The effect, however, is small for viewing angles below 50°. Harris *et al.* were concerned with the Along Track Scanning Radiometer (ATSR) and not the AVHRR.

One can consider the question of how necessary it is to take specific action to take into account the departure of ε from a value of unity. The answer is that if one is using, as NOAA does, a multichannel algorithm in which the coefficients have been determined by a least squares fit to *in situ* data then there is no need to make a specific correction for this effect. It is already taken into account by the fact that one is fitting measured surface temperatures to brightness temperatures. However, if one is using atmospheric data and software (for instance LOWTRAN) which is based on the solution of the radiative transfer equation, then one is converting satellite-received radiance into water-leaving radiance. When, at the final stage, one converts the water-leaving radiance into sea surface temperature then it is necessary to take account of the true value of ε, otherwise a significant error is likely to be introduced.

(ii)   *Point measurements versus pixel averages*   The AVHRR is gathering radiation from a large area. In reconstructing an image we think of the field of view as being a square of side about 1.1 km. We have noted in section 3.1.7 that this is not an accurate description of the situation; however, for present purposes it will suffice and we shall just regard the instantaneous field of view as an area of about 1 km². Thus any temperature calculated from the AVHRR data will represent an average temperature over this area. An attempt to measure the temperature by some field data collection method with a ship or buoy will give a value which is essentially a point measurement. If there are variations in the surface temperature throughout the field of view then there will usually be a real difference between a value from a point measurement and a value averaged over the field of view. In other words in comparing satellite-derived SSTs with values obtained from *in situ* values from ships or buoys one is not comparing like with like. One way around this problem is to make a two-stage comparison by using simultaneous airborne data as an intermediary. That is, one can use *in situ* data to calibrate airborne scanner data, with a field of view of a few metres by a few metres, and then use the airborne data to calibrate the satellite data. This has been done, for instance, by Singh *et al.* (1985) and Minnett and Saunders (1989).

(iii)   Skin effect and diurnal thermocline   This is a point which has attracted considerable attention. The problem is that, as seen by a radiometer, it is the temperature of an extremely thin layer on the surface of the sea that is being observed. This temperature is likely to be different by a significant amount from the temperature not far below the surface and it is this difference which is called the skin effect. Moreover, the surface temperature that is measured by a radiometer is not the temperature that is usually measured by an oceanographer, nor indeed is it necessarily the temperature that is actually of interest to an oceanographer. Most *in situ* measurements, using bucket, engine-room temperature of intake water, thermistor chain on buoy, etc., do not measure the temperature of this surface film.

Most oceanographers are interested in SST more representative of the upper few metres of the ocean, commonly referred to as the bulk SST. At present most satellite SST retrieval methods fail to recognise the importance of the difference between skin and bulk temperature. The most common method of correcting or calibrating satellite SSTs is to use bulk temperatures measured from drifting buoys (McClain 1981, 1989, McClain *et al.* 1985) which typically have their sensors mounted between 0.5 and 1.0 m below the surface. This depth varies as the buoy bounces in the turbulent environment of the sea surface. The earlier reviews and textbooks of Robinson *et al.* (1984), Robinson (1985) and Stewart (1985) give some discussion of the existence and importance of this problem and an early attempt to measure the surface temperature *in situ* was described by Constans *et al.* (1984). An experimental investigation of the skin effect is described by Hepplewhite (1989). This involved using measurements from an infrared radiometer mounted on a ship and from a conventional rubber bucket with a mercury-in-glass thermometer. Data were collected over an area from the tropical Atlantic to Antarctica. Values of a skin effect in the range −0.3 K to +1.2 K were found. Root-mean-square differences between buoy and satellite SST range from 0.5 to 1.5 K and are quoted by McClain *et al.* (1985).

The difference between skin and bulk temperatures contributes an added level of uncertainty to the satellite retrieved SST calibrated in this manner. Calibration of AVHRR data with buoy data would eliminate the skin effect if the magnitude of the skin effect were to be constant; however, that is extremely unlikely to be so. Skin temperature estimates (Schluessel *et al.* 1987) from the satellite data are still subject to atmospheric contamination and noise in the measurements but do not require correction for the bulk–skin temperature difference. Differences (ΔT) between the bulk temperature measured more than a metre below the surface and the surface skin temperature are of the order of 1 K. While this temperature difference appears to be quite small, it is significant in terms of its contribution to the determination of sea surface temperatures from infrared satellite data for studies of the global climate. If we are ever to develop satellite remote sensing techniques to attain the 0.2 K SST accuracy required for climate studies, we must account for the bulk–skin temperature difference in the calibration of the satellite infrared data. Once correctly calibrated as surface skin temperature, we can then investigate the relation between satellite-sensed skin temperature and the ocean bulk SST to determine the suite of transforms needed to relate these fundamentally different measures of SST.

Schluessel *et al.* (1987) proposed algorithms intended to determine the skin SST, one of which involved AVHRR data and one of which involved both AVHRR data and HIRS data. The AVHRR-only algorithm has the same form as the MCSST algorithms

$$T_{\text{S}} = a_0 + a_1 T_{\text{B4}} + a_2 T_{\text{B5}} \tag{4.2.17}$$

but where (i) the values of the coefficients depend on the scan angle (see Table 4.3(a)) and (ii) there is no difference between the day-time and night-time values of the coefficients. The algorithm involving both AVHRR and HIRS data can be written as

$$T_{\text{S}}^{\text{s}} = a_0 + a_1 T_{\text{B4}} + a_2 T_{\text{B5}} + a_3 T_{\text{H11}} + a_4 r \tag{4.2.18}$$

where

$$r = \frac{T_{\text{H8}} - T_{\text{H6}}}{T_{\text{H8}} - T_{\text{H7}}} \tag{4.2.19}$$

Table 4.3(a) AVHRR-only   SMSST   coefficients (Wick *et al.* 1992)

Scan angle	a_0 (°C)	a_1	a_2
$\theta < 5°$	-1.25	4.013	-3.016
$5° < \theta < 15°$	-1.26	4.039	-3.043
$15° < \theta < 25°$	-1.30	4.102	-3.105
$25° < \theta < 35°$	-1.38	4.219	-3.221
$35° < \theta < 45°$	-1.48	4.243	-3.241
$45° < \theta$	-1.84	4.512	-3.497

Table 4.3(b)   AVHRR/HIRS SMSST coefficients (Wick *et al.* 1992)

Scan angle	a_0 (°C)	a_1	a_2	a_3	a_4 (°C)
$\theta < 5°$	3.25	4.472	-3.567	0.117	-1.594
$5° < \theta < 15°$	3.37	4.504	-3.601	0.120	-1.634
$15° < \theta < 25°$	3.58	4.587	-3.690	0.127	-1.682
$25° < \theta < 35°$	3.92	4.753	-3.867	0.143	-1.736
$35° < \theta < 45°$	2.93	4.791	-3.893	0.133	-1.266
$45° < \theta$	0.55	5.215	-4.312	0.142	0.0

and T_{H6}, T_{H7}, T_{H8} and T_{H11} are the brightness temperatures calculated from HIRS data in channels 6, 7, 8 and 11, respectively. The values of the coefficients a_1, a_2, a_3, a_4 are given in Table 4.3(b). The temperature calculated in this way was called the satellite-measured surface-skin temperature (SMSST). Emery and Schluessel (1989) used GAC data to compare the SMSST and the MCSST for one two-week period in October 1982. Further work was done to make similar comparisons of the MCSST, CPSST and SMSST algorithms for several other two-week periods by Wick *et al.* (1992).

A very thorough recent investigation of the skin effect is described by Schluessel *et al.* (1990) and we shall summarise that work without citing all the references they cited. Schluessel *et al.* briefly recount the history of the first predictions of this skin effect in the early 1940s and its early observation in the 1960s and 1970s. The temperature of the skin is generally several tenths of a degree lower than the temperature measured just a few centimetres below the surface. While the thickness of this skin layer is always less than a millimetre, its actual thickness depends on the local energy flux through the sea surface due to molecular transports. The sharp temperature gradient, characteristic of the molecular sublayer, persists at windspeeds up to about 10 m s^{-1}; at higher windspeeds the skin layer is destroyed by breaking waves. Studies have shown, however, that this skin layer reestablishes itself within 10 to 12 s after the cessation of the destructive influence.

The persistent existence of the skin layer has been explained by theoretical and empirical models which demonstrate that the surface skin layer is needed to resist and regulate the fluxes of longwave radiative energy as well as the sensible and latent turbulent energy fluxes across the sea surface. It is the conductive surface skin layer that effects the molecular exchange of both the radiative and turbulent energy fluxes across the air–sea interface. While above and below this thin skin, turbulent eddy exchange mechanisms carry the heat, within the ocean and the atmosphere, to or from the interface, these eddy transports cannot carry the heat across the ocean's surface (except by spray bubbles); instead this is accomplished by molecular processes within the skin layer. Schluessel *et al.* (1990) reported experimental work that explored the character of the skin layer, its relation to external heat fluxes and its relation to coincident measurements of bulk temperatures.

During the day, solar heating warms the upper metre of the ocean, creating under low wind conditions an afternoon temperature maximum (Figure 4.8(a)) a few millimetres below the surface. Below this shallow temperature maximum is the diurnal transition to the lower temperatures of the ocean's mixed layer, while above it the temperature drops from the shallow maximum to create the cool skin of the ocean. The existence of this diurnal thermocline is another factor, in addition to the skin effect, which complicates the relation between the satellite-observed sea surface temperature and the value measured *in situ* at some depth. Observations of this diurnal warming of the sea surface in the Mediterranean with AVHRR data are described by Böhm *et al.* (1991). At night the surface will cool and this maximum is erased by vertical mixing creating the typical night-time profile as shown in Figure 4.8(b). This profile exhibits the cooler skin layer which is maintained at night by the longwave and turbulent fluxes. These diagrams in Figure 4.8 are intended only to display some very simple idealised cases. The great variety of atmospheric and oceanic conditions that actually take place may produce near-surface temperature profiles that do not exactly fit either of these cases, but in any event the thin skin of the ocean persists as a feature, both during the day and at night.

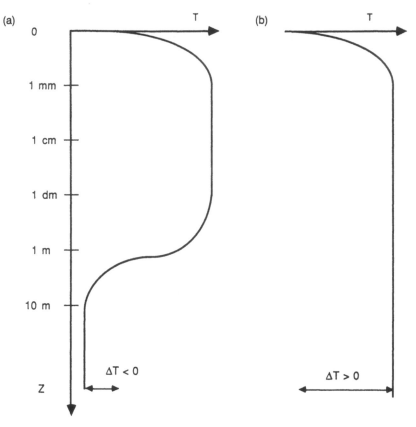

Figure 4.8   Schematic representation of possible near-surface temperature profiles (a) day-time profile with diurnal thermocline and (b) night-time profile (Schluessel *et al.* 1990).

Schluessel *et al.* (1990) describe the results of a six-week set of measurements taken from the German research vessel *Meteor* in the northeast Atlantic from 20 October to 28 November 1984. Radiometric measurements of skin temperature were made from the research vessel along with coincident measurements of subsurface bulk temperatures, radiative fluxes and meteorological variables. Over the entire six-week data set the bulk–skin temperature differences (ΔT) ranged between -1.0 and 1.0 K with mean differences of 0.1–0.2 K depending on wind and surface heat flux conditions. The bulk–skin temperature difference varied between day and night as well as with different cloud conditions, which can mask the horizontal variability of SST in regions of near-horizontal temperature gradients. An example of results for a two-day period is shown in Figure 4.9. The largest values of the magnitude of ΔT and of the heat fluxes occur at night and under clear skies. The mean values of $|\Delta T|$ over the whole six-week period were found to be 0.11 K for the day-time data and 0.30 K for the night-time data. Schluessel *et al.* (1990) used their data to derive parametrisations of ΔT in terms of heat fluxes and momentum fluxes (or their related variables) (for details see the appropriate section of their paper). They also studied the statistical distribution of values of ΔT that they obtained.

One important question which arises is that of whether the skin temperatures, seen by satellite radiometers, are representative of the temperatures at deeper levels

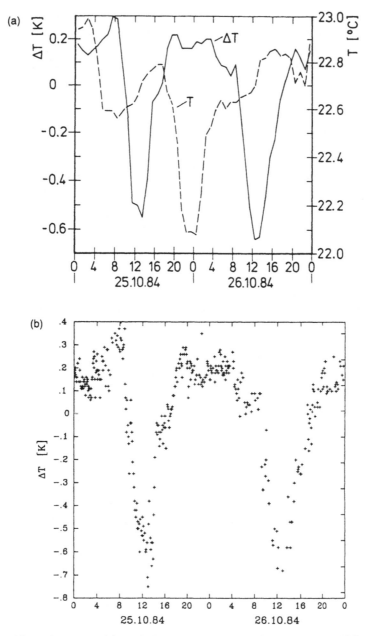

Figure 4.9 (a) Hourly means of ΔT (solid line) and 2 m depth bulk temperatures (*T*) (broken line) for 25 and 26 October 1984 and (b) five-minute means of ΔT for the same days (Schluessel *et al.* 1990).

of the ocean. Many of the well-known ocean circulation features are clearly visible in infrared satellite imagery. The foregoing discussions have shown that for small scales the changing skin effect might hide variations in bulk SSTs when observing the sea surface radiometrically. This could perturb the calibration match-ups done between buoy measurements at a single location and the satellite-derived SST from a coincident single pixel.

Insight into the coupling between skin and bulk temperatures at different spatial scales can be gained by looking at the correlations between these two parameters for ship tracks of different lengths. Schluessel *et al.* (1990) studied the mean cross-correlation between skin temperature and the bulk temperature at a depth of 2 m by dividing their data record from the moving ship into linear sections of between $\Delta x = 20$ km and $\Delta x = 640$ km. Values of the squared coherencies between skin temperature and bulk temperature, as a function of distance, were plotted for six segments of ship track (the longest being nearly 3500 km in length). The mean cross-correlation was found to rise from 0.5 for the short scales to over 0.9 for the longest tracks (see Table 4.4). Thus at longer scales one can expect the surface skin temperature variations to represent the ocean variability reflected by the bulk temperatures. At shorter scales, skin and bulk temperatures are only weakly correlated.

The general conclusion was that, taking a coherency of >0.7 (corresponding to a squared coherency of >0.49) as a threshold designing a good correspondence between skin and bulk temperatures, for length scales greater than 150 km the skin temperature is representative of the bulk temperature ocean variability even though individual skin–bulk temperature differences (at shorter wavelengths) show fluctuations of the order of 0.3 K. This threshold length scale might shift to lower values in those areas where strong horizontal gradients of SST overwhelm the skin effect or the temperature variations due to diurnal thermoclines. However, there is an important difference between a satellite image and *in situ* data from a ship track. The former is gathered almost instantaneously (over a period of a few minutes in fact) whereas data from a length of ship track are gathered over an interval of several hours or even several days. Thus even small horizontal temperature gradients might be detected in satellite images at large but constant bulk–skin temperature differences due to nearly homogeneous horizontal distributions of heat and momentum fluxes through the surface. These conditions can often be observed in infrared images where large areas are synoptically scanned and which do not show the temporal variation of ΔT as seen in the time series taken by the ship measurements.

The implications of the very detailed studies of the magnitude of the skin effect and of its day–night variation by Schluessel *et al.* (1990) for SST retrieval from satellite data are most important when observing diurnal warmings of the surface and when calibrating SST retrieval techniques with *in situ* buoy measurements. For the latter they recommended night-time comparisons to avoid the diurnal thermocline and to include meteorological variables in a prediction model for ΔT which should give the best match-up between satellite-derived skin temperatures and bulk temperatures measured from buoys. At night, SST retrieval can be carried out with

Table 4.4 Mean correlations between T_S and T_2 and their standard deviations (SD). (Schluessel *et al.* 1990)

Δx (km)	Correlation	SD
20	0.5	0.36
40	0.62	0.33
80	0.74	0.29
160	0.85	0.21
320	0.92	0.02
640	0.92	0.01

the aid of channel-3 data for cloud detection and filtering. Another type of calibration could be achieved by comparing remotely sensed SSTs with *in situ* skin measurements; however, this would require a more sophisticated *in situ* experimental set-up which is not practical for operational use.

(iv) *Cloud cover* If a scene is completely cloud-free there is no problem. If there are large areas of solid cloud then again there is no serious problem; it is just necessary to mask the clouds (manually or with some automatic process) and to reject the data, see section 3.2. The problem arises in the intermediate case which is – in practice – very common. If there is scattered or broken cloud then it is important to ensure that one's cloud masking method eliminates not only areas that are obviously cloudy but also the pixels from nearby areas that are likely to be cloud-contaminated. What is particularly difficult to handle are areas of thin haze, areas of thin dust clouds and areas containing subpixel clouds, i.e. small (cumulus) clouds which are smaller than the instantaneous field of view of the AVHRR. As we have already seen in section 3.2, it is exceedingly difficult to try to deal with these problems simply within one data set. The clouds are colder than the sea surface and therefore the presence of a cloud lowers the average temperature for the complete pixel. Thus the satellite-derived SST for such a cloud-contaminated pixel will be lower than it would be if the cloud were not there. Thus cloud-contaminated pixels are, perhaps, best detected by making comparisons with the SSTs computed for nearby dates for the same area and identifying and rejecting anomalously low values.

(v) *Sunglint* It is conventionally argued that the reflected solar infrared radiation at the wavelengths of channels 4 and 5 of the AVHRR is negligible and that all the radiation leaving the surface of the Earth at these wavelengths is emitted radiation. In general this argument is valid, as can be demonstrated by simple calculations based on the use of the Planck distribution function. Therefore, in deriving SSTs from channel-4 and channel-5 data along the lines described in this section, the effect of reflected radiation has always been neglected and the general success of SST retrieval from channel-4 and channel-5 day-time AVHRR data implies that diffusely reflected solar radiation can indeed be neglected. However, this appears to have led people to assume that this is also true for specular reflection, i.e. in areas of sunglint on the surface of the sea, as well. However, order of magnitude calculations indicate that in an area of sunglint the sea-surface-leaving radiance will include a significant component of reflected solar infrared radiation within the range of wavelengths 10–12 μm and that regarding this as part of the emitted radiation will therefore lead to a significant error in the retrieved SST. The SST in an area of sunglint will therefore appear too high unless an appropriate correction is made. Experimental evidence is given, for instance, by Saunders *et al.* (1981), Khattak *et al.* (1991) and Narendra Nath *et al.* (1993); a method of correcting AVHRR-derived SSTs in areas of sunglint has been proposed by Cracknell (1993) and tested by Narendra Nath *et al.* (1993). A more general discussion of sunglint in AVHRR data will be given later (see section 7.9).

4.3 OCEAN CIRCULATION

So far in this chapter we have largely been concerned with the determination of sea surface temperatures in absolute terms, i.e. in physical units of degrees Celsius or kelvins. The attention has been focussed on trying to find, with as much absolute accuracy as possible, the temperature of the sea for each pixel. We have also, however, seen briefly that satellite data can be used to study sea surface temperatures synoptically and to obtain regional or global coverage with a frequency and with a quantity of detail that would be quite unobtainable by more conventional methods. It is not, however, just a question of obtaining numerical values for the sea surface temperature on a wide network of grid points. These are various features whose existence had been established before satellite images of them became available and for which there was some evidence of their general stability or instability in terms of their location. But satellite images enabled synoptic views of such features to be obtained at frequent intervals. Consequently the location and stability of these features can now be observed in considerable detail. In this section we shall describe three classes of problem in ocean circulation that can be addressed with the aid of thermal infrared data from the AVHRR, but in which the absolute value of the surface temperature on a pixel-by-pixel basis is not important. The synoptic nature of the data and the frequency of acquisition mean that if an oceanographic feature or phenomenon has a temperature 'signature' associated with it then it is possible to study that feature or phenomenon, and its temporal evolution, in far greater detail with satellite data than had been possible previously. We have already encountered one example briefly in section 4.2 and Figure 4.2 in the case of the eddies or rings on the side of the Gulf Stream. The three things that we shall consider in this section are (i) flow patterns of water masses, (ii) boundaries, or fronts, between different water masses, the stability (or otherwise) of these boundaries and eddies that form at these boundaries and (iii) the determination of current vectors.

In terms of the processing of AVHRR thermal infrared channel data or images, this means that geometrical rectification of the images is important but that atmospheric corrections are not generally necessary. It is, of course, necessary to be able to be sure that a feature that one observes is an ocean surface feature and not an atmospheric feature associated with thin cloud, fog or haze; however, it is usually possible to separate atmospheric features from ocean surface features because atmospheric features are much less long-lived than ocean surface features. It is, of course, also important to choose cloud-free data for the area being studied, otherwise the sea surface will be obscured by cloud and the sea surface features will not be visible. The general question of accurate detection and enhancement of oceanic features in AVHRR data has been addressed by Simpson (1990).

4.3.1 Flow patterns

Quite often the different water masses have significantly different temperatures and then the boundary will be clearly visible in a thermal infrared image of the area. This means that the temperature, as determined from the thermal channels of the AVHRR, can be used as a tracer or indicator of flow patterns. One example is provided by some work by Jönsson (1989) on the Strait of Öresund. The strait of Öresund is approximately 100 km long, of width 10–20 km and connects the Baltic with Kattegat, see Figure 4.10. The flow is rather complex but most of the time it could be considered as estuarine, with less saline Baltic water (salinity $S = 8$–$10‰$) creating a northbound surface flow and a denser southbound bottom flow ($S = 32$–$34‰$). Kattegat is a shallow sea with an average depth of 23 m and is strongly

Figure 4.10 Geography of the Baltic, Öresund (The Sound) and Kattegat (Jönsson 1989).

stratified with the surface layer being of the order of 15 m deep. The shallowness of Kattegat, especially in the southern part, and the relatively deep surface layer cause the bottom water masses to be rather limited. Oxygen-consuming processes in the bottom waters could thus diminish the oxygen contents in these waters considerably with detrimental effects on the bottom ecology.

In recent years this oxygen problem – caused by the growth and later sedimentation and destruction of algae – has shown a tendency to become a more and more serious problem. Skälderviken and Laholmsbukten have been affected for a number of years with sometimes a total loss of oxygen in parts of the bays. The intense growth of algae is stimulated by inorganic nutrients in the surface layer and the most important substance is assumed to be nitrogen. A knowledge of circulation patterns and flow processes is necessary for understanding the problem and for taking remedial measures. Jönsson (1989) investigated the use of AVHRR thermal infrared data to study some aspects of the flow in southern Kattegat and how this information could be used in understanding the water quality situation, especially in relation to the way in which water from Öresund discharges to Kattegat and interacts with the bays which we have already mentioned. The flow was studied by visual interpretation of suitably contrast-enhanced images of the thermal infrared AVHRR data.

A visual interpretation of a single image from 9 August 1989 and of a sequence of three images, at intervals of approximately 24 hours, from 23, 24 and 25 May 1989 led to a number of general conclusions:

- A distinct coastal stream seems to exist along the Swedish coast in southern Kattegat at northbound Öresund flow. The Coriolis force might be the cause.

- In the widening section of northern Öresund the flow seems to separate from the Danish side. The reason is not known.

- The coastal stream does not penetrate Skälderviken very much.

- The temperature changes in the Öresund surface water are more probably caused by entrainment of bottom water into the surface water.

In relation to the oxygen depletion problem in southern Kattegat, it was argued that the existence of a northbound coastal flow along the Swedish coast of water originating from the Baltic might be one explanation for the water quality situation in this area as opposed to the generally held view that Swedish river effluents are the main cause.

Another example of ocean circulation studies using AVHRR thermal infrared data is described by Sousa and Fiúza (1989). Several hundred AVHRR thermal infrared images of the Atlantic off northern Portugal, prepared from data received for more than 10 years, were analysed by visual interpretation for investigating the filaments of cold upwelled waters which extend hundreds of kilometres offshore in that region during the upwelling season. Four such filaments were found which reappear each year at nearly the same places along the northern coast of Portugal and off western Galicia (NW Spain), after the first strong event of upwelling-favourable northerly winds in late spring or early summer. Their surface signature is detected in the satellite thermal imagery till the end of the summer. These structures are apparently separated from each other by warm mesoscale eddies with diameters of 80–100 km. The location of the cold filaments seems to relate with bottom topography, as they systematically occur over the ridges between the submarine canyons which indent the continental margin in the region. No relation was found with the morphology of the coast. There are other examples of similar studies from various parts of the world. These include the Norwegian coastal current (McClimans 1989, Lønseth and Bern 1991), large filaments transporting cold coastal water as far as several hundred kilometres offshore from the coast of northern California (Randerson and Simpson 1993), plumes of cool surface water flowing into the Black Sea, the Sea of Marmara and the Aegean Sea (Buttleman 1988), the evolution of a cyclonic eddy in the Tyrrhenian Sea (Viola and Böhm 1991) and the Leeuwin current, Western Australia (Prata and Wells 1990). Upwelling, which is more commonly studied with visible wavelength data (see section 7.3), can also be studied with thermal infrared data; for example upwelling off the West African coast has been studied with GAC data from 1981 to 1989 by Nykjaer (1993).

4.3.2 Fronts, eddies and rings

Because the oceans are dynamic systems with large masses of water circulating in them, one finds situations in which rather sharp boundaries exist between different water masses which, despite their close proximity, have not become mixed. We have seen one example already in Figure 4.2 where quite sharp boundaries exist to the Gulf Stream. What is usually of interest in this situation is two things (i) the location of the boundary and (ii) the temporal variations at the boundary. The absolute values of the temperatures of the different water masses and the value of the temperature difference across the boundary are not of great importance.

Some oceanographic boundaries or fronts change with time. For instance the boundaries on the Gulf Stream move, large eddies or rings form on the side of the Gulf Stream, they grow, they break away as separated rings and eventually they are lost by coalescence with the Gulf Stream or dissipation by mixing with the colder surrounding water. The life of a warm core ring is of the order of 3–8 months. These rings form in different places at different times and the pattern is constantly changing. A quite extensive study of the movements of warm core rings and the front on the north side of the Gulf Stream off the east coast of the USA has been described by Zheng *et al.* (1984) who also quote a number of related references. Measurements of AVHRR images (and indeed of VHRR images from NOAA-5 in 1977–1978) were made to determine the drift velocity and angular velocity of the rings, see Table 4.5.

Table 4.5 Satellite measurement values of ring rotation (Zheng *et al.* 1984)

Time	Orbit	Rotation speed ($\times 10^{-6}$ rad s^{-1})	Period (day)	Radius (km)	Tangent speed (cm s^{-1})
1977					
1–2 Apr	N5 3037, 3056	5	14	60	31
1978					
11–16 Jun	N5 8440, 8502	3	29	80	20
21–22 Sep	N5 9696, 9715	5	16	48	23
1979					
19–20 Nov	TN 5675, 5689	5	16	58	27
11–12 Dec	TN 5985, 5999	7	11	58	40
1980					
9–12 Sep	N6 6246, 6282	4	20	119	44
31 Oct	N6 6993, 7000	3	24	126	38
1981					
16 Feb	N6 8529, 8536	11	7	88	98
10–11 Apr	N6 9290, 9297	10	8	105	100
23–24 May	N6 9895, 9902	3	23	91	28
25–26 May	N6 9923, 9930	6	13	112	63
4–5 Nov	N7 1896, 1910	4	21	105	37
1982					
29 Jan	N7 3103, 3110	12	6	84	100
9–10 Mar	N7 3654, 3668	10	7	84	87
14–16 Mar	N7 3731, 3759	6	12	119	70
29–30 Mar	N7 3950, 3957	5	14	77	41
4 Apr	N7 4081, 4088	8	9	70	59
15–16 Apr	N7 4183, 4190	7	11	56	36
Total average		6	14	86	52
SDEV		3	7	24	28

It was found that (i) the angular velocity of the rings exhibits a marked seasonal variation with the maximum occurring in the winter and the minimum in the late autumn, (ii) the surface area of the rings increases monotonically with time and (iii) the Gulf Stream front at the north wall migrates seasonally and attains its northernmost position in the late autumn. There have been various studies of eddies using AVHRR data. These include studies of eddies in the Sea of Japan and the East China Sea (Huh and Shim 1987), on the Oyashio front, to the east of Japan (Vastano and Borders 1984), in the North Atlantic (Vaughan and Downey 1988, Viehoff 1989, Glenn *et al.* 1990), in the Norwegian coastal current (McClimans 1989) and near Pt Conception, California (Sheres and Kenyon 1990a,b). The shift of the California current closer to the shelf and the disappearance of the Davidson current at the surface in late April was studied by Foerster (1993) using 53 AVHRR images from a seven-year period to investigate the link of this behaviour with the wind direction. Airborne infrared scanner data and infrared scanner data from other satellites have also been used in various similar studies.

On the other hand some ocean features are long-lived and relatively stable in position. For instance, there is a front stretching between Iceland and the Faeroe Islands. This marks a relatively sharp boundary between warmer North Atlantic water and colder water of more polar origin. Robinson (1985) describes the study of the location of this front over a period of nearly a year using between 30 and 40 AVHRR images. It was possible to study the long-period seasonal movements of this front and also the short-period meanders, see Figure 4.11. A survey of sea surface temperature fronts on a global basis is given by Legeckis (1979).

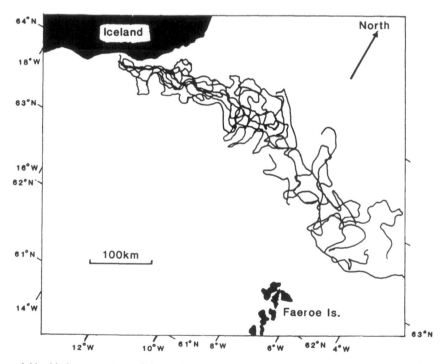

Figure 4.11   Various positions of the Iceland–Faeroe front from about 40 infrared images obtained between November 1978 and July 1980 (Robinson 1985).

Shelf fronts provide examples of features which have a thermal signature and which have extremely stable positions. These fronts are boundaries across which there is a significant difference of temperature (and also of salinity). It was observed that in middle latitudes these fronts appear in the same locations each spring, persist through the summer at these locations and disappear in the autumn. In a period of very stormy weather during the summer they might disappear, but after a few days they would reappear again in their usual locations. The question of why they appear at particular locations and why their positions are so relatively stable is obviously of interest and importance. The theory leading to an expression that enables the location of the front to be determined is a rather nice example of the application of simple but fundamental physical ideas based on energy considerations. An early account of the theory is given by Simpson (1981) and we shall indicate briefly what is involved in that theory.

In the summer significant amounts of energy from the Sun fall on the surface of the sea and some of this energy penetrates to different depths beneath the surface of the sea. In deep water a thermally stratified situation is established. In shallow water, where there is considerable tidal action, such stratification is not set up and the temperature is constant throughout the water column, see Figure 4.12. What is perhaps at first sight surprising is that as one goes from deep water, with a thermally stratified structure, to shallow water, with a mixed regime, there is not a gradual change from one situation to the other but a sharp transition at a particular location. This sharp boundary between these two regions shows up very clearly on infrared satellite images, Figure 4.13.

One starts by calculating the difference between the gravitational potential

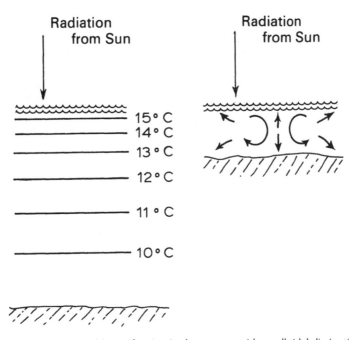

Figure 4.12 Diagram illustrating (a) stratification in deep water with small tidal dissipation and (b) tidal mixing in shallow water with large tidal dissipation.

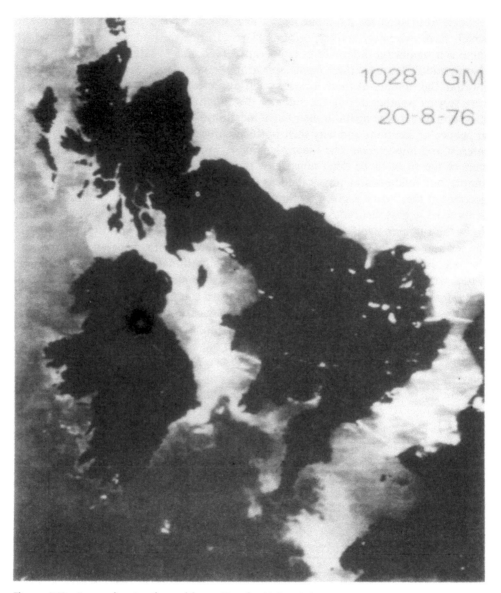

Figure 4.13 Image showing thermal fronts (Dundee University).

energy of a stratified water column shown in Figure 4.14(a) and that of the homogeneous tidally mixed column in Figure 4.14(b). The potential energy of the tidally mixed state will be the larger and so this state will only exist if there is adequate energy available from the tidal action to provide this extra potential energy. The detailed working leads to the condition that mixing is possible if

$$\varepsilon k\rho\,|u|^3 \ge \frac{\alpha g h}{2C_p}\,\dot{Q} \tag{4.3.1}$$

where

ε = fraction of the tidal energy that is available to provide mixing

Figure 4.13   (Continued)

k = constant in expression for whole tidal energy $k\,\rho\,|u|^3$
ρ = density of sea water
u = tidal stream velocity
α = coefficient of thermal expansion of sea water
C_p = specific heat of sea water
g = acceleration due to gravity
h = depth of water and
$\dot{Q}$ = rate of supply of heat from the Sun.

This condition can be arranged in the form

$\qquad P > 1$ stratification
$\qquad\quad < 1$ mixing

$\hfill (4.3.2)$

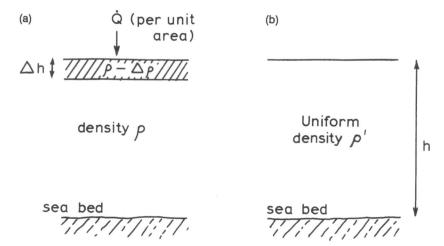

Figure 4.14 Schematic diagram of (a) unmixed surface layer on top of a uniform water column and (b) fully mixed water column.

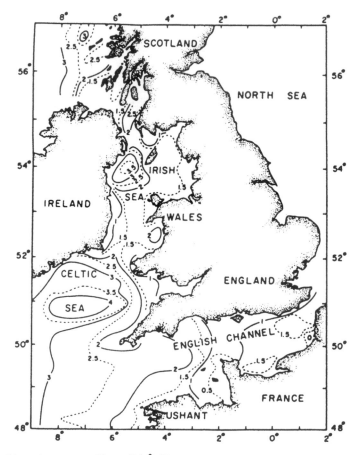

Figure 4.15 Map of contours of $\log_{10}(h/u^3)$ (Simpson 1981).

where

$$P = \frac{3\pi\alpha g}{8\rho C_{\mathrm{p}}\varepsilon}\left(\frac{\dot{Q}h}{u^3}\right). \tag{4.3.3}$$

Most of the quantities which are included in P are fixed and so, essentially, it is the parameter (h/u^3) that determines the location of fronts between stratified and mixed regions. It is therefore possible with a bathymetric chart and a tidal atlas to study the contours of (h/u^3) and to identify the locations at which fronts mark the boundaries between regions of stratified and tidally mixed water. Contours of $\log_{10}(h/u^3)$ in the Irish Sea, the English Channel and the Celtic Sea are shown in Figure 4.15; the contour $\log_{10}(h/u^3) = 1.85$ was found to be in approximately the same positions as the shelf sea fronts. In the early 1980s very considerable success was obtained in locating the positions of fronts in thermal infrared band images from the AVHRR. A very considerable measure of success was obtained in explaining the locations of these fronts in terms of the theory involving the condition (4.3.2).

More recent work has been devoted to questions of the temporal stability of the fronts, both in terms of their seasonal appearance and disappearance and in terms of the spatial variability in their location. It should be noted that in addition to tidal mixing there are other mixing mechanisms, especially mixing by wind action. There are also complications which were not taken into account in the theory which we have just mentioned; one of these is that if there is a flow of fresh water from the land into a region of coastal waters there will be differences of density which will need to be accounted for in the theory.

4.3.3 The determination of current vectors

In addition to studying general overall patterns of circulation and the evolution of features such as eddies, including the determination of the velocities and angular velocities of eddies, one can also attempt to determine the values of the current velocities at the sea surface. This can be regarded as an extension of a method that is used for the determination of cloud velocities. The use of data from geostationary satellites for studying cloud motions and thence deducing the wind speed at the corresponding height in the atmosphere is well established. Basically geostationary satellites generate visible images of large areas of the Earth's atmosphere and surface at half-hourly intervals. A particular cloud can be identified in two successive images but the locations of the cloud will be different. The displacement of the cloud between one image and its successor can be measured. The time interval between the acquisition of the two images is known (30 minutes) and therefore the velocity of the cloud can be calculated. This is assumed to be the velocity of the wind at that level. To be very useful it is necessary to be able to estimate the height of the cloud in order to determine the height at which the wind speed has been calculated. For example, wind vectors are now routinely generated from Meteosat data at the Meteosat Operations Centre at Darmstadt. In general the frequency of acquisition of AVHRR data is too low to be very useful for determining wind vectors. An exception is that cloud tracking for wind velocity determination has been carried out in polar regions (for which geostationary satellite data are not available) using AVHRR data. In polar regions the AVHRR passes are much more frequent than they are in equatorial regions (although the time intervals are 90–100 minutes rather than the 30 minutes of the geostationary satellites). For a description of this work see Turner and Warren (1989). The feature-tracking method can also be applied to the study of the motion of sea ice; the discussion of this problem will be postponed until section 7.1.

Until recently the only way to obtain sea surface velocity over large areas was to rely on measurements from ships and drift bottles. This is far from ideal, since measurements can only be made from one point at any particular time and such measurements are expensive. The use of satellite data has the advantage, along with other uses of remote sensing data, that it generates data of a quantity, spatial distribution and frequency quite unobtainable with traditional *in situ* methods.

Several workers have made use of satellite data to determine sea surface velocities from displacements of surface features in a time sequence of thermal infrared images. Legeckis (1975) determined the cyclonic eddies' drifting speeds on the inshore side of the Gulf Stream by examining successive images and we have already mentioned the work of Zheng *et al.* (1984) on eddies in the same area. Many investigators have now used the same technique to estimate the motion for mesoscale features and surface velocities in various parts of the world's oceans. Thus Tanaka *et al.* (1982) obtained the velocity vectors of the Kuroshio current and Vastano *et al.* (1985) derived a surface current map over an oceanic region near Georges Bank from sequential satellite images. Their results were shown to be in agreement with current charts based on sea surface measurements. Vastano and Borders (1984) used AVHRR data to determine the velocities of motion along the Oyashio front and for a large anticyclonic warm core eddy in the same area. A number of other examples are quoted by Simpson and Gobat (1994). Cracknell and Huang (1988) studied the surface current off the west coast of Ireland with data from the spring of 1984. The general direction of motion is in a northeasterly direction and one cyclonic eddy, centred at 57° 04′ N 10° 59′ W, was present at that time and it was studied in some detail. A very pronounced feature is the sea surface front between mixed coastal water and stratified Atlantic water along the whole west coast of Ireland. The front is characterised by cyclonic and anticyclonic eddies which have time scales of the order of 1–7 days and length scales of the order of 30 km.

As an example we quote from the work of Cracknell and Huang (1988). This was for the area shown in Figure 4.16(a) and used the following NOAA-8 AVHRR data supplied by the Dundee receiving station.

Date	Time (GMT)	Orbit	Lapse (hours)
25 April 1984	0919	5596	—
26 April 1984	0857	5610	23.63
27 April 1984	0836	5627	23.65

The data from these three scenes used were geometrically rectified using standard methods with a set of ground control points identified manually in the image and on a map of the nearby land areas. An accuracy of ± 0.8 km in the location of each pixel was obtained; thus the accuracy of the speed of sea feature movement can be estimated as ± 1.6 km/24 hr $= \pm 1.8$ cm s^{-1} in the case of the images taken with a 24 hr difference between their orbits. A total of 29 sea surface features were selected. By determining the displacement of each feature between two scenes and dividing by the time interval between the scenes the surface current velocity can be calculated.

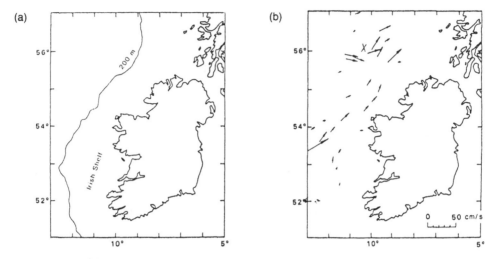

Figure 4.16 (a) Study area and (b) surface current vectors determined by the feature-tracking method (Cracknell and Huang 1988).

By applying this feature-tracking method to these features, 29 current vectors have been obtained, see Figure 4.16(b).

One of the important advantages of the feature-tracking method is that it enables us to investigate the sea surface front and eddies associated with it. The images used in the work just described show the Irish shelf front between mixed coastal water and stratified Atlantic water. The front is characterised by cyclonic and anticyclonic eddies. Using the feature-tracking method and analysing other infrared images, it was possible to obtain some characteristics for the Irish shelf front (for further details see Cracknell and Huang 1988).

There are two methods for computing surface velocities from sequential satellite images. One method, called the Maximum Cross Correlation (MCC) method or objective method, uses cross-correlations between sequential images to compute surface velocities; this is operator-independent (for details see Emery *et al.* 1986, Kamachi 1989, Jönsson 1993, Simpson and Gobat 1994). The feature-tracking method, which we have just described, needs subjective selection and tracking of surface features in the images and this is operator dependent. There is one impor-tant difference between these methods and this is the fact that the simple form of the objective method cannot estimate rotational motion and motion along the surface front while the subjective method can (Emery *et al.* 1986). Both methods have a great advantage over ordinary velocity measurements taken by ships and drift bottles. They give a synoptic and spatial view of the velocity distribution over a large area. However, they have a disadvantage in that they measure the drifting of sea surface features. The image-derived velocities represent a combination of both geostrophic mean and shorter-term wind-driven surface current.

Perhaps we should make some general comments about the feature-tracking method.

(i)   There is clearly an interesting pattern-recognition and computing problem that has to be developed for automatic tracking (a little has been done but not a great deal).

(ii)   It will not give current vectors regularly over a neat square or rectangular grid; it will only give current vectors where features happen to occur.

(iii) It is dependent on features remaining at the surface and not being lost or suddenly appearing as a result of vertical motions of the water.

(iv)   The most serious problem is that, although there are several passes per day of the NOAA satellites over a given area, if there is cloud around then one will not see the sea surface at all but will only see the clouds instead. This is a very serious limitation.

(v)   Tidal effects are, fortuitously, more or less eliminated because the scenes used are almost exactly 24 hours later and therefore at similar, though not identical, states of the tidal cycle.

(vi)   Finally, it is possible to use an improved geometrical rectification method (see the description of subpixel accuracy in section 3.1.6) with the aim of reducing the error from about ± 0.8 km to about ± 0.3 km in the rectified position of a pixel. This would enable us to reduce the error on the derived current vectors from about ± 1.8 cm s^{-1} to about ± 0.7 cm s^{-1}.

4.4   LAND SURFACE TEMPERATURES

Thermal infrared data obtained from satellites have been used for some time to study heat islands associated with cities, first of all qualitatively and then, more recently, quantitatively. A review will be found in section VIII-1.4 of Rao *et al.* (1990). Brightness temperature differences of between about 2.5°C and 6.5°C were found between cities and nearby rural areas in the USA. However, urban areas form only a very small fraction of the total land surface area of the Earth and we need to consider the question of the determination of (atmospherically corrected) land surface temperatures more generally.

We discussed the use of thermal infrared satellite data for the determination of sea surface temperatures before turning our consideration to land surface temperatures; this was because the case of land surface temperatures is considerably more difficult than that of sea surface temperatures. There are several reasons for this (see, for example, Becker and Li 1990).

■ The first reason applies to bare rock, bare soil or desert sand surfaces as well as to vegetated surfaces and this concerns the emissivity. Whereas the emissivity of the sea surface is that of water (albeit salt water) and its value is constant (and therefore known), for rock, soil or sand the value of the emissivity is much more variable than that of water. Thus for any land surface pixel, one has two unknowns, the temperature and the emissivity, whereas for a sea surface pixel there is only one unknown, the temperature.

■ Most land surface areas are much less homogeneous than the sea surface. Even non-vegetated land surfaces, apart from some desert or salt flats areas, are often very inhomogeneous on the scale of the IFOV (or pixels) of the AVHRR. For a vegetated surface it may even be quite difficult to define what is meant by the temperature of the surface of the land. One could consider the temperature of the underlying soil, or the temperature of some part of the vegetation (trunk, stalk, leaf, flower or fruit), or the temperature of the air that is trapped among the vegetation.

- The air surface temperatures just above the land surface are usually quite different from the land surface temperatures and this may weaken the assumptions behind the split-window method which is widely used for the sea surface (see section 4.2).

As a result of these problems, which make the determination of land surface temperatures so much more difficult than the determination of sea surface temperatures, relatively little work has been done on the determination of land surface temperatures from infrared data from satellites, in spite of the fact that the AVHRR generates an enormous amount of such thermal infrared data. Research on the retrieval of land surface temperatures from AVHRR data is thus still at an early stage. It will, therefore, only be described briefly and will cover two areas, namely the use of split-window (multichannel) methods for land surface temperature determination and the concept of thermal inertia. Before that we simply mention that the use of TOVS sounding data for atmospheric corrections of land surface temperature data has been described by Reutter and Olesen (1991).

4.4.1   Split-window methods

We saw in section 4.2 that it is possible to calculate atmospheric corrections to brightness temperatures using TOVS data in an atmospheric model calculation and that this has been used for sea surface temperatures. Although this approach has been used for land surface temperatures by Uspensky and Scherbina (1993) it has not been widely used so far for land surfaces. We did, however, also see in section 4.2 that the split-window method has been very successful in the determination of sea surface temperatures from AVHRR thermal infrared data. The question of adapting this method to the situation of land surface studies, where the emissivity is both spatially variable and its value differs significantly from unity, has been studied theoretically by Becker (1987). In that work he obtained an expression for T_S, the surface temperature, in terms of the two brightness temperatures in channel 4 and channel 5 and in terms of the emissivities ε_4 and ε_5 of the surface for the wavelengths of these two channels (see equation (20) of Becker 1987). Another way of looking at this adaptation of the split-window method is to consider the error ΔT that would arise in the calculated land surface temperature (LST) by taking $\varepsilon_4 = \varepsilon_5 = 1$ rather than using true values; Becker showed the value of ΔT is significant and is given by

$$\Delta T = 50 \, \frac{(1 - \varepsilon)}{\varepsilon} - 300 \, \frac{(\varepsilon_4 - \varepsilon_5)}{\varepsilon} \tag{4.4.1}$$

where $\varepsilon = (\varepsilon_4 + \varepsilon_5)/2$. Typical values of ε_4 and ε_5 for land surfaces give values of ΔT as high as 6 or 7 K. Therefore, while it is possible to derive a formal expression for T_S, the (land) surface temperature, it cannot be used in practice unless one has a reasonably accurate set of values of ε_4 and ε_5 for the land surface area being studied. Such information is generally not available.

However, the situation for the determination of land surface temperatures is not completely without hope. Since the split-window method has, in practice, proved to be so successful for sea surface temperatures (see section 4.2) and because it has the great advantage of simplicity, several workers have attempted to extend the split-window approach to the determination of land surface temperatures. This problem has been addressed, for example, by Becker and Li (1990), Li and Becker (1991) and Uspensky and Sutovsky (1991). For the sea surface the split-window method is based on the result, which can be justified theoretically, that the surface temperature, T_S, can be expressed as a linear combination of the channel-3, channel-4 and channel-5 brightness temperatures T_{B3}, T_{B4} and T_{B5}, respectively, i.e.

$$T_S = a_3 T_{B3} + a_4 T_{B4} + a_5 T_{B5} + a_6 \tag{4.2.6}$$

where a_3, a_4, a_5 and a_6 are assumed to be constants, on a global scale; in particular their values do not depend on the local atmospheric conditions. During the day a similar relation, but excluding channel 3, applies. Thus, as we have seen in section 4.2, once the values of these coefficients have been determined using *in situ* data from a relatively restricted set of data buoys, then these values of the coefficients are applied to the AVHRR data on a global basis. Becker and Li (1990) showed that for the land surface it was also possible to use an equation of the form of equation (4.2.6) in which the coefficients are still independent of the local atmospheric conditions. However, the coefficients a_3, a_4, a_5 and a_6 are no longer constants with values that are valid globally. Instead the values of these constants become functions of the emissivity. This leads to the replacement of a universal split-window method that is used so successfully for sea surface temperatures by a local split-window method for land surface temperatures. In the formulation of Becker and Li (1990) and considering the day-time situation so that channel 3 is excluded

$$T_S = A_0 + P \frac{T_{B4} + T_{B5}}{2} + M \frac{T_{B4} - T_{B5}}{2} \tag{4.4.2}$$

with

$$A_0 = 1.274$$

$$P = 1 + 0.15616 \frac{(1 - \varepsilon)}{\varepsilon} - 0.482 \frac{\Delta\varepsilon}{\varepsilon^2}$$

and

$$M = 6.26 + 3.98 \frac{(1 - \varepsilon)}{\varepsilon} + 38.33 \frac{\Delta\varepsilon}{\varepsilon^2}$$

where $\varepsilon = (\varepsilon_4 + \varepsilon_5)/2$ and $\Delta\varepsilon = \varepsilon_4 - \varepsilon_5$. These numerical values are only valid for NOAA-9, since the spectral response of the radiometer is used to compute these coefficients. Slightly different values of the coefficients will be required for AVHRR data from other satellites in the series.

The local split-window method therefore involves using the AVHRR data with equation (4.4.2) in association with a land-use map obtained from the visible and near-infrared channels, when effective emissivities of various surfaces are known. Errors in the assumed values of ε will lead to errors in the retrieved land surface temperatures, but these errors will be much smaller than the errors in ΔT mentioned above that would be obtained by using $\varepsilon_4 = \varepsilon_5 = 1$ (see Table 6 of Becker and Li 1990). This method has been used with data from two test sites by Kerr and Lagouarde (1989) and Kerr *et al.* (1992); in the latter work Kerr *et al.* demonstrated an accuracy of better than 1.5 degK. As an alternative to using the expressions for A_0, P and M in terms of ε and estimating the value of ε, one can regard A_0, P and M simply as parameters to be determined empirically by a least squares fit to a set of *in situ* data. This has been done, for instance, by Vidal (1991). However, this method is of limited usefulness because to be so reliant on *in situ* data considerably negates the benefits of using remotely sensed data. Attempts have also been made to use the success in studying water surface temperatures to investigate relatively homogeneous nearby land surfaces (salt flats or sand), e.g. in some work on AVHRR data for Lake Eyre in Australia (Barton and Takashima 1986). Further work on the development of the theory of the localised split-window method, and on the validation of their previous approach, was described by Li and Becker (1993); this involved extending the ideas to the inclusion of channel-3 AVHRR data in addition to using the channel-4 and channel-5 data. Estimating the land surface temperature emissivity difference between channels 4 and 5 using data from the Valencian area in Spain has been carried out by Coll *et al.* (1993). Land surface temperatures have been studied using both AVHRR and Meteosat data by Olesen *et al.* (1993); the spatial resolution of the AVHRR and the temporal resolution of the Meteosat data complement one another.

One situation in which the determination of land surface temperature from thermal infrared data is easier than for most cases is when the surface is snow covered. The surface is then fairly homogeneous and the emissivity is more homogeneous and its value is more well known than for most other land surfaces. Some studies of land surface temperature for snow-covered surfaces using AVHRR thermal infrared data are described by Collier *et al.* (1989) and McClatchey (1992).

There have been various attempts to apply thermal infrared data from Earth-orbiting satellites – and particularly AVHRR data – to studies of agricultural problems. These include studies of a tall grass prairie by Cooper and Asrar (1989), of a wheat field by Prata and Platt (1991) and of frost damage in orange groves in Spain by Caselles and Sobrino (1989). The economy of Valencia, Spain, depends largely on orange production, but when a frost occurs, this production is lost. Therefore, the assessment of the extent of frost damage is of great interest in the area. Caselles and Sobrino (1989) developed a theoretical model relating the temperature of the radiative system (the orange grove) and the temperature of each part (ground, orange tree side, and orange tree top). This relation depends on the proportions of ground, orange tree side and orange tree top as well as on orange tree and ground emissivities. By analysing the thermal differences among ground, orange tree side and orange tree top, they established a procedure that permits the estimation of the temperature of orange groves from the satellite measurement. This model is a simplification of the model developed by Becker (1981). It is assumed that an orange tree is a homogeneous medium as far as emissivity is concerned and therefore one does not distinguish between the orange tree side and top emissivities. So the radiance reaching the radiometer, L, is the weighted sum of the radiances coming from the orange tree top, L_t, the ground, L_g, and the orange tree side, L_s (see Figures 4.17 and 4.18):

$$L = P_t L_t + P_g L_g + P_s L_s \tag{4.4.3}$$

where P_t, P_g and P_s are, respectively, the proportion of orange tree top, ground and side viewed by the radiometer, which are given, in terms of the solid angles illus-

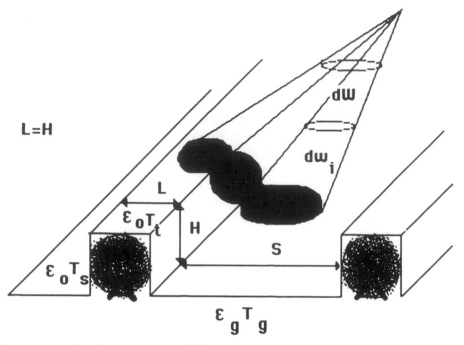

Figure 4.17 Schematic representation of the contribution of the different parts of an orange tree field to the radiance measured with a radiometer (Caselles and Sobrino 1989).

trated in Figures 4.17 and 4.18, by

$$P_i = \frac{d\omega_i}{d\omega} \tag{4.4.4}$$

with $i = t$, g and s. These quantities are strongly dependent on the viewing geometry, surface geometry and sensor characteristics (Sobrino *et al.* 1989), but in any case the relation

$$P_t + P_g + P_s = 1 \tag{4.4.5}$$

must be satisfied.

After some detailed analysis Caselles and Sobrino (1989) arrived at an expression for the effective temperature

$$T = aT_t + bT_g + cT_s \tag{4.4.6}$$

where T_t, T_g and T_s are the temperatures of the orange tree tops, the ground and the orange tree sides, respectively, and where the parameters a, b and c depend on the orange tree emissivity, the ground emissivity, the geometrical factors of the orange grove and P_t, P_g and P_s. What is important is that the a, b and c do not depend on the values of the temperature of the various parts of the system and they do not depend on the meteorological conditions. Thus, as in the theory of Becker and Li, we have a local split-window algorithm. Substantial tests of this model with AVHRR data are described by Caselles and Sobrino (1989) in their work on frost damage studies in orange groves in Valencia, Spain, and satisfactory results were obtained (Sobrino and Caselles 1991).

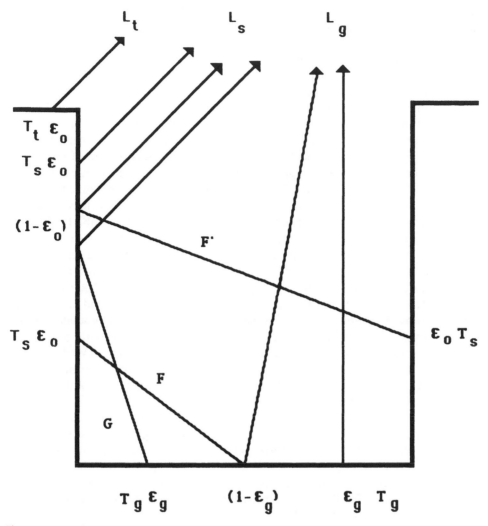

Figure 4.18 Different terms which compose the radiance observed by a radiometer (Caselles and Sobrino 1989).

4.4.2 Thermal inertia

In the consideration of thermal infrared data so far in this chapter we have been concerned with inverting the satellite-received data to determine surface properties, principally the temperature but also, for land surfaces, the emissivity. To study the subsurface conditions at sea and relate them to the surface temperature is a complicated problem. It is complicated because heat transfer, both horizontally and vertically, is largely accomplished by transport of the sea water in surface currents or by convection; the study of this would take us into lengthy oceanographic studies having no very close connection with the AVHRR. For the land, however, the transport of heat horizontally in the surface layer of the land is of no importance and the transport of heat below the surface vertically is by conduction only. A study of the surface temperature of the land may therefore be able to provide indirectly some information about the material below the surface and it was to assist with this that the concept of thermal inertia was introduced and has been used.

The thermal inertia, P, is a lumped parameter equal to $(k\rho c)^{1/2}$, where k is the thermal conductivity, ρ is the density and c is the specific heat capacity of the material. Thermal inertia cannot be measured directly in the field and its value must be inferred from measurements of the temperature variation during the diurnal cycle, combined with a knowledge of the heating processes which occur during the cycle and of the visible and near-infrared reflective processes during the day.

In the early 1970s the development of high quality visible and near-infrared sensors for satellite missions was followed closely by the construction of improved sensors in the thermal infrared region (10–12 μm) of the electromagnetic spectrum. The satellite for the Heat Capacity Mapping Mission (HCMM) was launched on 26 April 1978. This was designed to acquire 500 m resolution thermal infrared data. The HCMM gathered data in support of studies to determine the feasibility of using thermal infrared data to compute the thermal inertia for the Earth's surface. The data were obtained at about twelve-hour intervals at times when the temperature was presumed to be a maximum or a minimum. The satellite was in a Sun-synchronous orbit with nominal ascending equator-crossing time of 2.00 p.m. Local Standard Time to provide north mid-latitude crossing times of 2.30 and 13.30 Local Standard Time. A mission product of 'apparent thermal inertia' was produced for its potential value in determining surface characteristics from the satellite data. The scope of the experimental Heat Capacity Mapping Mission (HCMM) has been described by Price (1977). The theory underlying the apparent thermal inertia product is quite simplified (Price 1977) and the possibly limited value of the apparent thermal inertia was recognised.

There is a fairly extensive literature related to thermal inertia, but it is nearly all concerned with HCMM data and their applications in geological studies. We simply concentrate on the principles of thermal inertia studies with satellite data and indicate how AVHRR data could be used. While the archive of HCMM data is useful for historical studies, the satellite no longer operates. It is therefore worthwhile to attempt to use the AVHRR to determine thermal inertia values (Xue and Cracknell 1992, 1993, Cracknell and Xue 1996).

The origins of studies of thermal inertia for the surface of the Earth are derived from a one-dimensional model of heat conduction with periodic boundary conditions originally developed by Jaeger (1953) to study the surface temperature of the Moon. Let us think of the value of the temperature of the surface of the Earth as the response to the forcing mechanism of the incident solar radiation. The Sun rises in the morning and it shines more or less equally on different parts of the surface of the Earth, at least on a regional basis. But the response in terms of surface temperature varies enormously. At the end of the day the Sun sets and, again, the response, in terms of cooling varies enormously; generally those areas which experience large temperature rises during the day will suffer large temperature drops at night. Then the cycle starts again the following morning. The amplitude of the temperature variations of the surface, over a 24 hour period, varies enormously for different types of surface. The most extreme differences are between, say, dry sand and bare rock on the one hand and deep water on the other hand. One sees this very clearly in satellite images in the thermal infrared; at night the land is generally colder than the sea, but during the day the land becomes warmer than the sea. The amplitude of the temperature variations of the land surface is likely to be several tens of degrees C (the desert can be surprisingly cold at night) whereas the sea surface variations are only of the order of a few degrees (even this can be reduced by the action of waves or tides).

In qualitative terms, the thermal inertia is a quantity which is defined to give a measure of the reluctance of the surface of the Earth to respond to a given supply of heat energy (from the Sun). An over-simplified explanation of the difference between the response of the land and of the sea is to say that it is because the specific heat of water is very large. We recall that for the absorption of heat

$$Q = mc\theta \tag{4.4.7}$$

where Q is the heat supplied, m is the mass, c is the specific heat capacity and θ is the rise in temperature. However, the conductivity is also involved:

$$\dot{Q} = -kA \frac{(\theta_2 - \theta_1)}{(x_2 - x_1)} = -kA \frac{d\theta}{dx} \tag{4.4.8}$$

where $\dot{Q}$ is the rate of heat transfer, k is the thermal conductivity, A is the area of cross-section and $(d\theta/dx)$ is the temperature gradient. A fixed amount of heat arrives at the surface of the Earth; some of it raises the temperature of the surface layer and some of it is conducted away to layers beneath the surface.

The concept of thermal inertia was introduced to try to take into account both the specific heat and the thermal conductivity in identifying a quantity that would indicate the reluctance of the surface of the Earth (or of any medium) to respond to a given heat input. That is, large thermal inertia corresponds to small amplitude oscillations or variations in the surface temperature over a 24 hour period. Let us consider initially the effects of specific heat capacity and thermal conductivity separately. For a given heat input, a large specific heat implies a small temperature change (from equation (4.4.7)) and a large thermal inertia. For a given heat input and a large value of the thermal conductivity (i.e. for a good thermal conductor) a large fraction of the heat is taken away to lower layers and so the surface temperature rise is small. That is, a large thermal conductivity implies a large thermal inertia. What is therefore done is to define the thermal inertia, P, by

$$P = (k\rho c)^{1/2} \tag{4.4.9}$$

so as to take into account both of these factors. ρ, the density, is introduced because what matters here is really the specific heat capacity per unit volume, whereas c, the specific heat capacity, is usually defined as the heat capacity per unit mass.

Let us try to understand what the concept of thermal inertia means in practice. Figure 4.19 shows a very idealised situation. The amplitude of the oscillation is related to (in fact is inversely proportional to) the thermal inertia. Thus to determine P, the thermal inertia, we have to measure the amplitude of the oscillations $\Delta\theta_0$. By doing this one can thus get a measure of P, the thermal inertia.

Let us leave aside the question of the emissivity problem for the moment and suppose that we can resolve that. There is still a quite serious problem that satellite measurements, for any given area, are not carried out right through the day and night. They can only be made when the satellite is in range and this is determined by its orbit. With Landsat TM, which has a thermal channel, the data obtained once in 16 days at about 10 a.m. local Sun time is of no use for determining thermal inertia. The HCMM was devised to generate data at what were supposed to be the times of maximum and minimum surface temperature. Obviously one can, if one

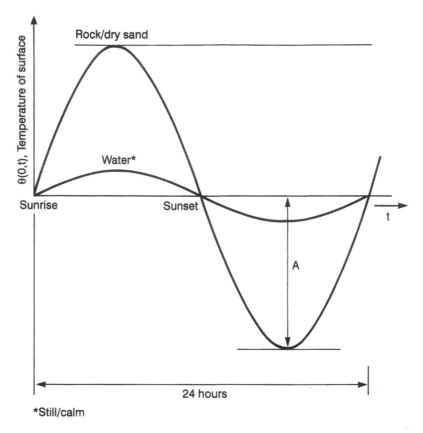

Figure 4.19 Idealised diurnal surface temperature oscillations for regions of different thermal inertia values.

wishes, work with that data set. But it was a one-off experimental mission. Like Seasat it was useful on a proof of concept basis, but it has all the limitations of such a data set:

- One cannot now go back and do retrospective fieldwork simultanously with the satellite data acquisition.

- The data are not available on an on-going operational basis and so one cannot use them for present or future environmental studies.

On the other hand, the AVHRR has some possibilities of being useful:

- The AVHRR has thermal infrared channels giving data of very high quality.

- It is an operational system which gathers data frequently and regularly, although almost certainly it does not gather data at the times of maximum and minimum temperature.

Most of the applications research involving HCMM data was concentrated in the fields of geology, hydrology, soil science and on the modelling of heat transfer. Cracknell and Xue (1996) give a comprehensive introduction to the applications of HCMM data. Several reports (Kahle *et al.* 1981, Watson and Hummer-Miller 1981, Watson 1982, Watson *et al.* 1982) have appeared describing the application of

HCMM data to thermal inertia calculations for the delineation of regional lineaments and rock types. As far as geological information about the nature of the near-surface rocks is concerned, once one has determined the value of the thermal inertia of such an area there is little point in making repeated measurements; once one has found the thermal inertia of that material its value is not going to change. However, one objective of NASA's HCMM was to evaluate the feasibility of using HCMM data to assess soil moisture effects by observing temperatures near the maximum and minimum of the diurnal temperature cycle. When the soil surface is wet, evaporation is the major factor controlling the surface heat loss. As the surface layer dries and the soil water supply cannot meet the evaporative demand, soil temperature is largely influenced by thermal inertia. Thus the diurnal range of surface soil temperature can be an indication of soil water content. Idso *et al.* (1975) found a significant relation between the diurnal range of surface soil temperature (bare soil) and surface soil water content, and reported that the relation was a function of soil type. Pratt and Ellyet (1979) presented a method for residual soil type discrimination and soil moisture determination. There is thus a potential need for ongoing determinations of thermal inertia and therefore an opportunity for using AVHRR data to determine thermal inertia data. Xue (1989) developed a method for soil monitoring with thermal inertia obtained from AVHRR data using theoretical relations among soil moisture, soil density and thermal inertia. A soil moisture map of the North China Plain was generated (Ma and Xue 1990). In the Group Agromet Monitoring Project (GAMP) in 1981–1985 such methods were, for the first time, applied in a systematic and quasi-operational way to monitoring a test area in Mali during a complete growing season (Rosema 1986a,b). The use of AVHRR data in connection with RIVERSAT, an interactive information system for real-time monitoring of Polish river catchments in relation to studying land surface temperature and soil moisture was described by Baranski and Mrugalski (1991).

The concept of apparent thermal inertia, which was generated as a mission product from HCMM, arises from a simple model of the absorption of solar radiation at the surface, developed by Watson (1973, 1975). In this simple thermal model it is assumed that the Sun causes periodic heating of the Earth's surface and that the ground losses of heat are only by radiative transfer. A very simple relation to compute apparent thermal inertia, which is only an approximation to the actual thermal inertia value, from the surface albedo (A) and the diurnal temperature difference (ΔT) values was chosen as follows

$$\text{ATI} = \frac{(1 - A)}{\Delta T}. \tag{4.4.10}$$

A generalised formulation to calculate thermal inertia from global remote-sensing data was given by Price (1977), which included various parameters like diurnal temperature difference, albedo, solar constant, atmospheric transmittance at visible wavelengths, the angular speed of the Earth's rotation, the ratio of the heat flux density transferred by the surface to the air to that transferred into the ground, and a parameter to account for solar declination and latitude. This was used with HCMM data in a simplified form as (Short and Stuart 1982)

$$\text{ATI} = NC \frac{(1 - A)}{\Delta T}, \tag{4.4.11}$$

where N is a scaling factor and C is a constant to normalise for solar flux variations with latitude and solar declination.

Kahle (1977) applied a finite difference technique to compute thermal inertia, based on the heat balance equation, utilising empirical and theoretical expressions and data from meteorological observations, topographical data and remote sensing day and night thermal data.

However, the apparent thermal inertia, as just defined, is only an approximation to the thermal inertia defined in equation (4.4.9). For the determination of thermal inertia one needs to solve the equation of thermal conduction, which governs the (vertical) flow of heat within the region below the surface. This is the diffusion equation

$$\frac{\partial^2 T(x, t)}{\partial x^2} = \frac{\rho c}{k} \frac{\partial T(x, t)}{\partial t} \tag{4.4.12}$$

where $T(x, t)$ is the temperature at depth x below the surface and time t; the quantity $D \,(= k/\rho c)$ is called the thermal diffusivity of the material. To solve this equation requires the use of the boundary condition on the temperature gradient $(\partial T(x, t)/\partial x)$ at the surface of the Earth $(x = 0)$. This boundary condition is given by the energy balance condition at the surface. To solve this problem one needs to make some approximations and different levels of approximation can be adopted. For instance Xue and Cracknell (1992) adopted the surface boundary condition

$$-k \left.\frac{\partial T(x, t)}{\partial x}\right|_{x=0} = (1 - A)S_0\, C_t \cos z' - R_{\text{Earth}} + R_{\text{sky}} - H - LE \tag{4.4.13}$$

where

S_0	= the solar constant
C_t	= the atmospheric transmittance in the visible spectrum
z'	= the local zenith angle for an inclined surface
R_{Earth}	= the Earth-emitted radiation
R_{sky}	= the downward sky longwave radiation
H	= the sensible heat flux to the atmosphere and
LE	= the latent heat flux to the atmosphere.

With this model it is possible to determine the thermal inertia from time-lapse sequences of AVHRR data where there is only one extra unknown, namely the phase angle of the diurnal variation of the surface temperature. The value of this parameter can be determined from the time of maximum temperature in the daytime and this information is easily obtained from a meteorological station in the study area. The thermal inertia can then be calculated from AVHRR thermal infrared data on a pixel-by-pixel basis (Xue and Cracknell 1992). From this model a second-order approximation thermal inertia model over areas with variable soil moisture with a surface temperature range 280–310 K has also been developed (Xue and Cracknell 1995b). Radiation balance investigations using AVHRR data have been described by Frouin et al. (1988) and Rimóoczi-Paál (1993) and latent heat flux determination using AVHRR and SSM/I data has been carried out by Schulz (1993) for the North Sea. The ground heat flux from the land surface has been estimated for Iceland using AVHRR data by Cracknell et al. (1989) and Mackay et al. (1991).

As already mentioned the interest in the potential use of AVHRR data as a source of thermal inertia data is primarily in connection with soil moisture studies.

The possibility of the use of AVHRR thermal infrared channel data to study soil moisture in an agricultural homogeneous region in France was investigated by Cheevasuvit *et al.* (1985). Five AVHRR scenes spread over the period February–September 1982 were used and encouraging results were obtained. However, far more work on soil moisture with AVHRR has been concerned with the use of vegetation indices derived from channel-1 and channel-2 data than with thermal infrared data (see chapter 5). AVHRR data from a few cloudless days in 1992 and 1993 were used by Csiszár *et al.* (1993) to study the daily amplitude of the variation of the land surface temperature for agricultural and forested areas of Hungary. Good correlation was found between this amplitude and both the vegetation index and the surface radiation balance.

CHAPTER FIVE

Vegetation

5.1 BACKGROUND

The use of AVHRR data in the study of vegetation is surely the most successful of the unexpected applications of AVHRR data. The use of the data in meteorology was the main application originally envisaged for the data. At least some of the marine applications of the data that we have described in chapter 4 had also been envisaged at the time that the system was originally planned and designed. It is probably fair to say that at the time of the planning and designing of the AVHRR (i.e. in the mid-1970s) no one had any idea that the data that it would generate would ever find any significant use in the study of vegetation.

Traditionally vegetation had been studied using (false colour) infrared photography. In the mid-1970s Landsat MSS (MultiSpectral Scanner) data were beginning to be used quite widely in vegetation studies. There have been many pieces of work done with Landsat data and these were mostly concerned with classification. Attempts were also made to try to study change or dynamic phenomena with Landsat. For instance attempts were made to predict crop yields in the LACIE (Large Area Crop Inventory Experiment) in the USA. But one problem was that the Landsat data can only be obtained relatively infrequently and one or two cloudy days which happened to be Landsat overpass days could wreck an entire season's predictions. The cost of Landsat data started to climb also. People began to realise that, in spite of its poorer spatial resolution, AVHRR data might be useful in this connection. Some of the major problems of using Landsat data are:

- infrequent coverage (once in 18 or 16 days) – orbital considerations
- further reduced coverage due to cloud cover
- too much detail for large-scale projects
- high cost of data.

Large-scale projects found not only the cost of Landsat data but also the large amount of such data that would be needed to cover the area being studied to be a problem. The idea of using AVHRR data to complement, or interpolate temporally between, the higher-spatial-resolution Landsat data began to gain ground. For instance the EEC now has a big Agricultural Project for Europe which is using AVHRR data in this way, not with any hope of looking at individual fields but looking at the general state of the vegetation in an area.

The TIROS-N, NOAA-6, NOAA-7, . . . series of satellites carrying the AVHRR began to command attention soon after the launch of TIROS-N in 1978 because, unlike their predecessors (NOAA-2 to NOAA-5) that carried the VHRR, the resolution is better and the output is digital. In fact the spectral response of channels 1 and 2 of NOAA-6 onwards is very similar to that of the Landsat MSS bands 5 and 7 (or 2 and 4 in the new notation), see Figure 5.1. While this was almost certainly accidental it does happen to be most fortunate as we shall see in the next section. The advantages of the AVHRR which, by now, have become widely apparent include the following:

- frequent coverage (several times per day)
- good calibration
- cheap data
- adequate spatial resolution for many projects
- direct readout within the horizon of the ground station (free data).

In section 2.4.2 we have already mentioned the normalised difference vegetation index as being a product generated by NOAA NESS and by various other AVHRR ground stations, in the same way that they generate sea surface temperature products. There is one big difference however. There is a sharp contrast between the principles involved in the study of vegetation, which is the topic of this chapter, and the study of the Earth's surface temperature, which was the topic of the last chapter. The sea surface temperature is a clearly defined physical parameter; the vegetation index is not. In the last chapter we were able to take the approach of a physicist. The physicist seeks to establish fundamental laws describing the operation of the physical universe and to represent these physical laws by mathematical equations. Any physical system is then studied by writing down the appropriate mathematical equations and attempting to solve them. The determination of the values of the parameters and of the solution of the equations may be difficult but the formulation of the problem is clear cut. In the previous chapter the equations in question were Planck's equation, to describe the emission of radiation by the Earth's surface, and the radiative transfer equation, to describe the passage of the radiation through the atmosphere before it arrives at the satellite. In the present chapter we are concerned with more complicated systems for which it is not realistic to attempt to formulate the problem in terms of the basic physical processes that occur at the Earth's surface.

There is a rather apt quotation from the philosopher A. N. Whitehead given at the beginning of the paper by Perry and Lautenschlager (1984):

> The aim of science is to seek the simplest explanation of complex facts. We are apt to fall into the error of thinking that the facts are simple because simplicity is the goal of our quest. The guiding motto in the life of every natural philosopher should be, 'Seek simplicity and distrust it'.

The use of multispectral scanners for the study of vegetation, especially using some form of multispectral classification scheme is extremely well established and has been very successful. However, the spatial resolution of the AVHRR is very

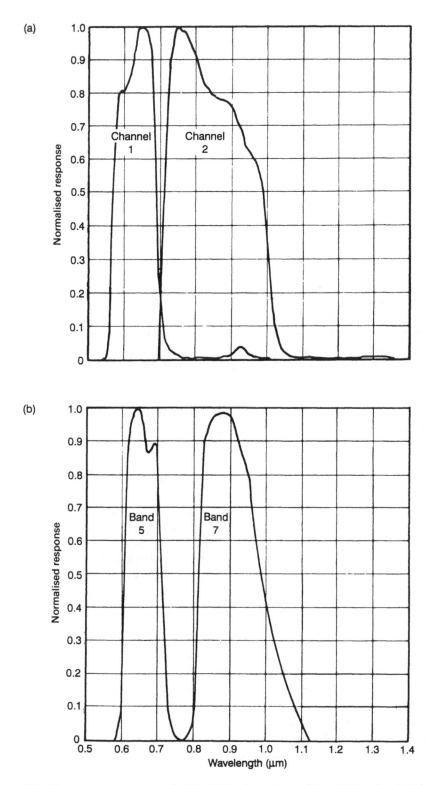

Figure 5.1 Spectral response curves for (a) AVHRR channels 1 and 2 and (b) Landsat MSS bands 5 and 7 (or 2 and 4 in the new notation) (Schneider *et al.* 1985).

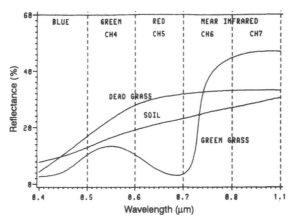

Figure 5.2 Idealised reflectance patterns of herbaceous vegetation and soil (Deering *et al.* 1975).

poor when compared with the sizes of parcels of homogeneous land cover in many parts of the world. Most of the successful land cover classification work has been done with high-resolution satellite data from SPOT and Landsat or with data from airborne scanners. While AVHRR data can be used in work on multispectral classification of land areas, their use is somewhat limited. We shall consider this, to some extent, in section 5.5 but this is not the most important way in which AVHRR data are used to study vegetated areas. One most frequently uses a vegetation index and so this chapter is mostly concerned with vegetation indices. It is therefore useful to start by reviewing the history of vegetation indices.

5.2   VEGETATION INDICES

The spectral response or reflectivity of a vegetated surface is generally high in the near-infrared wavelength range, see the curve labelled 'green grass' in Figure 5.2. For such a surface the signal generated in band[1] 7 (old notation or 4 in the new notation) of the Landsat MSS or in channel 2 of the AVHRR will therefore be high. At different stages in the growing season the reflectivity spectrum will be different. At a stage of vigorous growth in the middle of its growth cycle the reflectivity in the near-infrared will be higher than at other stages in the growth cycle. The concept of a vegetation index came about in connection with data generated by scanners such as the Landsat MSS with only a very small number of spectral bands. It can be used as a classifier to try to distinguish vegetated surfaces from non-vegetated surfaces. It can also be used to study the variations during the development cycle of a given crop. The early work was carried out with Landsat MSS data and the use of AVHRR data for vegetation studies only developed much more recently. Consequently, the early work on vegetation indices is formulated in terms of Landsat MSS bands. Perry and Lautenschlager (1984) noted that there were about four dozen formulae which have been proposed by various workers to define vegetation

---

[1] The term 'band' rather than 'channel' is most widely used for Landsat. For the AVHRR, usage is mixed but we have chosen the word channel as being probably the more commonly used term.

Table 5.1 Nominal wavelength bands of Landsat MSS scanner

Wavelength range (μm)	Band number (old notation)	Band number (new notation)
0.5–0.6	4	1
0.6–0.7	5	2
0.7–0.8	6	3
0.8–1.1	7	4

indices. In this section we shall refer to the four bands of the Landsat MSS in the old notation (i.e. as 4, 5, 6 and 7) because that is the notation that was used when the various different vegetation indices were first defined. Since most of them have not been widely adopted and they are therefore only of historical interest now, it seemed that there would be little point in translating all the definitions into the current notation (bands 1, 2, 3 and 4, respectively, see Table 5.1).

We start from the following ideas. In today's complex world it is important to obtain current and accurate information on a global basis regarding the extent and condition of the world's major food and fibre crops. Traditional sampling techniques for estimating crop conditions, based on field collection of data, are time consuming, and costly; they are often not generally applicable to foreign regions and it is not practical to attempt to use these methods to gather good data on a global scale. It became apparent that satellite remote sensing provided a method of gathering data frequently on a global scale, initially using Landsat. But the problem is that the satellite-flown instruments do not provide direct measurements of the parameters that are required, namely vegetation characteristics such as species, leaf area, stress and biomass. These parameters have to be deduced from the remotely sensed data, in the case of Landsat MSS from the signals recorded in the four spectral bands. The signals received are influenced by the vegetation characteristics, soil background and atmospheric conditions. The problem is to invert the satellite data to determine the characteristics of the vegetation, etc. Various workers have developed techniques for qualitatively and quantitatively assessing the vegetative canopy from spectral measurements. The objective has been to reduce the four bands of MSS data to a single number for predicting or assessing such canopy characteristics as leaf area, biomass and percentage ground cover. We shall follow the history of vegetation indices as given by Perry and Lautenschlager (1984) but would refer the reader to that paper for a full citation of references.

Idealised reflectance patterns for herbaceous vegetation and soil are compared in Figure 5.2. Dead or dormant vegetation has a higher reflectance than living vegetation in the visible spectrum and lower reflectance in the near-infrared. Soil has a higher reflectance than green vegetation and a lower reflectance than dead vegetation in the visible, whereas, in the near-infrared, soil typically has lower reflectance than green and dead vegetation. There are, by now, numerous studies of the reflectance spectra of different samples of vegetation in different states. The older work is described by Knipling (1970) and in chapter 33 of the second edition of the *Manual of remote sensing* (Colwell 1983).

The principle used in defining vegetation indices is based on the very steep slope of the spectral curve for green vegetation at about 0.7 μm (see Figure 5.2). The contrast between the near-infrared and visible AVHRR reflectances is an indicator of the amount and state of the vegetative cover. Numerous vegetation indices have been used to make quantitative estimates of leaf area index, percentage ground cover, plant height, biomass, plant population and other parameters. Most formulae are based on ratios or linear combinations and exploit differences in the reflectance patterns of green vegetation and other objects like those sketched in Figure 5.2. The very steep rise in the reflectance at the red–near-infrared boundary region of the spectrum is sometimes called the red edge. Technically the red edge position is defined as λ_{pr}, the value of the wavelength for which the slope of the reflectance curve is a maximum. The position of the red edge is not rigidly fixed; even for vegetation of a given species it shifts as a result of various factors including rainfall, temperature and season (see e.g. Miller *et al.* 1991).

Some workers have used the digital counts from the individual MSS bands (denoted by MSS4, MSS5, etc.) to estimate percentage ground cover and vegetative biomass. Others have used ratios of the MSS digital counts to estimate and monitor green biomass, etc. The 12 pairwise ratios (six of which are inverses of the other six) will be denoted by R45 = MSS4/MSS5, R46 = MSS4/MSS6, etc.

Rouse *et al.* (1973) proposed using the normalised difference of MSS7 and MSS5 for monitoring vegetation, which will be referred to as ND7. Deering *et al.* (1975) added 0.5 to ND7 to avoid negative values and took the square root of the result to stabilise the variance. This index is referred to as the transformed vegetation index and will be denoted by TVI7. Similar formulae using MSS6 and MSS5 were proposed:

$$\text{ND6} = (\text{MSS6} - \text{MSS5})/(\text{MSS6} + \text{MSS5}) \tag{5.2.1}$$

$$\text{ND7} = (\text{MSS7} - \text{MSS5})/(\text{MSS7} + \text{MSS5}) \tag{5.2.2}$$

$$\text{TVI6} = (\text{ND6} + 0.5)^{1/2} \tag{5.2.3}$$

$$\text{TVI7} = (\text{ND7} + 0.5)^{1/2}. \tag{5.2.4}$$

It is ND7 which eventually has became so widely adopted and is now almost universally known as the normalised difference vegetation index NDVI. It was, however, far from obvious at the outset that it would achieve such fame and importance and that most of the other proposed indices would sink into oblivion.

Perry and Lautenschlager (1984) argued that the addition of 0.5 does not eliminate all negative values and suggested the following formulae:

$$\text{TVI6} = [(\text{ND6} + 0.5)/\text{ABS}(\text{ND6} + 0.5)] \times [\text{ABS}(\text{ND6} + 0.5)]^{1/2} \tag{5.2.5}$$

$$\text{TVI7} = [(\text{ND7} + 0.5)/\text{ABS}(\text{ND7} + 0.5)] \times [\text{ABS}(\text{ND7} + 0.5)]^{1/2} \tag{5.2.6}$$

where ABS denotes the absolute value and 0/0 is set equal to 1(!). Perry and Lautenschlager showed that these formulae are equivalent for decision making to the basic ratios R65 and R75.

Various transformations of data in the four Landsat MSS bands are given by Perry and Lautenschlager (1984) taken from a variety of sources. One involved the use of the technique of sequential orthogonalisation underlying the Gram–Schmidt process to produce an orthogonal transformation of the original Landsat data space to a new four-dimensional space. This was called the 'tasselled cap' transformation and the four new axes were named brightness (soil brightness index, SBI), greenness (green vegetative index, GVI), yellow stuff (YVI) and nonsuch (NSI); it was credited to R. J. Kauth and G. S. Thomas. These new indices were given by

$$\text{SBI} = 0.332 \text{ MSS4} + 0.603 \text{ MSS5} + 0.675 \text{ MSS6} + 0.262 \text{ MSS7} \qquad (5.2.7)$$

$$\text{GVI} = -0.283 \text{ MSS4} - 0.660 \text{ MSS5} + 0.577 \text{ MSS6} + 0.388 \text{ MSS7} \qquad (5.2.8)$$

$$\text{YVI} = -0.899 \text{ MSS4} + 0.428 \text{ MSS5} + 0.076 \text{ MSS6} - 0.041 \text{ MSS7} \qquad (5.2.9)$$

$$\text{NSI} = -0.016 \text{ MSS4} + 0.131 \text{ MSS5} - 0.452 \text{ MSS6} + 0.882 \text{ MSS7}. \qquad (5.2.10)$$

An alternative to this was obtained by applying principal component analysis to MSS data; this was credited to a group from IBM. The structure of the resulting transformation and the interpretation of the principal components are similar to those for the Kauth–Thomas transformation:

$$\text{MSBI} = 0.406 \text{ MSS4} + 0.600 \text{ MSS5} + 0.645 \text{ MSS6} + 0.243 \text{ MSS7} \qquad (5.2.11)$$

$$\text{MGVI} = -0.386 \text{ MSS4} - 0.530 \text{ MSS5} + 0.535 \text{ MSS6} + 0.532 \text{ MSS7} \qquad (5.2.12)$$

$$\text{MYVI} = 0.723 \text{ MSS4} - 0.597 \text{ MSS5} + 0.206 \text{ MSS6} - 0.278 \text{ MSS7} \qquad (5.2.13)$$

$$\text{MNSI} = 0.404 \text{ MSS4} - 0.039 \text{ MSS5} - 0.505 \text{ MSS6} + 0.762 \text{ MSS7} . \qquad (5.2.14)$$

The similarity of these two results is remarkable because the ideas and techniques underlying the two processes are quite different. With principal component analysis the experimenter imposes no prior order or physical interpretation on the principal directions. Gram–Schmidt orthogonalisation, however, gives the freedom to establish indirectly a physical interpretation by choosing the order in which the calculations are performed. Another transformation was also proposed, based on the idea of spectral brightness and contrast. The first two components of the resulting transformation are similar to the first two components of the two preceding transformations:

$$\text{SSBI} = 0.437 \text{ MSS4} + 0.564 \text{ MSS5} + 0.661 \text{ MSS6} + 0.233 \text{ MSS7} \qquad (5.2.15)$$

$$\text{SGVI} = -0.437 \text{ MSS4} - 0.564 \text{ MSS5} + 0.661 \text{ MSS6} + 0.233 \text{ MSS7} \qquad (5.2.16)$$

$$\text{SYVI} = -0.437 \text{ MSS4} + 0.564 \text{ MSS5} - 0.661 \text{ MSS6} + 0.233 \text{ MSS7} \qquad (5.2.17)$$

$$\text{SNSI} = -0.437 \text{ MSS4} + 0.564 \text{ MSS5} + 0.661 \text{ MSS6} - 0.233 \text{ MSS7} . \qquad (5.2.18)$$

Richardson and Wiegand (1977) used the perpendicular distance to the soil line as an indicator of plant development. The soil line, which can be considered as a two-dimensional analogue of the Kauth–Thomas SBI, was estimated by linear regression. Bare soil spectra in the visible and near-infrared (NIR) are characterised by their spectral response, bidirectional reflectance distribution function (BRDF, see section 5.7) and polarisation properties (Bowers and Hanks 1965, Condit 1970, Coulson and Reynolds 1971, Stoner and Baumgardner 1981). For a given set of wavelengths, soil spectral signatures vary in two ways: there are brightness differences associated with the magnitude of the reflected flux and there are spectral curve form variations attributed to specific mineralogical and organic absorption features. It was found in practice that the major difference or primary variation among soil spectra is in the intensity of the reflected flux and that changes in the general shape of the curve are of much less significance. Moisture, roughness and decomposed organic matter all decrease the reflectance of the soil, producing a darker soil surface. The less significant feature of curve-form variations arises from variations in soil biogeochemical absorption properties. We imagine an idealised situation in which we consider a large number of different soil samples and find that the spectral reflectance curves of all the samples are of the same shape (within the wavelength range of the two red and near-infrared channels) but that the magnitudes of the reflectances (for a given wavelength) were different. That is, the reflectance $R(\lambda)$ could be represented by the relation

$$R(\lambda) = \alpha f(\lambda) \tag{5.2.19}$$

where α is a scale factor which takes different values for the different soil samples. That is, the reflectance curve for any one soil sample is related to that of any other soil sample by a simple scaling of the reflectance curve. Then if one plots the reflectances in channel 2 and channel 1 against one another for all these samples the points will all lie on a straight line and this is called the soil line, see Figure 5.3. Departures from this idealised straight-line behaviour will arise if there are differences in the shapes of the spectral reflectance curves. It is found in practice that departures from this straight line are relatively small. Jackson *et al.* (1980) plotted a wide range of soil types in NIR–red space and reported a slightly curvilinear relation. Huete *et al.* (1984) reported that the 'global' soil line actually consists of numerous non-parallel soil lines describing soil moisture variations for specific soil types. Two perpendicular vegetation indices were proposed by Richardson and Wiegand (1977):

PVI6

$$= [(-2.507 - 0.457\ \text{MSS5} + 0.498\ \text{MSS6})^2$$
$$+ (2.734 + 0.498\ \text{MSS5} - 0.543\ \text{MSS6})^2]^{1/2} \tag{5.2.20}$$

and

$$\text{PVI7} = [(0.355\ \text{MSS7} - 0.149\ \text{MSS5})^2 + (0.355\ \text{MSS5} - 0.852\ \text{MSS7})^2]^{1/2}. \tag{5.2.21}$$

These formulae are computationally inefficient and do not distinguish right from left

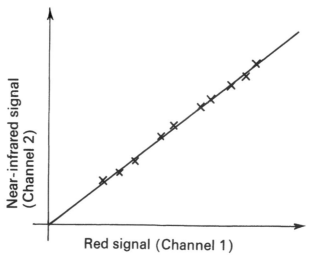

Figure 5.3   Sketch to illustrate soil line.

of the soil line (i.e. water from green stuff). The standard formula from analytical geometry for the perpendicular distance from a point to a line solves this difficulty, leading to

$$PV16 = (1.091\ MSS6 - MSS5 - 5.49)/(1.091^2 + 1^2)^{1/2} \qquad (5.2.22)$$

$$PVI7 = (2.4\ MSS7 - MSS5 - 0.01)/(2.4^2 + 1^2)\ . \qquad (5.2.23)$$

Other vegetation indices quoted by Perry and Lautenschlager include the DVI, given by

$$DVI = 2.4\ MSS7 - MSS5 \qquad (5.2.24)$$

the AVI, given by

$$AVI = 2.0\ MSS7 - MSS5 \qquad (5.2.25)$$

and the GRABS given by

$$GRABS = GVI - 0.09178SBI + 5.58959\ . \qquad (5.2.26)$$

Further details and the citation of references are given by Perry and Lautenschlager (1984).

Kanemasu *et al.* (1977) regressed winter wheat leaf area measurements of MSS band ratios and produced the following regression equation:

$$ELAI = 2.68 - 3.69\ R45 - 2.31\ R46 + 2.88\ R47 + 0.43\ R56 - 1.35\ R57$$
$$+ 3.07\ [R4 - (0.5\ R47)(R45)]\ . \qquad (5.2.27)$$

The leaf area index (LAI) is defined as the ratio of the area of the leaves to the area of the ground; the LAI can take values much greater than 1. Pollock and Kanemasu (1979) later used a larger data set plus stepwise regression and obtained another regression equation

$$CLAI = 0.366 - 2.265\ R46 - 0.431(R45 - R47)(R45) + 1.745\ R45 + 0.57\ PVI7\ .$$
$$(5.2.28)$$

Separate regression equations were also obtained for CLAI values above and below 0.5:

$$LAI = 1.903 - 1.138\ R56 - 0.071\ (R45 - R47)\ R45 - 0.016\ PVI6 \qquad (5.2.29)$$

if CLAI is less than 0.5 and

$$LAI = -5.33 + 0.036\ PVI7 + 6.54\ TVI6 \qquad (5.2.30)$$

if CLAI is greater than 0.5.

In areas where the vegetation canopy does not provide complete cover of the ground the soil background both contributes a reflected signal apart from the vegetation and also interacts with the overlying vegetation through multiple scattering of radiant energy. Arid and semi-arid regions are characterised by sparse vegetation cover, and the soil-background-reflected signal usually overwhelms the relatively small vegetation-reflected component. The use of vegetation indices for arid and semi-arid areas will be discussed in section 5.6. The question of generating a soil adjusted vegetation index (SAVI) has been considered by various authors (Huete 1988, Major *et al.* 1990, Huete and Tucker 1991). The soil adjusted vegetation index

(SAVI) is defined as

$$SAVI = \frac{(NIR + l_2) - (RED + l_1)}{(NIR + l_2) + (RED + l_1)}$$

(5.2.31)

which essentially corresponds to a translation of the origin in an MSS7–MSS5 plot (or channel-2–channel-1 plot for AVHRR) to a point $(-l_1, -l_2)$ so that the soil line passes through the origin.

Thompson and Wehmanen (1979) proposed a technique utilising transformed digital data for the detection of agricultural vegetation undergoing moisture stress. The MSS data are rotated into the Kauth–Thomas vectors (SBI, GVI, YVI, NSI) (see equations (5.2.7)–(5.2.10)) to screen out clouds, water, bare soil, etc. Each vector is evaluated and any vector having values considered unreasonable for agricultural data is discarded. The remaining pixels are considered the good pixels. 1% of the pixels with the lowest GVI values are then discarded. The lowest GVI value remaining becomes the soil line. A green number is then computed for each pixel by subtracting the soil line from GVI. The green index number (GIN) is then an estimate of the percentage of pixels in the scene with a green number greater than or equal to 15:

$$GIN = \frac{\text{number of pixels with a green number of 15+}}{\text{number of good pixels}} \times 100 .$$

(5.2.32)

Before leaving the account of Perry and Lautenschlager (1984) we should mention two other points. First, these authors describe a number of pieces of work aimed at establishing empirical relations among some of the vegetation indices that have been mentioned above. Secondly, they themselves study various formal equivalences among different vegetation indices. However, nearly all of these vegetation indices are of historical interest only. As mentioned above, one of them, ND7 or the NDVI, is now routinely calculated at ground stations, is very widely used and appears to be extremely successful in practice in numerous areas of application. The NDVI has proved to be extremely popular and is very widely used. With hindsight it is easy to see that there are several reasons why this should be so:

(i)   it is very simple to calculate and (as we shall see in section 5.4.3) it is easy to develop a very quick technique to eliminate data corresponding to cloudy areas

(ii) its definition is based on very simple, but very general and very sound, ideas basically involving exploiting the features of the curves shown in Figure 5.2 for healthy and senescent vegetation and noting that these curves are of very general applicability

(iii) the effects of extraneous factors (i.e. non-vegetation effects) are reduced by the normalisation, i.e. by dividing the difference (MSS7 − MSS5) by the sum (MSS7 + MSS5)

(iv) it has proved to be very successful in practice.

Point (iii) is very important. The extraneous factors and influences which are largely eliminated by the normalisation process include:

■ variations in the solar illumination depending on the day of the year or time of day

- variations in the atmospheric conditions
- variations in the optical path length of the atmosphere through which the radiation has travelled, arising from variations in the angle of observation
- variations in the reflectivity of the surface as a function of the angle of observation.

Definitions of vegetation indices were made more complicated to try to give better descriptions of special situations; perhaps the GIN is the culmination of this process. Another example of a rather complicated index is the vegetative sponge index VSI (Greegor and Norwine 1981, Norwine and Greegor 1983) which escaped the review of Perry and Lautenschlager (1984). However, as their definitions became more complicated these vegetation indices lost their physical simplicity and they lost the ability to be of general use in a widespread number of totally different applications. The NDVI has proved to be extremely versatile and has been applied successfully to studies of dense tropical vegetation, of temperate agricultural lands and of semi-arid and desert areas.

We have already mentioned briefly that (i) the idea of vegetation indices was developed originally with Landsat MSS, (ii) the AVHRR was never originally intended to provide data for vegetation studies but (iii) in the 1980s the application of the idea of vegetation indices to AVHRR data and the use of AVHRR data for vegetation studies became very widespread and very successful. Meteorological satellite systems other than the AVHRR could in theory be used for monitoring vegetation cover, including geostationary satellites like Meteosat. However, these satellites have not had the selection of bands to allow calculation of a ratio such as the NDVI. A review of the work in the early stages of the use of the AVHRR in the study of vegetation is given by Hayes (1985) who cites many references. A discussion of the early stages of the work leading to the development of an NDVI product by NOAA NESDIS is given by Yates et al. (1984). More recent surveys and discussions of the prospects for the use of vegetation indices derived from AVHRR data are given by Goward et al. (1991) and Gutman (1991). Some points from these sources are worth noting.

In 1984 the amount of work that had been done on vegetation studies with AVHRR data was so little that Hayes (1985) was able to summarise or review it in two printed pages citing only slightly over a dozen references. Now, about ten years later, a survey of similar depth of all the published work in the area would occupy a sizeable monograph. Moreover, nearly half of those early references quoted by Hayes were concerned with work that was making comparisons between vegetation indices derived from MSS and AVHRR data for common geographical areas and acquired as nearly simultaneously as possible.

The comparison between vegetation indices derived from Landsat MSS data and those derived from AVHRR data is very interesting and important. The principal differences between the two series of platforms and their sensors may be seen in terms of (i) radiometric resolution, (ii) spatial resolution or instantaneous field of view (IFOV), (iii) spectral resolution and (iv) frequency of coverage.

(i) *Radiometric resolution* The AVHRR has 10-bit radiometric resolution allowing 1024 levels of discrimination in measured response, while the MSS has 7-bit resolution in bands 4, 5 and 6, and only 6-bit resolution in band 7, all of which are converted to 8 bits with the aid of sensor-dependent calibration tables.

(ii) *Instantaneous field of view*   The AVHRR has a nominal ground resolution of 1.1 km at nadir, while that of the MSS is 79 m by 56 m. Approximately 200 MSS pixels represent the area covered by a single AVHRR pixel. For image processing, this represents a significant reduction of CPU time per unit area on the ground. This saving becomes a trade-off for the comparative loss of ground resolution, 1.21 km^2 at nadir for the AHVRR as compared to 0.004 km^2 for Landsat MSS.

(iii) *Spectral resolution*   The AVHRR has two channels measuring reflected radiation; the Landsat MSS has four. The saturation thresholds of the MSS sensors are significantly lower than those of the related AVHRR sensors, affording the AVHRR particular advantages in the analysis of highly reflective surfaces.

(iv) *Frequency of coverage*   The NOAA platforms provide daily coverage of the whole of the Earth's surface, while Landsat provides repeated coverage of the same point once or twice for one-satellite or two-satellite systems, respectively, in every 18 days (or 16 days). For land surface studies the occurrence of cloud severely limits the amount of usable imagery received, especially as for some areas a satellite overpass coincident with a cloud-free sky is rare. The daily availability of AVHRR data significantly improves on this situation. In addition, there is significant overlap of the ground area imaged in successive orbits, providing data of the same area from up to three successive passes in mid-latitudes.

There are, however, limitations on the daily availability of full-resolution data. A large part of the Earth's surface does not fall within the reception area of existing ground receiving stations and on-board recording facilities are limited to about 10 minutes per orbit, accommodating data from approximately 4500 km along the orbital track. Complete coverage, then, exists only for global-area coverage (GAC) data with a nominal resolution of 4 km, being resampled and averaged full-resolution data, although there is considerable effort now being devoted to the provision of daily full-resolution data for the entire globe (see section 7.9.6).

The importance of establishing the essential equivalence of NDVI values determined from MSS data and from AVHRR data was at least two-fold. First, it was to provide an *a priori* justification for the use of AVHRR data for vegetation studies; this was important in the early days. Secondly, its importance is that this equivalence provides the opportunity to use the high-frequency, but low-spatial-resolution, AVHRR data to interpolate between successive MSS sets of data, which for any given area are necessarily very infrequent. This combination of data from the two sources is now quite common. One example of this work is that of Townshend and Tucker (1984) who set out specifically to test the validity of using AVHRR data for land cover mapping by making comparisons with much higher resolution Landsat MSS data. Two approaches can be adopted in establishing relations between the two sets of satellite data. In the first, MSS data are registered to the AVHRR data set, thus deriving averaged MSS spectral values for each AVHRR pixel. Consequently, by obtaining correlations between the two sets of data, one obtains an indication of the extent to which the AVHRR data represent the areally integrated spectral response of the ground surface at a resolution of 1.1 km. One example of such a comparison is shown in Figure 5.4. A more stringent approach is to register the AVHRR data set to the MSS data, thus deriving an AVHRR value for every MSS pixel. In this case, the resultant correlations will give an indication of how well

NORMALISED DIFFERENCE VEGETATION INDEX

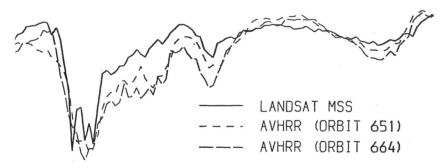

——— LANDSAT MSS
– – – – AVHRR (ORBIT 651)
——— AVHRR (ORBIT 664)

Figure 5.4 Comparison between NDVI derived from Landsat MSS and NOAA-7 AVHRR HRPT data (Hayes and Cracknell 1984).

the AVHRR data display the spatial variability of the ground surface as represented by MSS reference data with a spatial resolution almost 200 times finer in areal terms.

It was the latter, more rigorous, approach which was adopted by Townshend and Tucker (1984) (see also Tucker *et al.* 1984a). The AVHRR images were registered to the MSS images to an accuracy of approximately half an AVHRR pixel. Three areas were chosen for analysis. The first was in the Imperial Valley, California, including terrain around the north of the Salton Sea. Sample areas included both unvegetated desert surfaces, amongst which are alluvial fans, stone pavements, thinly veneered bedrock and bare bedrock surfaces, and vegetated areas of irrigated land characterised by large fields, many MSS pixels in area. By contrast, the second area was in the southern part of the Nile Delta in Egypt where, although irrigated land is found, the field sizes are very much smaller. Adjacent desert surfaces include alluvial sediments, stone pavements and thinly veneered bedrock. Thirdly, part of southern Italy in the provinces of Puglia and Basilicata was selected. This area includes local areas of intensive agriculture, but is predominantly covered by various forms of degraded natural vegetation such as macchia, as well as oak and beech woodlands. The internal spatial variability of much of the area is extremely complex. Although the pairs of images from the AVHRR and MSS were not captured at exactly the same time, their dates were sufficiently close to permit valid comparisons. For the Imperial Valley sites the AVHRR images were obtained on 17 June 1980 and the MSS images for 10 May 1980. For Egypt both images were obtained on 8 August 1981 and in the case of southern Italy the dates are 26 July 1981 and 24 July 1981. The results obtained by Townshend and Tucker (1984) strongly indicated that, despite the very coarse resolution of the AVHRR data compared with conventional MSS data, they are sufficiently strongly correlated to suggest that the former have significant potential for land cover mapping, especially at small scales and for large areas. This result is very important in terms of the potential use for the AVHRR to provide vegetation index data to interpolate temporally between the rather infrequently available data from Landsat. This conclusion is now widely accepted but it took several years and several pieces of work such as this to establish it.

5.3 SCALE

For the moment we suppose that for the AVHRR the NDVI is defined by

$$\text{NDVI} = \frac{X_2 - X_1}{X_2 + X_1} \tag{5.3.1}$$

where X_i ($i = 1, 2$) are the raw data output from channels 1 and 2 of the AVHRR. This is the direct analogue of equation (5.2.2) for Landsat MSS data. We shall turn later on (in section 5.4) to the question of the replacement of the raw data by the radiances in these two channels, but in this section we address the question of (spatial and temporal) scale. In particular we shall consider the extension of the use of multispectral classification in connection with agricultural studies from the initial situation using Landsat MSS data to the use of lower-resolution data, particularly AVHRR data, as well (Cicone and Metzler 1984). The use of vegetation indices, which involve data from only some of the spectral channels, is very important in this context. We shall illustrate this by considering LACIE and its successor, though other examples can be found. One of these is the Agriculture Project at the Joint Research Centre (JRC) of the European Communities; this is aimed at incorporating the use of remote sensing in the CEC's Agricultural Statistics System. The AVHRR is providing one important source of data for European projects (see, e.g. King 1989, Sharman 1989, Rossini *et al.* 1993). It is assumed that the reader is familiar with the principles of multispectral classification; they are described in numerous standard textbooks on remote sensing and will not be repeated here. We shall just note one or two points here. Initially classification was carried out to discriminate agricultural areas from non-agricultural areas and to distinguish one crop from another. Subsequently MSS data came to be used to study the health and state of development of a given crop. For this purpose the data from the visible (red, band 2, new notation) and near-infrared (band 4, new notation) proved to be particularly useful for monitoring the health and vigour of vegetation and crops.

The Large Area Crop Inventory Experiment (LACIE) was a joint project of NASA (the National Aeronautics and Space Administration), USDA (the United States Department of Agriculture) and NOAA (the National Oceanic and Atmospheric Administration). The LACIE programme was commenced in 1974, before TIROS-N with the first AVHRR was launched, and completed in 1978 and the background is as follows. While the United States was in the habit of publicising accurate forecasts of its wheat crop, many other nations either did not make reliable estimates of their crop or did not release their figures until annual purchases were completed. Such organisations as the UN Food and Agriculture Organisation (FAO) and the US Department of Agriculture (USDA) are chartered to provide information on global food production, but their reports had been heavily reliant on information generated by the various producing countries themselves.

The LACIE programme was designed to demonstrate that remote sensing from Earth-orbiting satellites can provide accurate and timely information on foreign commodity production, and that this information is significantly more accurate than data generated by conventional data-collection methods. The description of the project is given in a number of NASA reports that are not particularly widely available; however they are summarised in the chapter on 'Remote sensing applications in agriculture' in the second edition of the *Manual of remote sensing* (Colwell 1983).

Accurate crop-production forecasts require accurate forecasts of the area for harvest, its geographical distribution and the associated crop yield as determined by local growing conditions. Both crop area and yield are sufficiently variable from year to year and within a year to require periodic monitoring. These variations are created by slowly changing factors, such as irrigation, fertilisation and climate, as well as by rapidly changing factors, such as weather, market price and government policy.

To quantify the complex effects of these factors on crop production, both area and yield must be assessed at subregional levels (strata) where the limited ranges and simple interactions of the factors permit successful modelling and estimation. For example, a yield forecast stratum should be sufficiently homogeneous in soil type, crop variety, land use and climate to preclude the necessity for hopelessly complex yield-forecast models. Area and yield for significantly different crop sub-classes should be individually considered. For example, there are two major patterns of wheat production. Winter wheat, planted in the autumn, can have twice the yield of spring wheat, but it is subject to freeze damage during its dormancy period. Thus, despite its lower yield, spring wheat is often planted in severely cold regions. A single yield model cannot adequately describe the response of both types of wheat to such a wide range of weather conditions.

The ideal forecast system is one that can accurately assess current crop status and can detect and rapidly respond to changes in relevant conditions. During the first half of the crop year, area information is most important. The forecast accuracy of even the perfect yield model is limited by uncertainty about future events such as weather conditions and natural disasters. As harvest-time approaches, yield information increases in value because plant processes are closer to completion and the chances of a major perturbation by an unforeseen event are reduced. Thus, a crop forecast system should aim to produce reliable crop area information early in the season and then concentrate on increasingly accurate yield and production forecasts as the crop nears harvest.

In the LACIE design great emphasis was placed on an objective and quantitative evaluation of the technology under as many representative global agricultural conditions as possible. Goals were established for timeliness and accuracy. The LACIE experimental inventory system was designed to achieve monthly at-harvest production estimates that would converge to within 10% of the true estimate at the national level with a confidence of 90%.

LACIE was conducted in three phases, each covering a global crop year. In phase I, beginning in November 1974, the existing remote sensing technology was tested for the USA over the nine-state Central Plains region. Test results were sufficiently encouraging to expand testing in phase II to include wheat regions in the former USSR and Canada; in this period technology problems uncovered in phase I were also addressed. In phase III, a second-generation technology, developed in phases I and II, was used to forecast the 1977 Soviet wheat crop at the country level. Evaluations were continued in the Central Plains region of the USA where detailed ground observations and crop estimates were available for comparison. A limited amount of ground data were also collected in Canada as well. The project also conducted exploratory studies with data from India, China, Australia, Argentina, and Brazil.

The remote sensing crop-forecast system developed and evaluated by LACIE used Landsat MSS data to identify crops and estimate their area for harvest and it also used global weather data from the WMO ground network to forecast yield for the harvested areas. Instead of complete coverage by Landsat, a 2% random sample was employed. The error introduced by only using a small sample was shown to be quite small. An important segment of the LACIE program involved the development of machine-assisted image-analysis procedures. Meteorological quantities useful for agricultural assessment and potentially derivable from environmental satellite data include daily precipitation, maximum and minimum temperature, insolation and weekly snowcover. These quantities were used in soil moisture and crop yield models, crop calendars and, in the case of snow cover, winterkill (Yates *et al.* 1984). These observations are also used to provide early warning of damaging environmental situations such as drought and freeze events. The various types of data were processed for yield forecasting and crop maturity stage-estimation in LACIE and yield forecasts were input to the production forecasts.

Although ground data on crop identification and crop condition were not used directly to estimate crop area in the LACIE system, such data were used to develop techniques and to assess the accuracy of LACIE crop forecasts. To support the development of techniques, spectrometer and other field measurements were acquired at several intensive study sites.

An extensive accuracy-assessment programme was conducted to evaluate the performance of LACIE. Three years of intensive evaluation of LACIE estimates for the US crop and two years of experience in estimating the Soviet crop indicated that accuracy commensurate with USDA performance goals for foreign wheat-production forecasting was achievable in regions where fields are sufficiently large to be resolved by Landsat. In a 1977 quasi-operational test, the LACIE total wheat estimates for the former USSR were within 6% of the final Soviet figures six months before their release. The coefficient of variation of the LACIE total wheat estimate for the USSR was 3.8%, well within accuracy goals established for the project. More details of these results and of the evaluation of their accuracy will be found in chapter 33 of Colwell (1983).

We have given some discussion of the LACIE project for a number of reasons. It was carried out at a time when the only high-resolution satellite data available were Landsat MSS data, before the first AVHRR data were even available and at a time before the massive expansion of interest in vegetation indices occurred. The following question then arises. If the system is reasonably successful in relation to wheat production in the USA and in the former USSR how does it stand in relation to one of the LACIE design objectives, which was 'to demonstrate the technical and cost feasibility of a global agricultural monitoring system'? The key to the answer to this question lies in several features. The first is the part of the conclusion noted above that performance goals were achievable in regions where fields are sufficiently large to be resolved by Landsat MSS data. This condition holds for the two areas in the USA and the former USSR which were studied in LACIE. However, there are many other areas to which this condition does not apply. For instance, in Europe fields are very much smaller and much more irregular in shape than they are in the USA and a simple transfer of the LACIE methodology to the European context would not be expected to be successful.

The original idea of using satellite data in crop forecasting was to use high-resolution data (initially from Landsat MSS and later from other high-resolution systems) for classification purposes (to study areas planted, the health of the crop, etc.). In addition to this requirement there is the requirement for meteorological information. Meteorological observations with the desired coverage and timeliness are not available from conventional sources; thus crop production forecasters are constantly seeking improved data. Operational meteorological satellites provide rapid, continuous global coverage and are therefore potential sources of the required information. With the passage of time this distinction between the roles of the high-resolution satellites and of the meteorological satellites has become much more blurred. The high-resolution data have ceased to be amost exclusively used for the measurement of the area planted with each crop and became involved, principally by the use of vegetation indices, in the study of crop health and crop development. At the same time, the lower-resolution meteorological satellite data have come, in some cases, to be used in the early classification stage to assist in determining the area planted with a given crop.

As time passed the role of Landsat MSS or other high-resolution data for the classification role of the type used in LACIE began to be questioned for the reasons that we have mentioned already in section 5.1. The first problem is associated with the infrequent orbital revisit period (once in 18 days for the original Landsat MSS) coupled with the problems of cloud cover. A few unfortunately occurring days of poor viewing conditions can render a data set useless. This can be a particularly significant limitation in areas of frequent cloud cover, e.g. in tropical regions, and this was one of the factors leading people to consider not only AVHRR but also Nimbus-7 CZCS data as an alternative to high-resolution (but low-temporal) frequency data from Landsat, SPOT, etc. Another aspect is the quantity of data that would need to be processed. With the increase in population pressure and the continued exploitation of renewable and non-renewable resources, all nations are increasingly in need of accurate and current information about natural resources and agricultural production. For these areas, fine spatial resolution is often much less critical than data-processing cost and time. For example, to image the North American continent requires more than 1700 Landsat MSS scenes. Analysing this quantity of data is an enormous task given all the attendant computer processing, personnel and equipment requirements. In LACIE this problem was solved by using a 2% random sample; in other countries that might not be so successful. If similar information could be obtained at a reduced spatial resolution, this would relieve some of the processing burden and represent a benefit to many developing nations which are badly in need of natural resource information for decision making. This makes the AVHRR with its large area coverage and relatively coarse spatial resolution (1.1 km compared to 80 m for MSS) particularly attractive for certain applications. Coupled with this problem of handling enormous quantities of high-resolution data is the question of cost. It has been calculated (Quarmby *et al.* 1992) that a study similar to the LACIE project for Europe would require approximately 400 cloud-free Landsat MSS scenes. However, complete coverage of Europe is available for the comparatively low cost of 0.2% of the MSS cost by using AVHRR data; the cost is that of 20 Standard Family HRPT Archive Request Product (SHARP) scenes, each of which contains a four-minute slice from an AVHRR pass. Consequently, AVHRR data, analysed using an appropriate technique, offers a low-cost, and therefore more operationally viable, route for crop area estimation.

AVHRR data have come to be important in crop monitoring in two situations, (i) those situations in which the spatial resolution of the AVHRR is not a serious problem and (ii) those situations in which the lower-resolution data can usefully be used as a basis for interpolation between the relatively infrequent sets of high-resolution data.

The Foreign Crop Condition Assessment Division (FCCAD) of the US Department of Agriculture's Foreign Agricultural Service is the operational outgrowth of LACIE. Formed in 1978, the FCCAD is responsible for assessing and monitoring crop condition in selected areas of the world, with the ultimate goal of quantifying the assessment. This responsibility is carried out through analysis of satellite images, drawing upon all available crop, meteorological, soils, and other supporting information. Philipson and Teng (1988) describe the inclusion of AVHRR data among the data sources used in the FCCAD's work.

The principal image data used by the FCCAD analysts are Landsat MSS and NOAA AVHRR data. Computer-compatible tapes of the four channels of MSS data and the first two channels of AVHRR data are received respectively from EOSAT and NOAA within one week of acquisition. During peak periods, some analysts receive as many as 20 images per day. Before the satellite images are viewed by the analysts, each scene is processed to calculate vegetation indices which are averaged over pixels in geographically referenced grid cells. The cell averages are entered into a database. In FCCAD terminology the average value of the NDVI for a grid cell of pixels is called a cell VIN. The FCCAD's operational demands for obtaining timely information on a worldwide basis, with a limited staff, often precluded the adoption of image processing or analysis techniques which might otherwise be useful and the problem arose of how to incorporate the information from the database.

In the context of interpreting database values, the AVHRR portion of the database poses the more difficult problem. Because cropping in most agricultural areas of the world is neither monocultural nor in fields larger than 1.1 kilometre on a side, vegetation indices of nearly all cells of AVHRR pixels are derived from a preponderance of mixed pixels (i.e. pixels filled by more than one class of land cover). The question, then, is whether useful crop information can be reliably and efficiently derived from a time series of vegetation indices, when each index is the average vegetation index of a geographically referenced group of mixed AVHRR pixels. A procedure for interpreting and incorporating the AVHRR cell data should then be transferable to the MSS cell data.

The question addressed by Philipson and Teng (1988) was to evaluate the operational utility of the AVHRR database; that is, to determine if, despite its poor spatial resolution, the AVHRR cell data convey useful crop information, if they can be used efficiently and if their use improves crop interpretations. Philipson and Teng were able to demonstrate the potential usefulness of the database. They concluded that, despite the preponderance of mixed pixels, useful crop information can be obtained from the database and that its operational use could assist crop assessment. Similar conclusions have been reached at the Institute of Remote Sensing Applications at the Joint Research Centre of the European Communities; the particular interest there is the monitoring of European agricultural production and the prediction of European agricultural yields.

The procedure outlined by Philipson and Teng (1988) was as follows:

- preliminary steps, including collecting and analysing supporting information (e.g. crop calendars, crop area, yield and production data) and visual analysis of satellite images

- evaluation of cell VINs, including rejection of data from areas affected by cloud or haze
- crop monitoring and qualitative forecasting (for all or selected cells)
- modelling for quantitative production forecasts.

A number of possible problems were identified including the fact that vegetation indices may represent or be affected by non-crop vegetation which may respond differently from crops of interest, sensor degradation may affect data values (particularly over different years) and episodic or extreme weather events may have a controlling effect on crop conditions.

When this approach was adopted by FCCAD analysts and they began to interpret the AVHRR cell data in concert with imagery, weather data and supporting crop and soil information, they reported an improved capacity of crop assessment. The principal gain is in the capacity to examine objective measures and trend of vegetative growth from numerous dates in a given table or graph.

The problem of handling fields which are small relative to the IFOV of the AVHRR has been addressed by Quarmby et al. (1992); this is done by a process known as mixture modelling. In linear mixture modelling one assumes that each field within the IFOV contributes to the signal received at the satellite sensor an intensity of radiation that is characteristic of the cover type in that field and is proportional to the area of that field. Thus the signal x_i in band i is given by

$$x_i = \sum_j f_j M_{ij} + e_i \qquad (5.3.2)$$

where f_j is the fraction of the area of the IFOV occupied by cover type j, M_{ij} are the cover coefficients (which are assumed to be independent of f_j) and e_i is a noise term. It is assumed that the M_{ij} are known from some other source so that the analysis of the AVHRR data involves using the measured x_i and the given M_{ij} to determine the proportions f_j of the IFOV occupied by cover type j. It should be mentioned that (i) this ignores the point spread function of the sensor and (ii) the basic idea is similar to that used with thermal infrared data in section 6.2.1 in determining the fraction of an IFOV occupied by a very high temperature source.

The realisation of the potential importance of AVHRR data led to several pieces of work aimed at making comparisons between the AVHRR and MSS (Hayes and Cracknell 1984, Gervin et al. 1985, Gallo and Daughtry 1987, Ormsby et al. 1987, Woodcock and Strahler 1987, Suits et al. 1988, Nelson 1989). The relative success of the AVHRR data and the MSS data depends on the typical size of a parcel of land with relatively uniform cover within the scene and how this typical size relates to the sizes of the IFOVs of the AVHRR and of the MSS. For example, Gervin et al. (1985) compared the classification accuracy of MSS and AVHRR data for land-cover mapping by comparing the accuracy of land-cover information for the Washington, DC, area derived from NOAA-7 AVHRR data with that from Landsat MSS data. Their results produced overall land-cover classification accuracies of 71.9 and 76.8% for AVHRR and MSS, respectively. While the accuracies for predominant categories were similar for both sensors, land-cover discrimination for less commonly occurring and spatially heterogeneous categories was better with the MSS data set. The AVHRR, however, performed as well as or better than the MSS in classifying large homogeneous areas. Gallo and Daughtry (1987) demonstrated with field experiments on corn canopies how two vegetation indices (the NDVI and the near-infrared/red ratio) for the Landsat MSS and TM and SPOT systems may be estimated with AVHRR data. The vegetation indices of all four systems were found to be associated with similar amounts of variation in the examined agronomic variables. Thus, it was concluded that under similar viewing conditions, the AVHRR may complement measurements of the other sensor systems for monitoring surface features of the Earth. It must be appreciated, however, that this study did not address such questions as the effect of off-nadir viewing at quite oblique angles by the AVHRR. The effect of variation in the land cover within the IFOV of the AVHRR was evaluated by Ormsby *et al.* (1987) by simulating AVHRR signals with MSS data. A comparison between AVHRR and SPOT data for land-cover mapping in the Sahel region of West Africa was reported by Marsh *et al.* (1992).

Since there is now a variety of Earth-observing polar-orbiting satellite systems available ranging from SPOT and Landsat TM to AVHRR, a user of digital satellite imagery for remote sensing of the Earth's surface now has a choice of image scales ranging from 10 m to 1 km (or even 8 km or more if one is restricted to GAC data). Townshend *et al.* (1994) attempted to define a hierarchy of scales and of temporal resolution, see Table 5.2. The question of choosing an appropriate scale, or spatial resolution, for a particular application depends on several factors and was considered by Woodcock and Strahler (1987). These factors include the information desired about the ground scene, the analysis methods to be used to extract the information and the spatial structure of the scene itself. To select an appropriate scale of data, it is important to understand the manner in which images of a scene change as a function of spatial resolution. Several authors have investigated the spatial structure of images, usually at one or two discrete resolutions (Craig and Labovitz 1980, Labovitz *et al.* 1980). The spatial structure of images is expected to be primarily related to the relation between the size of the objects in the scene and spatial resolution. Graphs of local variance in images as a function of spatial resolution may be used to measure spatial structure in images.

Calculation of the data for graphs of local variance against spatial resolution is accomplished by degrading the image to successively more coarse spatial resolutions, while measuring local variance at each resolution. Local variance is measured as the mean value of the standard deviation of a moving 3×3 window. In an image, each pixel (except around the edge) can be considered the centre of a

Table 5.2 A hierarchical approach for land-cover data sets (Townshend *et al.* 1994)

Level of hierarchy	Time frame	Spatial resolution	Temporal resolution	Data source
H1	1–2 years	~20 km global	Monthly	GVI
H2	2 years	~8 km global	Weekly	GAC
H3	5 years	~1–2 km global	Weekly	LAC/HRPT
H4	1–2 years	20 m–1 km local/regional	Variable	SPOT/Landsat, LAC, etc.

3 × 3 window. The standard deviation of the nine values is computed, and the mean of these values over the entire image is taken as an indication of the local variability in the image.

To measure local variance at multiple resolutions, the image data are degraded to coarser spatial resolutions. The algorithm used to degrade the imagery simply averages resolution cells to be combined into a single, larger resolution cell. This approach implies an idealised square wave response on the part of the sensor and ignores the point spread function of the sensor (as seen in Figures 3.9 and 3.11). While this is unrealistic, it serves to illustrate the general effects of changes of scale. One effect of the degradation of images is that the number of pixels decreases as resolution becomes more coarse. Thus, there are a limited number of times that an image can be degraded and still have a reasonable number of pixels to estimate local variance.

Simonett and Coiner (1971) addressed the effect of spatial resolution on image structure by overlaying grids on aerial photographs and counting the number of land-use categories that occurred in each cell. By using different size grids, the effect of changing spatial resolution was evaluated. They demonstrated that the number of pixels containing more than one land-cover type was a function of both the complexity of the scene and the spatial resolution of the sensor.

An example of a graph of local variance against resolution is shown in Figure 5.5. This was obtained by Woodcock and Strahler (1987) from a 1:15000 scale air photo of a forested area in South Dakota. The photograph was scanned with a microdensitometer to produce a digital image with a spatial resolution of 0.75 m. The area comprised a simple forest scene composed of trees on a relatively smooth background. From Figure 5.5 we see that the variance is low at the initial resolution of the image. Resolution cells are much smaller than trees, as tree crowns in this area are about 8 m in diameter. If a pixel falls on a tree, its immediate neighbours are also likely to be on a tree and have similar values. In this situation, the standard deviation of a 3 × 3 window is low. Naturally, some pixels will fall along the borders of trees and background, and as a result will have high local variance. However, the mean value of the local variances for the entire image is low.

As the size of individual resolution cells increases, the number of pixels comprising an individual tree decreases, and the likelihood that surrounding pixels will be similar decreases. In this situation the local variance increases. This trend continues until a peak in local variance is observed at about 6 m. As resolution increases past

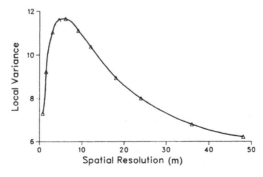

Figure 5.5 A graph of local variance as a function of resolution for an airborne scanner image of a forest area (Woodcock and Strahler 1987).

this peak, local variance decreases. This decrease is associated with individual pixels increasingly being characterised by a mix of both trees and background. As this mixing of elements occurs, all pixels look more similar and local variance continues to decrease. The image itself, at various stages of the spatial resolution will be found in the paper by Woodcock and Strahler (1987) who also consider examples of several other images. One methodological issue resulting from their results concerns the relation between the size of the objects in the scene and the spatial resolution of peak local variance. It was initially hypothesised that the peak would occur when the size of the resolution cells matched the size of the objects. However, in each of the graphs with a well-developed peak in local variance the peak occurs at a resolution-cell size somewhat smaller than the size of the objects in the scene. In the South Dakota forest image, for example, the peak occurs at 6 m for trees approximately 8 m in diameter. The peak in local variance occurs at 240 m for agricultural fields generally 420 m on a side in a second image based on TM data for an area of farmland near Dyersburg, TN.

Justice *et al.* (1991c) used NDVI images of Africa with a resolution of 8 km and studied the effect of successive degradation of spatial resolution to a final IFOV of 512 km square. Scale variance analysis was used to examine the spatial characteristics of the NDVI up to the scale comparable with that used for global modelling. This technique identifies the scales at which spatial variation is taking place and the relative magnitude of the variation. Both annual and seasonal images of Africa from 1987 were examined. The analysis revealed substantial differences within the continent on the scale at which spatial variation takes place. Commonly, for the annual image there was an increase in spatial variation with coarsening spatial resolution, although certain areas of complex surface conditions show markedly different patterns. There are substantial changes in the spatial characteristics of the NDVI with time. Analysis of monthly maximum value composites for September and February revealed different responses in scale variance as a function of spatial resolution. These spatial differences were most marked for areas where vegetation possesses strong seasonality. Interpretation of these results suggested that different factors appear to be controlling the spatial variation of the NDVI at different scales. Averaging at coarse grid cell sizes of 512 km as a means of representing surface conditions results in varied success in representing the NDVI. Averaging areas of transition and surface heterogeneity may result in a substantial over-simplification of surface conditions. Consideration needs to be given to the spatial characteristics of areas and their temporal variability if satellite-derived data are going to be applied validly to global models (see section 7.9).

A further aspect of the question of scale in relation to the use of AVHRR data for land cover classification is the question of whether to use the HRPT or LAC data with its 1.1 km spatial resolution or the lower-resolution GAC data (Townshend and Justice 1988). One factor involved is the relative time, and therefore cost, involved in processing the data set with the two different spatial resolutions. However, until recently at least, the choice between the two data sets may be determined by the availability of data. The GAC data (4 km resolution and only two spectral channels – see section 1.2.3) has the advantage that it has always been acquired throughout each orbit of the satellite and therefore gives global coverage more than once per day. For most users, however, these data are only available on an historical basis and not in near-real time. With a local APT ground station one can obtain data in near-real time for one's local area. The HRPT/LAC has the higher spatial resolution but, as we have also seen in sections 1.2.3 and 2.5 only about 10% of the data can be captured and archived by NOAA; for complete (temporal) coverage of an area one needs to have a direct readout station within range of the area in question. It is not, however, just a matter of having a direct readout station established and operating now. In many environmental studies there is a clear need for regular observation over a number of years.

Extensive archives of full-resolution data are far from widely available outside north America. By contrast, the global daily GAC archive dates back to June 1981 and continental-scale time series of these data are in regular use for environmental monitoring. So, for regional or continental-scale studies (outside Europe) requiring a historical perspective there is little alternative to using GAC data. The situation has only begun to change quite recently with attempts now being made to obtain global HRPT/LAC coverage, at least of all the land surface of the Earth (see section 7.9.6).

If quantitative use is to be made of GAC data it is necessary to establish its limitations. This is equally true whether information from the spectral or spatial domain is considered, and can be achieved by comparing the information content of the GAC with high-resolution data sets. Belward (1992) studied this problem using data relating to a study area in West Africa, from the coast of the Gulf of Guinea (approximately 3°N) to the northern border of Senegal (approximately 17°N) and from the Atlantic coast (approximately 18°W) to the Greenwich meridian. This question has also been considered in relation to fire detection, see section 6.2. Belward's approach was to use the technique used earlier by Woodcock and Strahler (1987) involving plotting graphs of local variance against resolution for GAC and HRPT data. Belward also considered a third data set which he called AVG and which consisted of digital HRPT data that had been degraded to the spatial resolution of GAC data. This was found not to have the same properties as the GAC data set. It was found that the mean of the local variance in the GAC data was much higher than in the AVG data set (or HRPT data set) for resolution up to 12 km. The difference should not be too surprising in view of the fact that the GAC data are generated by electronic subsampling or undersampling before they are tape-recorded on board. It has been suggested that the line skipping involved in the undersampling reduces the autocorrelation between neighbouring GAC pixels, thus introducing high local variance into the images (Belward and Lambin 1990). It is interesting to note that at resolution greater than 12 km the behaviour of the GAC local-variance/resolution graphs more closely parallels those of the AVG and the HRPT data. Taking the mean of blocks of 3×3 GAC pixels used to create the 12 kilometre resolution-cell size would appear to smooth out this effect, i.e. at coarse scales the GAC provides comparable spatial information to the HRPT. From the analysis of one set of data obtained during the dry season for this part of west Africa, Belward (1992) concluded that (i) structural variability, relevant to environmental studies, can be detected at regional scales using AVHRR data, (ii) HRPT data can provide spatial information at full resolution, but that spatial information is unreliable from full-resolution GAC data, and (iii) at resolutions coarser than 12 kilometres degraded GAC data are as good a source of spatial information as degraded HRPT data.

Plots of NDVI through the growing season, using AVHRR data at various different values of the spatial resolution, have been given by Malingreau and Belward (1989, 1992). For the routine estimation of range-land productivity in the ecologically sensitive Sahelian region, data from the NOAA AVHRR appear to offer the best trade-off in terms of spatial and spectral resolutions and frequency and area of coverage. However, due to the size of the field of view (of edge ~ 1.1 km) an accurate geolocation of images for a direct calibration of vegetation indices to above-ground phytomass requires impractically large areas to be covered. This has led to the proposal of a two-stage calibration scheme, whereby vegetation amounts are first related to Landsat MSS images and then to coregistered and (near) contemporaneous AVHRR data. The use of Landsat data in this manner for the study of rangeland vegetation in the Sahel has been described by D'Souza and Hiederer (1989).

Several attempts have been made to use either MSS or TM data to evaluate the validity of the use of AVHRR data for land-cover studies in forestry in cases where the feature size is smaller than the nominal IFOV of the AVHRR (Nelson *et al.* 1987, Iverson *et al.* 1989, Cross *et al.* 1991, Quarmby *et al.* 1992). Nelson (1989) used regression and ratio estimates to integrate AVHRR-GAC and Landsat MSS digital data to estimate forest area in the continental United States. Forest lands were enumerated for the 48 contiguous states using five different AVHRR-GAC data sets. The five GAC data sets tested were composed of different combinations of vegetation index and thermal infrared data acquired over the nine-month growing period in 1984. Twenty Landsat MSS scenes were selected countrywide and used to calibrate AVHRR forest estimates. Results indicated that the GAC and MSS forest estimates were not highly correlated; R^2 values ranged from 0.5 to 0.7. Although the ratio of means and linear regression corrections were, on the average, closer to national US Forest Service forest area estimates, these correction procedures did not consistently improve GAC estimates of forest area. GAC forest area estimates tended to be high in densely forested regions such as the northeast and low in sparsely forested areas. This fact, and the low correlation coefficients, indicated that AVHRR data should be used for primary stratification (with MSS as the second stage) and not as an auxiliary variable in a regression correction procedure.

5.4   CALCULATING THE NDVI

5.4.1   General considerations

So far we have supposed that the NDVI determined from AVHRR data is defined in terms of the raw data as in equation (5.3.1), which is the direct analogue of the original definition for Landsat MSS data. Thus we have

$$\text{NDVI} = \frac{X_2 - X_1}{X_2 + X_1} \qquad (5.3.1)$$

where X_1 and X_2 are the counts for channels 1 and 2, respectively. This is, indeed, what is done in the production by NOAA of their Global Vegetation Index (GVI) on an operational basis (Kidwell 1990). That index is calculated, it will be recalled, using raw data from the AVHRR (counts or 10-bit digital numbers on the scale 0–1023). In this section some developments in the calculation of NDVIs will be mentioned. For qualitative studies, or if one is concerned with data gathered over a relatively short period (such as one growing season), and if one is using data from the AVHRR on only one satellite in the series this is likely to be satisfactory. However, with recent developments in modelling of the behaviour of vegetation and

with the increasing interest in the use of long-term NDVI data sets for climatological studies, considerable attention is now being given to using the calibrated
radiances, rather than the raw digital output from the instruments when calculating
the NDVI. Thus the NDVI would be redefined as

$$\text{NDVI} = \frac{R_2 - R_1}{R_2 + R_1} \tag{5.4.1}$$

where the R_i ($i = 1, 2$) are the radiances in channels 1 and 2 given essentially by
equation (2.2.21):

$$R_i = M_i X_i + I_i \tag{5.4.2}$$

where M_i and I_i are the slope and intercept for channel i. The coefficients M_i and I_i
should be periodically updated because of the changes in the sensor sensitivity with
time discussed earlier (see section 2.2.6 on post-launch calibration).

The use of the *normalised* difference (in the NDVI) was intended to achieve the
elimination of (i.e. the cancelling out of) effects arising from differences of satellite
scan angle, satellite azimuth angle with respect to the target, solar elevation and
azimuth angles and variations in atmospheric conditions. It does not, however,
eliminate differences that arise from intrinsic differences between one AVHRR
instrument and another. The variations that may occur in the values obtained for
the NDVI if one uses data from different instruments are illustrated rather well by
some results of Gallo and Eidenshink (1988) who used AVHRR data from NOAA-9
and NOAA-10 for coincident sample locations. The data used by Gallo and Eidenshink (1988) were acquired over the southeast portion of the United States for the 6
December 1986 daylight orbits of the NOAA-9 and NOAA-10 satellites. These data
were selected because they were acquired for a similar area with nearly cloud-free
conditions over a large portion of the area. The near-coincident area of data acquisition for the two satellites, with nearly cloud-free conditions, is a relatively unusual
occurrence. The NOAA-10 image was registered to the NOAA-9 image and a total
of 38 coincident sample locations along two transects were selected for analysis
(Figure 5.6). No field data were gathered from the sample locations. The sample
locations included forest areas, agricultural land and water surfaces. The effects of
any possible ground tracking variations between the satellites, and differences in
original pixel size due to the scan angles, were minimised through the identification
of a three-pixel by three-pixel window that surrounded each initially selected sample
location. The mean value of the channel-1 and channel-2 response of each 3×3
window was used.

The channel-1 and channel-2 data of the NOAA-9 and NOAA-10 AVHRR
systems were calibrated to reflectance (often cited as albedo) values with the (pre-
launch) coefficients supplied with the data by NOAA (Kidwell 1991). The normalised difference vegetation index was computed from the data of each satellite for
each sample location. Solar zenith and azimuth angles at each of the sample locations were computed as were satellite scan and azimuthal angles. The data were
examined for the influence of the different solar and satellite orbital characteristics
on the visible, near-infrared and vegetation index values for the range of sample
locations.

The nadir paths and times of orbits of the two satellites result in a variety of
satellite and solar geometric differences between the data acquired with NOAA-9

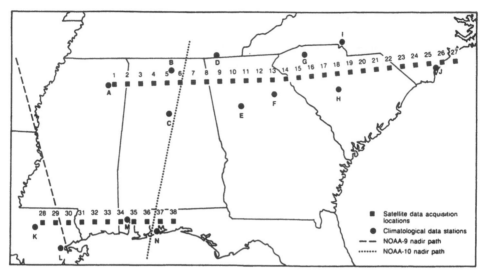

Figure 5.6 Selected sample locations, satellite nadir paths and climatological data stations included in the study of Gallo and Eidenshink (1988).

compared to NOAA-10. Figure 5.7 shows the satellite scan angle for each of the two satellites for sample locations 1–27, while Figures 5.8(a), (b) and (c) show the visible and near-infrared albedos and the NDVI for the same set of sample locations. Notice that for NOAA-10 there is no significant trend in the radiances in Figures 5.8(a) and (b) up to about sample location no. 12, for which the scan angle is about 15°. Beyond that, and for the whole set of sample locations with the NOAA-9 data (for which the scan angle is already about 15° at site 1), there is a trend to increasing response (arising from the high scan angle values). Notice, however, that no such trend is apparent in the NDVI values in Figure 5.8(c); this illustrates rather nicely

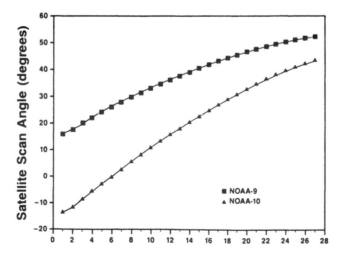

Figure 5.7 Satellite scan angle for the sample locations of the northern transect in Figure 5.6. Positive scan angles are in an easterly direction (Gallo and Eidenshink 1988).

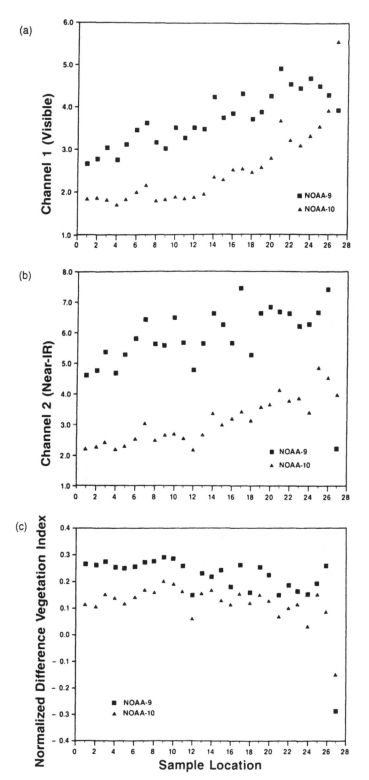

Figure 5.8 (a) Channel 1 albedo, (b) channel 2 albedo and (c) NDVI for the sample locations in the northern transect in Figure 5.6 (Gallo and Eidenshink 1988).

how the normalisation works in eliminating scan-angle effects. A similar set of results for the sample locations 28–38 is also presented by Gallo and Eidenshink (1988); these sample locations are all at low scan angles and, as expected, the radiance values in the individual bands therefore show no scan-angle effects. The conclusions from the NDVI values therefore include the following: (i) there are variations, from one sample location to another, in the NDVI values and these correspond to real differences in the nature and state of the vegetation on the ground, (ii) there is generally a strong correlation between the NDVI values calculated from the NOAA-9 data and the values calculated from the NOAA-10 data; these are strongly indicative of a real physical/biological significance, at least in the relative magnitudes of the calculated NDVI values, and (iii) however, there is a consistent difference, or offset, between the value of the NDVI calculated from NOAA-9 and that calculated from NOAA-10. This is attributable to the intrinsic differences of the two AVHRR instruments themselves. There are slight differences between one AVHRR instrument and another (see section 2.2.1) between visible (channel-1) and near-infrared (channel-2) wavebands and the response of the sensors within these wavebands. Channels 1 and 2 of the NOAA-9 and NOAA-10 AVHRRs have band widths, based on 50% relative response, of 570–700 nm and 715–982 nm, and 573–684 nm and 723–981 nm, respectively. However, the effect of these differences is expected to be very small. What is much more important is that these results suggest the need for post-launch calibration of channel 1 and channel 2 of both of the AVHRR instruments on NOAA-9 and NOAA-10 (see section 2.2.6).

Comparison of NDVIs calculated using X_i and R_i indicates that the reflectance-based NDVI yields systematically higher values than the one based on count values, see Figure 5.9. One should bear this in mind while comparing the vegetation indices derived from different sources of data. There are good reasons why the radiance, not the raw data, should be used in calculating the values of the NDVI. Instead of being used just in projects for individual areas, up to continental scale, serious attention is now being given to the use of the NDVI calculated from AVHRR data for studies of

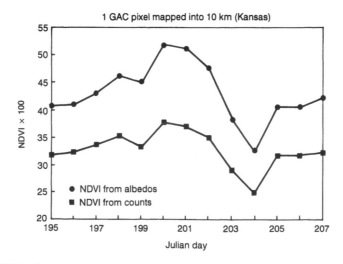

Figure 5.9   NDVI values calculated from digital values (counts) (■) and from albedos (radiances) (●) (Gutman 1991).

vegetation at global scale (section 7.9.6). After about ten years of serious use of AVHRR data for vegetation studies new demands are now being placed on the data. Researchers have related vegetation indices such as the NDVI to several vegetation phenomena including leaf area index, biomass, percentage ground cover and, most recently, absorbed photosynthetically active radiation (Sellers 1985, Tucker 1979, Jackson 1983, Daughtry *et al.* 1982, Curran 1983, Choudhury 1987, Peterson *et al.* 1987, D'Souza *et al.* 1996). These vegetation attributes are used in various models to study photosynthesis, carbon budgets, water balance and related terrestrial processes (Tucker and Sellers 1986, Nemani and Running 1989). Because vegetation foliage magnitude varies significantly in time and space, the high spatio-temporal observation frequency of the AVHRR is potentially of great value in global-scale biospheric studies. Knowledge of the AVHRR's capacity to provide consistent and accurate measures of land surface vegetation foliage conditions, at a high temporal frequency, is critical to the use of these data in global biospheric models. The importance of this is in connection with the radiation budget, that is the balance of upward and downward fluxes of energy at the Earth's surface. A recent special issue of the *International Journal of Remote Sensing* (volume 15, number 17, 1994) was devoted to the question of global data sets for land applications from AVHRR data (see section 7.9.6). This work, apart from needing global coverage, also requires data over extended time periods; therefore it places new demands on the quantitative aspects of the interpretation of the data and thus also on the calibration of the AVHRR instruments themselves. The importance of accurate calibration, not just of the various AVHRR instruments but also of other instruments on other satellites such as Landsat (TM and MSS) and SPOT which are used to study vegetation, is discussed by Price (1987a) (see also section 2.2). Long-term monitoring of land cover may require a transfer from using data from one satellite sensor to data from a different sensor having different spectral characteristics. Two general procedures for spectral substitution were described by Suits *et al.* (1988); they are a principal-components procedure and a complete multivariate regression procedure. They were evaluated through a simulation study of five satellite sensors (MSS, TM, AVHRR, CZCS, and SPOT HRV).

The early qualitative analyses of the relations between AVHRR spectral vegetation index measurements and terrestrial vegetation phenomena such as geography, phenology and annual primary production were exceptionally successful and encouraged great interest in the AVHRR sensor as a global vegetation observing system. The next logical step is to move from qualitative to quantitative interpretations of these satellite observations. The physical and biological linkages between the satellite spectral vegetation index measurements and biophysical phenomena occurring at the Earth's surface provide a basis to study biospheric phenomena at a global scale. Accomplishing this quantitative step, however, is complicated, not only by the question of the biophysical significance of the spectral measurements, but also by numerous intervening factors, such as instrument calibration, view geometry, atmospheric attenuation and cloud occurrence, which tend to obscure the relation between ground and satellite-observed spectral measurements.

Goward *et al.* (1991) and Gutman (1991) have reviewed the current understanding of the capacity and limits of the AVHRR sensor to produce consistent, high temporal resolution, global, surface-equivalent normalised difference vegetation index (NDVI) measurements. Goward *et al.* give an example to illustrate the importance of accuracy in NDVI values which are to be used in vegetation models. This is for the case of Monteith's production efficiency concept. This concept collapses the physiognomic and physiological complexities of plant processes to a single variable, the production rate, which is given by

$$P = \varepsilon \int_{t=0}^{n} I_t S_t \, dt \tag{5.4.3}$$

where

P = production (g m^{-2} y^{-1})
ε = production efficiency (g MJ^{-1})
I_t = percentage incident solar radiation intercepted
S_t = incident solar radiation (MJ m^{-2}) and
t = time.

It is possible to relate this to the NDVI and restate equation (5.4.3) as

$$P = \varepsilon \int_{t=0}^{n} (\text{NDVI})_t S_t \, dt \tag{5.4.4}$$

where $(\text{NDVI})_t$ is the value of the NDVI at time t. For $\varepsilon = 1.5$ g MJ^{-1} and incident photosynthetically active radiation PAR = 300 MJ m^{-2} month^{-1}, the error rate of P is 4.5 g per month for each 0.01 NDVI error (equivalent to a 1% error in estimating PAR absorption) incurred in the NDVI measurement. Because this error is cumulative through the growing season, small errors in the NDVI measurements can rapidly become significant and produce large errors in the calculated primary production. Errors can also accumulate in this production model if the temporal resolution of the observations decreases. The degree to which a given observation repeat cycle replicates the actual dynamics of vegetation foliage is dependent on the rate of foliar change through the growing season. Too infrequent measurements can seriously underestimate or overestimate the NDVI term in equation (5.4.4), particularly early and late in the growing season.

Remotely sensed measurement precision varies as a complex function of sensor design, irradiance and atmospheric attenuation. As a result, the capacity of a particular sensor to achieve desired measurement objectives can, and generally does, vary significantly in time and space. These same factors can also introduce serious errors in measurement extraction if not addressed in data processing and analysis procedures, particularly when indices such as the NDVI are applied to the data (Crippen 1988).

Goward et al. (1991) give an error analysis to study the effect on the calculated NDVI of errors $\Delta \rho_v$ in the channel-1 (visible) and $\Delta \rho_n$ in the channel-2 AVHRR (near-infrared) reflectances. The dependence of the error in the calculated NDVI on $\Delta \rho_v$ and $\Delta \rho_n$ is not particularly simple but some values are tabulated by Goward et al. (1991). For example, a sensor system with a precision of $\pm 0.1\%$ reflectance results in an NDVI precision which varies from better than ± 0.002 for bright targets to values exceeding ± 0.015 for dark targets. If the sensor precision decreases to $\pm 1.0\%$ reflectance, the NDVI precision decreases proportionally to ± 0.02 for bright targets and ± 0.2 for dark targets. These results are general and make no reference to what is the actual source of the errors $\Delta \rho_v$ and $\Delta \rho_n$. An extensive discussion of the actual sources of errors, as they affect the determination of the NDVI, is given by Goward et al. (1991). The sources of error are considered under the main headings of

- radiometry
- atmospheric effects
- spatio-temporal resolution.

These will each be considered in a little more detail. A rather similar review to that by Goward *et al.* (1991) was also published by Gutman (1991).

5.4.2 Radiometry

There are several separate sources of error here. These include

- variations in solar illumination, and
- radiometric calibration errors and drift.

We shall consider these in turn.

(i) Variations in solar illumination   The radiance received at the satellite depends on the reflectance of the Earth's surface and on the intensity of the solar illumination. If we ignore atmospheric effects for the moment (see section 5.4.3) what the sensor measures is the reflected radiance, which is the product of planetary reflectance and irradiance divided by π. This is given by

$$L = \rho^* \frac{E_0 \cos \theta_0}{\pi d^2} \tag{5.4.5}$$

where
L = reflected radiance
ρ^* = planetary reflectance
E_0 = solar irradiance at mean Earth–Sun distance
θ_0 = solar zenith angle and
d = distance between the Earth and the Sun (in astronomical units).

To study the reflectivity of the surface, it is necessary to separate the solar illumination factor and to make corrections to allow for its variation. There are two factors in the solar illumination, E_0 and $\cos \theta_0$. The solar irradiance E_0 (see section 2.2.2) does vary but the main factor in the variation of the solar illumination is the $\cos \theta_0$. We rearrange equation (5.4.5) and add subscripts λ to certain quantities, where λ is the mean wavelength of the channel in which the (satellite-received) radiance L_λ is measured. This gives

$$\rho_\lambda^* = L_\lambda \frac{\pi d^2}{E_{0\lambda} \cos \theta_0} \tag{5.4.6}$$

and what this means is that to obtain a reflectance ρ_λ^* from a satellite-received radiance L_λ using this equation we need to know the solar zenith angle θ_0. The correction, i.e. the departure from the value for vertical illumination ($\theta_0 = 0$), therefore involves $\sec \theta_0$, see Figure 5.10(a). This curve does not pass through the origin because the instrument has an offset of about 40 (for zero radiance input) and a full-scale response of about 976 instead of 1023. These margins are to allow for system drift. The solar zenith angle will vary throughout each orbit and will, of

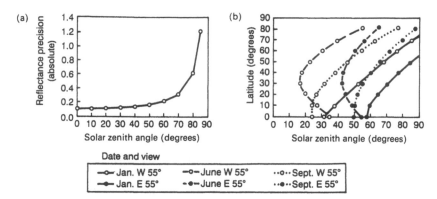

Figure 5.10 AVHRR reflectance measurement precision and observed solar zenith angles. (a) The variation in absolute measurement for the AVHRR as a function of solar zenith angle and (b) the variations in solar zenith angle, along the satellite track and across the scan swath, encountered in a single northern hemisphere pass of NOAA-7 for the summer and winter solstices and the equinox assuming a 14.30 h equatorial crossing (Goward *et al.* 1991).

course, vary according to the time of year. Examples of the variation of θ_0 along the satellite track, for the two extremes of the scan ($\pm 55°$), are shown in Figure 5.10(b) for a northern hemisphere pass for NOAA-7 for three different times of year. To make corrections for the solar zenith angle one would need to calculate θ_0 for each pixel in a scene and that would complicate the processing of the AVHRR data to produce the NDVI quite considerably. Another possibility, since the curve in Figure 5.10(a) is rather flat near the origin, would be to reject data with large values of θ_0. But, as can be seen from Figure 5.10(b) this would involve rejecting rather a large amount of data. For example, beyond a solar zenith angle of 80°, above 45°N latitude in the winter hemisphere for the afternoon orbit, the AVHRR's precision falls below $\pm 1.0\%$ and the NDVI measurements are unreliable indicators of surface conditions. In effect, observations are not available from the middle and high latitudes of the winter hemisphere for one month or more. Fortunately, with solar irradiance low, most photosynthetic activity is either dormant or small, so that loss of measurement precision, or loss of the measurements, in the winter hemisphere may not be too important in vegetation studies.

(ii) Radiometric calibration errors and drift The question of calibration of the AVHRR was discussed in considerable detail in section 2.2. We simply note one or two points here that are particularly relevant to the calculation of NDVI values. First, it is clear (Price 1987a) that vegetation indices computed from uncalibrated observations cannot be intercompared between differing sensor systems, even for those from the same family of instruments such as the AVHRR. Secondly, it is not adequate simply to use the original published calibration coefficients because the characteristics of an AVHRR may change, either after the pre-launch calibration but before launch or else after launch. The former can be handled by recalibration just before launch (since some of the instruments remained in store for a long time after calibration and before launch). However, it has been found that systems undergo further changes after launch. This degradation causes a decrease in sensor measurement precision with time and will cause significant errors in computed NDVIs unless suitable corrections are made. There are no on-board calibration standards employed in channels 1 and 2, at the visible and near-infrared wavelengths, of the

AVHRR, unlike in the case of channels 3, 4 and 5. Therefore post-launch recalibration is very important, see section 2.2.6.

Because the changes in gain and offset are not identical between the sensor channels or between the sensor systems, the failure to use post-launch calibration coefficients can introduce errors in excess of 20% for low to moderate NDVI values and lead to errors of interpretation of interannual variability assessments. These errors will be in addition to any other of the errors mentioned elsewhere in this section. Thus it is apparent that accurate assessment of interannual, interseasonal, and geographical changes in vegetation foliage with NDVI measurements requires accurate knowledge of sensor calibration changes. Thus the post-launch calibration work described in section 2.2.6 is extremely important.

5.4.3 Atmospheric effects and cloud cover

In section 4.2.3 we discussed atmospheric corrections to thermal infrared data for the determination of sea surface temperatures. We noted that a great deal of work has been put into atmospheric corrections and that now a high level of accuracy is routinely achieved in the generation of atmospherically corrected sea surface temperatures. By contrast, the atmospheric correction of NDVIs derived from AVHRR data is only in its infancy. There are two aspects or stages. The first is to demonstrate, for a given set of atmospheric conditions, that there is a difference between the NDVI, calculated with satellite received radiances, which can be denoted by $NDVI_\rho^*$, and the NDVI calculated from atmospherically corrected radiances at ground level, which can be denoted by $NDVI_\rho$. It has been shown (Duggin 1985, Fraser and Kaufman 1985) that the $NDVI_\rho^*$, the vegetation index which is usually calculated (i.e. without atmospheric corrections), can be as much as 30% lower than the equivalent ground-based (i.e. atmospherically corrected) values, see Figure 5.11. Some studies of atmospheric corrections to NDVIs have been undertaken by Singh (1986) and Singh and Saull (1988) for several different types of land area. These also clearly demonstrated the quite substantial changes made to the NDVI by making corrections for atmospheric effects. Since atmospheric conditions are known to be highly variable, in both space and time, it is clear that atmospheric corrections to AVHRR data used for NDVI calculations are very important. The second stage is to develop techniques that can be applied operationally to AVHRR for visible and near-infrared AVHRR data. This is far more difficult and work on this is only in its early stages; it is nowhere near the operational stage that has been reached with the thermal infrared and sea surface temperature determinations. For a discussion of what has been done so far see Mitchell (1989), Goward et al. (1991), Gutman (1991) and Singh (1992a). In addition to the factors that we have noted earlier in connection with atmospheric effects, there is an additional contribution arising from the surface topography. Differences in elevation of the land surface lead to differences in the geometrical (and therefore also the optical) path length experienced by the radiation (Teillet 1992).

The capacity of the AVHRR to observe landscapes at relatively high spatial resolution and at high temporal frequency is the primary factor which has led most temporal vegetation researchers to study AVHRR NDVI measurements. In theory the AVHRR provides daily repeat views of all regions of the globe. In reality, the temporal resolution of comparable surface NDVI measurements is lower than daily

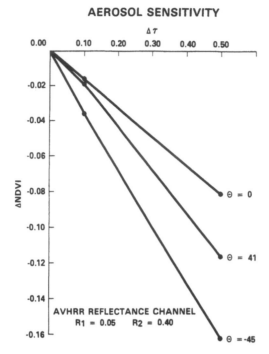

Figure 5.11 Change in NDVI as a function of aerosol optical thickness for three viewing angles of a dense green leaf canopy during the summer solstice (Holben 1986).

because of cloud cover. Cloud occurrence, for most locations on the globe, prevents daily repeat ground measurements; the magnitude of cloud impact in reducing temporal resolution varies significantly in time and space.

Although we mentioned in section 5.2 that the normalisation in the definition of the NDVI provides a significant amount of automatic compensation for various effects, it does not automatically compensate for clouds. It is estimated that on any given day about 50% of the Earth's surface is covered by cloud (Tarpley *et al.* 1984). A fully cloud-obscured observation (pixel) is a lost measurement; the ground is totally obscured and the vegetation index cannot be determined for that pixel. The question of developing cloud-detection algorithms so that cloudy areas can be identified and eliminated from data processing has been considered in section 3.2. The NDVI measurements from partially cloud-obscured pixels are significantly lower than from clear views and can therefore be substantially in error.

One could apply the cloud-detection algorithms described in section 3.2 routinely to all AVHRR data before they are used for the calculation of the NDVIs, as is commonly done when calculating sea surface temperatures. However, this is not usually done for NDVIs and there are several reasons for this. The first is that it is very expensive in terms of computing power needed. Secondly, in order to ensure the elimination of sub-IFOV-size clouds and of haze, quite harsh thresholding has to be applied and this may lead to the rejection of some cloud-free pixels. Thirdly, it has been shown that, subject to one or two precautions, cloud-contaminated data can be eliminated quite successfully by image compositing instead of by using the cloud-detection algorithms. Image compositing is a procedure in which geographi-

cally registered data sets, collected over a sequential period of time, are compared
and the maximum or minimum of a defined measurement (e.g. NDVI, temperature,
brightness) is selected to represent the conditions observed during that time period
(Holben 1986); this has already been described in connection with NOAA's operation vegetation index product in section 2.4.2. Compositing NDVI values
(retaining the maximum values) tends to select the least cloud-contaminated observations. The major advantages of this method are that it is simple, easily automated
and, for most regions of the world, quite effective in reducing cloud contamination
given enough daily observations. In addition to rejecting pixels that correspond to
total or partial cloud cover of the ground, there is also the problem that for large
viewing angles the field of view (on the ground) is very large and is viewed very
obliquely. It is quite common therefore to reject data from the edge of the swath of
AVHRR data. Thus, in producing NDVI maps one commonly rejects data from
various areas in different passes and uses the good parts from various different
passes to produce a mosaic. Thus what is presented to a user is sets of NDVI maps
(or digital data) on a regular basis, e.g. weekly, decadal (ten-days) or monthly.
NDVIs are often presented as density sliced and colour coded, see Figure 5.12 for
example (in colour section).

In addition to cloud masking and the elimination of data from pixels where small
(cumulus) clouds are present, it is necessary to carry out geometrical rectification of
every data set. This has been described already (see section 3.1). The major short-
coming encountered with image compositing is that, for any given time-composite
period and any given area, there is no assurance that any cloud-free observations
were recorded.

Selection of suitable time composite periods is still a topic of considerable discussion. For a random global 50% daily cloud-cover estimate, approximately seven
days of observations would be needed to produce a composite image that is 99%
cloud-free. In practice, cloud climatology displays both spatial and temporal persist-
ence which varies significantly across the globe. For example, monthly composites
of the Pacific Northwest region of the United States are generally not successful in
producing completely clear views of much of that region during the months of
November, December and January because cloud cover persists continuously
during this winter season. The case of a more typical mid-latitude summer cloud
climatology for the central Great Plains of the USA is described by Goward et al.
(1991). According to Justice et al. (1985) it may take more than a year of daily
observations to composite an entirely clear view of the intertropical convergence
zone. Only the use of a cloud-screening technique would guarantee that areas that
were always cloud-covered during the compositing period were eliminated. To
maintain a reasonable time resolution and also produce, as much as possible, cloud-
free surface measurements in very cloudy areas of the world requires the use of both
compositing and cloud screening.

Gutman (1989) determined statistically a relation between the maximum-value
composite and monthly mean normalised difference vegetation index (NDVI) using
data over the US Great Plains during 1986. The monthly mean NDVI was obtained
using a simple nine-day compositing technique based on the specifics of the scan
patterns of the NOAA-9 AVHRR. The results indicated that these two quantities
are closely related over grassland and forest during the growing season. It was suggested that in such areas a monthly mean NDVI can be roughly approximated by
80% of the monthly maximum NDVI, the latter being a standard satellite data

product. The derived relation was validated using data for the growing season of 1987.

As an alternative to the Maximum Value Composite (MVC), Viovy *et al.* (1992) proposed what they called the Best Index Slope Extraction (BISE). The MVC retains the highest NDVI value for a given location over a pre-defined compositing period, the assumption being that all contamination depresses NDVI values. The technique works best with a long compositing period, around 2–4 weeks. However, too long a period will distort the profile and may completely mask short-term changes in vegetation condition. Short compositing periods are more generally used; for example, the standard NOAA product, the Global Vegetation Index, is created on a weekly basis (see section 2.4.2) though this still retains a lot of noise. A smoothing technique to reduce the effects of radiometric and other disturbances on NDVI data without having a quantitative knowledge of the various interactions causing them has been proposed (Van Dijk *et al.* 1987, Van Dijk 1989). The method involves deriving composite weekly vegetation index values from the daily values for the area to be assessed followed by a smoothing of the weekly values over a selected period of time using statistical filters. The method was applied to an example of crop yield forecast. However, statistical filters generally do not remove noise and filtering after compositing severely restricts further analysis. An alternative is to use the original daily observations and attempt to isolate the true NDVI response of the surface from noise. This is what the Best Index Slope Extraction (BISE) attempts to do. The seasonal characteristics of vegetation are generally predictable, depending on ecoclimatic zonation and BISE accounts for the fact that plant growth and development is often asymmetric, i.e. periods of growth and senescence are not usually equal, and that in addition to somewhat gradual changes such as drought-induced stress, sudden changes can occur in the vegetation canopy, such as fire, deforestation or crop harvest. The BISE method was demonstrated by Viovy *et al.* (1992) using near daily geometrically corrected AVHRR Global Area Coverage (GAC) data for West Africa, covering the region from the Gulf of Guinea in the south (3°S) up to the northern border of Senegal (17°N) and from the Atlantic coast (18°W) to the Greenwich meridian. The typically noisy profiles can be seen in the original data shown in Figure 5.13. The BISE algorithm assumes (i) that cloud and poor atmospheric conditions will usually depress NDVI values, (ii) that the exception to this will be data transmission errors, such as line drop out, which can cause localised NDVI increases, though these will be abnormally high, and (iii) that decreases in NDVI relating to changes in vegetation status, while possibly sudden, will persist for a number of days, during which regrowth will be relatively slow.

From the first date of the time-series the algorithm searches forward and accepts the following point if it has a higher value than the first. Where the NDVI value from one day to the next decreases, this decrease is only accepted if there is no point in a pre-defined period of time (a 'sliding period') with a value greater than 20% of the difference between the first low value and the previous high value. If such a high value is encountered it is selected and the low point is ignored. The idea here is that sudden rises and falls in NDVI are not compatible with the gradual process of regrowth, but are a feature of moving from cloudy to clear sky conditions, or changing viewing angles. The threshold of 20% as an acceptable percentage increase in NDVI for regrowth during a sliding period was determined empirically for West African conditions. This eliminates high-frequency noise-related changes, but allows genuine drops in NDVI to be represented. The time period for decreasing cases is

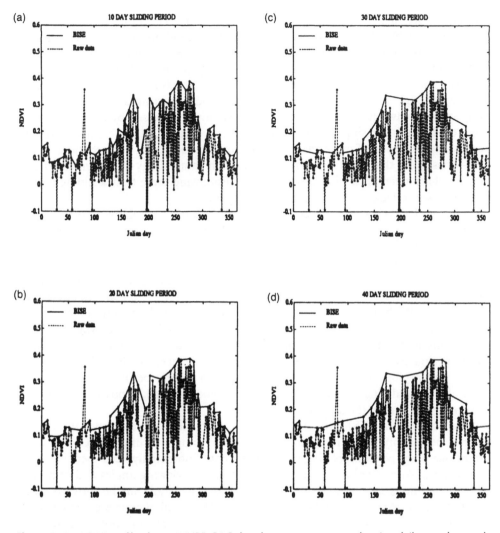

Figure 5.13 NDVI profiles from AVHRR GAC data for a savanna zone, showing daily raw data and cleaned profile using the BISE method with a sliding period of (a) 10 days, (b) 20 days, (c) 30 days and (d) 40 days (Viovy et al. 1992).

called a sliding period because the chosen point will now be taken as a new starting point for another iteration. Finally, points with a random increase greater than 0.1 are rejected; such a large fluctuation from one day to the next is impossible from natural surfaces and is therefore attributed to data errors.

One problem with the method is the choice of the length of the sliding period. Figure 5.13 shows four NDVI profiles obtained with sliding periods of 10, 20, 30 and 40 days. The results obtained from the BISE method are obviously very sensitive to the length of the sliding period; too short a period will retain too much noise, too long will tend to smooth out important changes.

The raw data were very noisy. Systematic drops in NDVI occur at intervals which match the repeat cycle of the NOAA satellites and are likely to be caused by changes in scanner viewing geometry (Gutman 1991). The more episodic irregularities are attributed to the effects of cloud and the random high value (day 80) to data transmission errors. Figure 5.13 shows how a BISE sliding period of 10 days reduces the noise introduced by variable scanning geometry but is too short for complete cloud filtering. A 20-day period (Figure 5.13(b)) is an improvement but still not long enough to account for the persistent cloud occurring at the height of the rainy season, between days 175 and 200. This is only suppressed by a 30 day period (Figure 5.13(c)). Increasing the period to 40 days (Figure 5.13(d)) offers no noticeable benefits, but a period of more than a month could hide natural perturbations in the time-series. Note that in all cases the spurious high value of day 80 is eliminated and that the sharp decrease at day 280 is retained. Following these tests a sliding period of 30 days was selected for final time-series preparation.

Viovy *et al.* (1992) also produced seven-day MVCs from the daily NDVI data for comparison with the BISE data set. The BISE was found to offer a number of advantages over the MVC. The MVC will always include spurious high values from data transmission errors such as for day 80 in Figure 5.13. BISE excludes these. MVCs made over longer periods are effective in reducing cloud and variable viewing condition influences, but the length of compositing period chosen affects the entire profile. With BISE, varying the length of sliding period only affects the profile where a sudden drop in NDVI is followed by a gradual increase; where NDVI sequentially increases or decreases the profile is unaffected. Nevertheless, the 30 day sliding period used by Viovy *et al.* will hide true decreases in NDVI if regrowth occurs very rapidly. In any case the selection of a sliding period should only be made after considering the general vegetation phenology and, if possible, the likely occurrence of cloud in the area of interest.

When comparing BISE with the MVC it is important to remember that BISE will not necessarily result in a reduction in data volume. Whilst MVC only ever retains one point per compositing period, BISE could potentially keep them all.

Because of surface anisotropy one finds that specific viewing angles and/or atmospheric conditions can produce high NDVI values, generally in the forward scattering direction (Pinty *et al.* 1989, Gutman 1991). These will of course be retained by BISE, though this criticism can also be levelled at the MVC and other methods of compositing. Dealing with directional effects is largely a matter of improving general AVHRR pre-processing; this will be discussed further in section 5.7.

Recently Pinty and Verstraete (1992) introduced a new vegetation index specifically designed to use the same spectral data provided by AVHRR, to be sensitive to the presence of vegetation and to be less sensitive than the NDVI to both soil brightness changes and atmospheric effects. This index, called the Global Environment Monitoring Index (GEMI) is defined as follows

$$\text{GEMI} = \eta(1 - 0.25\eta) - \frac{\rho_1 - 0.125}{1 - \rho_1} \tag{5.4.7}$$

where

$$\eta = \frac{2(\rho_2^2 - \rho_1^2) + 1.5\rho_2 + 0.5\rho_1}{\rho_2 + \rho_1 + 0.5} \tag{5.4.8}$$

and ρ_1 and ρ_2 are the reflectances in channels 1 and 2, respectively, of the AVHRR. This index was used with AVHRR data for July 1986 to estimate frational vegetation cover by Verstraete *et al.* (1993).

5.4.4 Developments in computational procedures

The problem of devising schemes to apply in the calculation of NDVIs to make corrections for the various effects described in this section has been addressed by Koslowsky (1989) and Phulpin et al. (1989). These procedures were developed and applied to AVHRR data for use in HAPEX-MOBILHY (Hydrologic Atmospheric Pilot Experiment – Modélisation du Bilan Hydrique). One of the main objectives of HAPEX-MOBILHY is to develop latent heat flux retrieval methods, especially in connection with global circulation model (GCM) studies and to study how satellite data can be used in this context. Phulpin et al. (1989) describe their automatic geometrical rectification, cloud screening and atmospheric correction schemes. In addition, the effects of precipitation and of haze were considered.

Although the calculation of the NDVI is not computationally demanding, Crippen (1990) argued that it was possible to use a faster, and more direct formula to produce results that are functionally and linearly identical to normalised differences. Computationally, the proposed index differs from the normalised difference vegetation index (NDVI) only in that the subtraction of the red radiance in the ratio numerator is eliminated. That is, one less step is needed than there is in the computation of the NDVI. The formula becomes

$$IPVI = \frac{NIR}{NIR + RED} . \tag{5.4.9}$$

This index has a finite positive range, from 0 to 1, thus avoiding the negative values which occur with the NDVI. It measures the percentage of near-infrared radiance in relation to the combined radiance in both the near-infrared and red bands and can therefore be termed the infrared percentage vegetation index, IPVI. The linear relation between this index and the NDVI is

$$\frac{NIR}{NIR + RED} = \frac{1}{2}\left(\frac{NIR - RED}{NIR + RED} + 1\right) . \tag{5.4.10}$$

Thus, the proposed index differs from the NDVI only by a gain of 0.5 after an offset of 1. Tests using task-specific software were run to demonstrate computational savings of 15–30% in overall computer time in using this proposed index instead of the normalised difference.

5.5 CLASSIFICATION AND PHENOLOGY

Previous sections in this chapter have been largely concerned with the definition and calculation of vegetation indices and with questions related to different spatial scales and the relation between feature size and the size of the IFOV. This section will be more concerned with some examples of classification of vegetation with AVHRR data, with the relation of the NDVI to the stages in the life cycle of vegetation and with relations between the NDVI and other parameters (such as leaf area index (LAI) for example) of a vegetation canopy. A review of many applications of the full-resolution, i.e. 1 km IFOV, AVHRR data to land-cover delineation and mapping was given by Ehrlich et al. (1994). The fact of the very small number of spectral channels in the visible and near-infrared wavelength ranges means that a multispectral classification of a land area using channels 1 and 2 for one AVHRR scene is not a feasible proposition. Looked at in another way, one can say that it is not possible to infer a specific vegetation cover type from a given value of a vegetation index such as the NDVI. It is also possible to make some use of channels 3 and 4 (or channels 3 and 5). This leads to the possibility of a four-channel classification scheme which has led to some success in distinguishing between forest, agriculture, grass and urban categories (see Kerber and Schutt 1986 for example), but this is still somewhat limited. However, while different types of vegetation may have similar values of the channel-1 and channel-2 reflectances (and therefore of the NDVI) at certain times of the year, they may have very different values at other times of the year, see Figure 5.14. Thus a considerable amount of information can be obtained by studying the form of the temporal evolution of the NDVI (Tucker 1979, Townshend and Justice 1986, Thomas and Henderson-Sellers 1987, Henderson-Sellers 1989, Arino *et al.* 1991, Derrien *et al.* 1993, Gond *et al.* 1993, Moulin and Fischer 1993, Sharman and Millot 1993, Vignolles *et al.* 1993). Classification of vegetation in France into 20 classes by considering not just the NDVI values but the profiles of the temporal evolution of the NDVI over the growing season has been carried out successfully by Derrien *et al.* (1993). Classification of vegetation into four classes for West Africa using time-profiles of the NDVI was carried out by Gond *et al.* (1993).

It is very useful to try to correlate the temporal evolution of the NDVI with the phenology of the vegetation canopy. Phenology is the study of the timing of recurring biological events, the causes of their timing with regard to biotic and abiotic forces and the interrelation among phases of the same or different species. Thus, phytophenological studies are concerned with the influence of seasonally varying environmental conditions, such as day length, air temperature and water availability, on the timing of plant development stages, or phenophases, including germination, flowering and senescence. The study of phytophenology is important in its own right for the insight it gives into the temporal organisation, evolution and

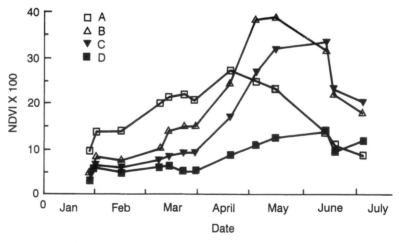

Figure 5.14   An example of the temporal evolution of the NDVI for four different sites in a study area in Spain. A: winter wheat, B: winter and spring barley, C: late spring barley and D: area suffering from frost damage and drought (Bradbury 1989).

functioning of ecosystems. Consequently, knowledge of the phenology of plant communities is relevant to estimating biological productivity, understanding land–atmosphere interactions and biome dynamics, modelling vegetative inputs into biogeochemical cycles, as well as for the management of vegetation resources. In the context of agriculture, phenology has been used in the planning of agricultural practices, the choice of optimum species for given bioclimatic conditions, the selection of optimum seeding dates and the prediction of harvest dates. Observed phenological changes are a valuable tool in the identification of vegetation cover and communities. Before the advent of satellite data, accurate quantitative information on the distribution and phenology of the world's vegetation formations was extremely limited. Yet such information is fundamental for the effective management of the Earth's resources. Such information is a basic requirement for the development of global biogeochemical and climate models and to develop an understanding of the dynamics of the major ecosystems. Before the use of AVHRR data the best estimates of the areal extent of the major vegetation formations of the Earth were obtained from compilations and syntheses of data from map sources or predictions of vegetation distribution based on controlling environmental parameters.

Although phytophenological observations have been made traditionally by a network of observers on the ground, such observations are now greatly augmented by remotely sensed data. One of the outstanding merits of remote-sensing procedures in the investigation of phenology relates to the spatially comprehensive overview of vegetation that is provided. Most early investigations surveyed the phenology of relatively small areas by the use of aerial photographs. Subsequently, the possibility of using satellite data, initially from Landsat but more recently from the AVHRR, has opened up the possibility of studying much larger areas. Previously, quantifying the seasonal activity of vegetation on a global scale had received relatively little attention because of the shortage of data on seasonal variations. We shall describe briefly two pieces of work. The first is by Justice *et al.* (1985) which is concerned with data for the entire globe and for several individual continents over the period from April 1982 to November 1983; it considers how the NDVI images display phenological changes at a very broad range of scales from subcontinental to global. That paper also cites an extensive number of relevant references. The second piece of work is by Lloyd (1989a, b) and is concerned with identifying particular parameters such as the start of the growing season, the time of maximum photosynthetic activity and so on in terms of the curve of NDVI versus time.

Figure 5.15 (in the colour section) (Justice *et al.* 1985) presents a global overview of vegetation activity which is represented in these images by the NDVI. Four different time periods were selected to show the main seasonal changes. A four-week compositing period was chosen to reduce cloud effects worldwide; this is longer than the compositing period (one week) used in the production of the NOAA GVI (see section 2.4.2). This composite was derived by selecting the highest value of the NDVI from four sequential weekly GVI products. For the purposes of display these products were sampled by a factor of 4 and this accounts for their somewhat grainy appearance. The NDVI is a bounded ratio with values from -1 to $+1$. However, values for land surfaces commonly range from -0.1 to $+0.6$. To facilitate comparison, the same colour coding has been applied to each of the NDVI images in Figure 5.15 by assigning blue, brown, green, red and purple colours to increasing NDVI values from -0.2 to $+0.6$. The colour coding and NDVI scale adopted for these images is provided in Figure 5.15(d).

In the first global image, covering the period from 12 April to 15 May 1982 (Figure 5.15(a)), the highest NDVI values are found near the equator, in the tropical rain forests and adjacent moist savannas. High values elsewhere are found only in the eastern USA, western Europe and eastern China. Low values are found not only in the desert areas but also in India, much of Canada, the West African grasslands, the north European plain and the former Soviet Union. At this time remarkably high values are found in the arid and semi-arid parts of Australia.

The second image derived for the period from 14 June to 18 July 1982 (Figure 5.15(b)) shows a dramatic change, especially for mid-latitude agricultural areas in the northern hemisphere. A belt of high NDVI values represented by the mauve and purple colours is almost continuous across the global landmass. The highest values are found in the eastern USA and Canada, in western Europe and in the former Soviet Union from Latvia almost to Lake Baykal and then again in the eastern-most areas north of Vladivostok. These areas correspond to the main grain-growing areas of the northern hemisphere. North of these agricultural areas there are marked, though less dramatic, changes of the NDVI found in the northern boreal forests. In Africa there has been a general northward shift of the main areas of high NDVI and a decrease in values in Angola and Mozambique. In South America the seasonal forest and tropical savannas of southeast Brazil show a decrease in the NDVI, whereas the tropical rain forest areas of the Amazon remain constant with overall high values.

The third image is for the period 13 September to 17 October 1982 (Figure 5.15(c)). Although the northern hemisphere mid-latitude agricultural areas continue to have higher NDVI values than in the April–May image, the values in most areas have declined significantly from the high values of June–July, presumably because of harvesting but also because other crops will have ripened and become senescent. High values remain, however, in the eastern half of the USA, probably because of the importance of crops such as corn which remain green until their harvesting in September and October. In contrast, the agricultural lands of Uruguay and southern Brazil show large increases in the NDVI with the onset of the southern spring. Southeast Asia shows the maximum values at this date. India has high NDVI values for most of its land area. This onset of greenness in fact occurred shortly after the June–July composite was obtained. In East Africa the NDVI values are low, corresponding to the dry season which lasts until the onset of the short rains later in the year.

On the image for 13 December 1982 to 16 January 1983 (Figure 5.15(d)) high NDVI values are found throughout East Africa and across the continent to the Bie Plateau of Angola. In South America the highest NDVI values are found south of the rain forest in a zone stretching from Bolivia through Paraguay to southern Brazil and Uruguay which had high NDVI values in the previous image. This zone now also extends southward to the grain-growing areas of the pampas of Argentina and the agricultural areas of Chile.

The last of these four images also displays some of the most marked atmospheric effects and it is appropriate at this stage to discuss some of the limitations of this product. Across the northern part of Figure 5.15(d) spuriously high values of the NDVI are found. Similar effects can be detected in the southern hemisphere in Patagonia, in the June–July image (Figure 5.15(b)). We have already discussed the importance of atmospheric effects in section 5.4.3. In this case one sees the consequence of low winter solar illumination levels and differential transmissivity in the red and near-infrared parts of the spectrum. In the winter months the solar illumination is low for high latitudes because of low solar elevations. Moreover, the radiation has to pass through a considerable thickness of atmosphere. The atmosphere absorbs shorter wavelengths most and hence the difference between the red and near-infrared recorded values is increased, resulting in spuriously high NDVI values. Poleward of the region of high NDVI area, no illumination is received and hence the NDVI values drop to zero. The September–October image shows an area south of the Taimyr Peninsula (in the former USSR) which would seem to be affected by cloud cover which has not been removed by the compositing process. Similar problems of very high frequency cloud cover also seem to be depressing the NDVI values over the islands of Indonesia on the same image and over the Amazon Basin in the December image.

It should be stressed that considering global data on this scale involves considerable spatial degradation of the AVHRR data. The GVI spatial resolution is poorer than that of GAC data (5 km × 3 km) which, in turn, is poorer than that of LAC/HRPT data (1.1 km × 1.1 km). If one studies vegetation indices over smaller geographical areas than the whole world one can afford to go to AVHRR data sets of higher spatial resolution than the GVI. Some examples are considered by Justice *et al.* (1985); see also section 7.9. Great use has been made of NDVI data over the last decade in the qualitative interpretive manner that we have just described; this has been both at continental scale and in a wide variety of more local regional studies. Some other examples of work at continental scales are given by Malingreau (1986) and Justice *et al.* (1989) for example and a further discussion will be found in section 5.6 since a great deal of the interest is in relation to semi-arid areas and the prediction of crop failures and potential famine. Some discussion will also be found in section 7.9.6, because of the importance of data on the reflective properties of the vegetation on the surface of the Earth in global climate modelling. Examples of regional studies involving the use of NDVI data from the AVHRR include studies of Tunisian grazing lands (Kennedy 1989), pasture production assessment in Niger (Wylie *et al.* 1991), assessment of millet fields in northern Burkina Faso (Rasmussen 1992), vegetation cover in Bangladesh (Ali *et al.* 1987) and in France (Derrien *et al.* 1992), forestry in several states of the USA (Spanner *et al.* 1990), and the silking stage in corn development in the USA (Gallo and Flesch 1989).

We now turn to the question of identifying particular stages in the plant life cycle in terms of the curve showing the temporal evolution of the NDVI. The work of Lloyd (1989b) was concerned with demonstrating that the vegetation index can serve as an indicator or measure of vegetation cover density, and of the health and vigour of the vegetation. Table 5.3 was produced by Lloyd (1989b) and it gives a phytophenological interpretation of the NDVI, or of the curve representing the temporal evolution of the NDVI through the growing season. A citation of references to justify the various entries in this table will be found in the paper by Lloyd (1989b).

A study of the phenology of the Iberian peninsular using NDVI data is described by Lloyd (1989a). On the basis of Table 5.3, Lloyd (1989b) developed a phytophenological approach to global vegetation cover mapping using NOAA GVI data. Seven-day maximum NDVI value composite GVI imagery for the period 2 July 1985–30 December 1986 was processed first to make eighteen 28-day and 35-day maximum NDVI value composite images of the world. These eighteen images, each of which corresponds approximately to a calendar month, were then processed to make three phytophenological variables listed in Table 5.3 as follows.

Table 5.3   List of phytophenological variables (Lloyd 1989b)

NDVI Definition	Phytophenological Interpretation
Date when minimum NDVI recorded	Time of minimum photosynthetic activity
Date when maximum NDVI recorded	Time of maximum photosynthetic activity
Date(s) when 0.099 threshold exceeded	Time of start of growing season(s)
Number of periods when NDVI > 0.099	Number of growing seasons
Length of period(s) when NDVI > 0.099	Length of the growing season(s)
Minimum NDVI value	Lowest level of photosynthetic activity
Maximum NDVI value	Maximum level of photosynthetic activity
Mean daily NDVI (time-integrated NDVI/ number of days in integration period)	Mean daily maximum potential photosynthetic rate
Mean daily NDVI for period when NDVI > 0.099	Mean daily maximum potential photosynthetic rate during growing season
Time-integrated NDVI	Gross primary production

- Time of maximum photosynthetic activity: this is defined as the month between January and December 1986 in which the maximum NDVI value was recorded.

- Length of the growing season: this is defined as the longest consecutive period between July 1985 and December 1986 when the NDVI value was greater than 0.099. An 18 month period was used in order to ensure that growing seasons which span calendar years (e.g. November to April) are described accurately.

- Annual mean daily maximum potential photosynthetic rate: this is calculated as follows. The NDVI value at each pixel of each monthly maximum NDVI value composite between January and December 1986 was multiplied by the number of days in the compositing period to create twelve monthly integrated images. The twelve monthly integrated NDVI values at each pixel were then summed and divided by the total number of days in the year. A classification scheme was devised using these three variables to generate 20 phytophenological classes (for details see Table 2 of Lloyd 1989b) and applied. The results for this 18 month data set were colour coded, or density sliced and colour coded, and represented as (subsampled) images, see Figure 5.16 (in the colour section). Enlargements of three parts of Figure 5.16 are shown in Figure 5.17 (in the colour section). Figures 5.16 and 5.17 are quite different from the conventional images that are

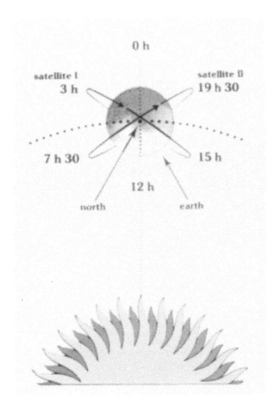

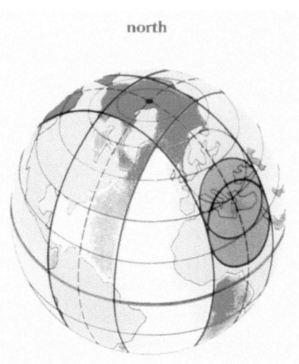

Figure 1.19 The morning and afternoon passes of the two operational NOAA series spacecraft.

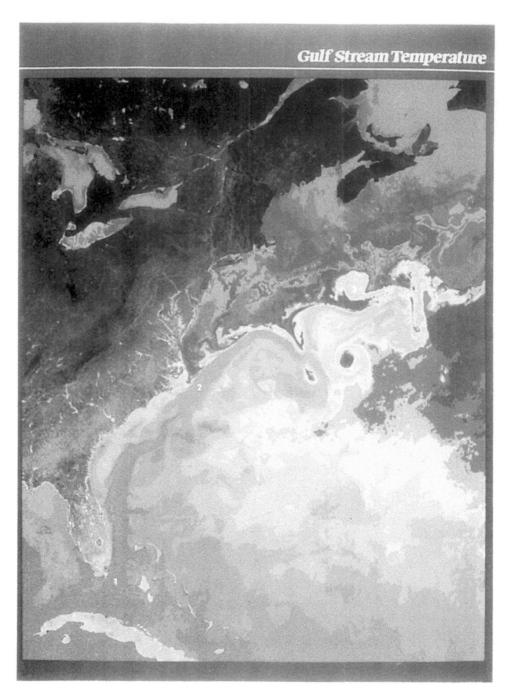

Figure 4.2 Eddies on the Gulf Stream.

Figure 4.5 Sea surface temperatures (National Remote Sensing Agency, Hyderabad).

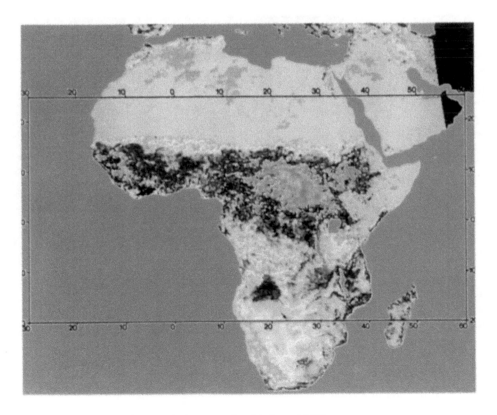

Figure 5.12 NDVI colour-coded composite (Holben 1986).

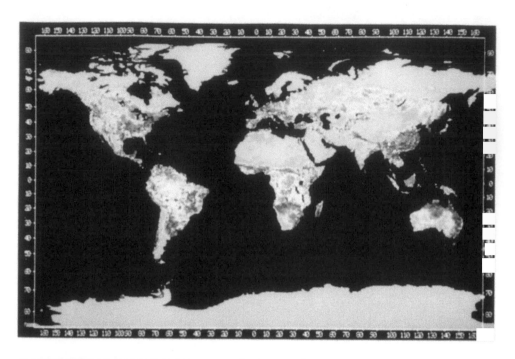

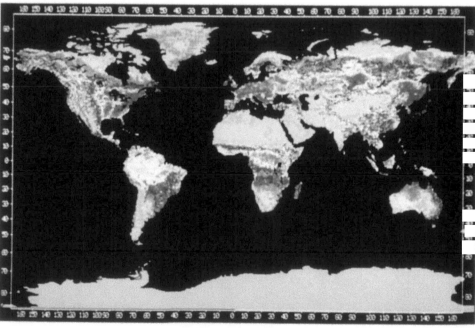

Figure 5.15 Global NDVI composites for (a) 12 April–15 May 1982, (b) 14 June–18 July 1982, (c) 13 September–17 October 1982 and (d) 13 December 1982–16 January 1983 (Justice *et al.* 1985).

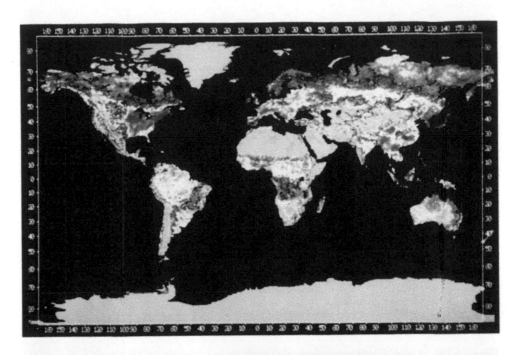

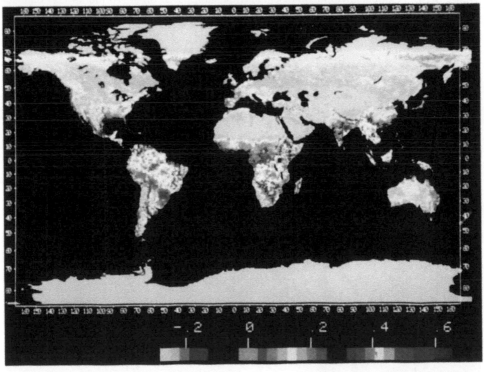

Figure 5.15 (*Continued*)

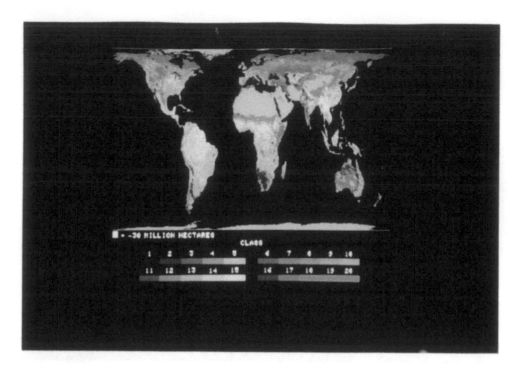

Figure 5.16 A global phytophenological classification based on an 18 month GVI data set (Lloyd 1989b).

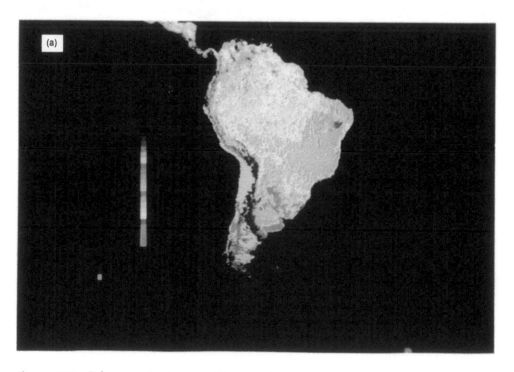

Figure 5.17 Enlargements of parts of Figure 5.16.

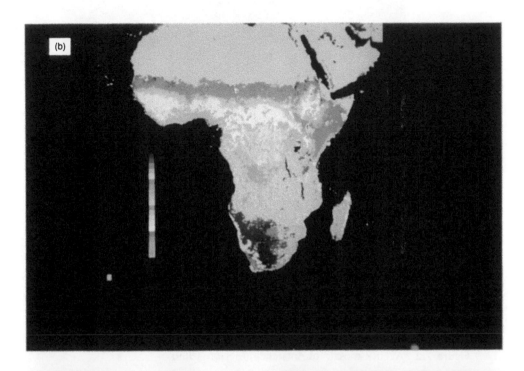

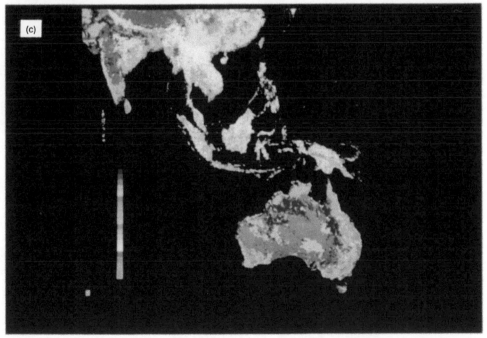

Figure 5.17 (*Continued*)

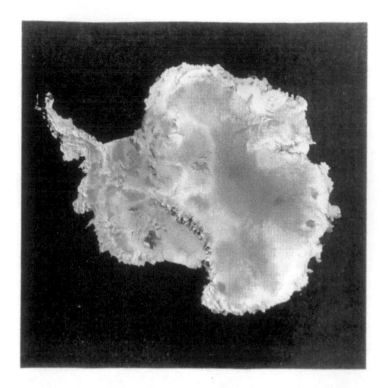

Figure 7.2 AVHRR mosaic of Antarctica (Merson 1989).

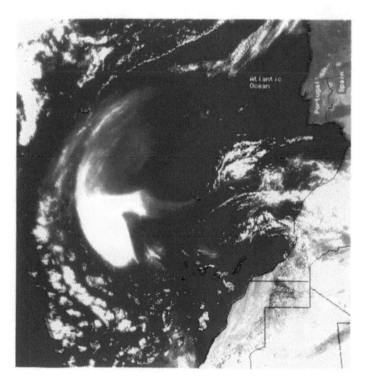

Figure 7.42 AVHRR image of a Saharan dust outbreak (Stephens 1995).

usually produced. What is commonly produced is a colour-coded image where, essentially, the NDVI is density sliced and a colour is attributed to each slice.

Achard and Blasco (1990) extended the idea of using the variation, through the annual cycle, of the NDVI to study the phenology of the vegetation canopy; they proposed the inclusion of the surface temperature, T_S, as well.

In addition to studies in which the NDVI is used in a largely empirical fashion in qualitative interpretations of NDVI maps, there have also been studies in which attempts have been made to establish correlations with other variables. In this way it is hoped to establish plausible mathematical relations with some local, or even global, semblance of generality. Such variables include (i) the quantity of vegetation (the biomass), (ii) the green leaf area index (GLAI, which is the area of green leaves per unit area of ground) and (iii) water availability (see, for instance Cihlar *et al.* 1991).

With regard to (i), Bradbury (1989) for instance studied the relation between NDVI and vegetation cover, biomass and crop growth for areas of cereal cultivation in central Spain. Figure 5.18 shows plots of NDVI against vegetation cover for the three months April, May and June 1989. For each individual month of field work there appears to be some correlation of vegetation cover with NDVI. However, no single linear correlation is apparent when the field data are considered as a whole. Fresh biomass values were found to have a poorer correlation with NDVI than the vegetation cover. The analysis of the relation between NDVI and crop growth stage showed that the NDVI increased up to the heading stage of crop growth, after which it declined.

Multiple regression analysis of the complete three-month data set indicated that NDVI is a function of crop cover and growth stage. The following equation resulted in an R^2 value of 0.787:

$$\text{NDVI} = (9.223 + 0.157\ \text{COVER} + 0.527\ \text{STAGE} - 0.0067\ (\text{STAGE})^2)/100 \quad (5.5.1)$$

where COVER is the vegetation cover in percent and STAGE is the mean Zadoks growth stage value (Zadoks *et al.* 1974).

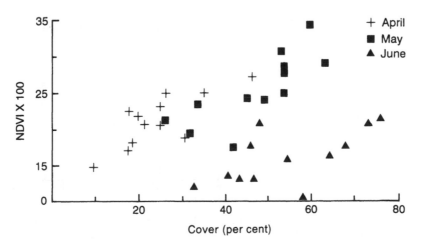

Figure 5.18 Plot of NDVI versus vegetation cover for a study area in Spain (Bradbury 1989).

Of the three variables biomass, leaf area index and water availability, it is probably (ii) the leaf area index, LAI, or green leaf area index, GLAI, which is the easiest to relate to the NDVI. The GLAI is defined as the area of green leaves per unit area of ground. To some extent the ideas involved have already been considered in what we have said in section 5.2 about a soil adjusted vegetation index, SAVI. But that approach was concerned with defining a different vegetation index to cope with situations where the vegetation canopy was not complete and some (soil) understorey was visible. Here we are concerned with relating the most conventional or most commonly used vegetation index, NDVI, to the fraction or proportion of the ground that is covered by leaves, i.e. to relate the NDVI to the GLAI.

For a very low value of the GLAI ($\ll 1$) there is a large amount of bare understorey (soil) visible, the near-infrared reflectance is low and the soil reflectance is reasonably high; the NDVI is thus low. As the GLAI increases the near-infrared reflectance increases and the red reflectance decreases; consequently the NDVI increases, see Figures 5.19 and 5.20. Eventually, for large values of the GLAI, where the canopy is several layers of leaves thick, the reflectances become almost unaffected by the addition of extra leaves and therefore the near-infrared and red reflectances, and therefore also the NDVI, tend to asymptotic values, see Figures 5.19 and 5.20. It can be seen from these figures that the curves are species-dependent. Given a calibration curve, for a given species, of NDVI versus GLAI one can use remotely sensed AVHRR data to estimate GLAI, provided the values of NDVI obtained are not in the region where the curve is flat. In other words, for a very thick canopy the reflected radiation is reflected near the surface and the reflectivity is insensitive to changes in the thickness of the canopy. Although the relation between NDVI and GLAI is a curve, one finds that the relation between the PVI (perpendicular vegetation index) and the GLAI is close to linear, see Figure 5.21 for example. This is not surprising since these two quantities are defined in rather similar ways (see section 5.2, also Curran 1983).

In some recent work Price (1992) has, effectively, applied the radiative transfer equation to the vegetative canopy to try to determine the amount of vegetation

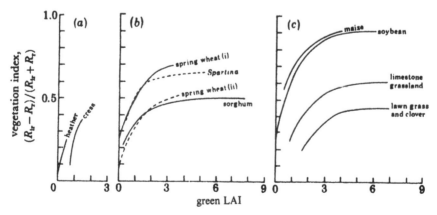

Figure 5.19 The relation between the NDVI and LAI (leaf area index) for different vegetation canopies (a) asymptote not reached, (b) asymptote reached at low LAI and (c) asymptote reached at high LAI (from a number of sources, Curran 1983).

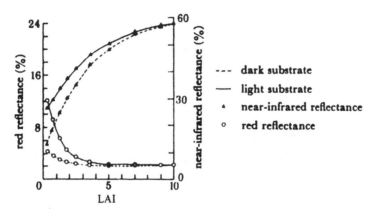

Figure 5.20 The relation between Leaf Area Index (LAI) and red and near-infrared reflectance for a vegetation canopy on a light-toned and a dark-toned substrate (Curran 1983).

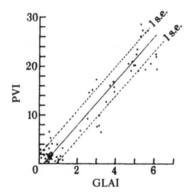

Figure 5.21 The relation between the PVI (Perpendicular Vegetation Index) and the GLAI (Green Leaf Area Index) for a *Pteridium-Calluna* association (Curran 1983).

from visible and near-infrared reflectance data. This involves reflection and transmission within the canopy and reflection by the underlying soil and assumes that variability within the structure of the canopy can be neglected, i.e. the canopy is treated as a homogeneous medium with an absorption coefficient, $\alpha(\lambda)$, and a scattering (reflection) coefficient, $\beta(\lambda)$. The downwelling radiation, I, and the upwelling radiation, J, will then satisfy the differential equations

$$\frac{\mathrm{d}I}{\mathrm{d}x} = -\alpha I + \beta J \tag{5.5.2}$$

$$\frac{\mathrm{d}J}{\mathrm{d}x} = \alpha J - \beta I \tag{5.5.3}$$

where x is (essentially) distance through the canopy but is actually expressed in terms of the leaf area index, LAI, of that thickness of canopy. By solving these two coupled (simultaneous) differential equations, and applying the appropriate boundary conditions at the upper surface of the canopy and at the underlying soil surface, an expression is obtained for the reflectance

$$R(\lambda) = \frac{[c(\lambda) - \alpha(\lambda)]/\beta(\lambda) + [(\alpha(\lambda) + c(\lambda))/\beta(\lambda)]D(\lambda)}{1 + D(\lambda)} \qquad (5.5.4)$$

where

$$D(\lambda) = \frac{[c(\lambda) - \alpha(\lambda)]/\beta(\lambda) + r_s(\lambda)}{[c(\lambda) + \alpha(\lambda)]/\beta(\lambda) - r_s(\lambda)} \exp(-2c(\lambda)\,(\text{LAI})) \qquad (5.5.5)$$

and where $c(\lambda)^2 = \alpha(\lambda)^2 - \beta(\lambda)^2$, $r_s(\lambda)$ is the reflectance of the soil and LAI is the leaf area index of the complete canopy. For details of the derivation of equation (5.5.4) see the paper by Price (1992).

Two problems arise. First, equations (5.5.4) and (5.5.5) give $R(\lambda)$ expressed as a function of LAI. This needs to be inverted since it is the channel-1 and channel-2 reflectances which are being measured with the purpose of trying to determine LAI, the leaf area index. This inversion involves mathematical manipulation of the equations. Secondly, the parameters $\alpha(\lambda)$, $\beta(\lambda)$ and $r_s(\lambda)$ are not readily available or easy to determine for any given species. What was done to cope with this second problem was to determine values of these parameters by fitting the data to published graphs of near-infrared reflectance to red reflectance, for various values of LAI; that is these curves are used as calibration curves. The final result (again for details see Price 1992) is

$$\text{LAI} = \frac{1}{2c_2} \ln(\{r_{\infty 2} - R_2 - [(r_{\infty 2} - R_2)^2 - 4(r_{\infty 2} - \alpha r_{\infty 1} - b)(R_1 - r_{\infty 1})]^{1/2}\}$$

$$\times [2(R_1 - r_{\infty 1})]^{-1}) \qquad (5.5.6)$$

where R_1 and R_2 are the channel-1 and channel-2 reflectances, $r_{\infty 1}$ and $r_{\infty 2}$ are the channel-1 and channel-2 reflectances for an infinitely thick canopy of the species in question ($\text{LAI} \Rightarrow \infty$) and a, b and c_2 are constants determined from the near-infrared/red reflectance calibration data.

The above argument is for a situation in which one assumes that there is a single species, density, etc. canopy in the whole field of view. In practice this is more likely to be relevant to SPOT or Landsat data. For the AVHRR, since the IFOV is very large (120 ha) it is likely that for any agricultural study one is not dealing with a pixel corresponding to a homogeneous canopy throughout the whole IFOV but that, instead, one has a mixed pixel corresponding to some area of bare soil and several areas of different crops at various stages of their development. In this situation, no general solution is possible from just two reflectance measurements in the visible and near-infrared, as two measurements are not sufficient to determine more than two unknowns. The situation can be simplified by assuming that the reflectance measurements correspond to the sum from two surfaces, a dense vegetation cover with fractional cover f, and bare soil with fraction $1 - f$. Such mixture models have been used frequently (e.g. Richardson and Wiegand 1991). Then by straightforward analysis one finds the value for f resulting from the heterogeneous surface:

$$f = \frac{R_2 - \alpha R_1 - b}{r_{\infty 2} - \alpha r_{\infty 1} - b} \qquad (5.5.7)$$

where the terms have been defined previously. Price (1992) also gives some discussion of the situation for sparse vegetation and discusses the relation of the results to several of the conventional definitions of vegetation indices which were considered previously in section 5.2.

Another recent development has involved attempting to use data from the thermal infrared channels of the AVHRR in conjunction with NDVI data in vegetation studies. This has proved to be very successful in studying the relation between the health of the vegetation and its supply of water. Vukovich *et al.* (1987) found a positive correlation between ground temperature and albedo values obtained from NOAA-7 AVHRR data for a semi-arid sub-Saharan region of Africa, Senegal, for various days in the period of September 1981 to October 1982; i.e. regions of high albedo were regions of high ground temperature and regions of low albedo were regions of low ground temperature. The highest values of ground temperature and albedo were found in a region characterised by sparse vegetation and in the dry season. The lowest values of ground temperature and albedo were found in regions characterised by dense vegetation and in the wet season. These results suggest that regions of high albedo in the sub-Saharan region of west Africa are regions of high ground temperature. Mechanisms for increasing the ground temperature under high albedo conditions were discussed. It was suggested that the increased ground temperature would increase the longwave radiation emitted to the atmosphere and the sensible heat flux at the surface, leading to a net heat gain in the local atmosphere which will result in a general rising motion. Seasonal evaporation has been related to green vegetation cover as measured by the NDVI (Kerr *et al.* 1989). Short and long term evaporation rates have also been calculated from the surface energy balance equation using the surface temperature, T_S, measured by space-borne sensors (Seguin *et al.* 1989). When these data sets from satellite sensors are combined, negative relations between T_S and NDVI are observed (Goward *et al.* 1985, Choudhury 1989b) confirming that fractional green vegetation cover is a major determinant of land surface evaporation. Variations in the slope of the T_S–NDVI relations have been related to increases in canopy resistance due to soil water deficit (Nemani and Running 1989) indicating that these relations contain information relevant to the regional estimation of evaporation.

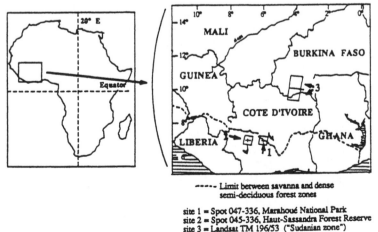

Figure 5.22 Study area of Achard and Blasco (1990) in West Africa.

In the West African study area of Achard and Blasco (1990), see Figure 5.22, the main original vegetation types on the two sides of the boundary limit are semi-deciduous dense forests in the south and various mixed vegetation types including a variety of savannas with scattered trees and shrubs in the north. No field work was done to classify the AVHRR data; the AVHRR data were classified by using classified SPOT and Landsat TM data degraded to 1 km resolution. The temporal evolution of the NDVI and of the surface temperature are shown in Figures 5.23 and 5.24 for the main types of land cover and for each of the three sites shown in Figure 5.22. Instead of plotting the temporal evolution of the NDVI or of the surface tem-

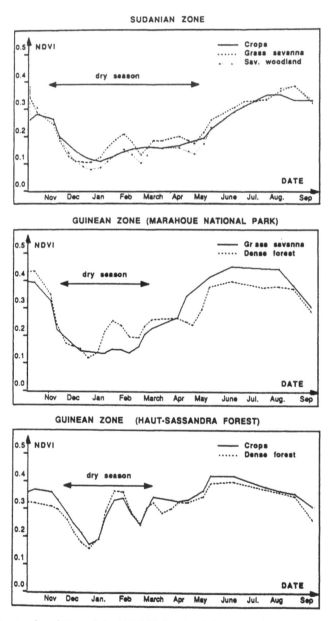

Figure 5.23 Temporal evolution of the NDVI (Achard and Blasco 1990).

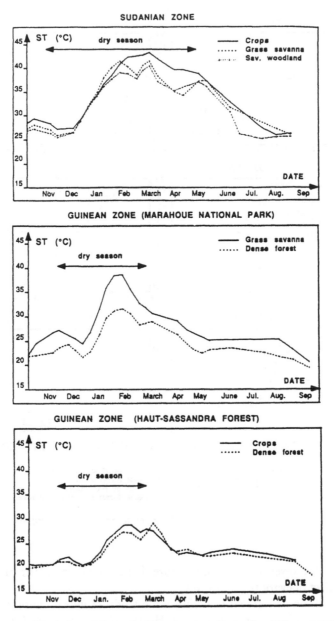

Figure 5.24 Temporal evolution of the surface temperature (Achard and Blasco 1990).

perature throughout the year, another method for presenting the data is to plot a succession of cluster diagrams in plots of surface temperature against NDVI for various months in the year, see Figure 5.25.

Figures 5.23–5.25 illustrate the capability of a time series of AVHRR data to discriminate different types of land cover in tropical areas. At the end of the dry season, it is possible to discriminate regions of natural savannas from wet forests. The savanna region is characterised by a low vegetation index corresponding to a relatively low photosynthetic activity and high surface temperature, whereas the forest region has a higher vegetation index due to a higher density of trees

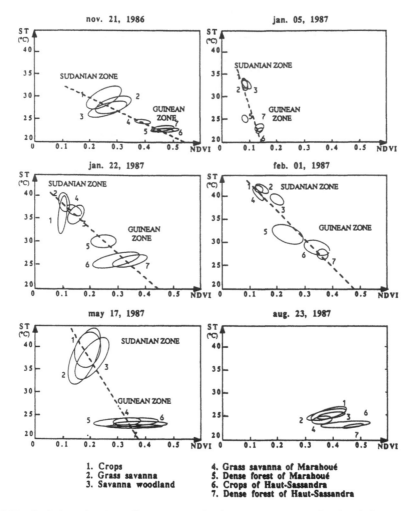

1. Crops
2. Grass savanna
3. Savanna woodland

4. Grass savanna of Marahoué
5. Dense forest of Marahoué
6. Crops of Haut-Sassandra
7. Dense forest of Haut-Sassandra

Figure 5.25   Evolution of various ellipses in NDVI/surface temperature (Achard and Blasco 1990).

(greenness), and a lower surface temperature due to the plant cover protection and the cooling effect of evapotranspiration.

Further work on soil moisture includes the study of wheat canopy resistance to stress induced by soil moisture deficiency (Hope 1988), corrections for soil moisture when estimating the LAI from vegetation index data (Clevers 1989) and also the study of forest and agricultural land in southeastern Australia (Smith and Choudhury 1990). As an example of the use of thermal infrared data in conjunction with the visible, near-infrared and mid-infrared data, we note that Williamson (1988) used airborne data and four spectral bands

RED	0.605–0.69 μm
NIR	0.76–1.05 μm
MIR	1.55–1.75 μm
TIR	8.50–13.0 μm

in two models for the GLAI for rough grassland. These are (i) using the mid-infrared

(MIR):

$$\text{GLAI} = -0.80 - (1.73 \text{ MIR}) + (0.10 \text{ NIR}) + (0.25 \text{ RED}) \qquad (5.5.8)$$

and (ii) using thermal infrared (TIR):

$$\text{GLAI} = 2.44 - (0.19 \text{ TIR}) + (0.17 \text{ NIR}) - (0.03 \text{ RED}) . \qquad (5.5.9)$$

The study of tropical deforestation by studying forest fires using AVHRR data from the mid-infrared (channel 3) and the thermal infrared (channels 4 and 5) will be discussed in the next chapter (see section 6.2). However, changes in forest area (after the fires have finished burning) can also be studied using data from the visible and the near-infrared channels of the AVHRR (Nelson and Holben 1986, Woodwell *et al.* 1987, Malingreau and Tucker 1988, 1990, Malingreau *et al.* 1989, 1990, Cross 1990, Paivinen *et al.* 1990, Vogelmann 1990, Cross *et al.* 1991). AVHRR data were also extensively used, for example by Dech and Glaser (1992), in studying the large number of fires that were started deliberately in early 1991 at the end of the Gulf War. Dech and Glaser were able to obtain qualitative information regarding the spatial distribution, height, thickness and some dynamic aspects of the smoke plumes. In addition to this, the investigation of satellite imagery during April, May, and June 1991 proved that large vegetated areas in the southern Mesopotamian region were obviously affected by the fall-out. The affected region belongs to the most important agricultural areas in Iraq.

It should by now be clear that there are two serious problems facing attempts to use AVHRR data in a conventional land cover classification. There are (i) the poor spatial resolution and (ii) the very limited number of spectral channels in the visible and the near-infrared. However, there is another approach to the use of AVHRR data for the classification of vegetation which it is useful to consider and this is in relation to large and relatively homogeneous areas of natural vegetation (Norwine and Greegor 1983). This is particularly important in relation to current work on the contribution of land areas to heat fluxes in relation to the modelling involved in global climate change studies (see section 7.9). The key to this lies in the fact that the distribution of types of natural vegetation over the globe is largely a function of climate, especially of spatial variations in energy and moisture budgets. As a general rule, climate is recognised as the pre-eminent control of natural vegetation at subcontinental-to-global scales. Ecologists and biogeographers have attempted to utilise climatic measures and indices in surveys, stratifications, and classifications of natural vegetation. It is commonly assumed that boundaries between areas of different types of natural vegetation correspond to climatic discontinuities and Norwine and Greegor (1983) incorporated this idea into their approach to natural vegetation classification. That is, one does not so much attempt to establish a direct relation between AVHRR data and vegetation but rather one attempts to relate the AVHRR data to the vegetation via the climatic parameters or conditions. This is not generally possible in agriculture or in managed forestry where species have been planted by human action. Norwine and Greegor's work thus involved trying to formulate a vegetation gradient model based on climatic and meteorological satellite data that would ultimately be appropriate for global vegetation stratification and monitoring. Norwine and Greegor (1983) made use of an approach developed by Trenchard and Artley (1981a, b); this involved the use of a climatological/meteorological variable called the hydrologic factor, HF, or sponge which is calculated from simple local meteorological data, namely the measured precipitation and the evaporation esti-

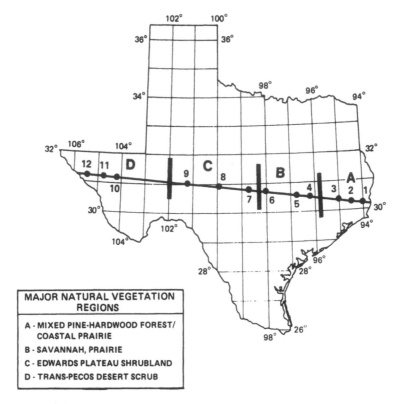

Figure 5.26 Sample locations and major vegetation regions in Texas studied in the work of Norwine and Greegor (1983).

mated from the measured maximum and minimum temperatures (for details see Norwine and Greegor 1983). The correlation between the NDVI and the hydrologic factor, HF, was investigated using AVHRR data. The correlation obtained between the NDVI and the HF is illustrated in Figure 5.26.

In Texas, the chosen study area of Norwine and Greegor (1983), there exists the most pronounced continuous, nonorographic, intra-state climatological gradient found anywhere in the United States. At least four distinct, first-order climatic types occur within the state (humid subtropical, tropical steppe, tropical desert, and mid-latitude steppe), with many more important subtypes (e.g. subtropical subhumid). In particular, there is an extraordinarily steep east–west moisture gradient, ranging from very humid in southeastern Texas (mean annual precipitation > 50 inches) to true desert in far western Texas (mean annual precipitation < 8.0 inches). This gradient strongly influences ecological patterns, and virtually controls the regional distribution of natural vegetation.

Mean annual HF values were calculated for 75 Texas locations and 12 sites were selected for comparison of HF and NDVI values, see Figure 5.26. AVHRR data from three 1980 scenes, 19 April, 10 July and 9 October, were used for the determination of NDVI values and these values are plotted against the corresponding HF values in Figure 5.27. In this figure there are thus three values of NDVI, corresponding to the three dates chosen, for each value of the HF. Some of the scatter in this figure arises from the fact that the NDVI varies throughout the growing season

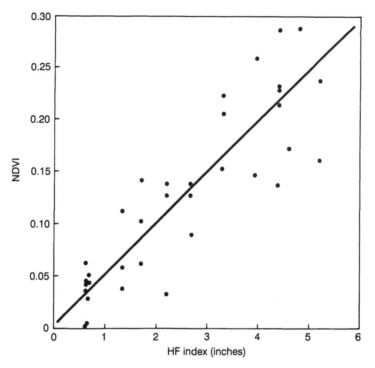

Figure 5.27 Plot of NDVI versus HF (hydrologic factor). Correlation = 0.86 (Norwine and Greegor 1983).

and it shows three instantaneous NDVI values while the HF plotted is a mean annual value. Having established a correlation between the NDVI and the HF, Norwine and Greegor proposed another index, the vegetation-hydrologic factor index, I_{vh}, where $I_{vh} = HF \times NDVI$. A table of values of HF, NDVI and I_{vh} and identifying the vegetation types is given in Table 5.4.

Table 5.4 Hydrologic factor, normalised difference, and vegetation-hydrologic factor index mean values for three dates in 1980 (19 April, 10 July, 9 October) for 12 sites along Texas transect (Norwine and Greegor 1983)

Vegetation regions	Site	Hydrologic factor	NDVI	Vegetation-hydrologic factor index
Pine-hardwood forest	1	5.20	0.22	1.14
	2	4.61	0.23	1.06
	3	4.40	0.25	1.10
Savannah/prairie	4	4.10	0.18	0.74
	5	3.31	0.19	0.63
Shrubland	6	2.70	0.12	0.32
	7	2.12	0.10	0.21
	8	1.71	0.10	0.17
	9	1.36	0.10	0.14
Desert/desert Scrub	10	0.70	0.04	0.03
	11	0.67	0.03	0.02
	12	0.63	0.03	0.02

5.6 EVAPOTRANSPIRATION, FIRE RISK AND SEMI-ARID AREAS

In this section we have gathered together several topics that relate to the loss of water by the soil–vegetation system or are associated with a shortage of water. A vegetation index, particularly the NDVI, is relatively easy to calculate on a regular basis, on a regional, continental or global scale, from AVHRR data. But it is not, of itself, an intrinsically useful quantity. It has to be related to scientifically or environmentally useful parameters; we have already seen some examples of this in section 5.5 and this section will provide some further examples.

5.6.1 Evapotranspiration

The large-scale distribution of soil moisture content and the evaporative processes over land are among the most important boundary conditions of the atmospheric system. Weather forecast models and climate models have the ability to predict these fields but it is necessary to be able to determine the values of these parameters from observed data for the purpose of model initialisation or model verification. Such diagnostics are also necessary to determine the inter-annual or long-term variability of the surface conditions over the continents. Until recently the only attempts to determine these large-scale fields have been made through water budget calculations, using a reconstruction of the potential evapotranspiration field (which is the evapotranspiration from a wet vegetated surface), from an estimation of the radiation balance at the ground or from an empirical approach relating it to the observed temperature and the length of the day. The verification of these computations, when possible, showed that they lead to realistic results. These methods based on ground data nevertheless suffer from important limitations: first, they do not take into account the variable properties of the vegetation and secondly they are based on data with a coarse spatial resolution that has a high degree of spatial non-uniformity and finally they are mainly statistical and apply essentially to the determination of long-term climatic distributions.

Remote sensing techniques which provide a global and almost continuous set of data at a resolution of a few kilometres are the only way to solve these problems. But their use is constrained by many difficulties. The first difficulty arises from the presence of clouds. However, even with cloud-free data there is the very important difficulty that remote sensing so often does not measure the actual physical or environmental quantities that one really wants to measure; some inversion scheme has to be developed to determine the values of the required quantities or parameters. One such example is provided by evapotranspiration. What we measure, with the AVHRR, is the satellite-received radiance in the five visible and infrared channels. Attempts have been made to relate the evapotranspiration either to these radiances directly or to a vegetation index such as the NDVI.

The latent heat flux, LE, which is an important parameter in climate modelling during day-time conditions, can be expressed as

$$LE = R_n - G - \rho C_p h(T_s - T_a) \tag{5.6.1}$$

where

R_n = net radiation
G = soil heat flux by conduction

$\rho\, C_{\mathrm{p}}\, h(T_{\mathrm{s}} - T_{\mathrm{a}}) =$ expression of the sensible heat flux H, with ρ and C_{p} constants (air density and specific heat, respectively)

h = turbulent exchange coefficient

$(T_{\mathrm{s}} - T_{\mathrm{a}})$ = difference between surface and air temperature.

The quantities on the right-hand side of this equation are not directly measured by remote sensing techniques.

Equation (5.6.1) is very widely used in various ways for crop water status determination on a local scale or surface flux estimation on a regional scale. Perhaps the main application of the instantaneous energy balance is to provide a basic tool for flux measurement methods or for soil–plant–atmosphere modelling studies, mainly working on a typical timescale of about one hour.

Concerning the objective of long-term monitoring of crop water conditions, it is necessary to consider longer periods, typically between 1 day and 10 days, in order to adjust to the timescale of agrometeorological processes and related models.

For local-scale application and for longer period applications Jackson et al. (1977) introduced a simple relation which has subsequently been used quite widely. This involves supposing that daily evapotranspiration, $\mathrm{ET_d}$, is related simply to R_{nd}, the net daily radiation, and the instantaneous value of $(T_{\mathrm{s}} - T_{\mathrm{a}})$ (the difference between the surface temperature and the air temperature) at midday. This is represented by the equation

$$\mathrm{ET_d} = R_{\mathrm{nd}} + a - b(T_{\mathrm{s}} - T_{\mathrm{a}}) . \tag{5.6.2}$$

For consistency the quantities ET and R_{nd} are expressed in common units, usually in mm of water (i.e. R_{n} is considered to be the latent heat of evaporation of the specified quantity of water). Various references to the subsequent use of this approach and to the question of the values of the coefficients a and b are cited by Seguin et al. (1991) for example. The values of the parameters a and b depend on the type and state of the vegetation; thus they are site-dependent as well as varying from year to year.

Equation (5.6.2) represents one day but it can very easily be extended to longer periods. Its interest is higher when applied to timescales larger than the daily basis, because the dispersion for individual values (of the order of ± 1 mm day^{-1}) is damped by passing to 5 days or 10 days. ET, the accumulation of daily values over n days, gives

$$\sum_0^n \mathrm{ET_d} = \sum_0^n R_{\mathrm{nd}} + na - b \sum_0^n (T_{\mathrm{s}} - T_{\mathrm{a}}) . \tag{5.6.3}$$

$\sum(T_{\mathrm{s}} - T_{\mathrm{a}})$ is thus linearly related to $\mathrm{ET} - R_{\mathrm{n}}$ (which corresponds to the accumulated value of the sensible heat flux H on the same period).

The quantity $\sum_0^n(T_{\mathrm{s}} - T_{\mathrm{a}})$ was named the stress-degree-day by Jackson et al. (1977) and it has been shown to provide a good description of crop water stress for irrigation or plant-breeding purposes. Equation (5.6.3) shows that this quantity $\sum_0^n(T_{\mathrm{s}} - T_{\mathrm{a}})$ has a physical basis (albeit mixed with the empiricism of the original idea behind equation (5.6.1)) allowing one to convert it into energy/water balance components considered in agrometeorological modelling for quantifying crop water stress and assessing its effect on final yield.

The accumulated form in equation (5.6.3) was used for Sahelian regions by Negré *et al.* (1988) and Seguin *et al.* (1989) using Meteosat data for the surface temperatures, for agricultural areas of France by Seguin *et al.* (1991) using both Meteosat and AVHRR data and also for France by Serafini (1987) and Lagouarde (1991), and for Poland by Dabrowska-Zielinska (1989) and Dabrowska-Zielinska and Grusczcynska (1993) using AVHRR data. Seguin *et al.* (1991) used five-day syntheses of Meteosat data for T_s, the surface temperature, for a three-year period (1985–1987) but used AVHRR data from about 40 dates per year to calibrate the Meteosat data since the calibration of the thermal data from AVHRR is better than that from Meteosat. The values of T_a were obtained from conventional measurements at ground meteorological stations. It was shown that thermal infrared data from meteorological satellites, if available on a regular basis, can be used to describe general trends in regional water conditions. Moreover, the possibility of relating surface temperature to evaporation by way of the energy balance equation permits one to obtain quantitative estimates of regional evapotranspiration. However, when applied to an area such as France the familiar problems of mixed pixels arises. The field size is usually much smaller than the IFOV of the AVHRR or of Meteosat and therefore the vegetation cover is non-homogeneous and the values of a and b will vary as well within the pixel.

The question of the validity of using only one instantaneous temperature difference each day when estimating the evapotranspiration, i.e. the validity of equation (5.6.1), has been examined by Vidal and Perrier (1989).

An alternative to using a relation that involves the instantaneous value of $(T_s - T_a)$ at midday, is to use T_a^{max}, the maximum air temperature during the day (Caselles and Delegido 1987, Caselles *et al.* 1992). This involves using an expression of the form

$$\text{ET}_d = A T_a^{max} R_{nd} + B R_{nd} + C \qquad (5.6.4)$$

where the values of the coefficients A, B and C are site-dependent and need to be determined empirically. This approach was applied to the Valencian region by Caselles and Delegido (1987) and to the La Mancha region in Spain by Caselles *et al.* (1992). T_a^{max} was determined using the AVHRR pass in the early afternoon, for instance the 14.30 GMT pass of NOAA-9. Values of the coefficients in equation (5.6.4) were determined using the daily values of R_{nd} and T_a^{max} over an extended period and fitting equation (5.6.4) to the values of ET_d calculated from evaporation data measured with a class A pan evaporimeter (see Figure 5.28). The values obtained for the Valencian region (39°35′ N, 0°20′ W) by Caselles and Delegido (1987) are

$$\text{ET}_d = (5.37 \times 10^{-4}) T_a^{max} R_{nd} + (1.57 \times 10^{-3}) R_{nd} + 0.37 \qquad (5.6.5)$$

and for Albacete (39°00′ N, 1°52′ W) in the La Mancha region by Caselles *et al.* (1992) are

$$\text{ET}_d = (5.99 \times 10^{-4}) T_a^{max} R_{nd} + (5.06 \times 10^{-3}) R_{nd} + 0.37 \ . \qquad (5.6.6)$$

It can be seen that the values of the first and third coefficients A and C are very similar for the two areas but that the values of the second coefficient, B, are quite different. This was attributed to the difference in altitude of the two sites, Albacete being 670 m above Valencia.

Given the various successes of using satellite-derived vegetation indices, the question of relating evapotranspiration to the NDVI naturally arises. It has been shown

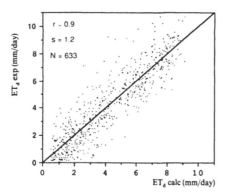

Figure 5.28 The correlation of experimental values of potential evapotranspiration (ET_d exp) versus calculated ones (ET_d calc) (Caselles *et al.* 1992).

empirically (Asrar *et al.* 1985) and theoretically (Sellers 1985, Tucker and Sellers 1986) that a nearly linear relation exists between the NDVI and the photosynthetically active radiation (PAR) absorption coefficient, which in turn can be related to the instantaneous dry biomass production and thereby to changes in the carbon cycle. As the biomass production is closely related to water losses, i.e. transpiration, it seems pertinent to analyse the relation between vegetation reflectance data and actual evapotranspiration (Running and Nemani 1988, Goward and Hope 1989). We consider some observed data from Senegal which show a strong relation between evapotranspiration and NDVI (Kerr *et al.* 1989).

Senegal is located between latitudes 12° N and 17° N in West Africa and covers three main climatic areas. From the south towards the north the climate is of Guinean, Sudanian and Sahelian type. Along the western coast the oceanic influence is present while the eastern part has continental characteristics. The mean annual rainfall ranges from over 1000 mm in the south to 150 mm or less near the Senegal river in the north. The rainy season usually starts in May (south of the area) and ends in October. The main crops are millet, cow pea and ground nut mixed with fallow. This area was selected by Kerr *et al.* (1989) for study because it offers a wide range of annual rainfall values across a relatively small area and because the ground network is relatively dense and reliable. AVHRR data and ground data spanning a period of four months from July to October 1986 were used in their study.

Because of cloud cover, and in order to reduce the viewing angle variations and to avoid problems with aerosols, only ten AVHRR scenes from this period were selected for processing. As the timescale that is mostly used in climatology is the five-day period, the data set was interpolated (maximum likelihood) so as to give the equivalent temporal sampling. This approach is valid because over a five-day period, the NDVI can be considered as constant. The computed NDVI were then extracted over a 3 pixel × 3 pixel window centred around a ground data station and the mean value was used. For the details involved in the determination of the evapotranspiration, ET, from the ground data see Kerr *et al.* (1989). What was computed was the integrated value over a five-day period and it was determined for several different types of vegetation (fallow, millet, ground nut and cow pea). Figure 5.29 shows plots of cumulative NDVI and cumulative ET for millet for three stations. It should be noted that the ET curve has been shifted by 20 days and fits the

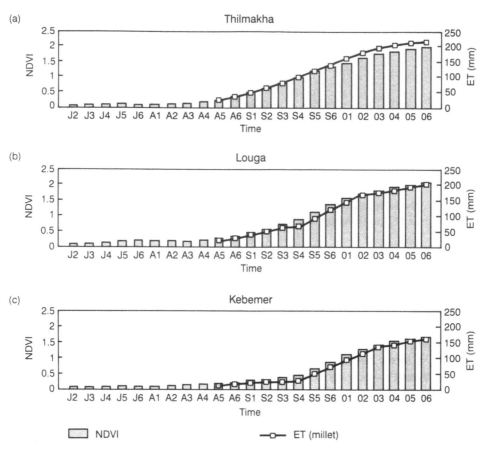

Figure 5.29   The evolution of the cumulative NDVI and the cumulative ET (shifted by 20 days) during the growing season for three locations (Kerr *et al.* 1989).

NDVI curve almost exactly. In other words, there is a time lapse of about 20 days between the occurrence of a physiological factor (i.e. ET) and its effect on the NDVI. The very good fit between the two data sets shows that the relation between the satellite-measured radiances and the water losses is almost linear.

We have already mentioned the relation between the NDVI and biomass production and therefore the carbon cycle. This has been observed experimentally using field measurements for various agricultural canopies and for the tidal wetland grass *Spartina alterniflora* (marsh cord-grass) (for some references see Bartlett *et al.* 1990). Reflectance measurements made with hand-held radiometers were used to compute the NDVI. Net CO_2 exchange was measured in a clear, climate-controlled chamber placed over square metre plots of the canopy. The NDVI was found to be nearly linearly related to the proportion of solar photosynthetically active radiation intercepted by green plant material. For the tidal wetland grass the relation was not as strong as was reported for growing agricultural canopies, but the regression model form persisted throughout the growing season (March–October) and was applicable to a wide variety of growth forms exhibiting different proportions of green foliage. The NDVI was also found to be nearly linearly related to net CO_2 exchange rates throughout the growing season and in plots subjected to varying levels of chronic stress. However, significant uncertainty in the relation was produced by changes in photosynthetic efficiency in response to varying environmental conditions. Estimates of CO_2 exchange based on NDVI were improved if the effect of ambient air temperature on net photosynthesis was incorporated. It was concluded (Barlett *et al.* 1990) that quantitative remote assessment of photosynthesis and net CO_2 exchange in natural vegetation is feasible, particularly if the analysis incorporates information on biological responses to environmental variables. Sellers (1985, 1987) combined models of radiative transfer and biological processes in plant canopies to provide a synthesis of the relations between vegetation indices, intercepted photosynthetically active radiation, photosynthesis and transpiration. In these models the dependence of the photosynthesis rate on intercepted photosynthetically active radiation produces a strong, nearly linear relation between vegetation indices and rates of canopy photosynthesis.

Tucker *et al.* (1986) tested the concept that remote measurements of vegetation can be related to global photosynthesis by comparing monthly, latitude-zonal averages of a satellite-derived NDVI with measured atmospheric CO_2 concentrations over a period of 34 months. The observed changes in CO_2 concentration were inversely correlated with changes in the NDVI, supporting the hypothesised link between vegetation indices and photosynthetic drawdown of atmospheric CO_2. Fung *et al.* (1983) extended this analysis by calculating CO_2 exchange from major, global biomes and incorporating these results into a global atmospheric transport model. The model output showed encouraging agreement with observations of seasonal changes in CO_2 concentration at several monitoring sites.

The work of Tucker *et al.* (1986) was based on the use of raw GVI data. Singh (1989, 1992b) applied a first-order atmospheric correction to the GVI data. Then the northern hemispheric CO_2 concentration cycle was compared with the GVI dynamics over the United States of America and the United Kingdom. The results showed that the GVI is a measure of the degree of photosynthetic activity in the terrestrial vegetation with an appropriate phase lag as was shown by Tucker *et al.* When southern hemispheric CO_2 concentration data were compared with the GVI dynamics over Australia and South America (south of the equator) then it was difficult to obtain the same conclusion. This is attributed to (i) the small amplitudes of CO_2 cycles arising due to the smaller amount of photosynthetically active vegetation and (ii) the occurrence of the *El Niño* phenomenon on the western coast of Peru which introduces distortions in the amplitudes of the CO_2 cycle.

The results of Sellers (1985, 1987), Tucker *et al.* (1986), Fung *et al.* (1983) and Hansen (1989) indicate the potential for remote monitoring of biophysical process rates in vegetation. However, as some of these authors point out, the approach is based on greatly simplified models of complex biological processes and the accuracy and repeatability of the analysis may depend strongly on the composition of the canopy and on environmental variables affecting the activity of plants. Evaluating the importance of these factors requires experimental studies at the level of individual ecosystems.

5.6.2 Drought and semi-arid areas

AVHRR data are now widely used in a variety of areas for drought monitoring and for making predictions of crop failure and therefore, in some areas, of famine.

AVHRR data for Oklahoma for the drought of 1980 were studied by Walsh (1987). This was based on attempting to estimate the Crop Moisture Index (CMI), the Drought Severity Index (DSI) and the Hydrologic Deficit (HD) by a combination of remote sensing vegetation indices. However the various vegetation indices which can be used are not all linearly independent. This work demonstrated that, while the AVHRR data clearly were indicative of the general effect of the drought on the crops, a correlation between the various drought indices and the AVHRR data was poor. The consequence, in terms of using satellite data, is that now one tends to work directly in terms of vegetation indices derived from the AVHRR data, rather than to proceed via the more conventional meteorological drought indices. This latter approach is illustrated by the work of the Foreign Crop Condition Assessment Division (FCCAD) of the US Department of Agriculture during the 1988 drought in the USA Corn Belt (Teng 1990). In this work, in which four states (Iowa, Illinois, Indiana and Ohio) were studied particularly, the FCCAD used the NDVI values as the primary data. From the NDVI values they calculated cell vegetation index numbers (VINs) which are the average NDVI values over all cloud-free pixels within cells with a nominal edge of 46.3 km on a geographical referenced grid. Using the VINs the FCCAD was able to detect the existence of drought early in the season, monitor changing conditions, and provide objective assessments of the drought's extent and severity. Field observations confirmed the image analyses, and underlined the importance of the timing of extreme weather events with respect to crop stages for interpreting VINs. The analyses were conducted in an operational environment, providing a unique test of the AVHRR data for large area, near-real-time crop monitoring. Because large area, operational remote sensing of crops is quite different from traditional, controlled, small plot research studies, more work is needed to link the two; this would improve crop assessment capabilities. Another point which was noted was the fact that problems may arise in using vegetation index data derived from different sensors (e.g. AVHRR on NOAA-9 and NOAA-10). Some work has been done for the FCCAD on empirically converting AVHRR VINs derived from one NOAA satellite to VINs from another; however, the conversions remain to be tested. A longer-term and more fundamental solution is the inter-calibration of the various sensors, see section 2.2.6. A specialised GIS known as a Drought Impact Information System (DIIS) involving the use of AVHRR data was described by Killmayer (1993).

In most semi-arid areas, vegetation growth is triggered by the first rains, and the completion of the growing cycle is dependent on water availability throughout the cycle. Water availability is, in fact, the main limiting factor for vegetation growth. As irrigation is not common practice in most areas, rainfall is a key parameter. It has been shown by various authors (see Hielkema et al. 1986 for instance) that primary production is directly linked with the amount of water available in tropical areas where annual rainfall ranges between 100 and 800 mm. Furthermore, it has been shown (Dancette 1983) that in such areas, rainfall is strongly related to yield. Consequently, for practical and economical reasons, it seems necessary to be able to monitor rainfall on a global scale. Looked at from another viewpoint, if rainfall, NDVI and biomass production or yield are all so closely related to one another in semi-arid areas, it is clear that studying the NDVI by remote sensing can be useful both as an indicator of (past) rainfall and of (future) expected yields.

The potential of NDVI data, obtained from the AVHRR, for monitoring grass-land production in the Sahel was recognised following work in Senegal in the early

1980s. Subsequently, a number of programmes have been carried out, and are continuing, that relate satellite-derived information to ground-collected data in Sahelian countries. References will be found in the various papers in one or two special issues of the *International Journal of Remote Sensing* (volume 7, number 11, 1986; volume 12, number 6, 1991) and in the paper by Diallo *et al.* (1991). One major reason for the interest is in relation to the possibility of using satellite data for famine early warning in the area, see for example Hutchinson (1991). The need for timely and accurate information on the status of agricultural crops and range-lands in Africa has received high priority since the onset of the current series of droughts in 1968. International relief responses to the famines that frequently accompanied drought were often late or poorly coordinated because famine could not be anticipated. Moreover, when famine occurred, information describing its extent and magnitude was often lacking, unreliable, or contradictory. This was due in part to the variable quality and long lag time in 'conventional' reporting such as meteorological observations or crop bulletins, that are compiled from incomplete or poorly managed ground observation networks. Since 1984, national and international agencies have sought to improve their ability to forecast famine in sub-Saharan Africa. A number of early warning systems have been implemented for this purpose that monitor physical and social variables that may indicate the likelihood and magnitude of famine. The advent of global meteorological satellite systems made available another source of comprehensive observations that, when combined with ground samples, could provide consistent descriptions of meteorological conditions for the entire continent of Africa. In the context of Africa considerable use has been made of AVHRR data to produce early warning of drought for the Famine Early Warning System (FEWS) of the US Agency for International Development (USAID), the Food and Agriculture Organisation (FAO) of the United Nations and other bodies (Johnson *et al.* 1987, Le Comte 1989). However, while these combined data sets could be used to describe more accurately weather patterns for large areas (of the order of 100 km^2), they are, as we have already mentioned elsewhere, inadequate to assess directly the general status of crops for relatively small areas. The lack of homogeneity arising from the existence of a chequerboard arrangement of vegetated and non-vegetated areas in the Sahel causes problems in the use of AVHRR data that are similar to the problems that arise in, for example, European agricultural areas where the field size is usually considerably smaller than the IFOV. Modelling of this situation was undertaken by Hanan *et al.* (1991).

Several famine early warning systems use satellite remote sensing data to supplement ground-based observations. These systems have demonstrated the advantages in timeliness and consistency of remote sensing data. Although user needs have not been clearly defined, experience gained in the operation of early warning systems and the results of related research suggest that (a) at the continental scale AVHRR GAC data offer many advantages over traditional ground data sources, (b) quantitative crop yield estimates might be improved through consideration of both photosynthetic activity of the vegetation and length of growing season, (c) qualitative comparisons of crop years have provided useful inputs to current early warning needs and (d) stratification of the region into coherent geographical areas would improve all estimates.

A careful analysis of field measurements of primary production in semi-arid grasslands in three Sahelian countries over a period of eight years was carried out by Prince (1991a) in conjunction with multitemporal sums of vegetation indices derived from AVHRR data. The results demonstrated a strong linear relation between the satellite observations of vegetation indices and the seasonal primary production in the range 0–3000 kg ha^{-1}. The confidence intervals of estimation of production were in the range ± 61–161 kg ha^{-1}. The question of different physiological types of plants and heterogeneous vegetation types was taken into account by Prince (1991b). AVHRR (NDVI) data were used, in conjunction with in situ data, by Diallo et al. (1991) to produce savanna primary productivity maps for Senegal north of the Gambia in 1987 and 1988. The use of NDVI for monitoring changing range-land conditions in Namibia using NDVI data has been described by Williams et al. (1993) and the assessment of herbage production in the arid range-lands of Central Australia has been investigated by Hobbs (1995). Desertification in Spain was studied with AVHRR by Roozekrans (1993). D'Souza and Hiederer (1989) used Landsat MSS data as an intermediate data source in relating AVHRR-derived vegetation indices to biomass production. Loudjani et al. (1993) described the estimation of Net Primary Productivity (NPP) from a semi-empirical model of Monteith which relates the amount of Absorbed Photosynthetically Active Radiation (APAR) to NPP. They used a model involving a linear relation between APAR and NDVI (GVI) and using incident global radiation derived from Meteosat data.

As work on the use of NDVI data for semi-arid areas has developed, so work has been done on the determination of atmospheric conditions, especially water vapour concentration and aerosol properties and concentration (Holben et al. 1991, Justice et al. 1991a, b, Soufflet et al. 1991, Faizoun and Dedieu 1993, Popp 1993, Tomasi et al. 1993). We have mentioned already (in section 5.4) the question of applying atmospheric corrections to satellite-derived radiances to produce ground-level radiances before using channel-1 and channel-2 AVHRR data to calculate the NDVI. We consider this briefly in the specific context of the Sahel. In the case of water vapour, channel 2 of the AVHRR contains a water vapour absorption band that affects the determination of the NDVI. Daily and seasonal variations in atmospheric water vapour within the Sahel affect the use of the NDVI for the estimation of primary production. This water vapour effect was quantified for the Sahel by Justice et al. (1991b) by radiative transfer modelling and empirically using observations made in Mali in 1986. In extreme cases, changes in water vapour were found to result in a reduction of the NDVI by 0.1. Variations of the NDVI of 0.01 would result from typical low atmospheric water vapour days within the wet season. If these conditions were to persist throughout the season it would lead to an overestimate of production of 200 kg ha^{-1}. This indicates very clearly the importance of attempting to determine atmospheric water vapour concentrations and to make the appropriate corrections to the NDVI and a procedure for making these atmospheric corrections was proposed by Justice et al. (1991b). As far as aerosols are concerned, Holben et al. (1991) described the establishment of a network of Sun photometers in the Sahel region of Senegal, Mali and Niger in order to monitor the aerosol characteristics needed for atmospheric correction of remotely sensed data. The aerosol optical thickness τ_a computed from the spectral Sun photometer measurements exhibited very high day-to-day variability ranging from approximately 0.1 to greater than 2.0 at 875 nm for both the wet and dry seasons. A gradient of decreasing τ_a from north-to-south latitudes in the Sahel for the wet season, July–September, was observed, which may be caused jointly by increased washout owing to the gradient of increasing precipitation to the south and the location of source regions for dust in the north. The Ångström wavelength exponent B (see equation (3.3.20)) was found to vary with the magnitude of the aerosol optical thickness, with values as high as 0.75 for very low τ_a, and values of 0.25–0.0 for high τ_a conditions. Analysis of τ_a data from this observation network suggests that there is a high spatial variability of τ_a in the western Sahel region. Statistical analysis performed on the wet season data showed that at a 67% confidence level the instantaneous values of τ_a can be extrapolated approximately 270–400 km with an error tolerance of 50%. Spatial variability in the dry season was of a similar magnitude. The ranges of variations in the NDVI in the Sahel region owing to commonly observed fluctuations in the aerosol optical thickness and aerosol size distribution were shown to be approximately 0.02 and 0.01, respectively.

The need, at least when working at a regional level rather than at a continental or subcontinental level, to take into account not only atmospheric effects but also the variations in other factors such as vegetation and soil types and terrain topography in crop yield assessment and forecasting has been considered by Maselli et al. (1992) using data from four rainy seasons in Niger. The problem of eliminating the effect of differences in geographical resources (climate, soil, vegetation type and topography) when the impact of weather on vegetation is considered is discussed by Kogan (1990). This involves a concept and a technique for eliminating that portion of the NDVI which is related to the contribution of geographical resources to the amount of vegetation. AVHRR data of the Global Vegetation Index format were used for the 1984–1987 seasons in Sudan. The procedure suggests normalisation of NDVI values relative to the absolute maximum and the absolute minimum of NDVI. These two criteria were shown to be an appropriate characteristic of geographical resources of an area. This modified NDVI was named the Vegetation Condition Index (VCI). Comparison between the VCI, NDVI and precipitation dynamics showed that the VCI estimates better portray precipitation dynamics than does the NDVI. According to Kogan (1990) the VCI permits not only the description of vegetation but also the estimation of spatial and temporal vegetation changes and weather impacts on vegetation.

The correlation between rainfall and NDVI, rather than between NDVI and biomass productivity, has also been considered by some workers. Justice et al. (1991b) made comparisons between rainfall estimates, based on cold cloud duration estimated from Meteosat data, and vegetation development depicted by NDVI data for part of the Sahel. Decadal data from the 1985 and 1986 growing seasons were examined to determine the synergism of the datasets for range-land monitoring. There was found to be a general correspondence between the two data sets with a marked lag between rainfall and NDVI of between 10 and 20 days. This time lag is particularly noticeable at the beginning of the rainy season and in more northern areas where rainfall is the limiting factor for growth. Areas of low NDVI values for a given input of rainfall were identified; at a regional scale they give an indication of areas of low production potential and possible degradation of ecosystems. Kerr et al. (1989) made comparisons between GAC data and ground-based rainfall measurements from 59 stations scattered over the country of Senegal for a four-month period from June to September in 1986, see Figure 5.30. The equation for the regression line is

$$CR = 3391.2 \, CGVI + 87.1 \tag{5.6.7}$$

where CR is the cumulative rainfall and CGVI is the cumulative GVI divided by the number of measurements.

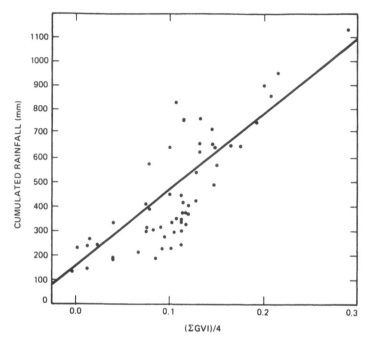

Figure 5.30 Scatter plots of cumulative rainfall versus GVI for 59 locations in Senegal during the period from June to September 1986 (Kerr *et al.* 1989).

The scatter of the points in Figure 5.30 is generally not very wide with some exceptions. Some values are underestimated, most probably due to the important rainfall which occurred in September (only one value of GVI is used per month and October data would have been very useful to account for these rains). Some values are overestimated, which can be explained by the green belts around towns, irrigation areas, or water bodies. Equation (5.6.7) was then applied to the cumulative GVI image to produce a satellite-derived rainfall map which, effectively, provides an interpolation scheme to cover the areas between the stations where the ground-based rainfall measurements were made. The map was much more detailed than any map produced by conventional means and gave useful information on parts of the country which are not very well covered by ground stations. The results also show that such an empirical method could be used with a satisfactory accuracy over quite a wide range of climatic conditions since the study area undergoes great variation in rainfall. The method, however, needs validation over several years. Further improvements of the results could be achieved if the whole rainy season was studied with finer time sampling, adapted to the cloud-over characteristics. Finally the method is not very efficient when vegetation is very sparse, in which case the use of passive microwave radiometry in conjunction with the NDVI might be necessary (see section 5.8).

5.6.3   Fire risk assessment

Fires in grasslands or forests present serious problems in many parts of the world. Observing and monitoring fires with data from remote sensing satellites, including

AVHRR data, is feasible and is practised (see section 6.2). However, since a vegetation index is an indicator of the health of a vegetation canopy one might expect that vegetation index values could be used to indicate areas of dry and highly combustible material, i.e. to indicate areas of high fire risks (Miller *et al.* 1983). Different methods have been proposed to convert this qualitative idea into a more quantitative approach. In this section we shall consider examples of the use of NDVIs, determined from AVHRR data, in fire-risk monitoring and assessment. Let us consider first the scale of the problem with reference to Spain (López *et al.* 1991).

Forest fires are one of the most important agents of environmental alteration in the Mediterranean area. Because of their high frequency and intensity, large areas of territory are affected by them. During the last decade there has been an improvement in the knowledge of forest fire phenomena, and a considerable increase in the resources available to fight against fire. However, the results in terms of protection have not been so favourable, since as time passes there are yet more forest fires that cause more damage. It seems evident that there is some failure in the fields of prevention, prediction and detection.

Spain is the country of the Mediterranean basin with the largest forest area burnt. Between 1980 and 1984 the total area burnt was, approximately, 1 000 000 ha, with 410 000 ha of this being tree-covered. Furthermore, in 1985 there were 13 000 individual fires affecting 500 000 ha of land. The economic damage suffered by Spain in the period 1980–1984 was estimated at about 25 000 million pesetas (US$210 M). It is much more difficult, but nevertheless important, to take into account the ecological damage produced by the destruction of forests. The protection of soil, the prevention of erosion, hydrological regulation, reservoir protection and the conservation of recreational areas could be seriously affected by this phenomenon. Moreover, we cannot forget the loss of human lives in the fight against fires and in the flooding caused by the removal of vegetation cover. One of the essential aspects in eliminating forest fires is the prediction of fire risk. Such a prediction demands an answer to the following questions at an acceptable error level. Where will fire break out? When will it occur? How will it develop?

AVHRR data were chosen, in preference to Landsat TM or SPOT data, for forest fire risk monitoring in Spain at national and regional level for the reasons that should, by now, be apparent to readers of this book. The high-spatial-resolution systems have serious limitations in terms of providing sufficiently frequent cloud-free images covering the whole country. From an operational point of view the NOAA AVHRR images have the following advantages:

- the whole country is covered by a single image
- the spatial resolution of the AVHRR HRPT data (1.1 km at nadir) integrates local variations and gives an average response that can be of high interest for large-scale projects
- the NOAA AVHRR images can be acquired daily and their cost is reasonable.

The application of NOAA AVHRR images to the study, monitoring and detection of forest fires is discussed elsewhere in this book; they can be studied once they are burning (see section 6.2) or in terms of assessing the area of damage afterwards as part of the general use of AVHRR data in studying land cover.

Before describing the fire-risk monitoring procedure by López *et al.* (1991) for Spain we turn to the consideration of the work which was done in the state of

Victoria, Australia, in relation to grassland fires. This involves a multitemporal analysis of AVHRR data through the spring and summer. It was found by experiment that for grassland there was an empirical relation between the fuel moisture content, FMC, and the vegetation index, see Figure 5.31 (Paltridge and Barber 1988). This takes the form

$$\text{FMC} = 250 \, \frac{V(t)}{V_{\text{N}}} \qquad\qquad (5.6.8)$$

provided FMC < 250, where FMC is expressed as a percentage of the biomass dry weight, V_{N} is the vegetation index in November (i.e. in the Australian spring) and $V(t)$ is the vegetation index at some time, t, after November. Thus the vegetation index is used to calculate the FMC and the FMC value is used to indicate the flammability of the grassland as shown in Table 5.5.

The FMC information is simply one parameter which must be considered by the Country Fire Authority (CFA) officers when assessing the fire potential on a particular day. The meteorology of the time is probably the most significant factor. The FMC information is supplied to the CFA in the form of six-colour ink-jet plots to distinguish the areas in the state which fall into the particular ranges identified above. The ranges were chosen somewhat subjectively, based on operational experience of the flammability of vegetation. The flammability of most grassland species

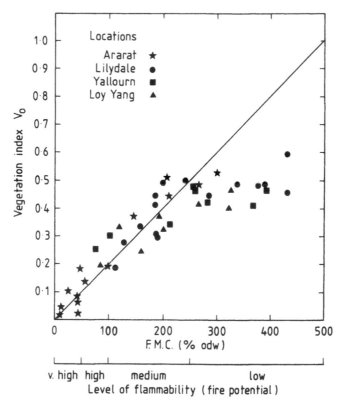

Figure 5.31 Full-cover satellite-observed vegetation index as a function of fuel moisture content at four locations in Victoria, Australia (Paltridge and Barber 1988).

Table 5.5 Fuel moisture content (FMC) and flammability (Paltridge and Barber 1988)

FMC range (%)	Flammability
0–25	extreme
25–50	very high
50–100	high
100–200	medium
200–	low

increases very sharply as their FMC drops into the 100–50% range. There is also evidence that for a given species there may be two significant values of FMC. The first corresponds to the point where the leaves become highly flammable. The second corresponds to the point where the stems and supporting structures become highly flammable. The hope was expressed that a more objective classification of the significant points of FMC would be developed in the light of experience. The system proved extremely valuable to the Country Fire Authority for continuous monitoring of potential fire-danger areas of the state. The extension of the technique, to take into account viewing angle corrections and atmospheric corrections, has been described by Paltridge and Mitchell (1990).

Let us now return to the case of fire-risk monitoring in Spain (López *et al.* 1991). The area studied is in the east of Spain, covering partially the provinces of Valencia, Castellón and Teruel. It is located between 39°03′ and 40°09′ N, and between 0°06′ and 0°55′ W, with an area of 8800 km², see Figure 5.32. The main reason for selecting that study zone was the occurrence of a forest fire in 1987 which began on 17 July and finished on 19 July, and was the largest forest fire that year in Spain, devastating 3600 ha (36 km²) of which 2810 ha were tree-covered. The fire began because of negligent use of fire when cleaning an agricultural field and a strong wind extended it to the rest of the area that was burnt. Six AVHRR scenes were used for the study, five from before the fire and one a month later: 7 May 1987, 13 June 1987, 23 June 1987, 2 July 1987, 12 July 1987 and 17 August 1987.

The general idea was to calculate relative decrements in the NDVI, on a pixel by pixel basis, and then to integrate those decrements over meshes of 10 km and 5 km grids and accumulate them from spring to summer. In this way a result was obtained that synthesised the NDVI variation with an information level equal to that used in standard calculations of forest fire risk index in Spain, over meshes of 10 km.

The quantity calculated was called the accumulated relative NDVI decrement (ARND) and it was calculated by subtracting, pixel by pixel, the NDVI between two dates and summing it over meshes of 10 and 5 km. We define

$$\text{RND}(N, \text{D1}, \text{D2}, I, J) = \sum_{L=J}^{J+N} \sum_{K=I}^{I+N} \frac{\text{NDVI}(\text{D2}, K, L) - \text{NDVI}(\text{D1}, K, L)}{\text{NDVI}(\text{D1}, K, L)} \quad (5.6.9)$$

where RND is the relative NDVI decrement, NDVI (D, K, L) is the NDVI calculated for day D for the pixel in row K and column L of the scene ($D = \text{D1}$ or D2), N is the mesh size, D1 is the image of a given date, D2 is the image of the following date, I is the row number (step N) of the first pixel of each reticular square, J is the

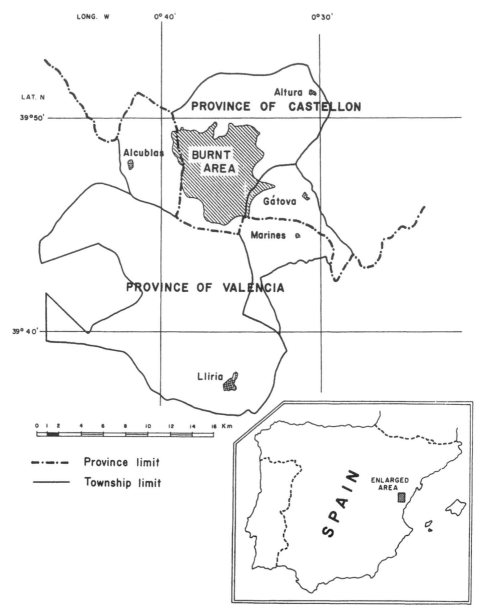

Figure 5.32 Location of study area for the work of López *et al.* (1991) on forest fires in Spain.

column number (step N) of the first pixel of each reticular square. The ARND is then defined as follows:

$$\mathrm{ARND}(N, d_1, d_t, I, J) = \sum_{h=d_1}^{d_t} \mathrm{RND}(N, \mathrm{id}_h, \mathrm{id}_{h+1}, I, J) \tag{5.6.10}$$

where id_h is the NDVI image of the date h, $d_1, d_2, \ldots, d_t$ are the dates of the available NDVI image, and I and J are the row number and the column number of each reticular square.

The ARND results were mapped on an electrostatic plotter, producing overlays at a scale of 1 : 200 000. A classification of the ARND into three groups was performed: high, medium and low decrements, in order to make an easier comparison with ground truth and vegetation maps. The results obtained from processing the NOAA data were compared with the field information from the study area and with auxiliary documents at a 1 : 200 000 scale such as vegetation and topographic maps, forest fire reports and meteorological data including precipitation, temperatures, speed and direction of winds and relative humidity. In considering the development of this technique to an operational level, some points need to be taken into account. It would be useful to have a weekly update of the NDVI images. This is feasible in Spain, where the number of cloud-free images is very high during summer. In an operational system these data should be included in a geographical information system with a digital terrain model for slopes and exposure models, and with climatic data updated by automatic stations in forest zones. Standard fire risk models may be integrated too. AVHRR data can be included in the calculation of fire risk index as an input to combustibility models.

5.7 DIRECTIONAL EFFECTS

5.7.1 Introduction

We need to distinguish between two different directional effects. One is associated with the consequence of the fact that the scan angle of the AVHRR is very large (about 55° on each side of nadir) and it is therefore important to consider various geometrical effects which it is not necessary to consider with high-spatial-resolution scanners such as Landsat or SPOT for which the scan angles are quite small. The second problem is associated with the reflectivity of the surface of the Earth. There are two idealised situations which may be considered in remote sensing of the Earth using reflected sunlight. One is that the reflection at the surface is specular, i.e. that the surface of the Earth behaves like a perfectly reflecting plane mirror. This situation is relatively unusual; generally it does not apply to the land but it does apply in the case of a calm water surface or of a water surface that is not too severely roughened (see section 7.8). A much more common assumption, and one that is frequently made about land surfaces, is that perfectly diffuse reflection occurs. In this case the surface is described as Lambertian and the reflected intensity is totally independent of the direction of the reflected radiation, i.e. of the direction of observation. In fact, however, the reflectivity is not simply a constant, or even just a function of wavelength, λ, but depends both on the direction of incidence of the solar radiation and also on the angle of reflection. In the case of Landsat or SPOT, where the total scan angle is small, the assumption that over a given scene the reflectance is independent of view angle is usually quite a reasonable assumption. However, when several scenes of a given area generated at different times of the year are concerned there will be significant differences in the direction of the solar illumination for the different scenes and these differences may be important. In the case of the AVHRR, where the scan angle is about 55° each side of nadir then variations in the refectivity of the ground with viewing angle within one scene are quite likely to be important; of course, variations in the solar illumination for different scenes will also be important.

5.7.2   View angle effects

The AVHRR scans over a large swath width, of about 2700 km, corresponding to about 55° on either side of nadir. This has a number of consequences which are essentially geometrical and which are not normally considered for high-spatial-resolution scanners like Landsat MSS or TM and SPOT since their angle of scan is quite small. In this section, therefore, we suppose that there is uniform ground cover over the whole scene that is being imaged and that the cover is a Lambertian surface, i.e. it reflects uniformly in all directions. Thus, for the moment, we can isolate the purely geometrical effects. It is important to appreciate that the nature of the optical system and the scanning mechanism of the AVHRR is such that

- the angular field of view is constant (solid angle $\delta\omega$)
- the scan mirror rotates at a constant angular velocity so that the scan angle, at the instrument, varies linearly with time
- sampling of the data, to generate pixel values, occurs at equal time intervals during the scan.

There are various consequences of this. Consider two situations where the view angles are θ_1 and θ_2 from nadir, see Figure 5.33, and the corresponding distances from the satellites to the ground are l_1 and l_2.

- A, the area of normal cross section from which radiation is received, increases as the scan angle θ increases. For θ_1 it is $l_1^2 \, \delta\omega$ and for θ_2 it is $l_2^2 \, \delta\omega$ where $\delta\omega$ is the solid angle of the beam.
- The corresponding area of the ground from which the radiation is reflected into the field of view of the AVHRR is further increased by the fact of increasing obliquity of view, i.e. the ground area imaged is $A \sec \theta$, which is equal to $l_1^2 \, \delta\omega \sec \theta_1$ at θ_1 and $l_2^2 \, \delta\omega \sec \theta_2$ at θ_2.
- The geometrical path length of the radiation is increased as θ is increased and consequently the optical path (= geometrical path × refractive index) is also

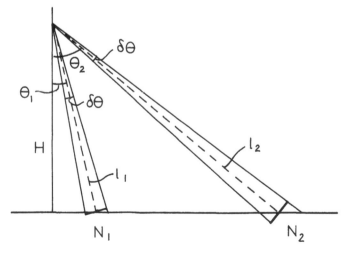

Figure 5.33   Instantaneous fields of view (IFOVs) for two different view angles, θ_1 and θ_2.

increased. In other words, the atmospheric effects suffered by the radiation will be increased as the scan angle increases.

There are further complications which we should perhaps stress. It will be recalled that in section 3.1.7 we mentioned the point spread function. Drawing a beam, as we have done in Figure 5.33, may lead one to suppose that the instrument responds uniformly to radiation originating from any point on the area A within the beam and that there is a sudden cut-off in response at the edge of the beam. In reality this is not the case as will be recalled from the discussion of the point spread function in section 3.1.7 (see Figures 3.9 and 3.11). Within the beam the instrument is most sensitive to radiation incident along the direction of the axis and less sensitive for angles approaching the edge of the nominal cone. Moreover, there is still some response for angles greater than the nominal angular width of the beam. In reconstructing an image from one channel of the data one then supposes that radiation received from a circular or elliptical area (see Figure 3.8) with fuzzy edges was actually received from a square area 1.1 km by 1.1 km; if the image is represented at 1 : 1 M scale for example, the corresponding square area 1.1 mm by 1.1 mm would be coloured a uniform shade of grey determined by the total amount of radiation received in that channel for that pixel. This description, however, ignores the fact that the scan mirror does not jump from one IFOV to an adjacent IFOV; it rotates continuously and the detector response is therefore a time integral over an interval during which the mirror is rotating at constant angular velocity. Thus the 'elliptical area with fuzzy edges' just mentioned is actually further smeared out as the point N moves a certain distance along the scan line.

As the scan angle increases, the IFOV on the ground increases in area (see Figure 3.8) leading to the need for linearisation when viewing the image (see Figure 2.4). The linear dimension of the IFOV in the scan direction at angle θ_2 will be given by $l_2 \, \delta\theta \sec \theta_2 = H \, \delta\theta \sec^2 \theta_2$. There is also some increase in the size of the IFOV on the ground in the direction of flight of the spacecraft; this is due to the effect of the ratio (l_2/l_1); but there is no increase due to a $\sec \theta$ factor in this direction. Figure 5.34 shows the variation in the pixel dimension along a scan line; it also shows the relative change in the air mass for the optical path of the radiation from the ground to the satellite as the view angle increases (i.e. l_1, l_2, etc., which go as $\sec \theta$ ($l_1 = H \sec \theta_1$, etc.) whereas the pixel dimension goes as $\sec^2 \theta$). No correction is introduced in the image to attempt to allow for a change in pixel size in the along-track direction because it would involve trying to disentangle overlapping scan lines. However, neglecting to make a correction to allow for this means that there is a greater contamination of the data in one scan line by radiation which has actually come from areas which correspond geographically to the adjacent scan lines. If the ground cover is changing from one scan line to the next, this may be an important source of error.

One way to overcome the problems of large-scan-angle effects is to restrict the use of data to that from the central part of the scan line and to reject all data from scan angles larger than some arbitrary cut-off angle. However, this is a draconian measure and it would seem to be better to attempt to make reasonable corrections for view-angle effects rather than simply to reject large amounts of data. This would seem to be especially important in those areas where frequent cloud cover already leads to a large loss of data.

Modelling of the effect of viewing angle on the NDVI has been undertaken, for

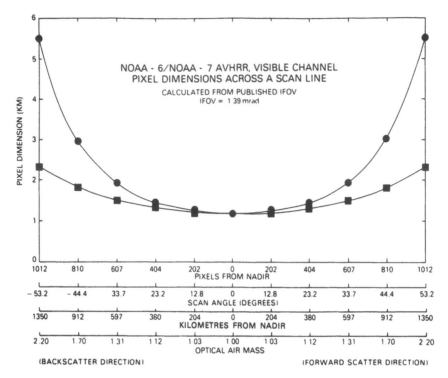

Figure 5.34 The across-track (along-scan) (●) and along-track (■) IFOV dimensions for the AVHRR (Holben and Fraser 1984).

example, by Holben and Fraser (1984) and Wardley (1984). The geometry is illustrated in Figure 5.35. Some of the results obtained by Holben and Fraser (1984) will be quoted to demonstrate that the effects are significant and thus to indicate that if one proposes to use data for large off-nadir scan angles one should be prepared to have to incorporate significant corrections. Let us look at it another way. The use of normalisation in generating the NDVI can be regarded as applying a first-order correction for view-angle and Sun-angle effects, making it thereby much more useful than most other vegetation indices. A demonstration of the first-order correction effect of the normalisation is given by Holben *et al.* (1986). However, following the huge initial success of the NDVI, it is becoming necessary to consider second-order effects and to make corrections for these effects. Holben and Fraser (1984) modelled various changes in channel-1 and channel-2 AVHRR radiances, and therefore of NDVI, under typical viewing and illumination conditions and including atmospheric effects using the atmospheric data of Dave (1978). In their model the following conditions were assumed:

- plane parallel surface at the point observed on the Earth
- no clouds
- no atmospheric polarisation
- Lambertian surface reflectance
- all orders of scattering
- a multilayered atmosphere.

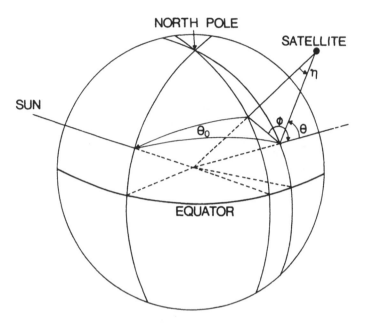

Figure 5.35 The solar zenith angle, θ_0, look zenith angle, θ, look azimuth angle, ϕ, and scan angle, η, for the AVHRR (Holben and Fraser 1984).

The radiances were calculated and stored in lookup tables as a function of θ_0, θ, ϕ, λ, model number and R_{sfc}, where θ_0 is the solar zenith angle, θ is the look zenith angle, ϕ is the look azimuth angle, λ is the wavelength of a discrete non-uniform spectral band, model number is one of five models each with unique atmospheric characteristics (Dave 1978), and R_{sfc} = Lambertian surface reflectance.

The radiances were integrated over Dave's spectral intervals to match the spectral bands of the AVHRR to simulate VIS and NIR radiances for all viewing and illumination geometries of the AVHRR. Holben and Fraser (1984) performed calculations for various different cover types and for various different illumination conditions (summer solstice, equinox and winter solstice), for latitude 30° (N). An example of their results is shown in Figure 5.36; further sets of results for various different types of land cover and different sets of atmospheric conditions will be found in their paper (Holben and Fraser 1984). Other calculations of the effect of viewing angle or of the solar zenith angle on the reflectance from vegetation and the satellite-derived vegetation indices have been undertaken by Bartlett *et al.* (1986), Singh (1988a, b, 1989), Gutman (1991) and Gupta (1992b) for example.

The correction procedure to compensate for off-nadir viewing developed by Gutman (1991) was demonstrated using an area of about 200 km by 200 km in eastern Kansas. The area is relatively homogeneous with tall-grass prairie as the predominant surface type. The data comprise 25 daily GAC samples (1 per each 40 km × 40 km map cell) during July 1986, which amounts to about 750 observations. It is worth mentioning that the cloud removal techniques needed to be adapted to the specific problem being studied by adjusting the values of the thresholds used in the various tests described in section 3.2.

It was assumed that this procedure eliminated most of the cloud and haze-contaminated pixels, but some verification was done using GOES images. It was

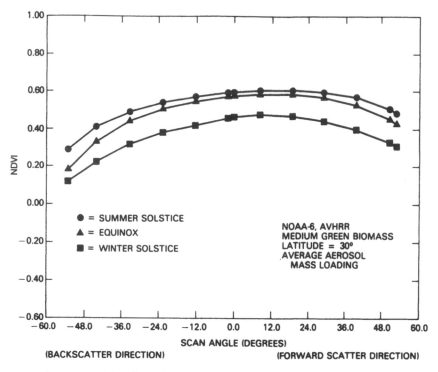

Figure 5.36   Illustration of the effect of viewing angle (scan angle) on NDVI values from model calculations (Holben and Fraser 1984).

assumed, for lack of any information to the contrary, that the reflectances in the area did not change significantly during the period (one month) during which the GAC data were acquired. Figure 5.37 shows the reflectances, R_1 and R_2, in channel 1 and channel 2, respectively, which remained after the application of the cloud removal procedure and they are plotted as a function of the viewing angle θ. The regular behaviour of the channel-1 data (squares) and channel-2 data (plus signs) as functions of θ enabled Gutman (1991) to fit the two sets of data, to a good approximation, by cubic polynomials for channel 1

$$f_1(\theta) = 7.4 - 0.08\theta + 0.001\theta^2 + 0.000001\theta^3 \tag{5.7.1}$$

and for channel 2

$$f_2(\theta) = 19.9 - 0.19\theta + 0.002\theta^2 + 0.000004\theta^3. \tag{5.7.2}$$

Corrections were then applied by using these polynomials to give

$$R_1' = \frac{f_1(0)}{f_1(\theta)} R_1 \tag{5.7.3}$$

and

$$R_2' = \frac{f_2(0)}{f_2(\theta)} R_2 \tag{5.7.4}$$

where R_1 and R_2 are the uncorrected reflectances. The corrected reflectances are also plotted in Figure 5.37 which demonstrates that R_1' and R_2' are independent of θ.

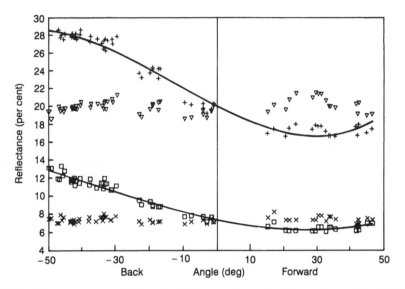

Figure 5.37 Reflectance over a 200 km by 200 km area in eastern Kansas during July 1986 after the application of cloud screening, observed channel 1 (□) and channel 2 (+) and angularly corrected channel 1 (x) and channel 2 (▽) (Gutman 1991).

The values of the NDVI calculated from the uncorrected radiances and from the corrected radiances are plotted, again as a function of viewing angle, in Figure 5.38. The NDVI calculated using the uncorrected data shows a similar decrease with increasing viewing angle as was calculated by Holben and Fraser and illustrated in Figure 5.36. Moreover, in both cases the decrease is greater in the backscatter direction than in the forward-scatter direction. The corrected NDVI in Figure 5.38 shows

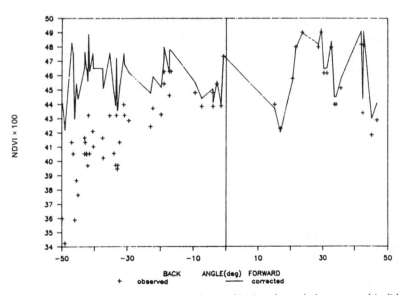

Figure 5.38 NDVI as a function of scan angle, observed (+) and angularly corrected (solid line) for the data in Figure 5.37 (Gutman 1991).

scatter, although the cover was assumed to be homogeneous over the entire area, but there is no obvious dependence on viewing angle. This method of correction is of general applicability. However, although we have presented this as a correction for viewing angle variations it should be noted that the variability in reflectances could partially be contributed by the change in solar zenith angle. The values of solar zenith angle vary from 25° to 45° in this data set. There is strong correlation between the viewing and Sun angles (see Gutman 1988) in AVHRR data, thus making it very difficult to distinguish between these two effects. Therefore, the correction which has been applied should be regarded as a correction for the combined effect of the viewing and Sun geometry for that particular range of Sun angles. It should therefore not be assumed that the values of the coefficients in equations (5.7.1)–(5.7.4) apply at all seasons of the year. One can expect the values of these coefficients to depend on the vegetation type and on the season.

Wardley (1984) carried out calculations of the effect of viewing geometry, i.e. off-nadir viewing, on the PVI (perpendicular vegetation index – see section 5.2). This index does not include a normalisation factor and therefore one would expect the effects of variation in viewing angle to be more important than for the NDVI. The calculations we have just mentioned were to demonstrate the effect of off-nadir viewing for a given direction of solar illumination. The complementary question, namely the effect of varying direction of solar illumination for a given viewing angle, actually nadir viewing, has been considered by Singh. The effect of varying solar zenith angle on the NDVI has been demonstrated by Singh (1988a) for various types of land cover. Calculations were made for four surface cover types, namely the high, moderate and low green-leaf vegetation densities and bare soil. The results showed that the NDVI for bare soil remains constant for solar zenith angles up to about 60°, then decreases for solar zenith angles above this. The NDVIs for high, moderate and low green-leaf vegetation densities remain constant up to about 30° solar zenith angle. However, for larger solar zenith angles these NDVIs were found to decrease smoothly and significantly. A method for the correction of GVI (Global Vegetation Index) data to allow for variations of solar zenith angle was devised by Singh (1988b, 1989). Goward et al. (1991), who devoted considerable attention to the question of whether the NDVI is calculated with raw data (digital numbers), radiances calculated using pre-flight calibration data or atmospherically corrected surface-leaving radiances, also calculated the NDVI as a function of solar zenith angle for various cover types. Similar results were obtained to those we have already quoted in Figures 5.36 and 5.37.

5.7.3   Bidirectional reflectance distribution function

As we have mentioned already in section 5.7.1, an assumption that is widely used is that of a Lambertian surface, i.e. that the intensity of diffusely reflected solar radiation is independent of the angle of scattering of the radiation at the surface. While the idea of a Lambertian surface is a convenient concept, and while it may have some approximate validity in certain situations, there are many circumstances in which the reflectance of a surface is very far from being Lambertian. One particularly important situation is that of a vegetated surface, for which not only does the reflectance depend on the angle of observation (relative to the surface normal) but it also depends on the angle of incidence of the solar radiation as well. In other words, the reflectivity is a function of two angles, the angle of incidence for the incoming sunlight and the angle of (diffuse) reflection, at which the surface is being observed. This leads to the idea of the bidirectional reflectance distribution function, BRDF (Kriebel 1976), $\rho^B(\theta_i, \phi_i; \theta_r, \phi_r)$ where θ_i and ϕ_i are the zenith and azimuth angles, respectively, of the incident light and θ_r and ϕ_r are the zenith and azimuth angles, respectively, of the reflected light. $\rho^B(\theta_i, \phi_i; \theta_r, \phi_r)$ is the most fundamental reflectance characteristic of a surface. If $L_i(\theta_i, \phi_i)$ and $L_r(\theta_r, \phi_r)$ are the incident and reflected radiances, respectively, as functions of angle, then $\rho^B(\theta_i, \phi_i; \theta_r, \phi_r)$ can be written as (Kriebel 1976)

$$\rho^B(\theta_i, \phi_i; \theta_r, \phi_r) = \frac{dL_r(\theta_r, \phi_r)}{L_i(\theta_i, \phi_i)\, d\Omega_i} .$$
(5.7.5)

This function is defined as the relation of that part of the total spectral radiance $dL_r(\theta_r, \phi_r)$ reflected into the direction θ_r, ϕ_r which originates from the direction of incidence θ_i, ϕ_i, to the total spectral irradiance $L_i(\theta_i, \phi_i)\, d\Omega_i$ impinging on a surface from the direction θ_i, ϕ_i. This particular bidirectional reflectance function is a unique characterisation of a surface and is not dependent on the irradiance distribution as are a number of other bidirectional functions presented in the literature. It may, of course, be a function of wavelength, λ.

To attempt to measure $\rho^B(\theta_i, \phi_i; \theta_r, \phi_r)$ directly is ambitious. For a rock, for a sand or soil sample one can imagine taking a sample into a laboratory and determining the value of $\rho^B(\theta_i, \phi_i; \theta_r, \phi_r)$ as a function of each of the five variables $\theta_i, \phi_i, \theta_r, \phi_r$ and the wavelength by measurement. For a smooth water surface one could calculate the reflectivity from first principles using the Fresnel reflection coefficients. For a vegetated area the measurement of the BRDF is not simple. It cannot be calculated from first principles, as can be done for water. Transportation of the material into a laboratory may not be feasible and results obtained with laboratory plants may not be applicable to field specimens, often of different species. Conducting the measurements actually in the field presents logistical problems, which may be more or less serious depending on the type of vegetation. Some work of this type has been done, e.g. by Kimes *et al.* (1985), Shibayama *et al.* (1986), Pinter *et al.* (1987), Huete *et al.* (1992), Cabot *et al.* (1993) and Flasse *et al.* (1993).

The two reflectance measurements which are most commonly reported in the literature for natural vegetation canopies are the bihemispherical reflectance, ρ^H, and the hemispherical-conical reflectance factor, ρ^C (Kimes *et al.* 1980). The bihemispherical reflectance, ρ^H, is defined as the ratio of the reflected exitance to the irradiance at the target surface. The hemispherical-conical reflectance factor, ρ^C, for a nadir-looking sensor having a field of view of less than 2π steradians, is measured as the ratio of the reflected flux of a surface in the direction of the sensor's field of view to the reflected flux of a perfectly reflecting horizontal Lambertian surface in the direction of the sensor's field of view. Mathematical definitions of these reflectances are given by

$$\rho^H = \frac{\int_0^{2\pi} \int_0^{\pi/2} L_r(\theta_r, \phi_r)\, d\Omega_r}{\int_0^{2\pi} \int_0^{\pi/2} L_i(\theta_i, \phi_i)\, d\Omega_i} = \frac{M}{E}$$
(5.7.6)

where $d\Omega_i = \cos\theta_i \sin\theta_i\, d\theta_i\, d\phi_i$, $d\Omega_r = \cos\theta_r \sin\theta_r\, d\theta_r\, d\phi_r$, M is the reflected exitance, E is the irradiance and

$$\rho^c = \frac{\int_{\phi_1}^{\phi_2} \int_{\theta_1}^{\theta_2} L_r(\theta_r, \phi_r) \, d\Omega_r}{\frac{E}{\pi} \int_{\phi_1}^{\phi_2} \int_{\theta_1}^{\theta_2} d\Omega_r} \tag{5.7.7}$$

where $\phi_1, \phi_2, \theta_1, \theta_2$ are the azimuth and zenith angle limits of the sensor's view.

From equation (5.7.5) it follows that

$$L_r(\theta_r, \phi_r) = \int_0^{2\pi} \int_0^{\pi/2} \rho^B(\theta_i, \phi_i; \theta_r, \phi_r) L_i(\theta_i, \phi_i) \, d\Omega_i. \tag{5.7.8}$$

As a consequence, the two commonly measured functions, ρ^C and ρ^H, are dependent on the bidirectional reflectance function of the surface and on the solar irradiance distribution.

The bidirectional reflectance distribution function, BRDF, $\rho^B(\theta_i, \phi_i; \theta_r, \phi_r)$ for any given surface is a function of a number of geometrical and optical characteristics of the constituents of the surface. For a vegetated surface one can expect the form of the BRDF to be particularly complicated, see for instance Kimes *et al.* (1980). There may also be diurnal variations in the BRDF, and therefore also in ρ^H and ρ^C, for a vegetated surface arising from variations in the orientations of leaves or the opening and closing of flowers as the day progresses. Some results of measurements of ρ^H and ρ^C for specific vegetation canopies are given, for example, by Kimes *et al.* (1980) and there are various other results to be found in the literature.

One may be forced to resort to modelling the canopy to obtain an expression for the BRDF. For example, Pinty *et al.* (1989) describe a model which yields an expression for the BRDF, as a function of the geometry of illumination and observation, and functions describing (i) the optical properties of the leaves (specifically, their single-scattering albedo and phase function), (ii) the statistical distribution of leaf orientations, (iii) the architecture of the canopy (which itself depends on the average distance between the leaves in the vertical and horizontal directions) and (iv) the contribution from multiple scattering occurring inside the canopy. The model is complicated and we shall not reproduce the details here (see Pinty *et al.* 1989). This work of Pinty *et al.* shows that (i) the NDVI does not depend only or simply on the intrinsic leaf properties and (ii) directional effects primarily due to the multiple scattering contribution in the near-infrared region always affect the vegetation indices. This, therefore, confirms theoretically what was indicated by the work of Holben and Fraser (1984), Singh (1988a,b, 1989) and Gutman (1991) with AVHRR data, which we have described already. Thus, it is necessary to correct for the angular variations in the NDVI, particularly when dealing with global scale observations where systematic trends in the relative geometry of illumination and observation do exist.

One should not fall into the trap of supposing that for a vegetation canopy the BRDF is something which is fixed and that all one has to do is to determine it once and for all. The value of the BRDF can be expected to depend on the general canopy geometry, the leaf orientation, the leaf area, the leaf reflectance and transmittance properties, the characteristics of other components of the vegetation canopy (stalks, trunks, branches, etc.) and soil reflectance, any of which may change as a function of time (Suits 1972a,b, Colwell 1974, Tucker 1980, Goel and Reynolds 1989, Brakke and Otterman 1990, Roujean *et al.* 1992).

If one is concerned with climate modelling it is total energy reflected by the surface which is the quantity of interest, not just the energy reflected in one particular direction (θ_2, ϕ_2). Thus it is the hemispherical reflectance (or albedo) which is of interest. In theory this could be determined by integrating the BRDF over θ_2 and ϕ_2. However, since in general the BRDF is unknown, this is not a practical proposition. Ways have therefore been devised to estimate the hemispherical reflectance (or albedo) from the nadir reflectance; it is not surprising that the results are not particularly accurate (Kimes and Sellers 1985, Kimes *et al.* 1987, Kimes and Holben 1992).

5.8 NORMALISED DIFFERENCE VEGETATION INDEX (NDVI) AND MICROWAVE POLARISATION DIFFERENCE TEMPERATURE (MPDT)

There has been an important development in recent years and it seemed to be worthwhile to devote a section to the topic. We have seen in this chapter that data from the visible and near-infrared channels of the AVHRR can be used to generate vegetation indices, especially the NDVI, and we have mentioned that, to some extent, data from the thermal infrared channels can be useful as well in some aspects of vegetation studies. However, in addition to emitting radiation in the (thermal) infrared part of the spectrum any object at ambient temperature emits some (quite small) amount of radiation in the microwave part of the electromagnetic spectrum. Such microwave radiation emitted from the surface of the Earth can be detected by microwave radiometers, such as the Scanning Multichannel Microwave Radiometer (SMMR) of which more than one example has been flown in space. One SMMR was flown on Seasat, but Seasat only operated for just over three months in 1979. Another SMMR which was flown on Nimbus-7 gathered data successfully for several years, from 1979 well into the middle of the 1980s. The spatial resolution of a passive microwave scanning radiometer is theoretically much poorer than that of a scanning thermal infrared radiometer. Diffraction effects, which are relatively insignificant for optical and infrared scanners, are important for microwave scanners. Thus to allow for this and to allow for the low intensity of the emission and give rise to a detectable signal, the IFOV is very much larger for a microwave scanner than for an optical or infrared scanner. For example, the IFOV of the SMMR is of dimension about 25 km. For a microwave radiometer, unlike an infrared radiometer, it is possible to detect and measure separately the horizontally and vertically polarised microwave signals. It has been suggested that the difference between the horizontally and vertically polarised signals could be used as an indicator of the state of the vegetation of the surface of the Earth (Njoku and Patel 1986, Choudhury *et al.* 1987, Choudhury and Tucker 1987a,b Choudhury 1989a, 1990). In particular, it was proposed to use the normalised difference of the brightness temperatures in horizontal and vertical polarisation measured at 37 GHz by SMMR on board Nimbus-7. This has been described as the microwave polarisation difference index, MPDI, or the microwave polarisation difference temperature, MPDT; we shall use the latter term here. Since the MPDT and NDVI are both connected to vegetation characteristics, there should be some correlation between them. This correlation has been analysed theoretically by Becker and Choudhury (1988) and by Choudhury (1989a).

A complete special issue of the *International Journal of Remote Sensing* (volume 10, number 10, 1989) was devoted to papers on the subject of the correlation between NDVI data from the AVHRR and MPDT data from Nimbus-7. Apart from the paper by Choudhury (1989a) that presented the theory we have outlined above, analyses of the two data sets for various geographical areas were presented. These included the Amazon Basin (Choudhury 1989a), some drought-affected areas of sub-Saharan Africa and northeastern Brazil (Tucker 1989), South America and Africa (Justice *et al.* 1989, Townshend *et al.* 1989a). Comparison of the data showed the MPDT response to differ considerably from the seasonal pattern exhibited by the NDVI. The MPDT showed a general decrease with increasing vegetation density, whereas the NDVI showed a positive correlation with the amount of green leaf material. In contrast to the NDVI, the MPDT showed a generally poor response to seasonal variations in green leaf material. Only those vegetation types exhibiting a substantial seasonal variation in the areal extent of vegetated cover showed strong seasonality in the MPDT data, e.g. annual grasslands. Several cover types with large proportions of trees and shrubs displayed seasonal variation in the NDVI but no equivalent change in MPDT.

Results of maximum-likelihood classifications applied to multitemporal data sets indicate that, overall, the NDVI data sets are substantially better than the MPDT data sets for land-cover characterisation. This is, perhaps, not surprising in view of the finer spatial resolution of the AVHRR. However, the greater sensitivity of the MPDT data in semi-arid areas results in their superior performance for some classes in these areas. The combined use of MPDT and NDVI data sets show clear synergistic benefits in using the two data sets. However, the evidence suggests that for most cover types, increasing the temporal frequency of the NDVI images is more advantageous than incorporating MPDT data sets. General conclusions were drawn by Townshend *et al.* (1989b). It was concluded that there was an important complementarity between these two quantities in environmental monitoring. Overall the NDVI was found to be more sensitive to green leaf activity, whereas the MPDT appeared also to be related to other elements of the above-ground biomass. Monitoring of hydrological phenomena can be carried out much more effectively by the MPDT.

To obtain more quantitative information from MPDT data it is necessary to do some modelling (Choudhury 1989a). It has been shown that ΔT, the difference between the brightness temperatures for the horizontally and vertically polarised radiation, can be expressed as

$$\Delta T = \tau_a(A, \theta)(T_S - T_{sky})\{R(\theta, H) - R(\theta, V)\} \qquad (5.8.1)$$

where $\tau_a(A, \theta)$ is the the atmospheric transmissivity, A is the height of the instrument above the ground, θ is the zenith angle of the direction to the instrument, T_S is the surface temperature of the land, T_{sky} is the sky temperature and $R(\theta, H)$ and $R(\theta, V)$ are coefficients for the horizontal and vertical polarisation, respectively, which contain land surface information. To make progress in terms of extracting useful information about the land surface from the measured values of ΔT it is necessary to develop a model that relates $\{R(\theta, H) - R(\theta, V)\}$, and therefore ΔT, to the land surface parameter that is of interest. In particular, expressions were developed (Choudhury 1989a) for net primary productivity and evapotranspiration.

It should be pointed out that a subsequent paper by Choudhury (1990) has led to some very considerable controversy (Tucker 1992) which is not yet totally resolved. Basically the argument concerned the quality of the correlation between some NDVI and MPDT values which has been claimed by Tucker (1992) to be less good than had been suggested by Choudhury (1990). It would not be very profitable to consider all the details of this dispute here; a paper by Choudhury and DiGirolamo (1994) contains a reply to the points made by Tucker (1992). This is obviously a subject area for further research.

5.9 THE STUDY OF SUB-IFOV SIZE OBJECTS

In satellite remote sensing we usually tend to think that the smallest object which we can reasonably expect to detect is an object which is comparable in size with the IFOV of the instrument being used. We also tend to assume that there is good contrast between an object and its surroundings. Thus for the AVHRR we would, initially, not expect to be able to detect objects with linear dimensions less than about 1 km. This would seem to rule out the study of any animals, birds or insects with AVHRR data. However, this need not necessarily always be the case (Cracknell 1991) and in this section we shall have cause to mention cattle, sheep, locusts, mosquitoes and tsetse flies. A single specimen of any one of the creatures just mentioned would not be detectable in an AVHRR signal. (We shall, however, see in section 6.2 that if the mid-infrared contrast between a sub-IFOV-sized object and its background is very great then the object may be directly detectable in channel-3 AVHRR data.) However, concentrations of small creatures sometimes do build up to be so large that these collections become directly detectable in AVHRR data. One example is provided by the case of blooms of blue-green algae and an example in which the evolution of such blooms was very successfully monitored with AVHRR data will be described in section 7.3.

An example of a collection of small creatures becoming sufficiently large to become directly detectable in AVHRR data is that of a swarm of locusts. The swarm is so large and contains so many locusts that it can be seen as a dark cloud in a satellite image. However, the direct observation in a satellite image of a swarm of locusts is not in practice very useful. By the time the swarming has occurred, let alone its existence been detected, it is too late to take any effective action to prevent crop devastation by the swarm. Likewise the detection, with AVHRR data, of the massive devastation caused to an agricultural area by a swarm of locusts is presumably possible because the effect in the image is large enough to detect. However, that too is of no use in trying to prevent the damage. What would be useful would be to use AVHRR to study the locusts' habitat and to detect the conditions that lead to swarming before it occurs. Then it is possible to take effective action, i.e. spraying, to control the locusts and prevent their swarming. A second example is provided by mosquitoes. These are smaller than locusts and never form clouds that would be large enough to detect with AVHRR data anyway. However, in this case too it has been shown to be possible to use AVHRR data to study the mosquitoes' habitat and to predict the behaviour of the insects from these studies. Thirdly, movements of larger animals such as cattle and sheep can sometimes be followed on large areas of open range-land by correlating their behaviour with the state of the vegetation as indicated by the NDVI. This has been done for sheep in the uplands of North Wales in the UK by Thomson and Milner (1989). This area is well covered with a variety of semi-natural grassland and heathland vegetation. Sheep are the only large herbivores that are present in large numbers in this area and so sheep grazing is the dominant biotic factor that controls these ecosystems. Different vegetation communities support different population desities of sheep and the soil, rainfall and altitude affect sheep distribution. These factors also affect the satellite-received radiance and therefore also the vegetation index as well. This work on North Wales was actually done with Landsat TM data, rather than AVHRR data, because of the rather small area involved. However, the general idea can be applied to large range-land areas for which AVHRR data would be more appropriate. For example, the monitoring of the phenology of Tunisian grazing lands using NDVI data from the AVHRR has been carried out by Kennedy (1989). We shall consider the examples of locusts and tsetse flies in a little more detail.

A considerable amount of work has been done on locusts using Landsat and AVHRR data. We only quote papers by Tucker *et al.* (1985a), Bryceson (1989) and Snijders (1989), but many other references are cited in these papers. As we have already indicated, what one has to do is to try to predict potential swarming of the locusts and take appropriate preventive action before damage to crops occurs. When the locusts are present in low numbers in their natural desert habitat or recession area they present no problem. When rainfall occurs the locusts multiply and live for a while on the ephemeral vegetation that appears after the rainfall. It is when the locust population has grown vastly and the local vegetation is exhausted that the locusts swarm and migrate, often over very large distances, in search of further supplies of vegetation, see Figures 5.39 and 5.40. The strategy of plague prevention is thus based on locating areas where rain has fallen, monitoring the locust population build up and then controlling the population by spraying where necessary. Given that the recession areas are large and generally not well supplied with meteorological stations on the ground, satellite remote sensing provides an important source of data for monitoring rainfall in these areas. It is not feasible to use Landsat data since approximately 700 Landsat scenes would be needed to cover

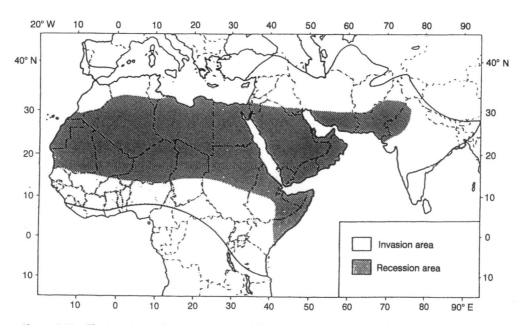

Figure 5.39 The invasion and recession areas of desert locusts in Africa (Tucker *et al.* 1985a).

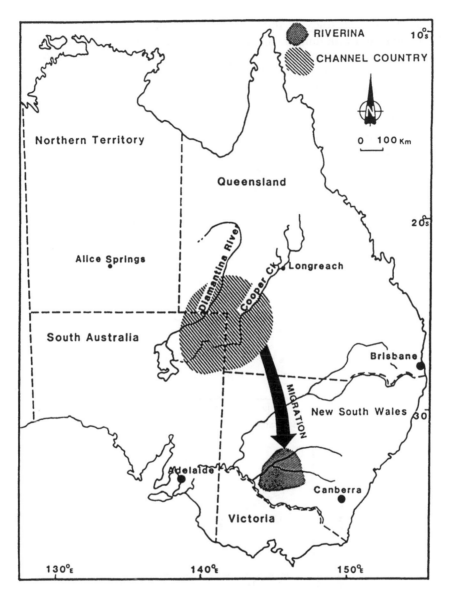

Figure 5.40   The recession area and migration pattern for locusts in eastern Australia (Bryceson 1989).

the recession area in Africa just once. Therefore the frequent, but relatively low spatial resolution, AVHRR data are far more appropriate. There are techniques based on the study of clouds in AVHRR images to locate precipitation (see section 7.6). One can also detect precipitation, after it has occurred, from its observable effects on the soil moisture and on the green-vegetation biomass. The study of soil moisture from satellite, though possible in principle using microwave or thermal infrared data, is not possible on an operational basis. Therefore, the use of visible-band and near-infrared-band data and of the vegetation index derived from these data is very important in this context and the first two references which were cited above are devoted to studying this; although they deal with quite different geographical areas and with different species of locusts, the general ideas have much in common. Tucker *et al.* (1985a) were concerned with the desert locust (*Schistocerca gregaria* Forsk.), while Bryceson (1989) was concerned with the Australian plague locust (*Chortoicetes terminifera* Walker) which affects areas of Queensland, South Australia and New South Wales.

The use of remotely sensed data in studying mosquito habitats has been described by Linthicum *et al.* (1987) and Wood *et al.* (1991) and tsetse fly habitats by Rogers and Randolph (1991, 1993) and Rogers and Williams (1993, 1994). The tsetse fly is a massive blight on life in Africa. Tsetse flies are a major constraint on animal production in about 10 M km^2 of Africa through their transmission of animal trypanosomiasis, while up to 25 million people are at risk from human trypanosomiasis, or sleeping sickness. The scale of the problem is vast. Africa contains about

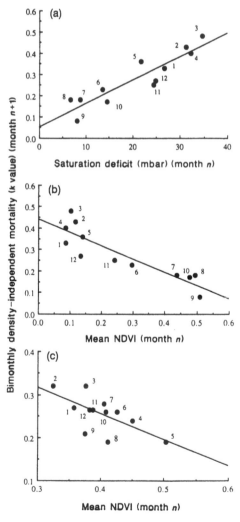

Figure 5.41   The relations between tsetse fly mortality rates and (a) ground based data, (b) satellite data for Nigeria and (c) satellite data for Uganda (Rogers and Randolph 1991).

20% of the world's pasture land, yet raises only 10% of the world's cattle and produces only 3% of the world's meat and milk. Indeed it has been argued that it was the fact that the animal trypanosomiasis, carried by the tsetse fly, precludes the use of draught animals for ploughing, which prevented the occurrence in tropical Africa of the agricultural revolution that occurred in the Middle East about 4000 BC. Ever since the link between the tsetse fly and the disease was discovered, countries affected have tried to control both. To control the tsetse fly over large areas requires a good knowledge of the flies' distribution so that elimination techniques can be most effectively targeted. A very interesting account of the biological background to the problem is given in the paper by Rogers (1991), including a detailed account of the role of climate in determining tsetse distribution. As Rogers puts it 'water relationships ... remain the Achilles heel of many insects and, directly or indirectly, are thought to limit insect distributions in dry environments'. A similarity was observed between whole-Africa NDVI maps and tsetse distribution maps and this led to the detailed study of the correlation between NDVI and tsetse fly distribution. Rogers and Randolph (1991) observed a correlation between tsetse fly mortality rates and the monthly mean NDVI (of the previous month) for two areas

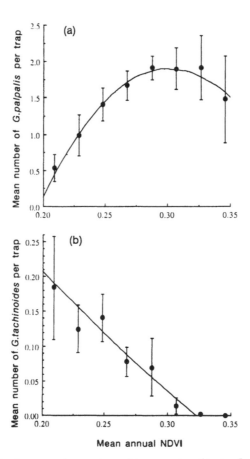

Figure 5.42 The relation between the numbers of two species of tsetse fly and the mean annual NDVI (Rogers and Randolph 1991).

(one in Nigeria and one in Uganda) (see Figure 5.41). They also found, using a large data set for the northern half of the Côte d'Ivoire for 1979–1980, a correlation between tsetse fly abundance and mean annual NDVI (see Figure 5.42). This work suggested that NDVI data can be used to predict tsetse fly abundance. The long-term aim is to produce maps of areas of high risk of trypanosomiasis transmission for the very large areas of tropical Africa where diseases carried by vectors such as the tsetse fly affect human and animal welfare. The use of the NDVI to provide information on two key parameters of vector-borne diseases, namely vector mortality rate and vector abundance, is a significant step towards this goal – but, as always, remote sensing does not provide the complete answer to a problem. Rogers and Williams (1994) described the use of a temporal Fourier analysis of the NDVI with examples from Africa. The analysis captures the important characteristics (i.e. the average, amplitude and phase) of the major annual and biannual cycles of vegetation growth. A strong association was demonstrated between the amplitude of the first term in the Fourier expansion (i.e. the amplitude of the annual cycle of vegetation growth) and savannah woodlands of Africa; a similar close association was found between the same features of the analysis and the areas infested by the tsetse *Glossina morsitans.*

CHAPTER SIX

Channel 3, the neglected channel

6.1 INTRODUCTION

It will be recalled that the AVHRR was designed primarily for meteorological purposes. We have mentioned in section 3.2 that channel-3 data play an integral part in multispectral classification schemes for clouds (see also chapter 4 of the book on clouds and satellites by Scorer 1986; see also Scorer 1989). However, one of the main reasons for writing the present book is to describe the enormous successes that have been achieved with AVHRR data in fields far removed from the field of meteorology for which it was originally designed.

Apart from a few notable exceptions, channel-3 data have not shared in this success. There are two main reasons for this. The first is the quality of the data. AVHRR data in channels 1, 2, 4 and 5 are of very good quality; they have low noise and they are free of the striping that is familiar, for example, to users of some of the older Landsat data. This is not the case, however, for channel-3 data. Channel-3 images often have a very pronounced herringbone pattern, see Figure 6.1. This noise is worse in data from some instruments in the series than from others; its removal was considered briefly at the end of section 2.2.6.

The second reason for the relative unpopularity of channel-3 data is the fact that in day-time the radiation received in channel 3 is a mixture of reflected solar radiation and emitted radiation. This makes the interpretation of the data more complicated than is the case for data from the other channels. In channels 1 and 2, except in one or two highly exceptional circumstances (e.g. gas flares or erupting volcanoes), one is dealing entirely with reflected solar radiation. This is in the visible and near-infrared region of the electromagnetic spectrum and so the interpretation of the satellite-received signals can be carried out in terms of the reflectivity of the surface at these wavelengths. In channels 4 and 5 one is dealing with almost exclusively (again except in very exceptional circumstances, in this case with sunglint) infrared radiation emitted by the surface of the Earth. The fact that channel-3 data contain something approaching a 50 : 50 mixture of reflected solar radiation and emitted radiation during the day-time, means that they are (not surprisingly) more difficult to interpret than the data from the other channels. At night there is of course only the emitted radiation and consequently night-time channel-3 data over sea areas are used in multichannel sea surface temperature algorithms (see section 4.2.3).

Figure 6.1   Channel-3 AVHRR image showing severe herringbone noise pattern (Dundee University).

We consider that there are two important areas of application of channel-3 AVHRR data; one of them is meteorological and one of them is non-meteorological. In accordance with the principles enunciated in the Preface, we shall not consider the details of the meteorological applications because we proposed to consider only non-meteorological applications of AVHRR data. We simply note that the behaviour of radiation at channel-3 wavelengths in relation to different types of clouds differs from that of radiation at channel-1, channel-2, channel-4 or channel-5 wavelengths. This means that channel-3 data play a very important role in the multispectral classification of clouds (see section 3.2 and the references cited therein, particularly the work of Liljas and Scorer). The application that we propose to discuss is concerned with small but intense sources of heat; this has proved to be very important and was quite unforeseen when the AVHRR was originally designed. Channel-3 AVHRR data have also been used in conjunction with data from other channels to study the forest clearing, in which burning plays a part, in various remote areas (Malingreau and Tucker 1990, Malingreau *et al.* 1990, Tucker *et al.* 1984b). In these cases channel-3 data are only some of the data used in a classification scheme.

6.2 SMALL INTENSE SOURCES OF HEAT

6.2.1 The principles of the detection of small high-temperature sources

A fire that generates a large smoke plume is very often clearly visible in channel-1 AVHRR data in which the smoke plume shows up very clearly, see Figure 6.2(a). In channel 3 the smoke plume will, most likely, be invisible but the site of the fire itself will probably show up very clearly, see Figure 6.2(b).

Croft (1977, 1978) presented views of worldwide agricultural burning, which is identifiable by its transient nature, using night-time imagery from visible-band imagery from the US Air Force's Defense Meteorological Satellite Program (DMSP). The DMSP sensors, which have sufficient radiometric sensitivity to pick out city lights, revealed, amongst other features, a huge band of agricultural/pastoral fires sweeping across central Africa. It was noted by Croft (1978) that the regions of poorest agricultural productivity were also the regions of greatest agricultural burning. However visible band imagery, whether from the DMSP or AVHRR instruments, is unable during the day-time to provide evidence of small fires, either from the smoke or from the flames of the fire itself.

It is possible, however, for fires that are considerably smaller than the IFOV (~ 1.1 km by 1.1 km) to be quite apparent in channel-3 data. Examples include gas flares, blast furnaces and agricultural straw fires. The ability of the channel-3 data to detect subpixel size high-temperature sources, or 'hot spots', has been well documented in studies of gas flares, forest and range fires, and steel production plants (Dozier 1981, Matson and Dozier 1981, Muirhead and Cracknell 1984a,b, 1985, Matson *et al.* 1984, 1987). Hot spots in channel-3 imagery can usually be distinguished from pixel error or highly reflective cloud (Muirhead and Cracknell 1984b) by their characteristic appearance.

The principle of what is involved in fire detection with AVHRR data is explained by Matson *et al.* (1987), see Figure 6.3. At typical room, land or sea temperatures, say ~ 300 K, the Planck distribution function peaks in the vicinity of channel-4 and channel-5 wavelengths, i.e. 10–12 μm. For a flame temperature, say 600–700°C or 900–1000 K, the peak moves to the vicinity of channel-3 wavelengths, i.e. 3.5–3.9 μm, see Figure 6.4. Kennedy *et al.* (1994) give some numbers to illustrate this. For a pixel corresponding to an IFOV of 1.1 km by 1.1 km, containing a single surface element of uniform temperature (300 K) the averaged radiant energy in channels 3 and 4 would be 0.442 W m^{-2} μm^{-1} sr^{-1} and 9.68 W m^{-2} μm^{-1} sr^{-1}, respectively, assuming an emissivity of 1 and no attenuation of the signal by the atmosphere. For a pixel containing two equal area elements with different temperatures, a fire at 800 K and the non-fire background of 300 K, the averaged radiant energy for channels 3 and 4 would be 670 W m^{-2} μm^{-1} sr^{-1} and 49.9 W m^{-2} μm^{-1} sr^{-1}, respectively. The radiance received by channel 3 increases by a factor of about 1500, compared to a factor 5 for channel 4. It is this differential response which is the basis for AVHRR fire detection. Figure 6.4 shows a typical plot of channel-3 and channel-4 brightness

(a)

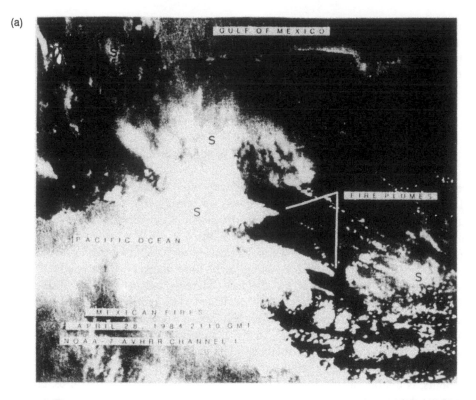

(b)

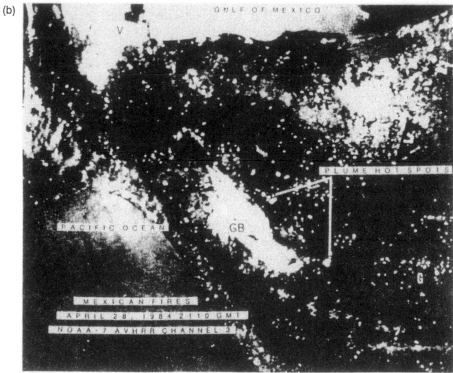

Figure 6.2 (a) Channel-1 and (b) channel-3 images of southern Mexico and northern Guatemala on 28 April 1984 (Matson *et al.* 1987).

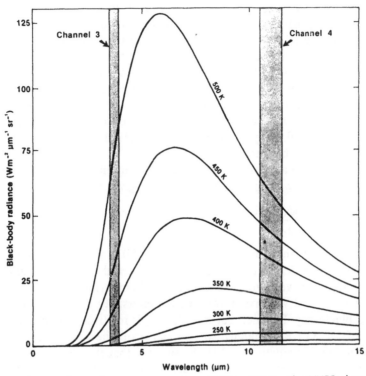

Figure 6.3 Blackbody radiation for temperatures from 200 K to 500 K with AVHRR channel-3 and channel-4 wavelengths indicated (Matson *et al.* 1987).

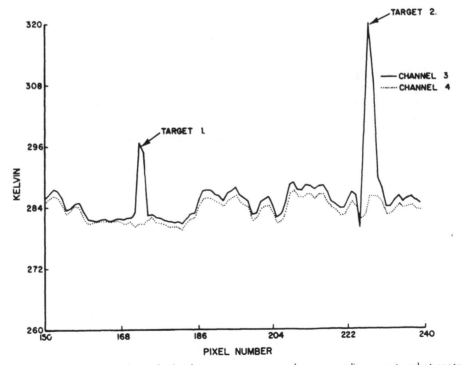

Figure 6.4 Channel-3 and channel-4 brightness temperatures along a scan line over two hot spots in Idaho (Matson *et al.* 1987).

temperatures over two high-temperature sources located in Idaho. Typical temperature differences between the two channels over land surfaces in general are usually about 1–2 degK. Target 1 in Figure 6.4 is a small controlled forest fire and target 2 is a phosphorus plant. At these sources, the channel-3 brightness temperatures are 16.2 degK and 33.9 degK higher than the corresponding channel-4 brightness temperatures. These targets are smaller than the IFOV and so these two higher temperatures correspond to pixel averages and do not give the hot target's temperature directly. One can, however, use an approach developed by Dozier (1981) and applied to fires by Matson and Dozier (1981) to estimate both the area and temperature of the hot target.

In the absence of an atmospheric contribution or attenuation, the upwelling radiance sensed by a downward-pointing radiometer is given by

$$L(T) = \frac{\int_0^\infty \varepsilon_\lambda B(\lambda, T)\phi(\lambda)\,\mathrm{d}\lambda}{\int_0^\infty \phi(\lambda)\,\mathrm{d}\lambda} \tag{6.2.1}$$

where $B(\lambda,T)$ is the Planck distribution function, $\phi(\lambda)$ is the spectral response of the detector (see section 2.2.3), the target area is assumed to be at a uniform temperature, T, and we have rewritten equation (2.2.18) in terms of wavelength λ, rather than frequency, ν, and we have introduced the emissivity ε_λ to allow for the target not being a blackbody. For most earth surfaces ε_λ is relatively independent of λ over the range of an AVHRR channel, so one can drop the λ subscript and move ε outside the integral in equation (6.2.1). We suppose that the channel-3 and channel-4 radiances are $L_3(T)$ and $L_4(T)$, respectively, as functions of blackbody temperature (i.e. $\varepsilon = 1$) and that the inverse functions of energy, E, are $L_3^{-1}(E)$ and $L_4^{-1}(E)$, respectively, for channels 3 and 4. For temperatures from 100 to 1000 K the expression in equation (6.2.1) was calculated numerically by Dozier (1981) using $\phi(\lambda)$ for the AVHRR on NOAA-6. The resulting values for $L_3(T)$ and $L_4(T)$ are shown in Figure 6.5. The shift in the peak of the Planck function towards shorter wavelengths, as temperature increases, causes the radiant contribution to channel 3 to increase more rapidly, at higher temperatures, than that to channel 4.

Suppose that we now have a mixed pixel corresponding to a hot target at temperature T_t which occupies a fraction p of the IFOV (where $0 \leq p \leq 1$) and a background temperature T_b which occupies the remaining fraction $(1 - p)$ of the IFOV. Strictly speaking, we should (a) specify the location of the hot target within the

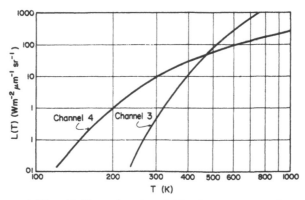

Figure 6.5   Radiances $L_3(T)$ and $L_4(T)$ as a function of T for the AVHRR on NOAA-6 (Dozier 1981).

IFOV and (b) use the point spread function for the AVHRR, but we ignore that just now. The brightness temperatures T_j ($j = 3, 4$) sensed by the AVHRR in channels 3 and 4 will be, in the absence of an atmospheric contribution or attenuation (and at night so that there is no reflected solar radiation in channel 3),

$$T_j = L_j^{-1}[pL_j(T_t) + (1 - p)L_j(T_b)]$$
(6.2.2)

where $j = 3, 4$. The background temperature can be estimated reasonably accurately from nearby pixels. This means that in the two equations (6.2.2) there are just two unknowns, p, the fraction of the IFOV occupied by the hot target, and T_t, the target temperature.

Using the theory described above with the data shown in Figure 6.4, Matson *et al.* (1987) were able to estimate the areas and temperatures of the two high-temperature targets causing the two peaks. For the two sources in Figure 6.4 they found the area and temperature were 0.28 ha and 430 K for target 1 and 1.7 ha and 483 K for target 2. As evidenced by the calculated target sizes and the detection of the phosphorus plant, target 2, it does not need a 1.1 km square target to cause a response in channel 3. (A 1.1 km square has an area of $((1.1 \times 10^3)^2/10^4)$ ha = 121 ha.) Thus we see an example which shows that small subresolution-scale high-temperature sources, such as fires, can be detected by channel 3. Very hot and small targets such as waste gas flares from offshore oil platforms have been detected as well as small fires from straw burning (see below). The question of the minimum size of the subresolution-scale high-temperature sources which can be detected will be discussed below (see section 6.2.4).

6.2.2 Difficulties in fire detection

It should be noted that the theory leading up to equation (6.2.2) neglects atmospheric effects (some discussion of this problem will be found in the paper by Dozier 1981) and, of course, it also assumes that there is no reflected solar radiation. It therefore applies only to night-time data, since during the day there is a significant reflected solar radiation contribution to the radiance detected in channel 3 of the AVHRR. The intensity of the reflected solar radiation is largest from cloud tops. Active cumulus cells and other highly reflective surfaces can reflect sufficient energy in the mid-infrared to create pseudo-hotspots. This problem, of course, vanishes at night.

Apart from cloud-top reflections giving signals that may be interpreted as hot spots, there is a second (and perhaps more obvious) consequence of clouds. Thick continuous layers of cloud obscure the surface of the Earth so that fires beneath the cloud will not be detected. Thus, it is important to remove cloudy areas from data that are to be analysed for fire-study purposes. Another possibility is that if a smoke plume rises almost vertically, as often happens with slash and burn in the humid tropics, the middle-infrared signal that is normally identifiable in channel-3 data as due to a fire can be obscured. 'Pyrocumulus' clouds from mass fires may obscure not only the generating fire event but also events downwind. On the other hand, if a plume is displaced by a strong wind a satellite may have a clear look at the burning area despite heavy smoke and receive a strong signal in channel 3 of the AVHRR. Plumes that shear at ground height and have dry optically thin smoke may be almost totally transparent in the infrared. A more extended discussion of these problems associated with the observations of fires in the infrared AVHRR data is given by Robinson (1991).

Another problem which arises is that the radiation from a fire is very intense; the contrast between fires and normal Earth temperatures is thus so sharp that fires easily saturate sensors instrumented for normal Earth surface observation (see also section 6.2.4). Conversely, when a sensor with a dynamic range appropriate for fire temperatures is directed towards the Earth, the resulting data for anything other than fires will be too granular to be useful for any but the crudest of applications.

Yet another problem is that the AVHRR does not provide continuous data gathering; it only provides one set of instantaneous data twice a day (or perhaps a few more times a day if one uses oblique views of an area). Thus when active fires, as opposed to post-fire char or scars, are being observed, measurements pertain to the instant of observation, not to the fire's average condition or ultimate magnitude. Thus, measurements made of a fire will vary considerably, both over its lifecycle and with the variations of the wind and the fuel bed. Moreover, fires typically manifest a strong diurnal rhythm, varying over time and space. Thus, fire measurements from a sensor platform with a fixed hour of observation may have strong biases that are difficult to eliminate.

Finally, low spatial resolution, though efficient, means a loss of contextual information. This loss may complicate the resolution of problems such as (i) signal attenuation by clouds, smoke or haze, (ii) artefacts from cloud or surface reflectance, (iii) effects of target and background emissivities on computations and (iv) biases imparted by the timing of the observation.

6.2.3   Fire studies

A useful summary of fire studies with AVHRR data is given by Robinson (1991). We shall follow Robinson's summary quite closely. The theory developed by Dozier (1981) which we have outlined above was for mixed pixels but included no analysis of AVHRR data for fires. Dozier's theory was applied to AVHRR data for fires by Matson and Dozier (1981) and later extended by Wan (1985) to arbitrarily dimensioned pixel fields overlaid by an N-layered absorbing and multiply scattering atmosphere.

Matson *et al.* (1984) examined fire appearances in AVHRR imagery, and studied the correspondence between fires sighted with 8-bit HRPT data and fire sightings by fire control agencies in the western United States. The experiment was affected by problems in the AVHRR tracking and receiving equipment. However, most of the fires sighted by the satellite were verified. Agricultural, slash and range improvement fires were dominant; a few wildfires, a peat-bog fire, gas flaring and mine smelts were also recorded. Matson *et al.* (1987) show channel-3 images of examples of fires in various parts of the world, Mexico, Rondonia (Brazil) and Mozambique.

An example of the observation of small intense sources of heat with data from channel 3 of the AVHRR is related to the identification and location of gas flares in the North Sea (Muirhead and Cracknell 1984a,b). During 1980, for example, gas was flared from oilfields in the British sector of the North Sea at an average rate of over 11 million m^3 a day. Using data from channel 3 of the AVHRR, one can identify gas flares which show up as pixels with very high temperature values. These pixels occur sometimes as isolated pixels, sometimes as a cluster of 2 or 3 pixels but never more than about 5 or 6 pixels. It is not that a flare is several kilometres in size but rather that its effect is seen in several pixels because its temperature is so high and because of the point spread function. There is a large number of platforms in the North Sea and some of them are clustered quite close together. The objective of Muirhead and Cracknell (1984a) was to see whether it was possible to carry out a geometrical rectification to sufficient accuracy to enable one to identify unambiguously the particular platform responsible for each flare observed in the satellite data. This work involved careful geometrical rectification of each scene, to an accuracy of better than 1 km; this was done for 11 scenes and it was possible to show that each flare could be assigned unambiguously to a platform that could be held responsible for it. An example of one of these images is shown in Figure 6.6. A section of the oilfield map is reproduced in Figure 6.7 with some of the satellite-derived flares superimposed on it. The number of flares identified, and the locations of these flares, varied from one scene to another, i.e. not all platforms burn gas in flares at any given time. In nearly every case it was possible to identify unam-

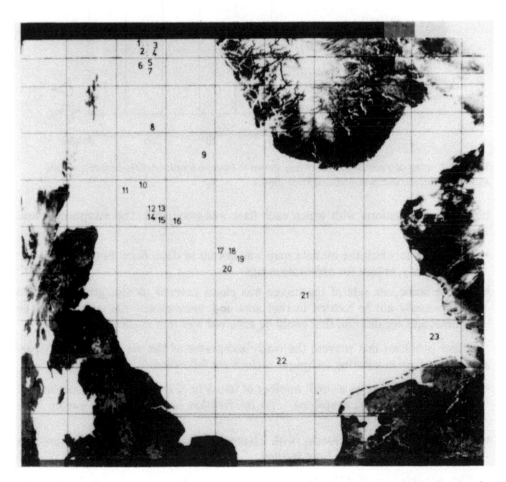

Figure 6.6 Channel-3 AVHRR image from NOAA-6 showing gas flares from oil rigs (Muirhead and Cracknell 1984a).

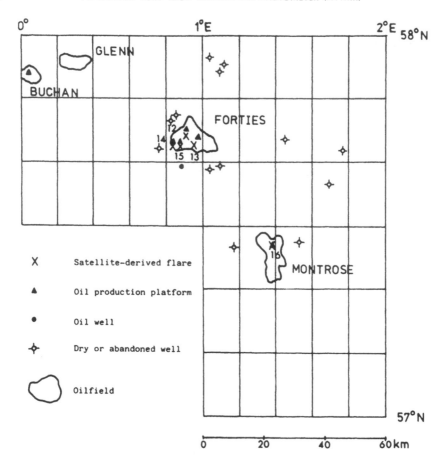

Figure 6.7 Map of a part of the North Sea shown in Figure 6.6 with satellite-derived gas flare locations marked (Muirhead and Cracknell 1984a).

biguously the platform with which each flare was associated. The exceptions arose from

- the extent to which the oil field map was not up to date, since things can change very quickly, at least for exploration rigs
- in one scene, one side of the image was cloud covered so that ground control points could not be located in that area and consequently the accuracy of the geometrical rectification that could be achieved was very much reduced.

Other problems that prevent the ready acceptance of the use of this as an operational gas-flare monitoring technique include the following:

- the satellite only gives a small number of (say 5 or 6) images per day corresponding to instantaneous snapshots – no information is provided in between these passes
- no attempt has been made, with channel-3 AVHRR data, to estimate the volumes of gas that have been burned.

The possible use of AVHRR channel-3 data in relation to straw burning by farmers has been studied by Muirhead and Cracknell (1985, 1986), Djavadi and

Cracknell (1986) and Saull (1986). As a result of changing agricultural practices, Britain finds itself with a large straw surplus with no established market. To dispose of it with the minimum financial outlay farmers have two realistic choices, namely either chopping the straw and ploughing it back into the soil (soil incorporation, a practice that is widespread in the rest of Europe) or burning it. In the 1980s many UK farmers resorted to the latter option in spite of the fact that soil incorporation is perfectly viable and costs little more than burning or that, with a modest investment of capital, large quantities of straw could be utilised as a fuel for domestic or greenhouse heating. As well as being wasteful, straw burning can occasionally prove hazardous. Fire damage to trees, hedges and even buildings often results from careless field burning, while road accidents have been known to occur as a consequence of the smoke produced. Muirhead and Cracknell (1985) used channel-3 AVHRR data to investigate the extent of straw and stubble burning across the country in the summer of 1984. The hot spots identified were not only attributable to straw/stubble burning; there are a number of high-temperature features that correspond to installations such as steel works and gas flares from oil refineries; these were excluded from the analysis. Figure 6.8, for Monday 20 August 1984, depicts about 350 burning fields and the distribution, as would be expected, corresponds roughly to the distribution of cereal production in Great Britain. Straw burning has subsequently been made illegal in Britain. However, as a continuous monitoring system the use of channel-3 AVHRR data would face several problems, some of which are common to the monitoring of gas flares and which we noted above. These include

- cloud cover
- the fact that images are only obtained at a small number of (predictable) times each day and so some fires may be missed
- inaccurate location of fires.

With regard to the last point, all the scenes were rectified using coastal ground control points and a least-squares approximation to quadratic polynomials giving r.m.s errors of the order of 1 km on the ground. This is not, in general, accurate enough to be able to identify unambiguously the actual farm involved in the burning. Thus the information gathered is only statistical and cannot generally be used to identify individual law-breaking farmers.

In both the cases of gas flares and straw burning discussed above, it was possible to demonstrate the feasibility of using AVHRR (channel-3) data to obtain relevant information. However, both projects failed in the sense that there was no willingness by the appropriate Government Department to give support to develop these applications to an operational state. This was for two reasons:

- basic unwillingness of civil servants to consider new methods and new technologies
- the genuine problem of incomplete (temporal) coverage, stemming from both orbital constraints and the frequent presence of cloud.

In other parts of the world it is forest fires rather than straw burning which is the problem, see also section 5.6.3. Forest fires, of course, arise from two distinct causes. Many of them occur naturally in the wild, and the forest in the areas in question regenerates naturally, as it has done for millenia. On the other hand, other fires

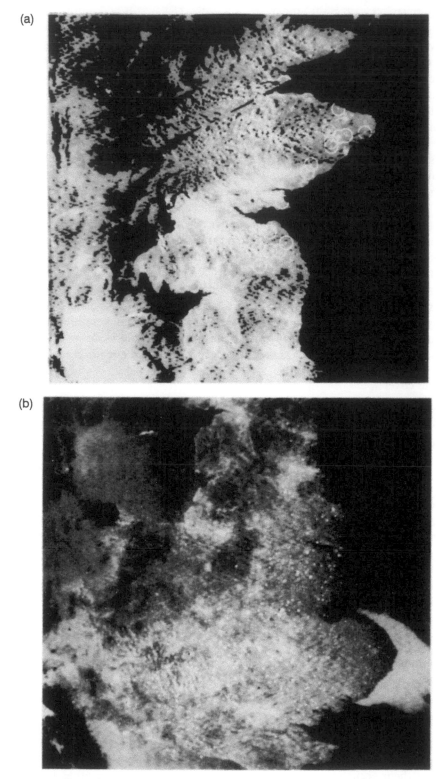

Figure 6.8   Channel-3 AVHRR image of (a) Scotland and (b) England and Wales from 20 August 1984 showing hot spots (white) (Muirhead and Cracknell 1985).

occur as a result of human activities; sometimes the fires are started by accident and sometimes they are started deliberately.

Early work on detecting subpixel-size fires had made use of the 1.1 km resolution HRPT or LAC data. However, the high-resolution data are only available if there is an appropriate direct readout ground receiving station or if the data capture has been scheduled for tape-recording on board the spacecraft. There are thus many areas of the world for which the high-resolution data cannot be obtained on a regular and frequent basis. Operational global coverage is, however, available with the lower-resolution GAC data. It might seem that such a coarse resolution would not be adequate for fire detection. However, Malingreau (1984) and Malingreau *et al.* (1985) examined the catastrophic Indonesian fires of 1983 in Kalimantan, Borneo, in Indonesia. GAC data were used because nothing else was available. Fifteen relatively cloud-free images and an unspecified number of cloudy images were investigated for the period from February to March 1983. Despite an initial expectation that the GAC resolution (4 km) would be too coarse for observing forest fires, many fires were in fact observed. They studied the correspondence between vegetation indices and hot spots as a function of date and location. Strong declines in greenness (NDVI) were associated with peak fire incidence. The observed chronology was interpreted in terms of the intensification of a drought leading to progressively higher rates of uncontrolled agricultural fires and, hence, to rampant uncontrolled forest burning. The testing and parametrisation of satellite observations were, however, limited by the absence of adequate ground information. Matson *et al.* (1987) also demonstrated that the spatial resolution of GAC data is not too coarse for those data to be used to study forest fires. Their paper shows a NOAA-7 visible band GAC image of northern Siberia, near Yakutsk, captured on 24 July 1984. The smoke plumes between 65–70° N and 105–125° E are a result of lightning-induced fires in this boreal forest area. Such fires are normal for this time of year and the resulting smoke had previously been routinely detected on NOAA satellite visible-band imagery. The paper also shows the coincident channel-3 image of the same area with many hot spots (white) on the image which are fires associated with the smoke plumes seen in the visible-band image. Even at the reduced GAC resolution, channel 3 responds to fire activity within the 4 km IFOV.

Forest fire monitoring in north America has been studied by Flannigan and Vonder Haar. All the HRPT data collected by the AVHRR receiving station in Edmonton, Alberta, in Canada during a severe nine-day fire outbreak were examined (Flannigan 1985, Flannigan and Vonder Haar 1986). Satellite observations were compared to daily fire reports from the Canadian Forest Service based on aerial reconnaissance. Five days were represented by data from both morning and afternoon passes and four days just by data from one pass each day. In total, 355 fires were observed. The cloud discrimination technique of Coakley and Bretherton (1982) was used to identify and discard cloud-contaminated pixels, etc., to avoid false positives from cloud-top reflectance. The Matson–Dozier algorithm was used to evaluate the sizes and temperatures of fires in the non-discarded pixels; the computed parameters were then compared with data in the fire reports.

About one third of all reported fires, but nearly 80% of fires not obstructed by cloud or smoke, were spotted on one or more occasion by satellite. Large (over 40 ha) fires were spotted 46% of the time, as opposed to 12–14% of the time for smaller (under 4 ha) fires. A fire's duration appeared to be as important as, or even more important than, its size in determining its probability of being observed. Less than

0.1% of all fires saturated the sensor. A greater percentage of unobstructed fire observations were identified during the day (87%) than at night (73%). The probability of identification increased rapidly for the first three passes since the outbreak of a fire and then tapered off.

Calculated temperatures were not reported by Flannigan and Vonder Haar, but they tended to be more indicative of smouldering than flaming combustion. Where more than five of the eight pixels contiguous to a target pixel showed signs of fire, it was assumed that the entire pixel was burnt. This appears to have resulted in over-estimation by more than a factor of three of mean fire size for fires under 400 ha. On the other hand, size estimates for very large (over 400 ha) fires averaged only about two-thirds of the reported size. Fire growth after the time of observation, emissivities below unity and atmospheric attenuation all probably contributed to this under-estimation.

A serious set of fires in the Yellowstone Park (USA) in the summer of 1988 was monitored using channel-1 and channel-3 AVHRR data; a black and white composite of data from these two channels for the afternoon of 7 September 1988 is shown by Hastings et al. (1988c). The estimation of the area of burning in this fire from AVHRR data was carried out by Cahoon et al. (1992); the burnt area was estimated as 3821 km^2. A number of workers have studied fires in the Amazon Basin in Brazil in connection with the problem of tropical deforestation. Matson and Holben (1987) applied the Matson–Dozier algorithm and made NDVI computations for all hot spots (defined as being all pixels for which the channel-3 brightness temperature exceeded 307 K) encountered in a $3° \times 6°$ area from a single LAC image centred near Manaus in the Amazon Basin. Of the 169 hot spots observed 143 (85%) saturated the sensor. The remaining 26 fires were analysed to determine their size and temperature, see Figure 6.9. Of these 26 fires, their areas were between 0.27 and 9.04 ha and their temperatures were within the range 376–609 K. The computed values seemed reasonable but no ground data were available to check the results. Typical observations of fires in Amazonia suggest that colonists and slash-and-burn cultivators generally burn 1–20 ha at a time. Smaller fires predominate. Flaming combustion occupies a large fraction, but usually less than one half, of the burn site during the hour or two of intense burning.

NDVI observations showed that pixels containing hot spots, and their nearest neighbours, tended to be less green than the general background. This was interpreted as fire damage to vegetation. Other possible factors include clustering of clearing (with the result that many clearings are next to land, such as secondary forest or pasture, that is less green than primary forest) and the prevalence of slash-and-burn at lake and river margins, resulting in frequent association of hot spots with adjacent water, flooded forest, and relatively bare ground found at low water.

The difference between the 85% of fires leading to saturation of channel 3 in the study of Matson and Holben (1987) in Amazonia and the less than 0.1% observed by Flannigan (1985) in Canada is large and unexpected. It is not clear whether the discrepancy is an artefact of the method or a reflection of a real difference in fire behaviour in the two areas. For a further discussion see the article by Robinson (1991).

Further work on the Amazon Basin was carried out by Malingreau and Tucker (1987), Pereira (1988) and Setzer et al. (1988). Malingreau and Tucker (1987), in a study of the southern Amazon Basin, studied fire points in AVHRR imagery. Daily LAC imagery was collected for the region between 6.5–15.5° S and 55–67° W over a

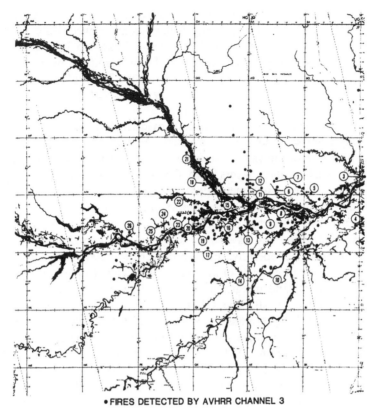

• FIRES DETECTED BY AVHRR CHANNEL 3

Figure 6.9   Map of the Manaus, Brazil, showing 26 fires (●) detected from AVHRR channel-3 data (Matson and Holben 1987).

two-year period. In 36 days of data from 1987, around 60 000 fires were counted. Using data from channels 1 and 2 of the AVHRR and methods developed by Ferrare *et al.* (1990), plumes were evaluated to determine the mass of the aerosol emitted from the observed fires. The aerosol mass, in turn, was used to estimate trace gas emissions.

Setzer *et al.* (1988) recorded and analysed HRPT data for fires in the seven states of the Brazilian Legal Amazon from 15 June to 2 October 1987. Observations were made in near-real time. As in the work of Pereira, channel-3 brightness temperatures over 317 K for the pixel were counted as fires. Assumptions were made, based on limited aerial and ground reconnaissance, about the average fraction of a pixel burned, the amount of double counting and the fraction of the observed burning that occurred in forested areas. These were used to calculate rates of deforestation. In total, nearly 140 000 fires were counted. The observed fires were estimated to have burned about 200 000 km^2 of forest in the seven states in question.

It is worth mentioning, in connection with large-area forestry studies, the role of GRID, the Global Resource Information Database. This is an environmental data support component of the Global Environment Monitoring System (GEMS) of the United Nations Environment Program (UNEP) (see, for example, Jaakkola 1990). The mission of GRID is to provide timely and reliable georeferenced information and access to a unique international GIS (Geographical Information System) service. GRID undertakes case studies with the international research community and maintains scientifically valid global or regional data sets in support of the concentration areas of UNEP. In this capacity, it has recently developed capabilities for global tropical forest cover assessment. The work is linked to UNEP's concentration area entitled 'Protection and management of land resources by, *inter alia*, combating deforestation and desertification and drought'. Through this work GRID plans ultimately to provide the world community with access to usable data on the condition and change of tropical forest cover.

In more detail, the services GRID aims to provide include (1) results of AVHRR/ LAC data analysis for selected areas of tropical forest, (2) integration and archiving in the GRID database of the various scientists' analysis results over tropical forests, (3) addition of ancillary data layers from GRID's global data sets, (4) assisting FAO's Forest Resources Assessment 1990 with the integrated data sets, and (5) giving expert advice and administrative support to the above functions.

There are currently a number of remote-sensing-oriented research projects dedicated to forest monitoring and assessment at global and continental scales. GRID has formal or informal cooperative links with various relevant organisations. To assist workers involved in projects of this type GRID aims to play an active role by systematically archiving and integrating relevant data in the GRID database and making them available to all. The GRID centres in Geneva and Nairobi are equipped with powerful digital image processing and GIS systems that are capable of the full range of operations necessary to accomplish data input, processing, archiving and distribution functions. In 1987–90, the forest-related work at GRID consisted of methodology development for (1) forest/non-forest delineation using AVHRR-satellite data over West Africa and (2) assessment of deforestation using satellite data over Amazonia. Data in the form of maps and satellite imagery were acquired from a variety of sources. After being processed and georeferenced, the data were stored in GRID's digital databases. During the methodology development phase in 1987–90, the focus of data acquisition at GRID was on purchasing AVHRR LAC data from NOAA. Landsat TM and SPOT scenes were acquired from EOSAT and SPOTIMAGE, respectively. Ground truth data were collected from various West African countries and from Amazonia in cooperation with local forestry authorities. The results of methodology development were encouraging enough to justify the continuation of the work at a global scale, and contributing to the development of operational systems for global tropical forest cover monitoring. Jaakkola (1990) also describes an ambitious forest database implementation phase consisting of the compilation and maintenance of a digital global forest cover map derived from AVHRR data. This was planned to be accompanied along with an expansion of the GRID network of centres and regional nodes.

Many forest fires are started deliberately in association with the clearance of forests. Thus the monitoring of fires is, in some cases, associated with studies of deforestation. Deforestation provides another example of a situation in which gaining knowledge by remote sensing may be possible, but governments may choose to ignore the results of the remote sensing studies. Several points should be made. First, deforestation is often spoken of as though it were only a problem of the tropics and, therefore, largely a problem of certain developing countries. It is often forgotten that various other areas of the world, e.g. parts of Europe, were once heavily forested but have been deforested over a period of many centuries. Admittedly the activity is more rapid and more dramatic in certain tropical areas at the present time. For example, the SEAMEO (SouthEastern Asian Ministers of Education Organisation)–France project was established to provide the four participating countries (Brunei, Indonesia, Malaysia and the Philippines) with the appropriate tools and methodologies for effective tropical forest monitoring with a short response time, based on the use of AVHRR data (Puyou-Lascassies *et al.* 1993). Secondly, there is a cautionary warning to be found in some work from the Philippines (Kummer 1992); the identification and quantification of a problem using remote sensing methods does not guarantee that the problem will be solved. The main points made by Kummer (1992) are worth noting, although most of the remote sensing work was done with Landsat and SPOT data, not with AVHRR data.

Deforestation has been widespread in the Philippines. Today approximately 20% of the total land area of 300 000 km^2 is forested as compared to 50% in 1950. Deforestation averaged 1570 km^2 per year from 1980 to 1987, and improper land use is the Philippines' number one environmental problem today. Remote sensing by satellite was suggested as a way of monitoring the rapid changes occurring in the extent of tropical forests and in the Philippines five forest and land use surveys using remotely sensed data have been conducted since the 1970s. Four surveys were conducted using Landsat data in 1973, 1974, 1976 and 1980 and one survey was conducted using SPOT data in 1987. The results of these surveys, in terms of the percentage of the total land area of the Philippines covered by forest, were

1973	38.0%
1974	29.8%
1976	30.0%
1980	25.9%
1987	23.7%.

The first Landsat survey of 1973 was computer based and was carried out jointly by the Philippine Bureau of Forestry and the General Electric Co.; all the computer work was done in the USA. It is now believed that the result of 38% over-estimated forest cover which was probably only about 30% in 1973. The Government of the Philippines had thought that forests covered 57% of the total land area in the early 1970s. But even before the Landsat survey the Government knew, from previous national surveys, that forest cover was less than 57%. However, notwithstanding this fact and the result of the Landsat survey, the Government continued to hold to this figure of 57% as part of its deliberate cover-up of the rate and extent of deforestation. Because the Philippine Government was not interested in rational management of its forests, the practical effect of this forest survey on Government policy was nil.

The other three surveys conducted using Landsat MSS data over the period 1974 to 1980 were all done visually by Filipino workers using black and white photomosaic images. The advantages of visual interpretation by Filipino photointerpreters are several: it is less costly, it can be done in-country, and it is conducted by people who are experts on the Philippines. The results, which have already been quoted above, seem to have been reasonably accurate in terms of the overall national percentage of land area covered by forest. However, although the results of these three surveys were more accurate they were no more effective than the original 1973 computer-assisted Landsat survey in actually leading to more effective forest management.

The rapid loss of Philippine forest cover since 1950 has been a tragedy of major proportions and, under the Marcos administration, there were deliberate efforts to mask the rate and extent of deforestation. The first four forest surveys mentioned above using remotely sensed data took place in a politically repressive environment and the results were ignored by the Philippine Government so that, in consequence, they had no impact on forest management. This, of course, was not the fault of the data or the researchers; at the same time, we also now know that the first Landsat survey seriously over-estimated national forest cover. The SPOT survey occurred one year after Marcos was forced out of office in 1986 but it also had some short-comings (for details see Kummer 1992), although its general conclusion (23.7% forest cover of the total land area) was reasonably accurate. In short, two decades of remote sensing research in the Philippines did not lead to any appreciable improve-ment in forest management, because political considerations from 1965 to 1986 con-sistently overrode the value of any data provided by remote sensing.

An unusual example of the observation of hot spots in channel-3 AVHRR data was provided by Dousset *et al.* (1993). Hundreds of fires were set alight in Los Angeles in a civil disturbance following the verdict in a court case on 29 April 1992. In the channel-3 AVHRR image shown by Dousset *et al.* from 03.47 PDT on 30 April 1992, approximately 10 hours after the start of the riots, an exceptionally large thermal anomaly was visible. It extended over more than 85 km^2 in south central Los Angeles where, on average, three new fires were started every minute during the three hours preceding the capture of the image.

6.2.4   Resolvability

We have now seen very clearly that channel-3 data from the AVHRR can be used successfully to detect hot spots that are very much smaller than the size of the IFOV of the instrument. The question arises as to what is the limit (in terms of both temperature and size), first for detection and secondly for analysis in terms of the Matson–Dozier theory to determine the size and temperature of the hot spot. As far as detection is concerned, Belward *et al.* (1993), by organising controlled fires in the savannah zone of Côte d'Ivoire, found that fires with a front longer than 50 m were detectable but smaller fires were not. However, a fire with a front length of 1 km that died out 10–15 min before satellite overpass could not be detected; this indi-cates that only flaming fires, not recently extinct ones, can be detected.

For the use of the Matson–Dozier theory for a fire that has been detected the sensor must not be saturated. This problem has been studied by Robinson (1991). The lower limit to the signal which can be interpreted to be due to a fire will be established by choosing a threshold for the signal in relation to the signals in nearby non-fire pixels. That is, in terms of Figure 6.4 one has to choose a threshold for the height of a peak; only peaks higher than that threshold will be interpreted as due to fires. Curves for three different choices of threshold (10%, 20% and 40% above the background (non-fire pixel value)) are plotted in Figure 6.10. If the size and tem-perature of a fire are such that it is represented by a point that falls below or to the left of the chosen detection threshold curve then that fire will not be detected. If the parameters of the fire correspond to a point that lies between the chosen detection threshold curve and the saturation threshold curve, then that fire can be detected and its parameters (size and temperature) can also be determined. If the parameters of a fire are represented by a point that falls above or to the right of the saturation

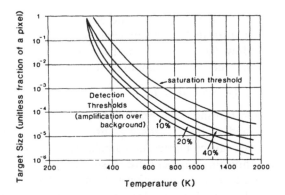

Figure 6.10 The envelopes of resolvable fires assuming a background temperature of 300 K, a saturation threshold of 325 K and detection thresholds of 10%, 20% and 40% above the background (Robinson 1991).

threshold curve, then this fire can be detected (provided it is not obscured by cloud or smoke); however, because the sensor has become saturated the area and temperature of the fire cannot be determined. To illustrate this, consider a fire at 600 K. If it occupies only 10^{-5} (0.001%) of the IFOV it will not be detected. If it occupies 10^{-3} (0.1%) of the IFOV it can be detected and its size and temperature can be determined. If it occupies 10^{-2} (1%) of the IFOV it can be detected, but the sensor will be saturated and the area and temperature of the fire cannot be determined.

The theory behind the determination of the saturation threshold curve is given by Robinson (1991) but can be reformulated in terms of the notation we have used previously in this section. Suppose we take equation (6.2.2)

$$T_j = L_j^{-1}[pL_j(T_t) + (1 - p)L_j(T_b)] \tag{6.2.2}$$

where $j = 3, 4$, and choose a value for the background temperature T_b and set T_j equal to the saturation threshold temperature for the appropriate AVHRR instrument taken from Table 6.1. This gives us the equation of a curve in the space of the two variables p (size) and T_t (temperature) of the target. A semi-automatic implementation of equation (6.2.2) with two channels of AVHRR data for savannah fire

Table 6.1 Calculated saturation temperature for AVHRR instruments (Robinson 1991)

Platform	Channel-3 saturation temperature (K)
TIROS-N	326.7
NOAA-6	325.3
NOAA-7	322.0
NOAA-8	324.2
NOAA-9	326.5
NOAA-10	331.3
NOAA-11	327.5

detection has been carried out by Langaas and Andersen (1991). Robinson (1991) simplified the use of equation (6.2.2) by replacing the spectral response $\phi(\lambda)$ by a simple rectangular window (i.e. $\phi(\lambda) = 1$ in the range 3.5–3.9 μm and $\phi(\lambda) = 0$ otherwise). Equation (6.2.2) then can be simplified to give

$$B(\lambda, T_j) = p\varepsilon_t B(\lambda, T_t) + (1 - p)\varepsilon_b B(\lambda, T_b) \tag{6.2.3}$$

where T_j is the saturation threshold temperature for channel 3 of the AVHRR. It is this curve which is plotted as the threshold saturation curve in Figure 6.11. The detection threshold curves can be plotted in a similar way by setting T_j in equation (6.2.3) as one's chosen threshold temperature. A similar treatment to that of equation (6.2.3) has been used in studying active volcanic craters, see section 7.7. We return briefly to the 26 fires studied by Matson and Holben (1987) in the Amazon Basin in Brazil and which we mentioned above, see Figure 6.9. Figure 6.11 shows a plot of size and temperature, given by Robinson (1991), for these 26 fires in relation to the envelope of resolvable events for a background temperature of 300 K, a detection temperature of 307 K and a saturation temperature of 325 K. As shown, there is some margin on the saturation side (perhaps caused by background brightness temperatures being below 300 K) and a few fires appear to be subthreshold, but in general the observations conform to the curve.

The question of saturation of channel 3 merits further consideration and we quote from a very interesting recent paper by Setzer and Verstraete (1994). Channel 3 supposedly saturates for Earth targets at about 47°C, or 320 K (Kidwell 1991), and also with sunglint in water bodies (Khattak et al. 1991, see also section 7.8) or sunlight reflected from some types of exposed soils (Grégoire et al. 1993). Calculations based on the use of Wien's displacement law predict that a fire of about 30 m × 30 m in extent and near the centre of a channel-3 IFOV should saturate it. This expectation is not supported by available measurements, however. For instance, large fires in forests and grasslands, sometimes occupying areas many times the size of the IFOV of the AVHRR, have been associated repeatedly with non-saturated values of channel 3 (for references, see Setzer and Verstraete 1994). For sunglint no actual data are apparently found in the literature. Brush (1993)

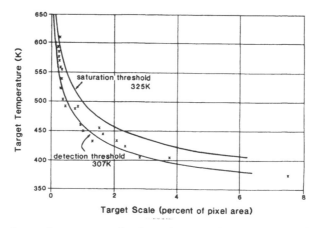

Figure 6.11   Fire sizes and temperatures for the 26 fires at the locations indicated in Figure 6.9 (Matson and Holben 1987, Robinson 1991).

showed a NOAA-12 AVHRR scene where the signal 'may' reach saturation; he also pointed to a 'transient' effect just entering the sunglint area, which could be explained by Setzer and Verstraete. This transient signal, which appears in the channel-3 images like an atmospheric halo/corona effect around the sunglint areas, whatever its size, as well as the non-saturation by very large fires, was claimed by Setzer and Verstraete (1994) to be the result of an on-board signal processing problem (see also Setzer and Malingreau 1993).

By using NOAA-11 AVHRR data from an area of sunglint, where the satellite-received radiance in channel 3 was clearly more than large enough to cause saturation, Setzer and Verstraete (1994) argued that there is a faulty on-board conversion of the analogue signal from the channel-3 sensor into DNs (digital numbers) output. This conversion should in principle take place monotonically, i.e. associate progressively lower DNs to increasing radiance measurements. Instead they postulated that beyond a specific threshold radiance, increasing radiances are transformed into increasing DNs. In this type of curve, schematically shown in Figure 6.12, a signal level higher than the level corresponding to a null DN results in DNs previously allocated to lower signal levels. This postulate was also supported by evidence from fires. Vegetation fires, no matter their temperature or size, provided the fire front is larger than ~50 m, do not saturate channel 3, contrary to calculations based on the emitted energy. Furthermore, actively burning fires were consistently associated with a DN of 45 in channel 3 during March 1990, which is precisely the same value found in the core of the sunglint by Setzer and Verstraete (1994). When examining histograms of channel-3 images containing many fires or sunglint a very steep change in their profiles was regularly found at a DN value of 45, indicating a discontinuity at this point. Thus it was concluded that the postulated engineering problem would also explain why hot fires over continents do not saturate channel-3 pixels. This could further imply that AVHRR channel 3 may actually be observing much higher temperatures than currently estimated on the basis of existing data, calibration coefficients and established calibration procedures given by Kidwell

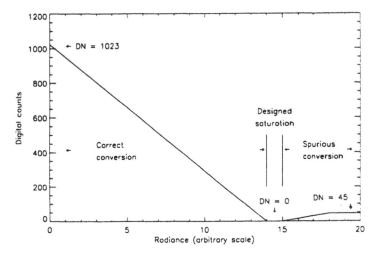

Figure 6.12 Hypothetical channel-3 conversion of radiance into counts (Setzer and Verstraete 1994).

(1991). Finally, the fact that at least NOAA-9 AVHRR exhibited the same pattern of non-saturation of channel 3 over active fires suggests that the same problem may affect several AVHRR instruments and not just the one on NOAA-11.

The reflective component of channel-3 AVHRR data presents problems in quantitative treatments of day-time fire data. Active cumulus cloud cells and other highly reflective surfaces can reflect sufficient energy in the middle infrared to create a pseudo-hotspot; this problem vanishes at night. Thus, if the diurnal distribution of fires were known and night burning were significant and well correlated to day burning, nocturnal data could be used to predict total burning. However, data on the diurnal distribution of fire are scarce, and the available observations indicate that in many regions night fires will be poor predictors of day-time burning. Alternatively, spectral filtering can be applied to eliminate colder reflective objects (Flannigan and Vonder Haar 1986). This carries a processing cost and, as mentioned above, is likely to eliminate observations of fire in pixels containing both clouds and hot spots.

CHAPTER SEVEN

Applications

7.1 SNOW AND ICE

7.1.1 Introduction

As was indicated in section 1.1.1, the AVHRR was designed in the mid-1970s as an instrument for a meteorological satellite, but since then the data from the series of AVHRR instruments have found an enormously wide range of applications in a whole variety of non-meteorological problems that were never originally envisaged. One of these, in the field of vegetation studies, has been described in some detail in chapter 5, while some special applications of channel-3 data were described in chapter 6. The present chapter is concerned with a wide range of other applications of AVHRR data.

The use of meteorological satellite data for monitoring snow and ice has been attempted since the launch of the first US polar-orbiting meteorological satellite in 1963. Some summaries of the early stages of the development of operational snow and ice mapping programmes are given by Barnes and Smallwood (1975), Rango and Peterson (1980), Schneider (1980), Rao *et al.* (1990), Foster and Hall (1991) and Walter (1991). The questions which can be addressed with the use of meteorological satellite data in general, and with AVHRR data in particular, include studies of

- the area (of the land surface) covered by snow or ice
- the depth of snow
- the area and depth of sea ice
- sea ice motions
- polar meteorology.

We shall address these topics in turn. However, in practice there are many problems. (a) Snow can be detected by its brightness in the visible channel, but cloud may be present and in that case it is important to be able to distinguish between cloud and snow. (b) It is generally not possible to determine the depth of snow from an image. (c) Normally sea ice can be discriminated from open water in channels 1 and 2 by its high reflectance and in the thermal infrared channels by its low temperature. But, as in the case of snow detection, the presence of cloud prevents the ice from being seen. Also determining the type and thickness of the ice that is present may be important to the user of the data. (d) In the early days snow and ice boundaries were identified by an operator using a cursor on a screen. For the routine generation of operational products it is preferable to have software available that will enable the boundary to be mapped automatically. Digital analysis of this imagery, however, is complicated by the presence of clouds and the mixed nature of the pixels. Such software should ideally be able to distinguish snow and ice from cloud and to distinguish between different types of ice.

Kidder and Wu (1987) have argued that the albedo difference between channel-1 and channel-2 data is a better indicator of the snow boundary than either channel alone. But with advances in cloud recognition techniques (see section 3.2) and the availability of high resolution (1 km^2) AVHRR data on a regular basis from ship-mounted receiving stations, as well as the rapid improvements in computing systems, the frequent acquisition of images that are at least partially cloud-free begins to outweigh the computational inconveniences.

Early approaches to snow-cloud discrimination (e.g. Barnes and Bowley 1968) have relied on pattern recognition techniques such as the conformity of snow cover to the terrain configuration and the uniformity of snow reflectances as compared to the uneven appearances of cloud, together with the identification of terrestrial features such as rivers, forest edges, roads and agricultural patterns. In addition, multitemporal analysis and techniques for the recognition of cloud shadows have been employed for snow-cloud discrimination.

Other methods have relied on the spectral signatures of snow, and have included visible and infrared thresholding techniques. Improved snow-cloud discrimination procedures have been developed using near-infrared wavelengths, for example the Landsat Thematic Mapper (TM) band 5 (1.55–1.75 μm), the US Air Force Defense Meteorological Satellite Program (DMSP) satellite (1.5–1.6 μm), and the Skylab S-192 sensor (1.55–1.75 μm) (Hoffer 1975, Crane and Anderson 1984, Dozier 1984). At this infrared wavelength snow appears relatively dark while ice and water clouds appear bright. However, there is at present no channel in the region of 1.5–1.7 μm on the AVHRR instruments. Separation of snow from cloud is also feasible using the 2.10–2.35 μm waveband although the discrimination is less sensitive than in the near-infrared (Ødegaard and Østrem 1977).

Improvements to snow-cloud separation procedures have been achieved using the near-infrared in conjunction with data from the visible and thermal infrared channels (Hunt et al. 1974, Kimball 1980, Bunting and D'Entremont 1982).

Using AVHRR data Gesell (1989) was able to separate snow and water clouds by developing an extension to the AVHRR Processing scheme Over Land, cLoud and Ocean (APOLLO) software package which was developed at the UK Meteorological Office to derive surface and cloud parameters from AVHRR digital data (see section 3.2). To find cloud-free and partly cloudy snow and ice pixels, Gesell (1989) developed a day-time algorithm which uses all five AVHRR channels as follows. The threshold testing of the reflected part of channel-3 radiance leads to a definite distinction between snow/ice and water clouds due to the clouds' much higher reflectivity at 3.7 μm. The detection of sea ice is based on threshold tests of visible reflectances and, in particular, of the temperature difference between channel-4 and channel-5. Snow is identified if a high visible reflectance is combined with a low reflectance in channel-3 and with a ratio of channel-2 to channel-1 reflectances similar to that of a cloud. The latter criterion is also mostly suitable to distinguish between snow-covered and snow-free ice areas. Some tests of this algorithm applied by Gesell (1989) to AVHRR data from the 1987 Baltic Sea ice season have shown reasonable classification results with the exception of a few areas with ice clouds or with ice topped water clouds.

Harrison and Lucas (1989) used an unsupervised multispectral clustering technique to separate cloud and snow-covered pixels in some work on snow-covered mapping for the UK in the winter of 1985–1986. The method also allowed the detection of several snow classes which vary according to depth, percentage cover, vegetation influences and snow condition. Cloud shadows and snow in forest could also be recognised to varying extents.

Various descriptions exist of the use of AVHRR data in regional snow surveys in various parts of the world. These include, for example, part of northwest China (Liu et al. 1986), the St Louis area (Kidder and Wu 1987) and the Sierra Nevada mountains (Schneider and Matson 1977) in the USA and the case just mentioned in the UK (Harrison and Lucas 1989); doubtless there are many others.

The problems associated with cloud cover can be eliminated by using passive microwave data. Data from the SMMR (Scanning Multichannel Microwave Radiometer) have been used quite extensively for ice mapping; the data have been processed routinely to provide daily sea ice concentration charts. These charts have been the exclusive source of information on cloud-covered ice, drifting in the polar oceans. However, (i) the SMMR has very much poorer spatial resolution than the AVHRR, the IFOV of the SMMR being about 27 km by 18 km for its highest frequency channel (37 GHz) and (ii) the SMMR is not operational but is an experimental instrument, which was flown on Seasat and on Nimbus-7. The use of AVHRR HRPT data for snow monitoring, in conjuction with data from another passive microwave scanner, the Special Sensor Microwave Imager (SSM/I), on the current family of DMSP satellites, is described by Barrett et al. (1991).

7.1.2 Area of snow cover

The basic idea behind the production of snow and ice charts is very simple. Areas covered by snow or ice can be detected and delineated from AVHRR data so that maps or charts showing the areas covered by snow or ice can be produced. This information can be augmented by data from Landsat and geostationary satellite scanners. The data need to be geometrically rectified to a map projection, the snow or ice boundary needs to be located and this information transferred to the map product. Snow coverage has been measured by NOAA since 1966 and snow charts are produced weekly by manual analysis from the most recent cloud-free images obtained during the week. These charts are digitised and aligned with a polar stereographic map. The resolution achieved is about 190 km by 190 km at 60° N. Data are archived on magnetic tapes. Due to the high cloud coverage the results are poorest in summer and autumn. The interests of users of the mapped snow and ice data are quite varied. For example, in agricultural activities it can be useful to have a map of the area covered by snow. This is important for instance for agricultural monitoring in northern areas of the USA; this is because snow is necessary in northern wheat growing regions to protect winter wheat from damage from low tem-

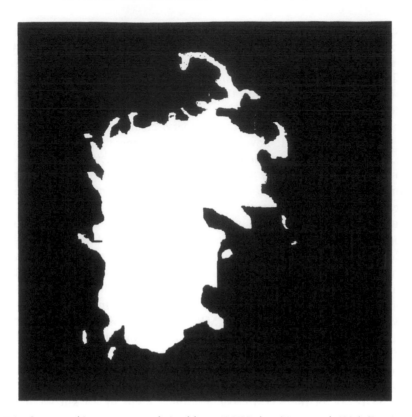

Figure 7.1   Snow- and ice-cover map derived from AVHRR data (Yates *et al.* 1984). (Reprinted with permission of Elsevier Science Inc.).

peratures (Yates *et al.* 1984). A comparison of the National Oceanic and Atmospheric Administration (NOAA), Air Force Ground Weather Centre (AFGWC) and US Navy Snow and Ice Charts has been made by Kukla and Robinson (1981).

Figure 7.1 illustrates a northern hemisphere snow and ice map for the period 22–29 March 1982, produced at NOAA, NESDIS. The input data are mapped seven-day minimum brightness composites of NOAA-7 GAC data. Snow and ice boundaries were drawn by analysts on a screen using a cursor and the mapped visible and infrared composites. This product has been prepared once a week and shipped to the US Department of Agriculture in Houston since 1 November 1982. This product defines an area of relatively long-lived snow cover. Snowfalls lasting only a day or two are overwritten in the seven-day compositing process. A rather famous AVHRR mosaic of Antarctica has been produced by Merson (1989), see Figure 7.2 (in the colour section). At the other extreme, regional snow-cover maps are produced for various river basins (Schneider 1980), see Figure 7.3.

NOAA NESDIS, because it receives GAC data, is able to produce on a regular basis maps or charts of snow and ice cover giving global coverage at frequent intervals. Other organisations that operate direct readout stations receiving HRPT/LAC data are able to produce snow and ice maps of higher spatial resolution but only within the restricted areas of visibility from their own ground stations.

Figure 7.3 Map showing the locations of operational snow-cover basins (Rao *et al.* 1990).

7.1.3 Depth of snow

While the area covered by snow (or ice) can be determined quite easily from visual inspection or from digital processing of AVHRR images, the determination of the depth of the snow is a much more difficult problem. One of the reasons for seeking to determine the depth of the snow is to be able to predict the melt-water runoff that can be expected from a given catchment area once the snow melts. The main interest in snowmelt runoff prediction is in relation to flood prevention and to the optimisation of the management of reservoirs, which should lead to an improvement in hydroelectric power production. Rather than attempting to determine the snow depth, on a point-by-point basis, what is done is to establish an empirical relation between the observed spring meltwater runoff and the (percentage) area of snow cover in a catchment area. Such a graph is, obviously, specific to a given catchment area and is of no relevance to any other catchment area. Figure 7.4 shows the relation between satellite snow-cover estimates and measured runoff over a six-year period for the Indus River, India. Correlations for the Indus were quite favourable (Rango 1975). This method has been developed in Norway and Greenland, and

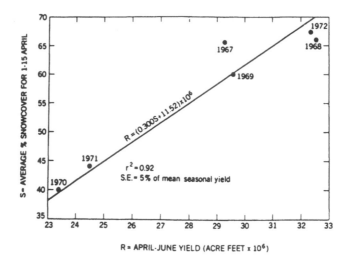

Figure 7.4 Satellite-derived snow-cover estimates versus measured runoff for the Indus River (Rango 1975).

more sophisticated versions have been developed for the Swiss Alps. A comparison of the three methods and models is given in Table 7.1 (Haefner and Pampaloni 1992).

In the case of Norway and Greenland the basins under surveillance are relatively large and the relief considerably less pronounced than in the Alps; consequently simpler methods could be applied successfully. The requirement for improved power production was the main driver for snow mapping and monitoring in Norway (Østrem *et al.* 1979, Østrem 1981, Andersen 1982, Faanes 1991). Based on using AVHRR visible data, a method was developed to evaluate the snow residue and to

Table 7.1 Comparison of the three methods for snow-melt runoff forecasting (Haefner and Pampaloni 1992)

Author	Østrem *et al.* (1979)	Søgaard (1983)	Martinec *et al.* (1983)
Satellite data	NOAA (Landsat)	NOAA	Landsat-MSS and TM, NOAA, orthophotos, etc.
Digital terrain model	No	Yes	Yes
Ground measurements	Runoff measurements	Snow-pack properties	Weather parameters
Snow-cover extraction	Density slicing; linear interpolation model	Computing pixel values into albedo values	Supervised digital classification
Water equivalent	None Extrapolation of snow area versus runoff from observation series	Yes Field measurements depending on altitude and easting	Yes Computed by a degree-day factor, (index of energy balance)
Runoff forecast	Seasonal runoff	Average weekly or monthly discharge	Average daily discharge

predict the corresponding melt-water volume for various basins. For both categories, 'totally snow-covered' and 'snow-free', a corresponding grey tone level was defined in selected test areas at various altitudes and a linear interpolation was established between them. The method is only applicable in almost vegetation-free basins after a part of the terrain has become snow-free. The expected runoff is then calculated from a curve which relates the extent of the remaining snow cover and the subsequent runoff for the melting period. The curve, developed from ground sampling, is an empirical curve which is specific to the basin for which it was determined. This procedure has now been used to estimate snow-melt runoff operationally for several years.

The snow mapping project in Western Greenland was also based on using AVHRR data (Søgaard 1983, 1986). In two major basins mapping of the snow cover and also of the water equivalent was elaborated between 1979 and 1984 and validated by field observation and hydrological simulation. The method used includes geometrical rectification, radiometer calibration to albedo and surface temperature, correction of slope effects (for visible channels) and atmospheric effects (for the thermal infrared channel) and computation of the snow cover from a linear expression, classifying the snow cover into five categories. In addition, the relation between snow cover estimated from the AVHRR and the remaining snow-pack water equivalent obtained from field observation, was investigated with a view to using the relation with later AVHRR data for hydroelectric power production.

The input data needed for the Snow-melt Runoff Model (Martinec *et al.* 1983) used in the Swiss Alps includes the air temperature and precipitation (if possible from a weather station centrally located in the study area) and the snow cover derived from the depletion curves obtained from snow mapping. The requirements of the users for operational forecasts, namely hydroelectric power companies, are for very high accuracy for rather small, often artificially influenced, catchment areas. On the other hand the European Alps are characterised by a very variable climate and by small and heterogeneous watersheds with rugged terrain. Consequently only sophisticated methods will provide the necessary accuracy. Table 7.1 summarises the most important parameters of the model. It has been tested in various basins in simulation form (Baumgartner 1987). Figure 7.5 provides the results of a simulation in comparison with the actual runoff. The model is not capable of estimating heavy rainfall; efforts should be undertaken to modify the model in such a way that it can also handle such events. Ongoing studies are determining the daily flow for the Tavanasa hydroelectric station in the eastern Swiss Alps (catchment area 215 km^2, 1277–3210 m above mean sea level) with the aim of carrying out the computation in near-real time for operational runoff forecasting (Seidel *et al.* 1989). A further development, the Alpine Snow-Cover Analysis System (ASCAS), involving the use of AVHRR data, was described by Baumgartner and Apfl (1993); it enables snow hydrologists and climatologists to quantify snow-cover variations, to determine snow-melt runoff and the influence of temperature variations on runoff as well as to derive climatic differences between regions and between years.

The above methods involve the determination of the surface area covered by snow from the AVHRR data. They accept that with visible and near-infrared data the direct determination of the depth of the snow is not possible and that the relation between area of cover and volume of runoff has to be determined empirically. Only microwave methods offer any hope for measuring the snow thickness directly.

An early paper by Lillesand *et al.* (1982) on the use of AVHRR imagery for

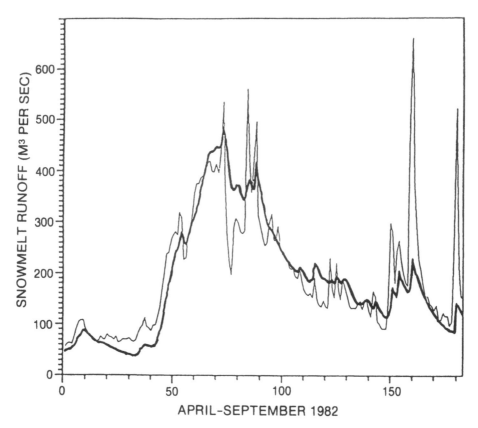

Figure 7.5 Simulation of snow-melt runoff for 1982 for Vorderrhein basin (thick line) and the natural reconstructed runoff (thin line) (Baumgartner 1987).

snow-cover mapping demonstrated the considerable problems associated with the masking influences of the various cover types in a heavily forested area.

While for the purpose of runoff predictions it is possible to use an empirical approach which circumvents any need to determine the depth of the snow, there may be situations in which it would be useful to determine the depth of the snow. Some attempts have been made to relate the intensity of the satellite-received radiation in the visible channel(s) to the depth of the snow. However, the results show very considerable scatter, see Figure 7.6. Depth, however, is not the only factor affecting the reflectivity of the surface of the snow. Strong *et al.* (1971) first suggested the use of multispectral satellite imagery to detect melting snow and ice packs by observing changes in the spectral brightness of these fields in the visible and infrared regions of the spectrum. Laboratory studies by O'Brien and Munis (1975) of various natural snow samples under different aging conditions showed that when the snow surface temperature is below the melting point, there is little difference in the reflectance over the visible and infrared range. Once the snow surface temperature is raised above the melting point, however, the reflectance drops dramatically at wavelengths greater than 0.75 μm. Refreezing of the melting snow surface does not restore the reflectance to its pre-melting levels. Such reflectance differences provide evidence on the condition of the snow pack that could be used by water-resource planners in their estimation of expected runoff.

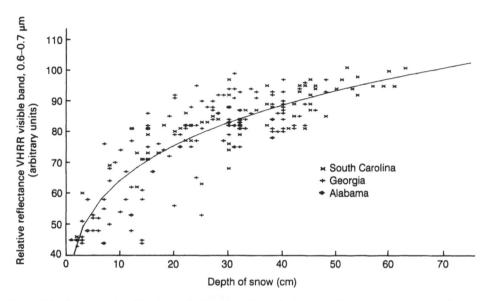

Figure 7.6 Relation of visible channel reflectance (from the VHRR) with snow depth (McGinnis *et al.* 1975).

7.1.4 Sea ice cover

Manual interpretation of AVHRR imagery has been routinely incorporated for a number of years into sea ice charts produced for the Arctic, Antarctica and Baltic regions. The main interest in this product is for navigational purposes, but it is also relevant to climate modelling. The US Navy/NOAA Joint Ice Centre has produced weekly sea ice charts of the Arctic since 1972 and of the Antarctic since 1973 by interpreting satellite images. All charts produced are archived at the National Geophysical Data Centre, NOAA.

NESS (later NESDIS) started operational ice analyses of the Great Lakes during the 1973–4 season using the visible and infrared data from the VHRR and then, later on, from the AVHRR. Analyses were produced twice weekly and distributed by facsimile and mail. Currently, the ice analysis of the Great Lakes is done by the Navy/NOAA Joint Ice Center at Suitland, Maryland, and distributed by facsimile. Although satellite data are the main sources of information for the analyses, reports from lakeside land stations, ship reports and aerial reconnaissance are also used. The Great Lakes ice analyses are done manually using photo-interpretation techniques. Ice concentration is expressed in per cent of the lake surface covered. Sea ice observation at the Berlin Meteorological Institute started in 1966; during the first years, without infrared data, gaps during winter-time were filled with information from sea ice charts from the UK Meteorological Office and other sources. Since 1983 monthly supplements to the *Berliner Wetterkarte* have been published, which show the position of the actual sea ice edge within the receiving area of the Institute's HRPT (High Resolution Picture Transmission) station and additionally from Canadian regions. Figure 7.7 shows the sea ice boundary and also the extents of the maximum and minimum ice coverage during the period from 1966 (the beginning of the Berlin Meteorological Institute's VHRR sea ice observation) to 1989. The ice

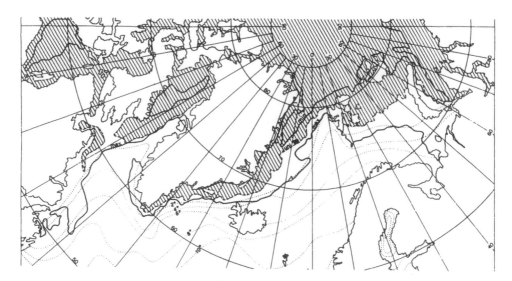

Figure 7.7   Sea ice border at the end of June 1989 showing the ice-covered area (hatched) and the ice boundaries at the times of minimum and maximum cover during the period 1966–1989 (Eckhardt 1989).

edge plotted represents the edge of the areas with 70% coverage and is valid for the last ten days of the month. The main interest of this product is for navigational purposes at sea. The variation in the sea ice coverage for the same period, derived from the Berlin Meteorological Institute's data, is shown in Figure 7.8. Such records are very valuable for climatological studies. A satellite-based sea ice monitoring system developed in the former USSR has been described by Nikitin (1991).

The development of automatic detection of sea ice, instead of using human interpretation, is described in some detail by Gerson and Rosenfeld (1975), beginning with air photos taken over Arctic waters and with early satellite images. Various sea ice mapping projects are described. The early satellite images were of poor quality. The scale of air photos is so large that the number of photos one needs to handle is vast; the information obtained from them was at best a summary of human visual scanning of the pictures. An overview of methods for deriving ice information from operational satellite data is given by Eckhardt *et al.* (1992), with special attention to AVHRR data; these include the use of single-channel and channel-difference images,

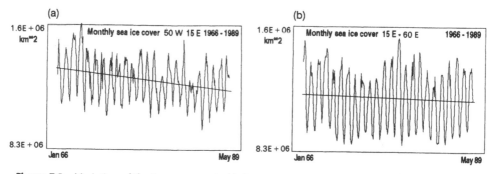

Figure 7.8   Variation of the ice coverage in (a) the western and (b) the eastern part of the region covered in Figure 7.7 (Eckhardt 1989).

minimum brightness and microwave data from the MSU on board the NOAA satellites. The automatic tracking of ice floes from AVHRR image sequences was described by Moctezuma *et al.* (1993).

Differentiation of ice from clouds by the use of computers has been done for some time. The methods were originally developed by those who were interested in the clouds and wanted to remove underlying ice. An early method was to make use of the fact that ice is relatively stationary compared to relatively fast moving clouds; thus a time-series of satellite images was overlaid and the stationary part removed. A method which evolved from the above is the use of minimum brightness charts. This is a computer process where the charts, or composites of satellite pictures, are compared for a period of five days and the darkest value for each point on the Earth's surface is retained. The bright points remaining are those permanently bright, i.e. ice.

An early attempt at the analysis of satellite images for sea ice mapping was to differentiate four classes in a satellite image, namely clouds, ice/snow, water and land, and to use a decision tree (Gerson and Rosenfeld 1975). Land is the most difficult class to differentiate. It is easiest to eliminate the land from consideration by using a map, in registration with the image, to determine which image points (pixels) are located on the land and which on the sea. There should be no difficulty in doing this with an accuracy of better than one pixel, even for high-resolution satellite imagery. However, it should be pointed out that coastal leads, which are often used for navigation of imagery, may be too narrow to detect on such imagery.

Once land has been eliminated from consideration, it will usually be easy to distinguish water from non-water (i.e. ice or clouds) by simply thresholding the image's grey level. Exceptions to this are new and weathered ice, which are probably indistinguishable from water on satellite images. (It would also be impossible to distinguish bare ice from snow-covered ice, but this distinction is not important for our purposes.)

It remains only to distinguish ice from clouds. This discrimination is the main topic of the paper of Gerson and Rosenfeld (1975) who describe the use of statistical features to discriminate sea ice from clouds, using five classes (three types of ice and two types of cloud) and claiming to achieve 90%, or higher, accuracy. It should be pointed out that even if the ice/cloud decision is in favour of cloud, ice may still be present under the clouds. In this case, we can use the *a priori* probability of ice (at the given location and season) to make a final ice versus not-ice decision. This decision can be based on the results of earlier ice reconnaissance of the area, as well as on ice concentration maps for the given time of year. (These maps give contours of ice probability, based on subjective estimates derived from historical data.) The use of a fuzzy logic technique for the accurate separation of sea ice from clouds and from the open ocean in AVHRR imagery of the Arctic is described by Simpson and Keller (1995).

Various algorithms for sea ice monitoring with AVHRR data have been presented by Burns *et al.* (1989). Algorithms were developed for three purposes: discrimination of cloud cover and ice-covered areas from open water areas, estimation of ice concentration, and measurement of the sizes of ice floes and open water areas (polynyas) within the ice field. For ice–water–cloud discrimination, a dynamic threshold technique was applied to six channels derived from combinations of the five spectral channels of the AVHRR. Ice concentration was obtained from the visible channel data by assuming a linear relation between pixel intensity value and amount of ice cover within a pixel area. Binary realisations of the ice concentration result then serve as the basis of the floe (polynya size algorithm), which applies a series of different sized masks to obtain size frequency distributions from the imagery. These algorithms were applied to AVHRR imagery obtained in June 1984 during the MIZEX (Marginal Ice Zone EXperiment) 1984 field experiment which allowed validation of the results with high-resolution aircraft and surface measurements of ice conditions. The extensive use of AVHRR images for studying mesoscale eddies in the Fram Straight in the MIZEX 1983 and 1984 campaigns has been described by Iohannessen *et al.* (1987a,b) and in the Norwegian coastal current by Ikeda *et al.* (1989) and Iohannessen *et al.* (1989b).

The use of Landsat MSS data in conjunction with AVHRR data for determining sea ice concentrations in the Barents Sea has been described by Kloster (1991). Some work has been done with SMMR data in conjunction with AVHRR data (Maslanik *et al.* 1989, Dey and Feldman 1990) using AVHRR thermal infrared data (in cloud-free situations) to test the validity of ice charts derived from SMMR data. These ice charts have a 50 km × 50 km resolution. For the purpose of these tests winter polynyas and fractures recorded on cloud-free thermal infrared AVHRR images, with a 1.1 km × 1.1 km resolution, were analysed and compared to the same polynyas and fractures recorded on the SMMR-derived ice concentration charts, processed for the same time and over the same area. It was concluded that, although there is a loss of detail, due to the coarser resolution of SMMR data, in all three cases, the ice concentration charts provided valuable information on openings in the ice. The SMMR information was found to be consistent with the location, orientation, and size of polynyas and fractures as recorded on the AVHRR images. Maslanik *et al.* (1989) merged five-channel AVHRR data, four channels of SMMR brightness temperatures and SMMR-derived ice concentration and ice type and co-registered the data to a polar stereographic grid. The merged data sets were used in combination with meteorological information for integrated studies of clouds and sea ice.

The use of neural networks for the classification of a merged AVHRR and SMMR data set has been described by Key *et al.* (1989). They used four surface classes, namely land (= snow-free land), snow (=snow-covered land or ice cap), water (= open water) and ice (=sea ice) and eight cloud classes. The results were compared with the results of manual interpretations and with those obtained from a supervised maximum likelihood classification procedure and some aspects of the advantages and disadvantages of using neural networks were discussed.

Gradually, over the years, the level of sophistication used in classifying ice has been increased. From a single class of ice as we have just described, attempts have been made to distinguish between several different types of ice/snow. Thus for instance Scharfen *et al.* (1987) defined four types of ice based on *in situ* and aerial observations and geographical designation schemes in the Arctic Basin, see Table 7.2, using DMSP data (which have many similarities with AVHRR data). Morass-utti (1992) introduced a 26-class scheme and used a cubic polynomial to fit the observed reflectance curves of the classes, namely

$$R(\lambda) = a + b\lambda + c\lambda^2 + d\lambda^3 \qquad (7.1.1)$$

$\lambda_{min} \leq \lambda \leq \lambda_{max}$, where $R(\lambda)$ is the reflectance at wavelength λ, a, b, c and d are empirically derived coefficients and λ_{min} and λ_{max} are the minimum and maximum wavelengths of the satellite channel (respectively 0.4 and 1.1 μm). Since most of the

Table 7.2 Surface classes and associated percentage areas of surface types defined by Scharfen *et al.* (1987) (Morassutti 1992)

Class	Description
1	Fresh snow cover of 95 per cent of ice
2	Snow covers between 50–95 per cent of surface with remainder being bare or ponded ice (spring – initial stage of snow melt)
3	(a) 10–50 per cent of ice surface is snow covered with numerous melt ponds (b) Following pond drainage, predominantly bare ice with snow patches and scattered ponds (only for August) (final stage of active snow melt)
4	Heavily ponded or flooded ice with less than about 10 per cent snow cover or bare exposed ice

reflectance curves exhibited 'noisy' trends, employment of polynomial fits of the third degree was deemed suitable. Observed spectral reflectance curve data were obtained from detailed field measurements and corrected for cloud effects. Equation (7.1.1) was fitted to each of the curves therein for the spectral range 0.4–1.1 μm. Parameter estimates and correlation coefficients of these generalised 'smooth' functions are given in Table 7.3. They serve as 'look-up' spectral reflectance functions for *in situ* or aerial observational studies. The idea was to calculate the broad-band (DMSP) reflectances in the visible and near-infrared channels for each of the 26 ice types; the result is given in Table 7.4.

We have seen that the visible wavelength data from the AVHRR can be used, either for cloud-free images or with a cloud-detection algorithm, to determine the area of the ocean covered by ice. However, while it is useful to determine the area of the ocean that is covered by ice, for navigation purposes one also needs to be able to obtain information about the thickness of the ice as well. If the ice is very thin, a ship can pass through it without the aid of an ice-breaker, at greater thicknesses it will need the assistance of an ice-breaker and beyond a certain thickness even an ice-breaker will be of no use.

There are different methods for distinguishing between first-year and multi-year ice. The most common methods relate to microwave signature, i.e. the emissivity as a function of frequency, or to correlations between various channels of the instrument. For the SMMR data the multi-year ice concentration can be determined by applying a standard algorithm from Svendsen *et al.* (1983). However, multi-year ice concentration can also be determined from surface temperature observations using the thermal infrared channel data from the AVHRR (Rees and Squire 1989, Key and Haefliger 1992, Christensen *et al.* 1994). All that can be measured directly by the AVHRR is the surface temperature of the ice. However, this temperature depends on the heat flow through the ice and on the energy balance at the ice surface (Cox and Weeks 1988). Simple physical arguments confirm that the surface temperature theoretically becomes lower the thicker the ice, given steady winter conditions. This means that differences in observed radiance in the thermal infrared frequency range can potentially be used to deduce ice thicknesses. Even with limited accuracy, the technique could potentially be used as a basis for distinguishing between first-year and multi-year sea ice. Indeed LeSchack (1981) exploited the same idea to deduce under-ice roughness from surface temperature observations based on thermal infrared radiometry. LeSchack's conclusions were based on studies of an area extending from the Fram Strait along the northern coast of Greenland to approximately

Table 7.3 Parameter estimates and correlation coefficients (cc) of cubic polynomials of spectral reflectance data from Grenfell and Maykut (1977) and Grenfell and Perovich (1984) for differing sea ice conditions. Coefficients are for the spectral range 0.4–1.1 μm. (Morassutti 1992)

Class/surface condition	a	b	c	d	cc
1 Bare sea ice	−0.202	4.236	−6.485	2.753	0.995
2 Blue ice	−1.398	9.989	−15.512	6.944	0.988
3 Frozen melt pond	−1.134	8.700	−14.076	6.515	0.970
4 Melt pond (7–10 cm deep)	−0.549	5.348	−9.118	4.343	0.937
5 Melt pond (> 12 cm deep)	−0.328	3.771	−6.498	3.131	0.948
6 Scraped milky ice	−0.441	5.948	−8.759	3.586	0.983
7 Refrozen melt pond – underlying water drained	−0.087	3.628	−5.911	2.551	0.984
8 Dry wind-packed snow	0.356	2.036	−2.088	0.404	0.983
9 Melting snow	0.747	0.073	0.766	−0.960	0.999
10 Dry packed snow over cold ice	0.218	2.810	−3.429	1.111	0.944
11 Snow covered first-year melting ice	0.448	1.567	−2.226	0.833	0.986
12 First-year ice with dusting of snow	−0.074	3.644	−5.090	1.995	0.986
13 First-year ice with crumbly surface layer	0.165	2.316	−3.359	1.313	0.989
14 Melting first-year ice with crumbly surface layer	−0.171	3.641	−4.615	1.540	0.973
15 Frozen multi-year white ice	0.172	3.299	−4.961	2.095	0.996
16 Melting multi-year white ice	−0.686	7.360	−11.050	4.751	0.997
17 Melting first-year white ice	−0.608	5.683	−8.171	3.439	0.993
18 Melting first-year blue ice	−0.667	5.958	−9.697	4.454	0.988
19 Dry snow	0.512	1.931	−2.630	0.995	0.995
20 Wet new snow over multi-year ice	−0.797	−0.427	−0.496	0.018	0.999
21 Melting old snow	−0.152	4.424	−6.126	3.388	0.995
22 Partially frozen melt pond	−0.887	8.442	−13.446	6.067	0.994
23 Early season melt pond with white bottom on multi-year ice	−0.137	5.510	−10.951	5.582	0.971
24 Mature melt pond with blue bottom on multi-year ice	−0.054	4.020	−8.140	4.188	0.964
25 Melt pond on first-year ice	−0.522	4.909	−8.441	4.072	0.946
26 Old melt pond on multi-year ice	0.085	1.471	−3.257	1.735	0.946

70° W. Ideally, the existence of a thickness-dependent surface temperature difference should be verified by field measurements. Numerous authors have described sea ice temperature profiles in the open literature, but simultaneous measurements in first-year and multi-year sea ice in the same area are sparse. One published study of interest is that by Blanchet (1988) who combined several reports and found that the representative average ice temperatures during winter were −10°C for first-year sea ice and −12°C for multi-year sea ice. The corresponding summer values were 1.5°C and −3°C, respectively. During steady winter conditions, the temperature profiles are largely linear with the ice bottom at the freezing point temperature. The 2°C difference in average temperatures implies a difference in surface temperature of the order of 4°C and such a surface temperature difference can be identified in AVHRR thermal infrared data.

The use of surface temperature and its correlation with ice type has been described by Steffen and Lewis (1988) with some work for Baffin Bay. The northern

Table 7.4 Spectrally averaged reflectances and component reflectances for sea ice. The class numbers refer to the surfaces in Table 7.3. (Morassutti 1992)

Class	Spect. avg. alb.	$R(<\lambda^*)$ visible	$R(>\lambda^*)$ NIR
1	0.520	0.568	0.312
2	0.420	0.433	0.081
3	0.320	0.328	0.050
4	0.230	0.240	0.054
5	0.210	0.219	0.047
6	0.680	0.744	0.362
7	0.400	0.438	0.174
8	0.810	0.854	0.673
9	0.800	0.852	0.622
10	0.790	0.832	0.660
11	0.750	0.784	0.649
12	0.640	0.684	0.486
13	0.590	0.629	0.455
14	0.600	0.645	0.390

region of Baffin Bay (74–79° N and 70–78° W) is known as the North Water Polynya – a region of inhomogeneous ice cover during the winter and spring months. During the winters of 1978–1979 and 1980–1981 a series of low-level aircraft flights were conducted over the North Water covering a length of 2300 km. Surface thermal infrared temperatures were measured and visual surface ice characteristics photographed with the aid of a searchlight attached to the aircraft. Correction procedures were applied to the data which produced a temperature accuracy of $\pm 0.15°C$. A matching was made between visual ice types and ice surface temperatures along with ice properties using the aircraft and ice/meteorological data collected at a base station near Resolute Bay, Northwest Territories, Canada. Also, surface temperatures outside the spatial extent of the aircraft measurements were extrapolated by means of NOAA thermal infrared data. From the two sets of data, regional surface temperature maps were constructed. Results indicated that greywhite ice for thicknesses of 0.15–0.3 m was the dominant type of ice especially during January and February with a marked increase of white ice percentage towards the end of the ice season. Ice-free areas or warm water actually constituted only a very small percentage of the total area for both winter seasons.

For summer conditions, the prospects for distinguishing first-year and multi-year sea ice in thermal infrared data are not as encouraging as for the winter. While Blanchet (1988) still found a 1.5°C difference in average temperatures, the profiles are no longer linear, and both first-year and multi-year sea ice surface temperatures are likely to tend toward the melting point. The net heat flux will change direction and be vertically downward for a period of about four months during the summer. Indeed, many multi-year ice floes are observed to have melt ponds on the surface during the summer. Blanchet found representative sea ice surface temperatures during summer of $-0.5°C$ for first-year sea ice and $-1°C$ for multi-year sea ice, a difference of a mere 0.5°C.

Summer warming is not the only process that may obscure surface temperature differences. A heavy snowfall will at any season tend to create a homogenous surface temperature field. After some time the differences will reappear, as a steady-state heat flow balance is reached. Consequently, for the mapping of multi-year sea ice concentrations, only images preceded by several days of steady conditions without precipitation should be used.

Christensen *et al.* (1994) performed a classification to distinguish between first-year ice and multi-year ice, using AVHRR thermal infrared data for an area off the northeast coast of Greenland. A serious problem encountered was that the observed surface has three components, namely multi-year sea ice, first-year sea ice and open water. To simplify the discussion and analysis, open water was disregarded. This is reasonable for winter pack-ice conditions in the study area, except for polynyas, of which a few are reported to form every year along the coast of northeast Greenland. Considering only first-year and multi-year sea ice, the observed surface was thus considered as a two-component mixture. Each of the two components was assumed to represent a distinct thickness category. Pixel values below approximately 140 were found to correspond to 100 per cent first-year ice and pixel values above approximately 190 were found to correspond to 100 per cent multi-year ice (higher pixel values correspond to lower (brightness) temperatures). If one assumes one uniform thickness for first-year ice and a second uniform thickness for multi-year ice then one can use a linear relation for intermediate pixel values to determine the relative areas of first-year ice and multi-year ice in mixed pixels (still assuming a total ice concentration of 100 per cent, i.e. no open water).

Natural sea ice, however, contains a continuous spectrum of ice thicknesses and therefore it possesses a continuous spectrum of surface temperatures as well. Consequently, for natural sea ice covers, an asymptotic relation should thus be expected, as shown by the dotted curve in Figure 7.9. However, it is not possible to distinguish directly between fraction and thickness effects. Christensen *et al.* (1994) describe the use of SMMR data to calibrate the AVHRR results. The validity of this calibration is limited to periods of the year when the ice (or snow) is dry. In the Greenland Sea and East Greenland Current, this is the case roughly from October to May, although even during this period the signatures change with season, calling for variable tie-points for the SMMR algorithm. Also, the applicability is limited to periods where the surface temperatures of the two ice types differ substantially. This means the winter period, where more exact dates depend on local conditions and

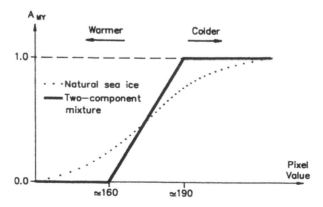

Figure 7.9 Multi-year sea ice concentration (A_{MY}) as a function of pixel values for natural sea ice and for a two-component mixture (Christensen *et al.* 1994).

must be determined from experience. The SMMR/AVHRR method combines the superior accuracy of the microwave frequency range with the superior spatial resolution of the thermal infrared frequency range, but it is only applicable under day-time, cloud-free, steady freezing weather conditions.

7.1.5 Ice movements

AVHRR data can be used in monitoring the movement of individual large icebergs. One example is described in some detail in the AVHRR news section of *Photogrammetric Engineering and Remote Sensing* (Hastings *et al.* 1988a). This involved the iceberg B-9 which broke away from the Ross Ice Shelf in October 1987. The dimensions of B-9 estimated from the AVHRR images were about 155 km in length and 35 km in width. The separation of B-9 from the Ross Ice Shelf and its subsequent movement from late October 1987 until January 1988 is illustrated by three thermal infrared AVHRR images of 27 October 1987, 25 November 1987 and 3 January 1988, see Figures 7.10–7.12, kindly supplied by G. K. Rutledge, NOAA/National Weather Service. An image showing the location of this iceberg on 1 July 1989 is given by Hastings *et al.* (1989c), see Figure 7.13. For such a large iceberg AVHRR data clearly provide the opportunity for operationally monitoring its position.

Time-lapse sequences of AVHRR images can be used to determine ice floe displacement vectors, and therefore ice floe velocities (Ninnis *et al.* 1986). This involves the same basic idea as is involved in cloud velocity determination from time-lapse sequences of geostationary satellite data and in sea surface current velocity determination from tracking features in the AVHRR thermal infrared images (see section 4.3). Collins (1989) described the analysis of a sequence of AVHRR images of the Beaufort Sea from June 1984 (centred at 72° N, 127° W) using an automated sea ice detection method to determine sea ice velocity. Moctezuma *et al.* (1993) described the automatic tracking of ice floes in the East Greenland Sea and in Terra Nova

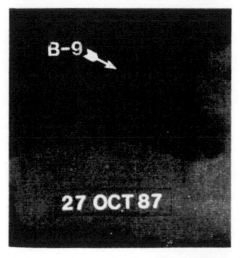

Figure 7.10 AVHRR channel-4 image from 27 October 1987 showing the large iceberg B-9 shortly after its separation from the Ross Ice Shelf (Glenn K. Rutledge in Hastings *et al.* 1988a). (Reproduced with permission of The American Society for Photogrammetry and Remote Sensing).

Figure 7.11   AVHRR image showing iceberg B-9 on 25 November 1987 (Glenn K. Rutledge in Hastings *et al.* 1988a). (Reproduced with permission of The American Society for Photogrammetry and Remote Sensing).

Figure 7.12   AVHRR image showing iceberg B-9 on 3 January 1988 (Glenn K. Rutledge in Hastings *et al.* 1988a). (Reproduced with permission of The American Society for Photogrammetry and Remote Sensing).

Figure 7.13 AVHRR image showing iceberg B-9 on 1 July 1989 (Hasting *et al.* 1989c).

Bay in the Ross Sea using sequences of AVHRR scenes. Dech (1989) describes the use of 148 sets of NOAA-9 and NOAA-10 AVHRR data from May 1988 for analysing sea ice motions in the area of the Fram Strait and the Greenland Sea. AVHRR channel-1 image pairs of subsequent data sets are taken for interactively computing ice floe displacement vectors. From the North Polar Sea, which is ice-covered all year round, the Transpolar Drift Stream delivers huge pack ice masses through the Fram Strait to the Greenland Sea where the drift ice is forced by the cold East Greenland Current with enormous speed (up to 50 cm s^{-1}) southwards. Along the western coast of Spitsbergen, the northernmost branches of tropical water, the West Spitsbergen Current (WSC) merges into the East Greenland Current and causes a significant northerly ice extent beyond 80° N. In the investigation period from 4

May to 25 May, the ice coverage in both working areas decreased continuously by about 20–25 per cent.

On the basis of 43 AVHRR image pairs, each of them representing a nearly 24 hour time shift, a total of about 4000 ice-floe displacement vectors for both investigation areas was calculated by Dech (1989). Depending on the special conditions for each floe (size, rotation and cloud contamination), one or other of the two methods mentioned in section 4.3 was used, namely (statistical) maximum cross-correlation and single-point (or feature) tracking. Using the cross-correlation procedure, the calculation of the correlation matrix can be described in vector notation: at time t_1, the target grid is a $k.l$ vector A. The search grid sections at time t_2 of the lag positions are $k.l$ vectors $B(p, q)$. The grey values of both image sections represent the

(a)

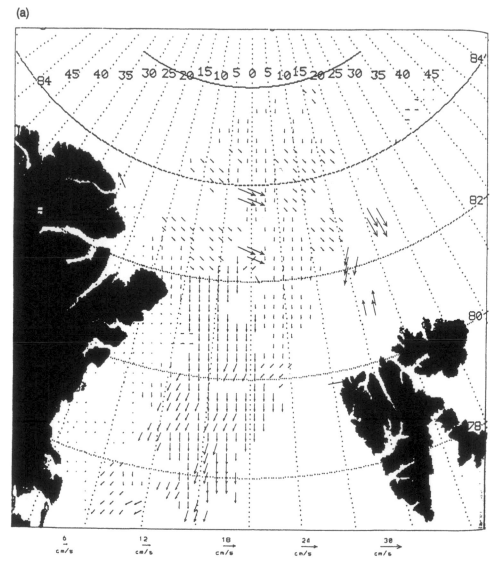

Figure 7.14   Ice velocities determined from a pair of AVHRR images determined (a) by feature tracking and (b) by maximum cross-correlation (MCC) (Emery *et al.* 1991).

vector components. Between vector **A** and each vector **B**(p, q) (p, q are indices for the lag positions) which are both normalised, a correlation measure is calculated. This correlation measure is the cosine of the angle θ between **A** and **B**(p, q). If $\cos \theta = 1$, the vectors point out to the same direction; this case indicates optimal (maximum) correlation. In order to validate the quality of correlation, the correlation matrix can be displayed on a screen. In earlier work a Fourier transform technique had been used, but that has the disadvantage that all the pixels have to be used and so one cannot apply a cloud mask to the data.

This maximum cross-correlation method was predominantly used if floe rotation did not exist. Otherwise, single-point tracking was used in cases of high rotation and highly structured cloud contamination. A useful way is to track the centre displacements of each floe. The vector data sets obtained allow the derivation of drift direc-

(b)

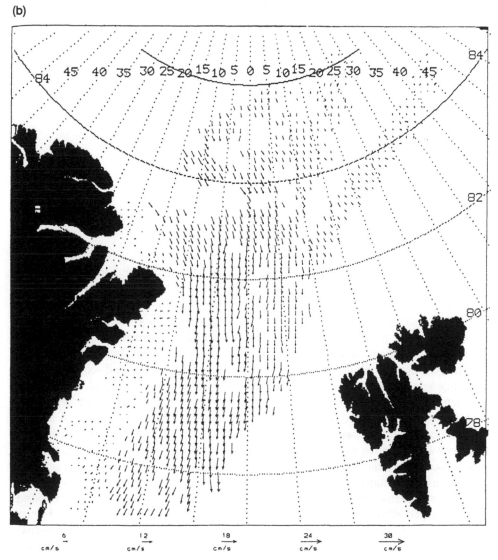

Figure 7.14 (Continued)

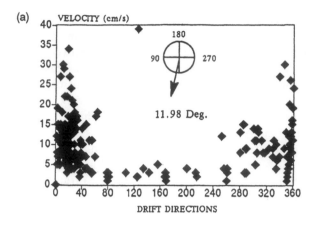

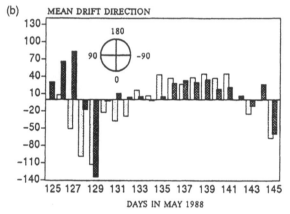

Figure 7.15   (a) Ice drift velocities and directions from 16 to 17 May 1988 and (b) mean ice drift directions from 4 May to 25 May 1988 (Dech 1989).

tions and velocities as well as of areas with ice divergence and ice convergence. Figure 7.14 shows some results from similar work by Emery *et al.* (1991) for the same area using six sets of AVHRR data from April 1986. Figure 7.14(a) shows ice velocities determined from the (subjective) feature tracking method and Figure 7.14(b) using the maximum cross-correlation method with a spatial filtering of the inferred ice motion vectors. Emery *et al.* (1991) analysed both near-infrared (channel-2) and thermal infrared (channel-4) AVHRR data and compared the resulting sets of velocity vectors obtained; good agreement was found between the two sets of vectors. The reason for the spatial filtering is to eliminate erroneous vectors arising from cloud contamination and other non-advective causes. A similar set of velocity vectors was determined by Dech (1989) for the May 1988 data. Furthermore Dech carried out a statistical analysis of the ice velocity vectors. Generally it was found that the actual drift velocities increase with decreasing ice thickness and decreasing distance to the marginal ice zone, whereas the mean drift velocities rise with decreasing sea ice concentration. Deviations from the mean drift are found to be enlarged with decreasing velocities, and especially at the ice edge, see Figure 7.15(a). The mean drift direction in the observation period is SSW which approximately agrees with the mean direction of the East Greenland Current. However, wind remains the dominant component. This is clearly recognised by the change of the drift direction to NE between 6 May and 9 May (Figure 7.15(b)). If the wind blows from northerly directions, i.e. in the East Greenland Current direction, drift speeds rise distinctly (to more than 40 cm s^{-1}).

7.1.6 Polar (Arctic and Antarctic) meteorology

The AVHRR is a particularly important source of meteorological data in polar regions for two reasons (i) because more conventional meteorological observations are very sparse in these areas and (ii) because the geostationary satellites do not gather images from polar regions. Although in the early days full resolution AVHRR data for the Arctic and Antarctic regions was only available from the tape-recorded LAC data, we now have coverage of Arctic regions by various land-based direct readout stations (in Alaska, Canada, Norway, etc.) while a number of HRPT stations have now been established at the various Antarctic research bases (see the Appendix). In the past the amount of LAC data recorded for polar regions was often quite small because of competition for tape-recording facilities by demands for data from other geographical areas. With the establishment of HRPT direct readout stations in Antarctica this has now changed. Although these stations can provide coverage of the whole continent, in practice data are not taken from all passes and only a selection of the data are retained for archiving. Archive facilities for the 1 km data outside the Antarctic are improving and some data are becoming available to the research community. For example, it is only recently that the data have been recorded at the American bases for research use, but an archive has been established at the Scripps Institute in California from which imagery can be obtained.

Turner (1989) presented a review of the applications of AVHRR data for operations and for research in the Antarctic. At present, meteorology and climatology are currently the major applications of AVHRR data in Antarctic research. Operationally the imagery is used as an aid in short-period forecasting since numerical models of the atmosphere give poor forecasts around the Antarctic. Generally the interpretation of Antarctic imagery is much more difficult than at lower latitudes owing to the similarity of the temperatures and albedos of the clouds and ice and snow covered surfaces. During the period when there is solar illumination, interpretation is easier as the infrared and visible channels can all be used. Thus textural information from the visible channels can be used together with the temperature field. However, when only the infrared channels are available and the clouds and surface have a similar temperature then discrimination of features can be very difficult indeed. During the middle of winter the imagery often appears inverted compared to mid-latitude data. This is because the cold air flows to the lower areas where it cools the surface, leaving the high ground with warmer surface temperatures.

Current Antarctic meteorological and climatological research is concerned with two broad areas of study: investigations of Antarctic meteorological phenomena and the assembly of data sets for use in climate studies. In the investigation of weather systems, AVHRR imagery has provided new insight into the evolution and structure of many phenomena on a range of scales from the synoptic to the mesoscale. This has been aided by the large number of NOAA satellite passes received at high latitudes. At the Antarctic coastal bases 10 or more passes per day can be received giving long time series of imagery. On the mesoscale, topographic drainage flow of air from the high Antarctic plateau has been detected on infrared imagery (Swithinbank 1973, D'Aguanno 1986, Bromwich 1992). These katabatic winds can be observed because of the adiabatic heating of the air as it quickly descends down steep valleys with subsequent warming of the underlying ice. The result is dark tongues at the base of valleys which can be observed regularly on the imagery. AVHRR imagery has also proved important in allowing the identification of developing weather systems in the very active coastal region. Here short-lived, mesoscale vortices are often observed to develop in the baroclinic zone at the boundary of the cold continental air and the relatively warm maritime air (Turner and Row 1989). Many of these vortices only last a day or two and are never detected by the synoptic scale observing network or represented in the operational numerical analysis systems. However, some do grow to be major mesoscale disturbances which can have a significant effect on Antarctic operations. A number of these vortices appear to be similar to northern hemispheric polar lows and intercomparisons of mesoscale systems from both hemispheres are under way (Turner and Warren 1988). When AVHRR imagery is used in conjunction with high-resolution sounder data it is possible to gain further insight into the mechanisms that are important in the dynamics of the Antarctic atmosphere through the study of combined cloud distribution and thermal data.

The availability of digital AVHRR data has allowed the production of a number of new meteorological products which are currently used in a research environment, but which could have applications in operational forecasting. One example is the determination of wind velocities from time-lapse sequences of images of clouds. For other geographical areas this is now routinely done with data from geostationary satellites. However, the geostationary data are not available for polar regions because they are out of sight from the satellites. Moreover, the frequency of acquisition of AVHRR data is much higher in polar regions than in equatorial regions. For these two reasons it is therefore worthwhile considering using time-lapse sequences of AVHRR images in polar regions for wind velocity determination. Turner and Warren (1989) described a technique for computing high-latitude wind velocities by following cloud tracers in sequences of remapped AVHRR images. Although polar clouds are more featureless in infrared satellite imagery than clouds at lower latitudes, owing to the limited convection over the ice, it was possible to find suitable tracers and apply many of the tracking techniques developed for use with geostationary satellite data and the method appears to be promising.

Kikuchi et al. (1992) have described some work by Japanese Antarctic Research expeditions in the Mizuho Plateau, East Antarctica. Data have been collected on the annual mean temperatures, which are inferred from 10 m depth snow temperatures, and on the prevailing wind directions, which are inferred from the snow surface reliefs. These two parameters, wind and temperature, are the most important factors in describing the weather and climate on an Antarctic plateau. The data from ground-based observations have been combined with remotely sensed data both from satellites and aircraft. Annually averaged NOAA channel-5 brightness temperatures were found to correlate well with the 10 m snow temperatures. The synthesised prevailing wind field shows a clear distinction between the katabatic wind system and that of the synoptic disturbances. Confluence zones were also identified but their presence seemed to be only intermittent. A slight but significant climate difference has been observed between ridges and troughs of the plateau due to the difference in katabatic wind forces.

A recent development has been in the use of AVHRR imagery in arctic marine mammal research. Remote sensing has traditionally been of only limited use in Arctic marine mammal research. This is primarily because (i) the summers are short and often cloud covered, while the winters are long and solar illumination is limited to a few hours of twilight, and (ii) orbital sensors operating on a repeat cycle of 16 days or so (i.e. Landsat and SPOT) simply do not offer the biologist a sufficiently reliable tool for habitat assessment studies. Compounding the data availability problem was the substantial cost involved in the acquisition and interpretation of the high-resolution digital remote sensing imagery. AVHRR data avoid many of these problems. Barber and Richard (1992) described the work which is being done with AVHRR data in support of marine mammal aerial survey programmes by the Marine Mammal Management section of the Department of Fisheries and Oceans, Central and Arctic Region of the Canadian Government. The Department of Fisheries and Oceans Canada, is actively involved in the management of marine mammals in the Canadian Arctic. The Government of Canada spends a considerable amount of money annually in an attempt to assess and monitor population abundance, and to obtain information necessary for the management of the various marine mammal species. The large expanse of the Arctic and the adverse climatic conditions make the task of data collection a considerable challenge. AVHRR imagery has proved to be useful in helping to develop a better understanding of marine mammal preferred habitat and as a source of *a priori* knowledge for the design of aerial surveys.

The primary focus of the Department of Fisheries and Oceans' remote sensing programme is to assess the population abundance, distribution and habitat preference during the spring and summer season of four species of Arctic marine mammals (see Figure 7.16) the narwhal (*Monodon monoceros*), beluga (*Delphinapterus leucas*), walrus (*Odobenus rosmarus*) and ringed seals (*Phoca hispida*). The narwhal is a deep-water species that inhabits the eastern Canadian Arctic. Narwhal summer in the fiords and inlets south of Lancaster Sound and winter in Davis Strait. The beluga is a coastal species preferring the warm shallow water of river estuaries in the summer. Eastern Canadian Arctic populations winter in Hudson Strait, Davis Strait and Baffin Bay. Western populations winter in the Chukchi Sea. Walrus occupy a variety of locations throughout the central and eastern Arctic. These animals remain in the Arctic throughout the year. In the summer they haul-out at traditional land sites (the Inuit name is Ooglit) and on sea ice pans. Ringed seals are ubiquitous throughout the Canadian Arctic in both the summer and winter seasons. Pressure on these marine mammal populations, both by Inuit subsistence hunting and by the potential threat of industrial activity, provided the stimulus for this research.

The following research questions are similar for each species.

- What is the population size in each management area?
- What is the distribution of the species within and between management areas?
- What is the relation between the animals and their habitat?

To address these questions AVHRR imagery is used together with data from an airborne remote sensing system. The airborne platform, primarily used to collect information on abundance, distribution and habitat, was developed over a three-year period and consists of three primary sensors and a fully integrated computer

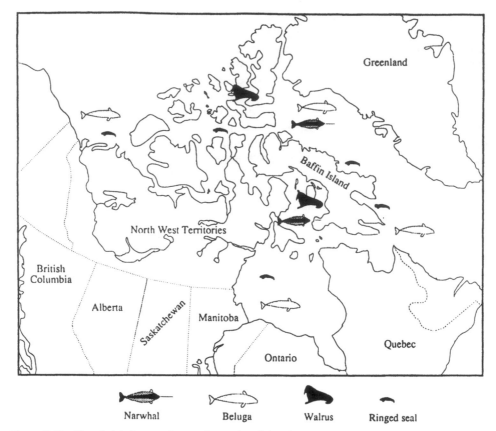

Figure 7.16   Narwhal, beluga, walrus and ringed seal distribution in the Canadian Arctic (Barber and Richard 1992).

annotation system mounted in a DeHavilland Twin Otter. A large format air survey camera (RC8) is mounted in the belly of the aircraft. This metric camera is used for population sampling (photo-counts) and photogrammetric measurements such as distances between members in a pod, length frequencies of particular animals, estimates of animal travel speeds in adjacent frames, etc. The other two instruments are mounted in a specially designed nose cone of the Twin Otter; these are a forward-looking infrared imager which provides surface temperature measurements, and a high-resolution video camera which provides a panchromatic image of the ocean surface, forward of the aircraft. Both these sensors record on to standard half-inch video tapes which are useful for qualitative interpretation or which can later be digitised, using a video frame grabber, for quantitative analysis. The data from the forward-looking sensors contribute both as unique high-resolution habitat data and as a means of checking the AVHRR habitat information. Navigation information, flight attitude, altitude, and data specific to each sensor are collected and stored in a portable on-board microcomputer.

The data collected with the system is the only practical method of obtaining specific information on population abundance and distribution of these marine mammals. The cost of the airborne platform makes this approach an expensive necessity. To maximise information extraction, inexpensive AVHRR satellite imagery is used. The combination of aerial and orbital data complement each other. The line transects provide high-resolution point-specific information; the AVHRR imagery provides more synoptic, spatially complete data. The AVHRR imagery is used for survey preparation and analysis of three habitat parameters, sea ice concentration (location and movement), sea surface temperature (specifically oceanographic upwellings and fronts) and sediment features (concentration and distribution). The paper by Barber and Richard (1992) describes in some detail the use of the AVHRR data in relation to the beluga and the study of their habitat in terms of these three parameters. The procedures are illustrated from the area of the Nelson River Estuary. It was concluded (Barber and Richard 1992) that the strength of AVHRR imagery in Arctic marine mammal research is a direct result of the timely coverage and relatively inexpensive cost. Digital image analysis systems and geographical information systems have recently made it possible to observe macro-scale features of marine mammal distributions which were previously undetected, due to limitations in our ability to sample adequately their entire distribution. It is the long-term commitment of NOAA to maintain the polar platforms and the improvements in image analysis systems and geographical systems which now allow this form of marine mammal remote sensing to take an active place among the existing tools of Arctic marine mammal researchers.

7.2 WATER BODY BOUNDARY MAPPING; BATHYMETRY

Data from the high-spatial-resolution Earth-observing satellite systems are widely used for cartographic purposes to cover areas of the world for which accurate maps have not already been made by more conventional surveying methods. For large areas in some developing countries the best map that is available is a satellite image. Mapping of such areas includes the use of satellite data to delineate water boundaries, for lakes, rivers, estuaries and coastal areas, including the possibility of flood monitoring. This is best done with image data in the near-infrared part of the electromagnetic spectrum; this is because the finite signal from the land is easy to distinguish from the near-zero signal which is received from the water since the reflectivity of water at near-infrared wavelengths is virtually zero. It is also possible to obtain bathymetric information from multispectral satellite data at visible and near-infrared wavelengths. This can be done in two ways. First, it can be done by delineating the water boundary, using the near-infrared data, in an inter-tidal zone or in a lake where the water level changes with time. Secondly, it can be done for shallow water by using the intensity of the reflected signal in the visible bands from the surface of the shallow water. In very many situations these methods are not appropriate for AVHRR data because (i) the spatial resolution of the AVHRR is too poor for this purpose and (ii) for reasonable success the method requires more spectral channels in the visible wavelength range than are possessed by the AVHRR. There are, however, a few exceptions.

7.2.1 Lakes

In principle the channel-2 (near-infrared) data from the AVHRR can be used to map the boundaries of water bodies. In practice, however, the poor spatial resolution makes this impossible for a very large number of water bodies, such as small lakes or narrow rivers. The use of AVHRR data to measure the area of a large lake was demonstrated by Harris and Mason (1989) for Lough Neagh in Northern Ireland which has an area of 390 km². This is of no practical importance for Lough Neagh because the area of Lough Neagh is (a) constant (at least to a much smaller variation than could ever be determined with AVHRR data) and (b) accurately known from conventional survey data. However, there are some large lakes whose areas vary greatly (often to total dryness) and which are in isolated locations so that multitemporal AVHRR can be very useful in studying the area of the lake and, hence, the volume of water in the lake.

One example of a lake which has a changing water level and an accompanying changing surface area and which is also large enough to be studied with AVHRR data, is Lake Chad. The use of AVHRR data to study this lake has been described by Schneider *et al.* (1985); they were concerned with monitoring the lake boundaries, vegetation growth and surface temperature patterns using AVHRR data over the period from November 1981 until November 1982. Initial pilot studies of terrestrial vegetation using AVHRR data from NOAA-6 and NOAA-7 were carried out for study areas in Senegal, Egypt and Algeria by scientists from the United Nations Food and Agriculture Organisation (FAO), the US National Aeronautics and Space Administration (NASA) and the US National Oceanic and Atmosphere Administration (NOAA). This work was extended in 1982 to include several new areas in the Indian subcontinent and Africa, including the Lake Chad basin. The impact of the 1970s drought on Lake Chad, and its location astride the borders of four north central African nations, made the area particularly interesting to study. The waters of the lake are crucial to agriculture in Nigeria, Chad, Niger and the Cameroons. The lake extent and vegetation patterns in the basin are extremely changeable and monitoring would require frequent time-consuming and expensive remapping if done by conventional, i.e. non-satellite, means.

Lake Chad is the largest lake in West Africa and the fourth largest lake on the continent. The lake lies 281 m above sea level in the zone between the desert and the savannah known as the Sahel. During normal periods, which can be several decades long, the lake water covers about 20 000 km² with maximum length and width of 200 km, and depths no more than 5–8 m. Lake Chad drains an immense 1 500 000 km² region. Annual rainfall ranges from 1400 mm in the southwest portion of the basin to less than 200 mm in the north. The rainfall is monsoonal, and more than 90% of the annual total falls in the four months from June to September. The level of Lake Chad has suffered an acute long-term fall from a severe drought that has affected the Sahel–Sudan zone in Africa since the mid-1960s. Between 1964 and 1974 the surface area of the lake was diminished by a factor of 3 and its volume by a factor of 4. In 1972 the flow of the Chari River, which provided in a mean year about 80% of the lake's water supply, reached a 100 year low. In July 1973 the lake reached the lowest level ever gauged. This devastating drought continued without surcease through the 1970s and into the 1980s. Recent measurements of precipitation and lake levels taken at the Bol gauging station on the east shore of the lake are shown, respectively, in Figures 7.17 and 7.18. Examination of fossilised plant spores in the lacustrine sediments shows that this drying of the lake is not unique in recent geological history and probably occurred half a dozen times in the last 100 years.

Landsat images had been used to study changes in lake extent during the 10 year

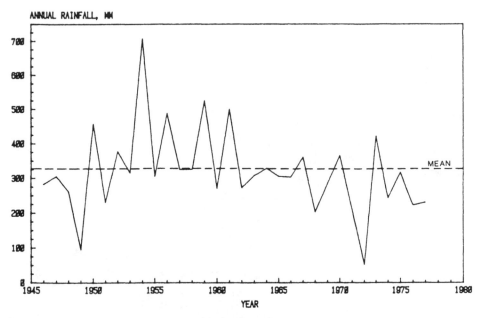

Figure 7.17 Annual precipitation (in mm) at the Bol gauging station on Lake Chad (Gaston 1981).

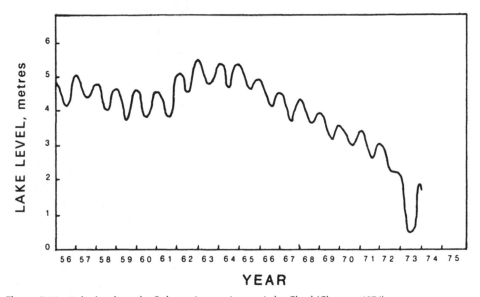

Figure 7.18 Lake levels at the Bol gauging station on Lake Chad (Chouret 1974).

period from 1972 to 1982. These images show that the central portions of the lake dried up between 1972 and 1973, leaving two separate areas of water in the northern and southern basins. Sixteen NOAA-7 AVHRR scenes from 5 November 1981 until 29 November 1982 were studied by Schneider *et al.* (1985).

Figure 7.19 shows the total lake area and the standing water area by using thermal infrared, channel-4 and channel-5, AVHRR data. Each character on the

(a)

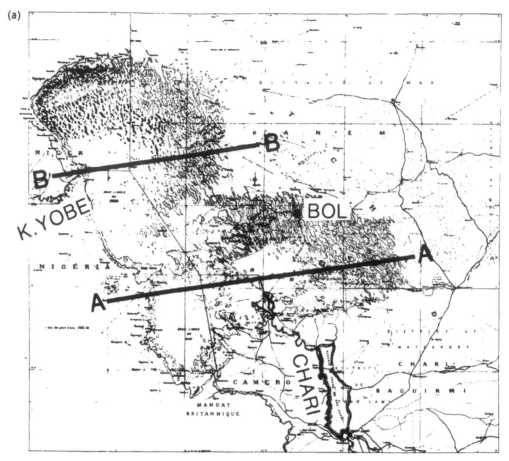

Figure 7.19   (a) Map of Lake Chad and vicinity showing the two profile lines A-A and B-B chosen for detailed study.

map is a temperature value in degrees Celsius averaged over an array of 3 × 4 pixels. Isotherms are printed out in intervals of 1°C. The boundaries of the lake are clearly manifested along its western and southern edges by quite sharp temperature boundaries; they are less clear, but still apparent, on the northern and eastern edges. Outside the lake the temperature is about 40°C everywhere, except in the flood plains of the two rivers, the Chari and the Komdugu Yobe, which feed in to the lake. Within the lake the temperature is lower (mid-20s to mid-30s °C) in the dry but vegetated areas and it is lower still (22–24°C) in the areas of standing water in the southern basin and in the small body of free water in the Komadugu Yobe delta. The temperatures in Figure 7.19(b) were calculated using a multichannel algorithm developed for oceanographic purposes (see section 4.2.3), i.e. an emissivity of 1.0 was assumed. As the emissivities in this area may vary from 0.9 (desert) to 0.98 (plants), the actual ground temperatures, but not the water temperatures, may be somewhat higher than those depicted in Figure 7.19.

   Channel-1 and channel-2 AVHRR data were studied in detail along two transects through the Lake Chad basin, see Figure 7.19(a). The first, which is shown as A-A in Figure 7.19(a), extends 175 km along a W-E axis at 13° 10′ N. It begins in Nigeria

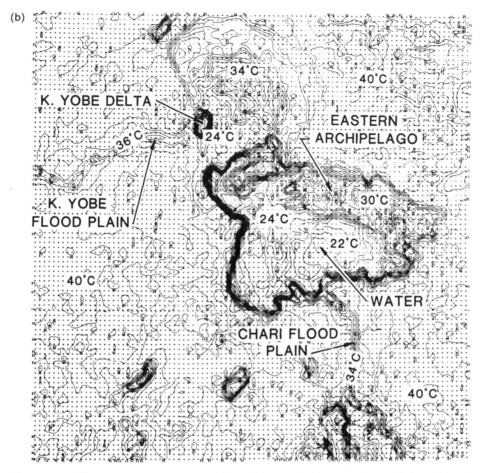

Figure 7.19 (b) Temperature map generated for Lake Chad using AVHRR thermal infrared data for 29 November 1982 (Schneider *et al.* 1985).

and passes through the southern Lake Chad basin, including the basin's only perennial body of free standing water, finally ending in Chad. Channel-1 and channel-2 profiles for this transect are given in Figure 7.20. The horizontal axis on each plot gives distance along the transect; the vertical axis gives albedos (uncorrected for solar zenith angle). In general, channel-2 exceed channel-1 albedos in vegetated areas, the gap between the two increasing directly with vegetation vigour. Channel 1 exceeds channel 2 only over water. The lake–basin boundaries are most distinct in the channel-1 profile, where highly reflective desert abruptly meets the darker vegetated regions. The brightening of the vegetation areas in the near-infrared combined with the less reflective water gives a better land–water interface in channel 2, making it easier to locate the lake boundaries and isolated pools of water. Although there are some variations through this 12 month period, the linear dimension of the lake along this transect remains at about 40 per cent of the full size of the lake along this transect. A similar transect, shown as B-B in Figure 7.19(a), across the northern part of the lake, which had almost completely dried out, was also studied.

A second example of a study of a large lake with AVHRR data is of Lake Eyre in Australia. The location, shape and depth contours of Lake Eyre are shown in Figure

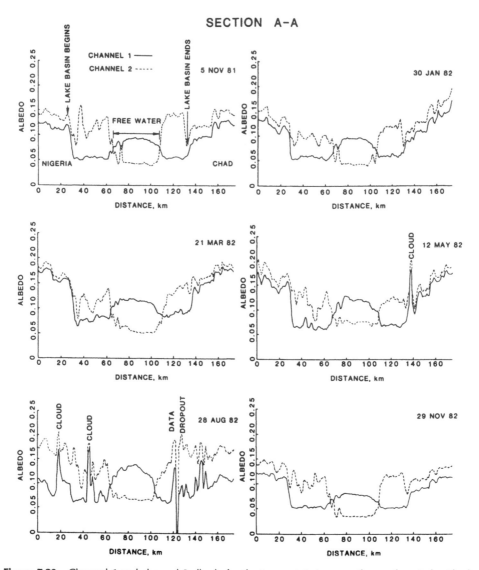

Figure 7.20 Channel-1 and channel-2 albedo for the transect A-A across the southern Lake Chad basin (see Figure 7.19) for six dates during the period from November 1981 to November 1982 (Schneider *et al.* 1985).

7.21. The lake drains an area of over 1 million km². The normal state of the lake is dry and it can remain so for decades until a tropical depression penetrates inland and causes extensive rainfall and filling (flooding) of the lake. It can take from a few weeks to a few months for the lake to fill, depending on the rainfall distribution. The last two major floodings were in 1974 and 1984 but records of floods go back nearly 100 years.

Some very extensive work was done by Prata (1990) on Lake Eyre using AVHRR data, to study the filling and emptying of the lake in 1984–1985. The spectral, spatial and temporal characteristics of the AVHRR are particularly well suited to the study of Lake Eyre because of its shallowness (up to 6 m), its large area (9690 km²) and because the hydrological timescales are long (days to months). The contour data

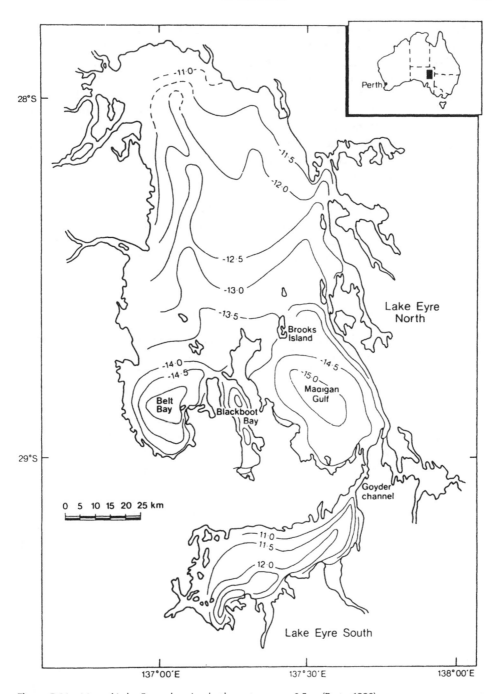

Figure 7.21 Map of Lake Eyre, showing bathymetry every 0.5 m (Prata 1990).

were derived from bathymetric surveys during the 1974 flooding. In the 1984 flooding heavy rain fell between 8 January and 15 January and the lake was observed to be filling by 17 January. It reached its highest levels in early February corresponding to about 30% of its full capacity. Flow from Lake Eyre South to Lake Eyre North occurred through the Goyder Channel and some salt transfer took place.

HRPT data were obtained from a receiving station in Perth, Western Australia; data from 28 passes between 17 January 1984 and 6 September 1985 were used.

The lake was contained within a 256 pixel by 256 pixel area so that it was not necessary to process the whole area from a pass. The steps involved in the processing of the data were (i) calibration of the thermal infrared channels, (ii) conversion of visible and near-infrared data to radiances, (iii) geometrical rectification, (iv) reprojection to transverse Mercator projection and (v) derivation of (a) water depth, (b) areal extent of standing water and (c) water surface temperature. Steps (i)–(iv) are operations which we have already discussed elsewhere, but it is appropriate to consider point (v) here.

(a) *Water depth estimation* As we indicated at the beginning of this section, it is relatively unusual to use AVHRR for bathymetric work; it is much more common to use Landsat or SPOT data. This is because usually one needs more spectral bands and better spatial resolution. But in this case the water is very shallow and the depth varies only very slowly in the horizontal plane. In Prata's paper it is shown that

$$\Delta a_i = \tau_{aw}\, \tau_{wa}\, \tau_a\, \rho(\lambda)\, \exp\{-\kappa(\lambda)[\sec(\beta_{Sun}) + \sec(\beta_{sat})]z\}\, \cos(\beta_{Sun}) \tag{7.2.1}$$

where Δa_i is the spectral albedo difference between the shallow water pixel and a deep water reference pixel, τ_{aw} is the air–water transmittance, τ_{wa} is the water–air transmittance, τ_a is the atmospheric transmittance, $\rho(\lambda)$ is the bottom reflectivity, $\kappa(\lambda)$ is the absorption, z is the depth, and the angles are indicated in Figure 7.22. A plot of intensity (D^*) versus depth is shown in Figure 7.23. The value of the absorption coefficient $\kappa(\lambda)$ derived from the best-fit straight line is 0.51 m^{-1}.

The results of the bathymetric studies using the AVHRR data were generally in agreement with the contours shown previously in Figure 7.21, with the exception that the satellite data for 1984 shows a shallower area on the northeastern side of

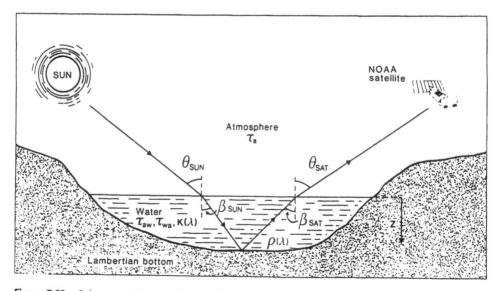

Figure 7.22 Schematic diagram showing the geometry used for the radiative transfer model (Prata 1990).

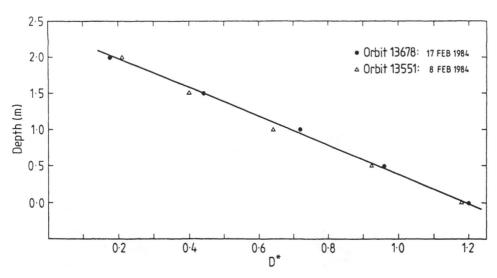

Figure 7.23   Plot of D^* versus depth, z, for two satellite orbits (Prata 1990).

the Madigan Gulf which arose from the flow of water and sand and silt from Lake Eyre South which we have already mentioned.

(b)   *Water surface area estimation*   This is an important step in the process of estimating the rate of evaporation. The water boundaries can be determined very easily using the channel-2, near-infrared, data and the results for some of the dates are shown in Figure 7.24. In fact Prata used the quantity

$$Q = \frac{\text{channel-2}}{\text{channel-1}} \tag{7.2.2}$$

rather than just the channel-2 data on their own to delineate the boundary of the water. The argument used for this is that the absorption by vegetation in channel 1 is stronger than in channel 2 and therefore channel-2/channel-1 is a better water/land discriminator than channel-2 data alone.

For the evaporation rate calculation one assumes that there is (a) no inflow (rainfall, river flow or surface runoff) and (b) no groundwater exchange, so that the only losses are due to evaporation. One then also assumes that there is a known relation between depth and surface area; actually there are two separate relations for the North lake and the South lake, namely

$$\text{South:}\quad h_S(t) = a_0 + \sum_{i=1}^{4} a_i(A_S(t))^i \tag{7.2.3}$$

$$\text{North:}\quad h_N(t) = b_0 + \sum_{i=1}^{4} b_i(A_N(t))^i \tag{7.2.4}$$

where $h_S(t)$ and $h_N(t)$ are the depths and $A_S(t)$ and $A_N(t)$ are the surface areas.

The coefficients were found by a least squares fit, see Figure 7.25. From the area and the change in depth, the volume of water lost by evaporation in a given interval, and therefore the evaporation rate, can be calculated.

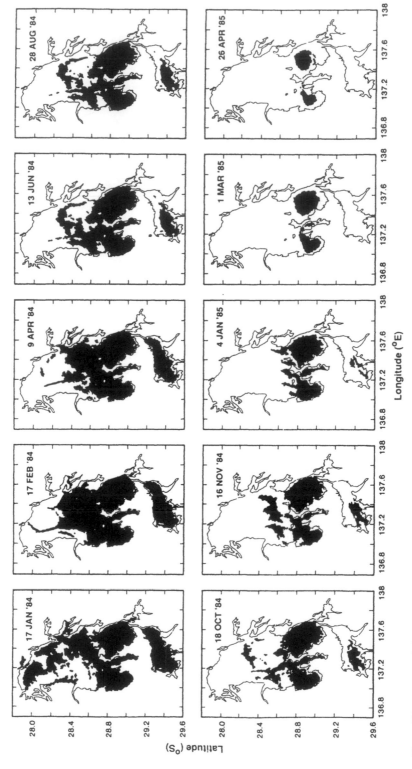

Figure 7.24 Maps of the Lake Eyre basin showing the area of standing water on various dates (Prata 1990).

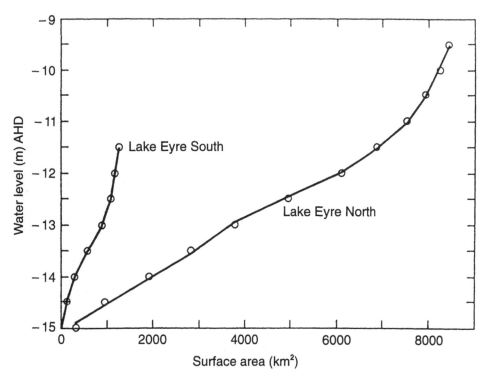

Figure 7.25 Water level against area in Lake Eyre (Prata 1990).

(c) *Temperature* The water temperature can be calculated, as we have already seen in section 4.2.2. This can be used as an alternative parameter in the calculation of the area of the lake, at least using night-time winter data based on the water remaining warmer than the land at night because of the high thermal inertia of the water.

One can appreciate this work on Lake Eyre as being an extensive and careful study carried out using AVHRR data to provide a knowledge of the behaviour of Lake Eyre that could not be obtained in such great detail by more conventional methods. However, one might be tempted to ask what was the practical value of such information about an arid or semi-arid area with sparse vegetation, animal life and human population. One important aspect of the answer lies in the question of climate research since precipitation and evaporation rates are important parameters in climate studies and we have seen how the work of Prata (1990) enabled the evaporation rate to be determined. There are many lakes which behave like Lake Eyre and have volumes which fluctuate in response to changes in the evaporation and precipitation rates within their catchment basins. On a global scale it would be a major undertaking to carry out similar studies to the Lake Eyre study for all the lakes that undergo large volume changes.

(d) *Volume estimation* Currently measurements of lake volume fluctuations have been performed on relatively few lakes using ground-based measurements of the levels and areas of the lakes. These methods rely on survey data to obtain the relation between lake level and lake area, which is then used in conjunction with one parameter (usually *in situ* values of lake level) to obtain the lake volume change.

It has been pointed out, however, by Mason *et al.* (1985) that remote sensing of levels and areas from satellite altimeters and images offers an alternative method for measuring lake volume change. In particular, closed lakes (those without surface outlets) display considerable volume changes which are relatively easy to model (Street-Perrott and Harrison 1985, Mason *et al.* 1985). In addition there are also variations in the seasonal cycles of some open lakes (those with an outlet).

The water balance of a closed sealed lake is described as follows

$$\frac{dV}{dt} = R - A(E - P) \tag{7.2.5}$$

where V is the lake volume, R is the runoff rate from the catchment basin, E is the evaporation rate over the lake per unit area, P is the precipitation rate per unit area over the lake and A is the area of the lake. Since $(dV/dt) = A\, dL/dt$ the equation can be rewritten in terms of lake level, L,

$$\frac{dL}{dt} = \frac{R}{A} - (E - P) \tag{7.2.6}$$

or, in terms of A,

$$\frac{dA}{dt} = \frac{dA}{dL}\frac{dL}{dt} = \left(\frac{dA}{dL}\right)\left[\frac{R}{A} - (E - P)\right]. \tag{7.2.7}$$

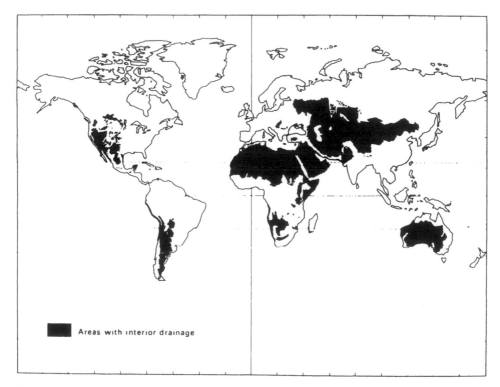

■ Areas with interior drainage

Figure 7.26   The areas of the world where inland drainage can occur, i.e. where closed lakes can exist (Street-Perrott and Harrison 1985).

It is useful to establish, for a given lake, the relation $A(L)$. $A(L)$ is a calibration curve like that illustrated already for Lake Eyre in Figure 7.25. For many lakes ground survey data are not available, but $A(L)$ could be obtained by simultaneous remote sensing of both $A(t)$ and $L(t)$. Satellite altimeter data are available, intermittently from 1978, from Seasat, Geosat and ERS-1, and AVHRR data are available continuously from 1978. Satellite altimetric measurements are now so accurate that it is feasible to think in terms of studying large ephemeral lakes by a combination of altimeter data to give $L(t)$, the height of the water surface, and AVHRR data to give $A(t)$, the surface area. Once the calibration curve $A(L)$ for a lake has been established, then it is not necessary to measure both $A(t)$ and $L(t)$. It will be possible to use the calibration curve to calculate lake volumes from one instrument alone (i.e. the AVHRR or an altimeter). Thus such remotely sensed data should provide a valuable global proxy indicator of climatic change from 1978 onwards. Rapley *et al.* (1987) estimated that, globally, there are about 1500 lakes of area greater than 100 km², of which about 200 are closed. It was to assess the accuracy of the measurement of the area with AVHRR that the work of Harris and Mason (1989) on Lough Neagh was performed.

This alternative approach can supplement the current ground-based methods and can also be tested against them. In particular, remote sensing offers the potential to monitor large numbers of lakes on a global scale. Many of the most climatically sensitive (closed) lakes occur in semi-arid regions for which there is little or no *in situ* data, see Figure 7.26.

7.2.2 Flood monitoring

Flood situations develop and evolve rapidly and to be able to provide useful information to the rescue services it is necessary to have access to the data quickly and to have a rapid time-lapse sequence of images available so as to be able to follow the development of the flood situation. Various attempts have been made to use high-resolution satellite data in this context, but they have often been frustrated by the difficulty of getting rapid access to the data and by the low frequency of capture of the data. To be of any assistance in alleviating the effects of the floods for humans or animals the data must be available and analysed in near-real-time. The AVHRR provides the possibility of having several images of a flooded area every day (subject to cloud cover); moreover, if the data are received at a direct readout ground station in the vicinity of the study area, the AVHRR data should be available quickly. The AVHRR does, as we have noted already in connection with the monitoring of lakes, have rather coarse spatial resolution. Therefore AVHRR data can only be considered for use in connection with very large-scale flooding. Examples which have been studied successfully with AVHRR data include the flooding of the Darling River in central eastern Australia in 1988 (Barton and Bathols 1989), flooding of the Yangtze and Huaihi river basins in China in the spring and early summer of 1991 (Sheng *et al.* 1993) and the floods that occur quite frequently in Bangladesh (Ali *et al.* 1987, 1989, Carey and Pritchard 1989).

For the flood in the Darling River in mid-1988 Barton and Bathols (1989) observed that the thermal channels at night gave better land-flood discrimination than the visible data during the day. Although the spatial resolution of the data is 1 km, useful estimates of flood area were made. The success of the AVHRR in this instance was assisted by the fact of the slow movement of the flood front which is typical of floods in central Australia, where the fall of the rivers from catchment area to the sea is small, and by the fact that there is typically very little cloud cover over the area. In the work of Sheng et al. (1993) the boundaries of the flooded areas were delineated by using the ratio, Q, of the channel-2 and channel-1 intensities (see equation (7.2.2)); this was found to eliminate or minimise the influence of clouds.

Bangladesh is well known for its frequent serious flood situations. Almost every year it is seriously affected by floods during the southwest monsoon season (June–September). People and livestock are killed, properties are lost, agricultural lands are destroyed, crops are ruined and general economic development is impeded. Bangladesh is located in the confluence of the lower alluvial valleys and the delta plain of the main drainage systems of the Himalayas, the Ganges and the Brahmaputra, see Figure 7.27. The Ganges rises in the southern slopes of the Himalayas, while the Brahmaputra rises in Tibet and flows round the eastern Himalayas before entering Bangladesh. These two rivers, while traversing hundreds of kilometres, are fed by numerous tributaries.

Flood waters in Bangladesh have two main sources (a) excessive rainfall, both locally and in surrounding countries, and (b) snow melt in the Himalayas. Ali et al. (1987, 1989) studied AVHRR data from late September 1984 showing summer flood conditions of Bangladesh, while images of the flooding in September 1988 were published by Carey and Pritchard (1989). The work of Ali et al. demonstrated how the imagery could be used for monitoring the progress of flooding and flood-damage assessment. The images at the time of the floods were compared with a low water image from the previous winter, namely from January 1984. In the January image the river channels are narrow and the water is confined to the deepest parts of the channels. In the September images the rivers were seen to be much wider than in January. The turbidity was also studied and the rivers were found to have much higher turbidity during the period of flooding. The widened rivers and the inundation of the flood plain were clearly visible in band 1 and band 2 of the AVHRR data. The areas were also studied in the thermal infrared bands of the AVHRR. In these channels only the main flow of the rivers was apparent; this is presumably because the thermal data reveal the temperature differences between the cooler deeper fast-flowing water of the main channels and the shallow, Sun-warmed and nearly stagnant overbank waters of the flooded areas. The work of Ali et al. should be regarded as a feasibility demonstration; it was carried out on a historical basis using AVHRR data obtained from NOAA/NESDIS after the event. Since then the direct readout station of the Bangladesh Space Research and Remote Sensing Organisation (SPARRSO) in Dhaka has become operational so that it should now be possible to acquire and analyse flood-related AVHRR data for Bangladesh quantitatively in near-real-time.

The question of flood monitoring in Africa using AVHRR data has been addressed by Legg (1989) with particular reference to the floods which occurred in Sudan in 1988. A specific requirement was to assess the extent of flooding in the areas of the country away from major population centres, where access problems prevented ground-checking of rumoured flooding. The high frequency of coverage which is possible with AVHRR can only properly be utilised by having a direct readout HRPT ground station which includes the study area within its range. No such station existed in North Africa at that time. At that time the proposed HRPT stations in Ougadougou and Nairobi were not operational. Most of Sudan is

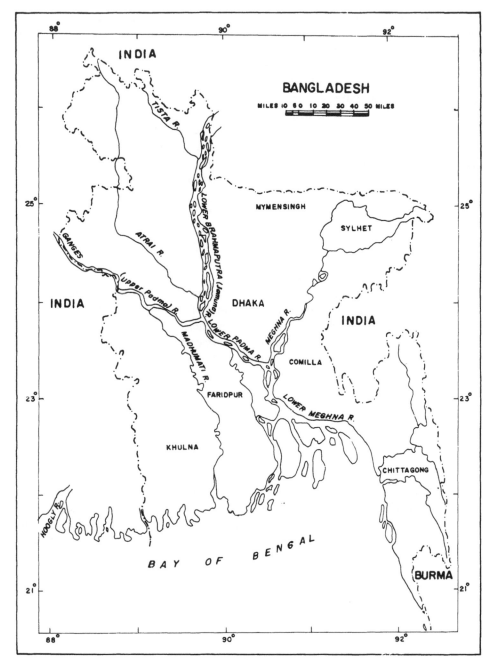

Figure 7.27 Bangladesh, showing major river systems (Ali *et al.* 1989).

covered by HRPT receiving stations in Saudi Arabia and North Yemen, but access to data from those stations could not be obtained, at least not on the near-real-time basis required for that project.

Attempts to obtain near-real-time cloud-free recorded AVHRR data of the study area from NOAA were unsuccessful. Imagery for an overpass in mid-September was largely cloud-covered, although the area immediately south of Khartoum was partially visible. A further three AVHRR images covering Sudan after the flooding emergency (November and December 1988, February 1989) were obtained from NOAA. The only near-real-time data that were actually received for this project with coverage of Sudan were from the geostationary Meteosat satellite. Consequently, the real-time part of this test study described by Legg (1989) was carried out only with Meteosat visible data acquired between 19 and 28 August by RAE Lasham. The very cloudy AVHRR data acquired in mid-September by NOAA, and relatively cloud-free imagery acquired in November and December 1988 and February 1989, were used to assess the potential of this sensor for this type of study, if real-time imagery were to be available from a suitably located receiving station.

This work was rather similar to the work already mentioned in relation to flooding in Bangladesh. It demonstrated that if full-resolution (HRPT) AVHRR data could be made available in near-real-time then they would be very valuable in monitoring the development of large-scale flooding. Operational use of AVHRR in disaster relief is critically dependent on having strategically placed HRPT receiving stations. A modern HRPT receiving station is not inordinately expensive and it is not very large. It would be possible to construct a portable station weighing about 500 kg, which could be transported to critical areas in times of emergency. Such a station could acquire usable data within 12 hours of installation and could provide real-time information to governmental and aid agencies.

Legg (1989) concluded with a set of recommendations as follows. If remote sensing is to be able to respond more rapidly and effectively to emergency situations such as the recent one in Sudan, a series of actions should be initiated:

- An AVHRR mosaic of Africa, which could be used as a reference image for change-detection, should be prepared, funded by United Nations agencies and/or national aid agencies. Sufficient cloud-free data are probably available in the NOAA archives.

- Priority should be given by international agencies to the installation of HRPT receiving stations in strategic locations for purposes of disaster monitoring and routine environmental studies. The existing regional centres in Ougadougou and Nairobi are preferred, since they already have a long-term involvement in remote sensing and operate on an international basis.

- Communication between existing HRPT receiving stations should be improved and mechanisms established for rapid exchange of data in times of emergency. Standardisation of data formats should also be agreed.

- The possibility of commissioning portable HRPT receiving stations for emergency use should be investigated.

- In the longer term, the feasibility of later-generation Meteosat-type geostationary satellites having improved sensors with wavelengths similar to AVHRR, and a resolution of one kilometre at nadir, should be investigated. Such a system could provide real-time data for monitoring natural disasters such as floods, fires, storms and volcanic eruptions, as well as allowing medium-to-long-term monitoring of vegetation change and desertification, at the same time as continuing to provide operational meteorological information.

7.3   WATER QUALITY MAPPING

7.3.1   AVHRR and other systems

The most common and most successful oceanographic applications of AVHRR data are undoubtedly based on the use of the data from the thermal infrared bands for studying sea surface temperatures (see chapter 4). The visible wavelength AVHRR channel (channel 1), on the other hand, is generally considered to be inappropriate for oceanographic applications because of its low radiometric sensitivity. Under most conditions, ocean volume reflectance (light backscattered from the water column and excluding water surface reflectance or sunglint) within the 0.58–0.68 µm range of channel 1 is less than 3 per cent even though the concentration of phytoplankton and other suspended particles may change by several orders of magnitude (Morel and Prieur 1977). Within the AVHRR visible band, these waters will appear quite dark and, apparently, homogeneous. The scanners that are generally used for ocean colour studies have visible bands that are more sensitive; they also have a greater number of bands, which are necessarily narrower in wavelength range and which are able to be used in multispectral classification or in multichannel algorithms for the determination of water parameters such as suspended sediment or chlorophyll concentrations.

From 1978 until 1986 the Coastal Zone Colour Scanner (CZCS), which was flown on the Nimbus-7 satellite, generated ocean colour data in several narrow bands in the visible and near-infrared region of the electromagnetic spectrum. The work done with the CZCS demonstrated the potential of satellite-borne ocean colour instrumentation for a wide, and still widening, range of marine applications (see, for instance, Gordon and Morel 1983). Data from the Landsat Multispectral Scanner (MSS) and Thematic Mapper (TM) instruments, from SPOT and from other high-resolution systems such as INSAT, are quite widely used for studying coastal waters, estuaries and small inland water bodies. Overviews of what can be done in this respect can be found in the literature (Berie and Cornillon 1981, Khorram 1981, Klemas and Philpott 1981, Smith *et al.* 1982, Gordon *et al.* 1983a,b, Sathyendranath and Morel 1983, Sturm 1983, Dwivedi and Narain 1987). Following the demise of the CZCS, which was only an experimental instrument, there was no immediate successor available and so, to cover the gap until data from SeaWiFs or other new systems become available, people have looked around for other possible sources of satellite data for ocean colour studies. The high-spatial-resolution systems are of little use over large areas of the ocean for several reasons (some of which we have already encountered in other contexts): (i) the data are available only relatively infrequently and too infrequently for some oceanographic work, (ii) accurate geometrical rectification based on the use of ground control points is not possible because there are no land areas in most scenes, (iii) coverage of a large area would involve constructing mosaics of many scenes and (iv) the data are very expensive. Before the demise of CZCS, AVHRR was used for a number of oceanographic purposes, but these were mostly based on using the data from the thermal infrared channels which are of far better quality than the thermal infrared data from the CZCS. The data from the visible and near-infrared channels of the AVHRR had been used for a few oceanographic purposes, e.g. for studying very large oil spills such as that from the IXTOC-1 blowout in the Gulf of Mexico in 1979 or from the *Exxon Valdez* in 1989. However, before the demise of the CZCS no very serious attempts were made to use AVHRR data for ocean colour studies. It was the frequency of data generation and the reliability of the AVHRR which led people to explore the possibilities of using this data source following the demise of CZCS. The possibility of using AVHRR data for ocean colour work, in spite of the fact that there is only one visible-wavelength channel, has been explored for a number of different sea areas and we shall consider some examples briefly. In this context the AVHRR has the advantages that we have already noted at various other stages in this book. A particularly strong point in favour of the AVHRR, given the timescale on which oceanic conditions may change, is the daily coverage (even twice a day if both the morning and afternoon satellites are used), giving access to the area of interest whenever a cloud-free day occurs, compared to the Landsat overpasses only once every 16 (or previously 18) days. Also, the costs and the potential real-time use of AVHRR data favour the AVHRR over the expensive Landsat and SPOT data, for which the accessibility problems can often, in practice, be frustrating. A further strong point is the fact that the AVHRR instruments are part of an operational meteorological satellite system that is guaranteed to continue for several years to come. A weak point of the AVHRR data is the low spectral resolution, especially in the visible wavelengths. On the other hand, the radiometric resolution of the AVHRR (10 bits) is much better than the resolution of most other (8 bits) satellite sensors. The parameters which are commonly studied using ocean colour data include such parameters as total suspended matter, chlorophyll concentration and floating patches of dense concentrations of algae (algal blooms) and some large oil pollution events.

The estimation of water quality parameters from space is based on the absorption and scattering of sunlight by suspended or dissolved material in the upper water column. These processes are wavelength-dependent and therefore so is the atmospheric interference of the signal received by the satellite sensor. For this reason the choice of wavelength bands for sensors is very important and the wavelength ranges of the channels of the AVHRR were not optimally chosen for ocean colour monitoring. Indeed, very large amounts of AVHRR data from the oceans contain no useful information whatever, although there are some noteworthy exceptions. However, as so often happens in Earth observation, one has to make the best use of the least inappropriate data available.

7.3.2 Suspended sediment concentrations

The use of AVHRR data to study suspended sediment concentrations has been explored by Strong and Eadie (1978) for the Great Lakes, by Lyon *et al.* (1988) with data for Sandusky Bay, Lake Erie, by Prangsma and Roozekrans (1989) and Spitzer *et al.* (1990) for the North Sea and certain large areas of inland water in the Netherlands, by Johannessen *et al.* (1989a) in Skaggerrak and off the coast of Norway and by Fiedler and Laurs (1990) for the Columbia River plume. We shall describe some of this work. The swath width (2700 km), instantaneous field of view (1.1 km) and radiometric resolution (10 bits) for the AVHRR are largely comparable with the corresponding CZCS characteristics (1650 km, 0.8 km and 8 bits, respectively) with daily day-time observations. On the other hand, the spectral resolution of the AVHRR is much less; there are only two channels in the visible (0.58–0.68 μm) and near-infrared (0.725–1.1 μm) bands for the AVHRR compared with five for the

CZCS. AVHRR imagery has been used (a) in conjunction with Landsat MSS data, (b) in conjunction with CZCS and (c) on its own (following the demise of CZCS).

Thus, for example, Lyon *et al.* (1988) used both MSS and AVHRR data, together with on-site sampling and hydrodynamic and water quality model simulations, to evaluate surface sediment concentrations in Sandusky Bay, Lake Erie. Two Landsat MSS scenes and two AVHRR scenes from the period 10–28 June 1981 were used together with *in situ* data from within the same period. Good correlation was obtained between the satellite data and the *in situ* measurements and the satellite-derived results were similar to the results of the hydrodynamic and water quality model simulations. Gagliardini *et al.* (1984) used AVHRR data in conjunction with Landsat MSS and Nimbus-7 CZCS data to study the Rio de la Plata Estuary. Sørensen *et al.* (1991) used a combination of AVHRR and Landsat TM data for water quality monitoring in some Norwegian fjords.

In their work on the Columbia River plume Fiedler and Laurs (1990) used AVHRR data in conjunction with CZCS data over the period 1979–1985. The Columbia River, with a mean flow of 7300 m^3 s^{-1}, is the largest point source of freshwater flow into the eastern Pacific Ocean. The Columbia River water forms a low-salinity plume extending outward from the river mouth above a shallow (<20 m) halocline. Seawater is entrained into this plume by wind-generated turbulence and by frontal mixing and secondary lateral flow induced by potential energy within the plume. River flow and turbulent mixing rates are such that anomalously low surface salinities ($<32.5\%$) may be detected over the entire shelf and slope and out to 400 km from the coast of Washington and Oregon (Figure 7.28). The spatial resolution of CZCS and AVHRR are very suitable for the study of this plume because it is large and because it is subject to large variations in its shape in response to coastal winds and wind-driven surface currents. In this region the coastal wind velocity shows strong seasonality, with northerly winds prevailing from May to September and strong southerly winds from October to April. The spring transition is marked by a change from prevailing southerly to northerly winds. The timing of the spring transition varies from year to year, with some nominal dates as follows: 8 May 1979, 20 April 1980, 23 June 1981, 22 April 1982, 4 April 1983, 26 May 1984 and 5 May 1985. Higher-frequency fluctuations in wind velocity, at the scale of about a week, appear as occasional relaxations or reversals of the seasonal prevailing winds. These fluctuations are caused by the passage of storm systems through the area.

In the work of Fiedler and Laurs (1990) both CZCS data and AVHRR data were used together. However, they were used in a complementary fashion. About 70–80 satellite images, acquired during the period 1979–1985, were analysed, including approximately equal numbers of AVHRR and CZCS scenes. The visible-channel data from the CZCS were used to determine pigment (chlorophyll) concentrations using the approach of Gordon *et al.* (1983b) which has been widely used with CZCS data. The thermal infrared data from the AVHRR were used, in the standard way, described in chapter 4, to determine sea surface temperatures. The orientation, shape, size, colour and relative temperature of the plume were found to vary in these images. A seasonal pattern of two basic forms is illustrated schematically in Figure 7.29. During winter, plume water is advected northward along the coast (Figure 7.29 (a)). The most obvious plume water is a semicircular mass of green water, colder than the surrounding coastal water, just north of the mouth. Water along the coast north of this mass is still relatively cold and green, but is probably a mixture of

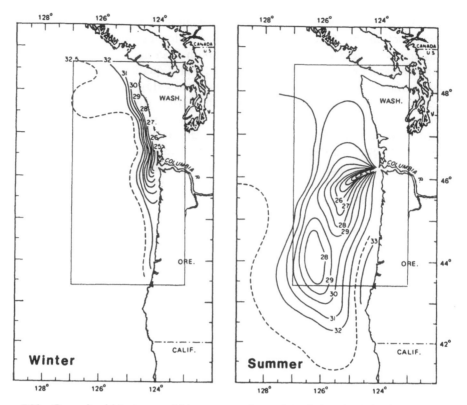

Figure 7.28 Generalised (a) winter and (b) summer surface salinity (‰) in the Columbia River plume (Fiedler and Laurs 1990).

Columbia River water, runoff from coastal rivers and oceanic surface water driven onshore. During summer, the plume is advected southward, usually separated from the coast (Figure 7.29(b)) due to offshore Ekman transport, but sometimes adjacent to it (Figure 7.29(c)). Plume water is warmer than the surrounding coastal water, which is cooled by upwelling. During the transition between the basic winter and summer forms, the plume is roughly circular at the mouth of the river (Figure 7.29(d)). More detailed discussion of the plume's characteristics and its seasonal variation will be found in the paper by Fiedler and Laurs (1990).

In the absence of CZCS data, Prangsma and Roozekrans (1989) used AVHRR data to study suspended sediment and blue-green algal concentrations along the coastal waters of the North Sea and in the Ijsselmeer in the Netherlands (see also Roozekrans 1989). Areas of cloud were eliminated using a standard algorithm package at the Royal Netherlands Meteorological Institute (KNMI), which is based on the APOLLO (AVHRR Processing Over Land cLoud and Ocean) package made available by the UK Meteorological Office Research Unit in Oxford, with additions and enhancements developed in-house (see also section 3.2). Some of these additions apply to cloud-clearing and others are transcriptions of known CZCS and TM algorithms adapted to the different spectral resolution of the AVHRR instrument. *In situ* measurements of total suspended matter were obtained from the Department of Roads and Waterways of the Netherlands Ministry of Public Transport, Rijkswaterstaat. These measurements have been taken in the tidal waters along the Nether-

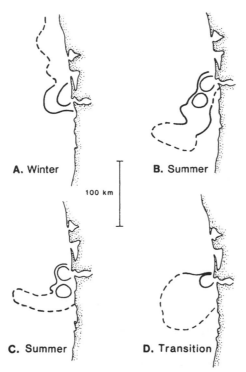

Figure 7.29 Schematic drawings of basic forms of the Columbia River plume. Solid lines represent strong colour and/or temperature boundaries and broken lines represent weaker boundaries (Fiedler and Laurs 1990).

lands coast providing a wide range of total suspended matter concentrations (1–120 mg l^{-1}). The measurements represent total suspended matter concentrations in the upper part of the water column. In relatively shallow waters, like the southern North Sea, the total suspended matter concentration increases towards the lower part of the water column, dependent on meteorological conditions (e.g. windspeed).

In channel 1 only about 20 per cent of the signal received by the AVHRR originates from the water column; 80 per cent is contributed by the atmosphere. In channel 2 the contribution of the water column to the signal is negligible since the water molecules absorb the near-infrared radiation. It is therefore extremely important to try to calculate the atmospheric correction to channel-1 AVHRR data as accurately as possible. This was done by Prangsma and Roozekrans by adapting the methods used for the atmospheric correction of CZCS data to the case of AVHRR data (Viollier *et al.* 1980, Singh and Cracknell 1986). Work based on the use of a radiative transfer model (Dirks and Spitzer 1987) indicated a log–log relation between the total suspended matter, TSM, and the atmospherically corrected water column reflectance ρ_1 from channel-1 AVHRR data:

$$\log(\text{TSM}) = a \log(\rho_1) + b \,. \tag{7.3.1}$$

In practice this is essentially an empirical relation, where a and b are parameters that are site-dependent and time-dependent. Thus a and b have to be computed from a regression between *in situ* total suspended matter observations and co-located satellite measurements of ρ_1 for every individual image.

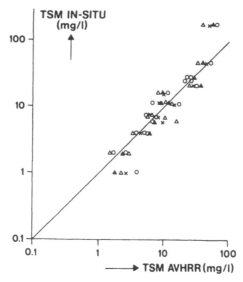

Figure 7.30 Comparison of *in situ* total suspended matter (TSM) concentrations with AVHRR-derived values from three different dates, 31 October 1984 (○), 1 November 1984 (△) and 2 November 1984 (▲), and average values (×) (Prangsma and Roozekrans 1989).

An example of results obtained from the above procedure is shown in Figure 7.30 (Prangsma and Roozekrans 1989, Spitzer *et al.* 1990) which is derived from *in situ* measurements along the Dutch coast for the period 28 October until 7 November 1984 and compared with satellite-derived data from three reasonably cloud-free AVHRR scenes from 31 October, 1 November and 2 November 1984. Correlation coefficients for the three AVHRR orbits vary between 0.84 and 0.89. In Figure 7.31 the corresponding spatial distributions of suspended sediment concentrations are presented for these three successive days. The results show the total suspended matter distribution to be fairly conservative on a day-to-day basis, although details may vary depending on, for example, meteorological conditions. The patterns are generally in line with what is commonly known about the area, but do contain more detail than is usually available.

Prangsma and Roozekrans (1989) also studied freshwater lakes in The Netherlands during the summer months which show large areas of floating blue-green algae. They chose to do this in terms of the NDVI which is sensitive to the presence of green vegetation on land. Since

$$\text{NDVI} = \frac{\text{channel-2} - \text{channel-1}}{\text{channel-2} + \text{channel-1}} = \frac{X_2 - X_1}{X_2 + X_1} \tag{5.3.1}$$

and since clear water is nearly a blackbody for near-infrared radiation, the value of X_2 will be zero and the NDVI will be negative for a clear-water surface – indeed the value of the NDVI will be close to -1. However, if blue-green algae are floating near the surface the reflectance in channel 2 will be dramatically increased, due to the large concentration of chlorophyll, and the value of the NDVI will be increased, probably enough to make it have a positive value. In August 1983 floating layers of blue-green algae caused considerable inconvenience in the marinas around the Ijsselmeer, an artificially enclosed sea in the Netherlands. The arrival of many persistent high-pressure systems during the whole month increased the occurrence of layers in

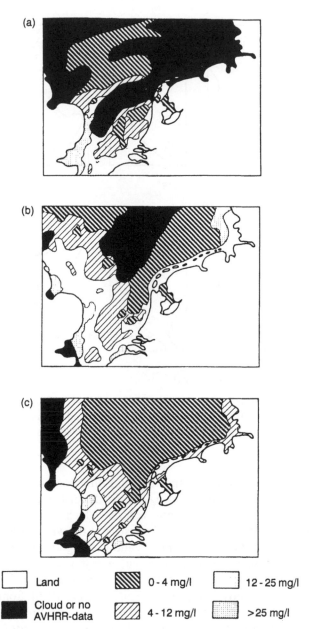

Figure 7.31 Distribution of AVHRR-derived total suspended matter concentrations in the southern North Sea for three successive days (a) 31 October 1984, (b) 1 November 1984 and (c) 2 November 1984 (Prangsma and Roozekrans 1989).

the algae-rich Ijsselmeer. These high-pressure systems also resulted in the availability of a convenient series of cloud-free AVHRR images for this period. Figure 7.32 shows the NDVI and water surface temperature for a period of five consecutive days, 18–22 August 1983, derived from AVHRR data. In Figure 7.32 the positive NDVI values seem to be indicative of the presence of blue-green algae at the surface. The surface temperature distribution confirms the strong surface heating by several degrees due to the presence of these biologically active absorbers.

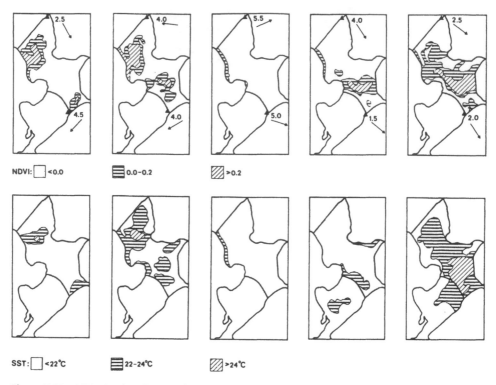

Figure 7.32   NDVI (top) and sea surface temperature (SST) (bottom) in the Ijsselmeer for five successive days (left to right), 18–22 August 1983. The numbers and arrows indicate the magnitude (m s^{-1}) and direction of the wind speed at the time of the pass (Prangsma and Roozekrans 1989).

When the wind speed increased and the wind direction changed by 180° during 20 August 1983, enhanced vertical mixing caused the distribution of the blue-green algae through a larger part of the water column which is also seen in the disappearance of the hot surface layer. On 21 August 1983 the wind speed in the southern part of the lake had dropped below 2 m s^{-1} and floating algae immediately became visible in the NDVI image again. Notice that it was not attempted to determine quantitative concentrations of chlorophyll from these data although some discussion of the factors involved is given by Spitzer *et al.* (1990). Some discussion of the question of atmospheric corrections based on using a multi-spectral approach is given, for example, by Sturm (1981) and Barale and Schlittenhardt (1993). Attempts have been made to make atmospheric corrections by using a two-look approach instead of a multi-spectral approach (Khosraviani and Cracknell 1984, 1987, Xue and Cracknell 1965a).

7.3.3   Algal blooms

The serious study of plankton blooms in the ocean using satellite data began with visible-band data from the Coastal Zone Colour Scanner (CZCS). This was an experimental instrument flown on Nimbus-7 and it generated data from 1978 until 1986. The data from that instrument tended to be analysed on an historical basis long after they had been generated. It has been widely used in research work but was hardly ever analysed in near-real-time for use on an operational basis.

Within coccolithophore blooms the reflectance of the water column may increase to 25 per cent or greater (Holligan *et al.* 1983, Ackleson and Holligan 1989). Coccolithophores are marine phytoplankton composed of a nearly spherical central cell, the *coccosphere*, and surrounded with one or more layers of calcium carbonate plates, *coccoliths*. The most abundant coccolithophore appears to be *Emiliania huxleyi*, for which the coccosphere is between 5 and 8 μm in diameter. The coccoliths are typically 2 μm in width, 4 μm in length and 0.5 μm in thickness. Under certain conditions, the coccoliths may be detached from the coccosphere and freely suspended within the water column. The process whereby coccolithophores produce and shed coccoliths is poorly understood, although various theories concerning buoyancy regulation, light conditions for photosynthesis, and the maintenance of internal CO_2 concentrations have been proposed. High concentrations of detached coccoliths, which at times exceed $78\,000\ ml^{-1}$, are believed to be responsible for the increased volume reflectance of the sea.

Figure 7.33 shows a NOAA-10 AVHRR image of an extensive coccolithophore bloom which occurred within the Gulf of Maine between mid-June and mid-July 1988. This scene is one of a sequence collected to monitor the development and dynamics of the bloom; a simple atmospheric correction has been applied to it, as a result of which land and clouds appear black. Cape Cod is shown along the lower left margin, Nova Scotia within the upper right portion of the image, and clouds occupy the lower right margin. The image has been linearly contrast stretched to enhance water features.

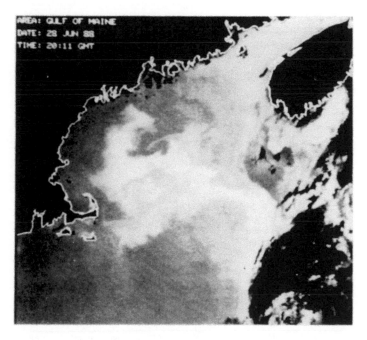

Figure 7.33 NOAA-10 AVHRR channel-1 image of the Gulf of Maine, from 28 June 1988, showing an intense bloom of coccolithophores (white areas). Black areas represent either land (upper and left margins) or clouds (lower right) (Ackleson and Holligan 1989).

The coccolithophore bloom (light grey and white areas) appears to occupy most of the southwestern portion of the Gulf, centred over Wilkinson Basin (bathymetric features are shown in Figure 7.34), and covers an area of approximately 25 000 km^2. Analyses of surface water samples collected within the bloom between 9 July and 12 July indicated a nearly monospecific *Emiliania huxleyi* population and coccolith concentrations in excess of 20 000 ml^{-1}. The boundary of the bloom is characterised by complex filament and eddy structures. The eastern boundary of the bloom follows closely the 100 m depth contour outlining Georges Bank, an area of shallow, well-mixed water located approximately 150 km east of Cape Cod. Along the northern tip of Georges Bank, a portion of the bloom appears to be exiting the Gulf of Maine through the Northeast Channel and flowing into the North Atlantic. This feature is indicative of surface water flow out of the Gulf and southwest along the shelf break and is consistent with the general circulation of surface water within the Gulf of Maine.

Blooms of this nature are episodic and do not necessarily appear every year. When they do appear it is not necessarily in the same region of the Gulf and they are temporally dynamic (their shapes change as a result of surface currents). The advantage of AVHRR data over the higher resolution satellite data is that their high frequency of coverage makes it possible to monitor the appearance, temporal evolution and decay of blooms of this type in great detail. By tracking small-scale eddies and other structures highlighted by coccolithophores from one image to the next, it is possible to chart surface water movements in the same way that similar features in the surface temperature pattern can be used (see section 4.3).

Coincident shipboard and satellite detection of species-specific phytoplankton blooms in the Chesapeake Bay during the spring 1982 season by Tyler and Stumpf

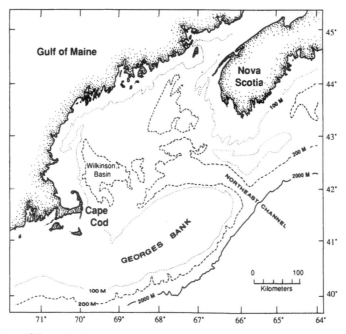

Figure 7.34   Map of the Gulf of Maine showing relevant coastal and bathymetric features (Ackleson and Holligan 1989).

(1989) documented the spatial and temporal variability of these red tides. Both AVHRR and CZCS data were processed and confirmed the kinetics of a 100 km^2 bloom of the dinoflagellate *Heterocapsa triquetra* including abundance changes due to diurnal migration, and axial and cross-stream spatial reconfiguration due to tidal excursion. Turbidity and thermal information gleaned from both satellite and ships emphasise the association of blooms with physical features of the system such as fronts. Thus, remote detection and rapid monitoring of bloom kinetics in turbid estuaries is feasible.

Tyler and Stumpf (1989) make the important point that satellite images, provided they are delivered in near-real-time, can be very useful to aid in the planning of a scientific cruise. The fieldwork carried out in the Potomac River was initially undertaken to examine the fine-scale bloom dynamics of estuarine dinoflagellates. The typical protocol for a ship involves a survey of the entire estuary and major tributaries for the highest concentration of cells and then a return to the highest biomass area to commence a 48-hour study of migration and physiology. In essence, one begins a cruise blind. The scientist can design a sampling programme, on the basis of past experience, to cover the areas where blooms usually appear and that is what was done in this case. Days of survey time might have been saved in 1982 if one had had in hand a satellite image enhanced to show the bloom. For a number of years the Dundee University Satellite Station has been supplying CZCS and AVHRR images of European waters for just this purpose to various oceanographers planning their cruises.

In the study of the Potomac River (Tyler and Stumpf 1989) the details of the upstream–downstream, as well as cross-stream, reconfiguration of the Potomac *Heterocapsa* bloom were clarified with the satellite data. Only with such synoptic coverage can the interplay between the physical dynamics and organism behaviour be thoroughly understood. Another concept that was verified is the patchiness of the system. Heterogeneous distributions of phytoplankton with patch sizes of a few kilometres are a standard feature of the estuary. Shipboard collection rarely resolves such fine-scale features.

While satellite remote sensing in estuaries can never replace *in situ* observations, it is becoming increasingly apparent that its use can improve the interpretation of the shipboard measurements. It should also be suitable for documenting seasonal and interannual variations to establish trends. This is a powerful readily available data set which should be a routine tool in any basic or applied research programme.

Another example of the study of the development of an algal bloom and of its spatial advance and retreat using AVHRR is provided by Johannessen *et al.* (1989a). Blooms of toxic algae in reservoirs pose a threat to drinking water supplies while blooms at sea in coastal waters present a serious threat to fish farms. Some blooms have been observed in the past with Landsat and with CZCS. But Landsat data are obtained too infrequently to enable one to track the development and decay of an algal bloom. CZCS was also infrequent and, anyway, CZCS is now no longer available. So people have turned to the visible band of AVHRR to see if it can be used in the study of blooms. The work described by Johannessen *et al.* (1989a) concerned a bloom of *Chrysochromulina polylepis* which appeared in Skagerrak in May 1988 and advanced a large distance northwards along the coast of Norway before halting and retreating. An interesting and important feature of this work was the fact that the AVHRR data were analysed in near-real-time and led to precautions being taken to reduce the damage to the Norwegian fish-farming industry. It is a very good example of a successful application of AVHRR data, not just in terms of historical studies of an event but in terms of providing really useful input in near-real-time into an environmental monitoring and protection programme.

During the last few decades an increasing number of epidemic red tides have been reported globally. The occurrences of these blooms were unexpected. Likewise no forecast or warning was provided prior to the bloom of the toxic *Chryso-chromulina polylepis* in Skagerrak in May 1988. The sea farming industry along the southern and southwestern coast of Norway suffered significant damage, and a total of 480 t of caged fish at a market value of about 30 million NOK were lost before any precautions could be taken. Several observations of dead wild fish and other fauna in the upper 30 m were also reported along the southern coast of Norway as well as in Kattegat and eastern Skagerrak. No single explanation is presently offered for this bloom. However it is thought to have been due to a combination of increasing pollution and a set of weather conditions and sea conditions that were particularly favourable to the development of the bloom (for details see Johannessen *et al.* 1989). The general circulation in the North Sea and Skagerrak is shown in Figure 7.35.

The first documented observation of the toxic algae was reported from the west coast of Sweden on 9 May 1988, while dead wild fish were first observed on 13 May 1988. The bloom along the west coast of Norway culminated towards the end of May. In the meantime maximum algal concentrations in the surface layer (upper 30 m) were measured up to a maximum of 90 million algal cells per litre of water. Concentrations between 5 and 10 million per litre at salinities of about 30‰ are sufficient to kill caged trout and salmon.

On 24 May scientists in Bergen, Norway, with backgrounds in different aspects of the marine disciplines, formed a team with the responsibility to coordinate the monitoring of the algal distribution and relevant environmental conditions. Daily forecasts of the algal front from 24 May to 3 June were distributed by radio, TV and newspapers. A private monitoring group was also established by the Seafarming Sale Association in Trondheim. Several research vessels, smaller speedboats, remote

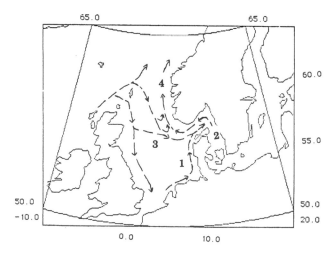

Figure 7.35 Schematic of the general circulation system in the North Sea and Skagerrak (Johannessen *et al.* 1989a).

sensing aircraft, and drifting buoys tracked by service ARGOS as well as near-real-time access to NOAA weather satellite data were at the team's disposal.

Advanced Very High Resolution Radiometer (AVHRR) data from the NOAA-9 and NOAA-10 satellites were received at the Tromsø ground station. Data tapes were sent by express mail to Bergen and received at the Nansen Remote Sensing Centre about 6 hours later and processed to obtain maps of the sea surface temperature distribution. Early on during the bloom, observations indicated a close correlation between the algal front and the satellite-derived surface temperature front. The spreading and advection of the algae were thus indirectly monitored by satellite thermal infrared data during cloud-free periods. A remarkable amount of warm surface water (temperature > 10°C) was found in central Skagerrak. While the warm water was primarily confined to a large cell in the eastern Skagerrak and along the east coast of Denmark on 28 April, it covered the entire eastern Skagerrak by 15 May followed by a westward growth between 15 May and 30 May. Except for a few cells or filaments of warm water along the Norwegian coast, no similar evolution was observed in the Norwegian Coastal Current and central North Sea where the mean surface temperature remained near constant (about 6–8°C).

The thermal infrared AVHRR images provided estimates of the abnormal warm frontal advection along the coast of Norway, thus indicating the propagation of the algal front, which was further verified by *in situ* observations. Between 15 and 21 May the front moved southwestward at an average speed of 5 km per day. On 21 May the major temperature front hugged the southeastern coast of Norway near Kristiansand at 8° E. Between 21 and 22 May the westward advection of the front rapidly increased to about 30 km per day. This is not an unusual speed for the Norwegian coastal current which varies considerably both in time and space. As expected, evidence of meanders and eddies of 30–60 km could also be seen in the Norwegian coastal current as it advected northwestward out of Skagerrak. A large meander southwest of Stavanger was noticeable on 22 May overlying a topographic ridge. This meander was used to initiate the numerical model on 22 May. On 30 May the image shows that the narrow coastal front reached the Boknafjord south of Karmøy, indicating a mean advection speed of about 25 km per day since 22 May.

Several types of *in situ* data were collected. These included mapping the three-dimensional velocity structure with an Acoustic Doppler Current Profiler, sea surface temperature measurements and CTD (conductivity and temperature profiles versus depth) measurements; so the main features of the current patterns in the area were identified (for further details see Johannessen *et al.* 1989a).

In addition to using the thermal infrared AVHRR data and the *in situ* data for making daily forecasts, these data were also used as input to, and to provide verification of, a numerical model (again for details see Johannessen *et al.* 1989a). A summary of the observed algal front, defined as a concentration of 0.5–1.0 million algal cells per litre, indicates the advance and retreat along the southern and south-western coast of Norway from 21 May to 3 June (Figure 7.36). The behaviour of eddies and filaments within the bloom behind the front was also presented. Few other severe environmental episodes have received as much attention in Norway as this toxic algal bloom. In addition to the public interest, the forecasts were primarily used by fish farmers along the southwest coast of Norway between Stavanger and Bergen, and led to the towing of about 130 seafarms (with estimated insurance value of about 1.0–1.5 billion NOK) to safety in low salinity water deep in the fjords along

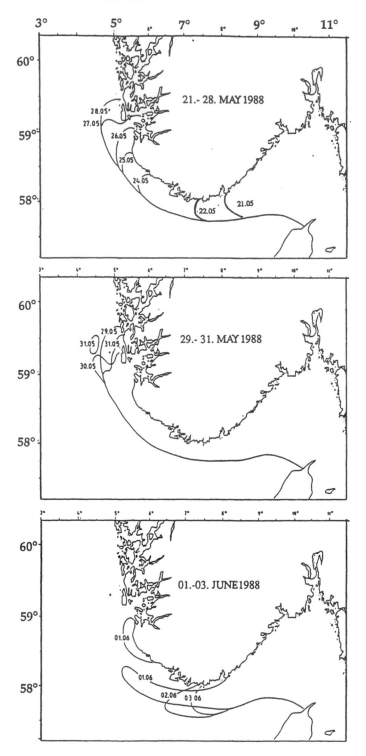

Figure 7.36 Summary of the observed algal front (including both advance and retreat) in the Norwegian coastal current from 21 May to 3 June defined by 0.5–1.0 million algal cells per litre. Note that the algal front locations of 21–22 May are assumed to be overlapping the temperature fronts inferred from the NOAA satellite IR images (Johannessen *et al.* 1989a).

the west coast. There is one very important point which comes out of this piece of work. This is the fact that the AVHRR data were only useful because they were obtained and processed very quickly. It would have been an interesting academic exercise to process the AVHRR data six months or a year after the event and show that one could map the development of an algal bloom with the data. However, what was of practical use is that people did use the data in near-real-time and did monitor the development of the bloom successfully and did predict its movements.

This particular piece of work was enormously successful in minimising the losses sustained by the fish-farming industry and the use of AVHRR data played a key role in the operation. That it was successful was largely a result of the fact that the AVHRR data were made available and were processed in near-real-time. This is dependent on having a reliable and efficient direct readout ground station within range of the disaster area being studied. Monitoring the development of a bloom in near-real-time is clearly a great improvement on just studying it on an historical basis months after the event and long after there is any chance to reap economic benefits from the study. It also indicates that very useful information can be extracted from the AVHRR data, notwithstanding the fact that quantitative mapping of the concentrations of the algal blooms is not possible without simultaneous *in situ* calibration data. However, as mentioned before, this algal bloom occurred unexpectedly. The next stage, in which there may or may not be a role for AVHRR data, will be to predict when and where toxic algal blooms are going to occur.

7.3.4 Oil pollution

We did mention very briefly earlier in this section that the large oil spill associated with the IXTOC-1 blowout in 1979 was able to be observed in AVHRR data. Landsat data were also available at that time. Although both Landsat and AVHRR data were used there were delays in the delivery of the data, in the analysis of the data and in the delivery of the analysis of the data to the people involved in the monitoring of the spill and in attempting to clean it up. At that time the results were rather disappointing and the contribution of remote sensing satellite data to the clean-up operation was rather minimal. Ten years later, at the time of the *Exxon Valdez* disaster, things were better and, at least, an accurate historical record of the disaster was obtained. But some of the logistical problems in the near-real-time delivery and analysis to enable the use of satellite data within the timescale required in the clean-up operation still remain to be solved. An account of the use of remotely sensed imagery of the *Exxon Valdez* disaster is given by Stringer *et al.* (1992). The *MV Exxon Valdez* ran aground on Bligh Reef, Alaska, on 24 March 1989 (see Figure 7.37). As a result over eleven million gallons of crude oil circulated through the western Prince William Sound region of Alaska, oiling many of its beaches. A good deal of this oil subsequently entered the Gulf of Alaska and some appears to have been transported even beyond Kodiak Island, 500 km to the southwest (Figure 7.37). Immediately following the spill an effort was undertaken to capture satellite imagery that might be of value in monitoring the spill's location and movement. Tapes of AVHRR, Landsat TM and SPOT satellite imagery were acquired from dates prior to the oil spill and throughout the spring and summer when the oil was in Prince William Sound and the Gulf of Alaska. Later, as part of a process to verify

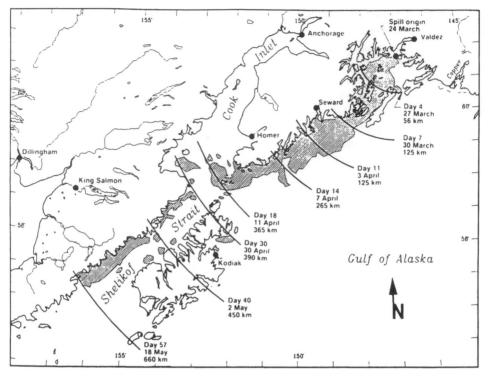

Figure 7.37 Generalised extent and progression of oil spilled from the *Exxon Valdez* as seen from airborne observation from 24 March to 18 May 1989 (Stringer *et al.* 1992).

oil identification, data were obtained that had been acquired of the study area just prior to the spill. The images were digitally analysed to distinguish spectral signatures related to oil and to remove noise. These results were then recorded on film and reproduced as photographic prints.

Numerous AVHRR images were acquired from the NOAA Command and Data Acquisition (CDA) station at Gilmore Creek, Alaska, and from the NOAA satellite data services division at Camp Springs, Maryland. However, because of cloudiness on many of the images, only seven were digitally analysed covering the period 16 March–30 April 1989. The Gilmore Creek data are 8-bit (i.e. 256 grey levels) and are available almost immediately after the satellite pass while the Camp Springs data are 10-bit but are only available retrospectively. Subtle patterns in the water were observed in areas where spilled oil was located on several of the AVHRR images in channel 1, channel 2 and channel 4. Channels 3 and 5 were not used because the oil-related signature was not as distinct as in channel 4. Channel-4 images exhibited more distinct patterns than channel-1 or channel-2 images. What was particularly interesting is that in earlier work on the IXTOC-1 spill it was found (Hayes 1980, Cracknell *et al.* 1983) that while the oil could be detected very clearly in channel-1 AVHRR data, the oil was apparently almost invisible in the thermal infrared channel data. At that time this invisibility was taken to have arisen as a consequence of an accidental compensation effect between the oil–water emissivity difference and the oil–water temperature difference. There was (a) no possibility of checking this supposition and (b) no reason to suppose that this compensation would be a general phenomenon for oil spills. The area of interest was subsectioned from the data and the land was masked out. The remaining data were contrast stretched so that sea surface temperature patterns were distinct. Colours were assigned to specific ranges of digital numbers to delineate sea surface radiant temperature patterns. The near-real-time 8-bit AVHRR imagery from Gilmore Creek was acquired the same morning as the satellite pass. The contrast was enhanced on the thermal infrared band (channel 4) and the dominant sea surface radiant temperature pattens were mapped to four grey levels, since it was found that four grey levels provided the most information that could clearly be discerned on a telefaxed AVHRR image. These maps were transmitted to the Alaska Department of Environmental Conservation at Kodiak, Alaska, and from there to the University of Alaska Research Ship *Alpha Helix* before the end of the day. This proved to be useful in inferring the movement of surface water from sea surface temperature patterns. However, such low-resolution data are not conducive to detecting spilled oil except when the oil is still in a large assembly. Since the 10-bit AVHRR data from Camp Springs were only available retrospectively they could only be processed on an historical basis. Several TM and SPOT scenes were analysed for the period of the spring and summer following the oil spill, but this could not be done in near-real-time. If the handling of Landsat TM or SPOT high-resolution imagery could be such that same-day reception, analysis and field transmission could be achieved, very critical tracking and monitoring information would be available.

The images from AVHRR, Landsat TM and SPOT were analysed to help to ascertain the extent of the spill and to monitor its trajectory along the Alaskan coast (Stringer *et al.* 1992). Digital image processing techniques were utilised to highlight spectral response related to oil on the water surface and on the beaches. In addition to the satellite imagery described here, airborne visual observations with hand-held oblique photography, vertical aircraft mapping photography, side looking airborne radar (SLAR) as well as shipboard sightings and samplings were acquired to monitor the position and extent of the oil and its effect on the marine environment. Turbidity and sea surface temperature data were enhanced to provide information on the circulation and distribution of surface water bodies for other research units.

The analysis and interpretation of the data on the *Exxon Valdez* oil spill given by Stringer *et al.* (1992) are very detailed but are not central to the theme of this book. We turn to those authors' conclusions, in the light of their experience with data related to the *Exxon Valdex*, about the use of satellite imagery in relation to oil spill detection and monitoring. At this time imagery provides an excellent historical record but, largely because of logistical constraints, it does not provide timely information at high resolution to assist in clean-up operations. It was concluded that for satellite imagery to be very useful during the active phase of spill monitoring and clean-up, several criteria will have to be met: (1) near-real-time availability of imagery to field operations, (2) high resolution to detect windrows, (3) one- to two-day repeat satellite coverage and (4) cloud penetration. Spilled oil can be detected on satellite imagery, especially in infrared wavelengths. The imagery provides the precise positions of windrows more accurately than an airborne observer's position estimates on a map. The satellite imagery also shows the precise shape and extent of an oil windrow. Furthermore, there is considerably less chance that windrows will escape detection because of areas not evaluated. However, the presently available imagery does not appear to record sheen very well. Further research needs to be performed to identify precisely what characteristics of an oil spill the presently available satellite sensors detect.

The requirement for one-day to two-day repeat satellite coverage with sufficient spatial resolution to detect the oil is not inconceivable. SPOT with its pointable sensor partially addresses this need. A single satellite probably cannot provide all the data coverage required but several satellites could. SPOT, in conjunction with Landsat and other platforms to be launched in the future, could very likely provide the time resolution required for evaluating a dynamic, short-term environmental crisis. Recently a second SPOT system was launched. Two functioning SPOT satellites can provide nearly daily data coverage at Alaskan latitudes.

The fourth criterion, cloud penetration, can only be met with active microwave sensors which are now beginning to be launched and evaluated (Almaz, ERS-1, JERS-1, Radarsat). Data at these wavelengths are capable of detecting spilled oil as indicated by SLAR imagery and as shown on imagery from Seasat, the first satellite radar platform, which was active in 1978. Radar is sensitive to sea state (wave heights). Oil dampens the capillary surface waves thus providing a means of detecting its presence. However, we are still a very long way from having near-real-time delivery, analysis and interpretation of active microwave data from satellites.

It is by now clear from the *Exxon Valdez* experience that techniques and procedures improved markedly since the IXTOC-1 blowout of ten years earlier; however the transition to operational use depends on political and economic considerations. With an appropriate technological investment, imagery can be received, analysed and sent to the field on the same day the satellite records the data to assist clean-up operations. The increase in the number of imaging satellites scheduled to be launched in the 1990s in conjunction with cloud-penetrating radar will provide even better data sources to assist monitoring and clean-up of future oil spill catastrophes.

Data from the thermal infrared channels of the AVHRR were successfully used by Cross (1992) to monitor an oil slick that was produced during the Gulf War off Kuwait. This was interesting in several ways. First, the discharge of oil occurred in a war zone; this hindered the routine use of dedicated airborne reconnaissance aircraft. Details of the extent and progress of the resulting slick were inevitably sparse during the hostilities. Consequently public domain satellite sensor data became an important source of information. Secondly, the slick was very large so that the coarse spatial resolution of AVHRR was not expected to be a serious limitation because of the large extent of the slick. The high frequency of AVHRR coverage provided a better opportunity for obtaining cloud-free data than could be expected from Landsat or SPOT. Contemporary reports quoted Saudi Petroleum Ministry estimates that the release involved between 5 and 12 million barrels of oil, making it the largest slick in history. The discharge was believed to have commenced on 25 January, several days after the start of the Gulf War, and to have emanated from the Sea Island oil terminal located 12 km offshore of Mina Al Ahmadi, south of Kuwait City. The resulting pollution posed a threat to 1 million birds and their habitat, 600 dugongs (sea cows), offshore coral islands, green turtles, shrimping grounds, fisheries and humpback dolphins.

Twenty AVHRR LAC scenes acquired between 17 and 29 January 1991 were previewed and checked for cloud cover and for view angle over the study area. Five (three day-time and two night-time) scenes were selected and converted from 10-bit to 8-bit radiometric resolution using a linear look-up table that maintained the original dynamic range over the area of interest. The five scenes were all recorded from the NOAA-11 satellite. The Earth location coordinates were extracted from the raw data, and subsets obtained over the area of interest. These subsets were then geometrically corrected to a geographical projection with a nominal 1 km pixel resolution using the supplied Earth location data. The images were inspected visually in the region of the slick location given in the press and other reports. It was apparent that all of the thermal channels (3, 4 and 5) offered a ready differentiation between the sea and a series of irregular objects, believed to correspond to the slick, along the western shore. The raw data values were converted to brightness temperatures. Because absolute measurements were not required it was decided not to incorporate atmospheric corrections.

It should be stressed that in the absence of extensive reliable ground data it was necessary to take steps to identify the objects detected to be oil. In the day-time images the objects were seen to have higher brightness temperatures than the surrounding water, while the sense of contrast was reversed at night. Notwithstanding the complexity of the relative spectral response of oil and water, this observation was considered to be consistent with the fact that oil has a lower heat capacity than water and would, therefore, be expected to warm up during the day, and cool down at night, more rapidly than the surrounding water. The contrast between the supposed slick and the water was found to be much greater in the scenes recorded

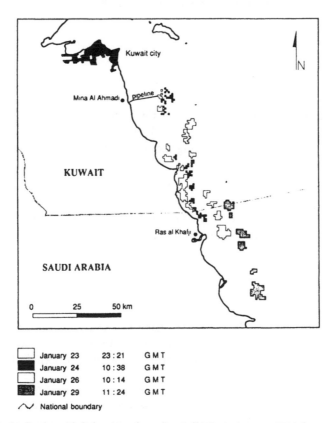

☐	January 23	23 : 21	G M T
■	January 24	10 : 38	G M T
☐	January 26	10 : 14	G M T
▨	January 29	11 : 24	G M T
∿	National boundary		

Figure 7.38 Progress of an oil slick arising from the Gulf War in January 1991 determined from AVHRR thermal infrared data (Cross 1992).

during the day. Interpretation of one of the night-time scenes was particularly difficult, because of the risk of confusion between a series of slick-like objects off the
Kuwaiti coastline and a line of clouds detected just inland of the coast. In an
attempt to resolve this, two of the cloud detection tests of Saunders and Kriebel
(1988) (see section 3.2) were carried out. Although it was recognised that the thresholds given by Saunders and Kriebel were not necessarily applicable to the latitudes
in question, the slick-like objects passed the tests as cloud-free. Further evidence
that the objects did indeed correspond to the slick was obtained when the digitised
course of the oil pipelines and the terminal at Sea Island were overlain with the
imagery. In the case of the first three scenes the northern limit of the supposed slick
was precisely located at the terminal.

Channel 3 was not considered further because of the confusion between reflected
and emitted signals during day-time acquisitions. Channels 4 and 5 appeared to
contain similar information to one another, and the former was arbitrarily chosen
for tracking the slick's progress. The third scene in the sequence (night-time on 24
January) showed what was presumably a large plume of smoke emanating from the
Sea Island terminal. This scene was rejected because the smoke obscures the sea and
the nascent slick in the vicinity of the terminal. A threshold was applied interactively
to the channel-4 values over the region of the slick in the remaining images. The
threshold selection was guided by visual interpretation, taking cognisance of the
reported slick location and the known southward drift during the period in question. The movement of the slick is shown in Figure 7.38.

7.4 FISHERIES

Remote sensing can, potentially, do several things for the fishing industry, but not
all of them are based on the use of AVHRR data. The problem is that, as a result of
other technological advances, the fishing industry is facing a serious situation. The
problems include

- over-fishing
- lack of sound management principles and proper legal superstructure
- increased international competition for resources
- increased pollution
- drilling and dredging operations
- coastal development.

Stocks of fish in many areas have been depleted or completely destroyed and many
breeding grounds have been damaged or destroyed. In many areas we see increased
competition and, in some cases, political problems among fishermen from different
countries.

Fishermen can no longer expect to travel only short distances from port and rely
on conventional detection techniques. The increased distances they must go involve
greater travel time, greater fuel consumption and greater risk of damage and loss of
gear, crew and ships.

Remote sensing can be involved in helping fishermen in three ways, namely by providing

- better weather information
- better sea-state information
- help in locating fish.

As far as weather information is concerned, remote sensing information in general, and AVHRR data in particular, represents only one of a very large number of sources of data that provide input into weather forecasting systems. Infrared data by themselves do not get you very far, even if they are available in near-real-time. The input of remote sensing into providing sea-state information is similarly marginal. In so far as sea state is related to the weather, and in particular to the wind speed, this is really the same situation as with weather forecasting. Direct observation of sea state is possible with remote sensing techniques, but this involves active micro-wave, i.e. radar, techniques rather than AVHRR data.

In connection with the question of locating fish, AVHRR data can provide useful information in certain circumstances. With remote sensing from aircraft or satellites one is, clearly, not attempting to see individual fish; from satellites one is not even attempting to see shoals of fish. Rather, one is trying to study indirect indicators that point to fish locations and fish migration patterns. The relevant parameters are (i) sea surface temperature and (ii) phytoplankton blooms/chlorophyll concentrations. The relevance is that where the fish choose to spend their time will be determined by the nature of the environment and by the availability of food supplies. In the expression 'nature of the environment' we include water density which, in turn, relies on salinity and temperature and we also include the level of dissolved oxygen, which also depends on temperature. One cannot determine salinity from satellites (though perhaps one can from aircraft) but we have already seen in chapter 4 that one can determine sea surface temperature from satellites. To be of any use to fishermen, in an operational situation, the data on sea surface temperatures have to be received, processed and passed on to them very quickly, that is, in a matter of a few hours from the time of capture of the data by the satellite.

The study of phytoplankton blooms and chlorophyll concentration is based on the colour, rather than the temperature, of the water and we have already discussed this in section 7.3.3. Although satellite techniques for fish location are rather indirect there are compensating advantages. The coverage, in terms of area, by satellite data is enormous and this may in fact be necessary. Some species of fish move very rapidly over large distances. For instance, one species of fish that is of interest to Scottish fishermen is the blue whiting and this species appears to move from the vicinity of Iceland in late January or early February down past the western isles of Scotland to reach an area off the west coast of Ireland in late March or early April. Another example is provided by the migration of the species Sardinella aurita off the west coast of Africa, which has been studied using data from the thermal infrared channel of Meteosat (Domain et al. 1980). A study of sea surface temperature using AVHRR data with relevance to fisheries off the coast of Namibia has been described by Brown et al. (1993). The migration of mackerel is of interest to Canadian east coast fishermen. In addition to information which is published, there is a suspicion that many other examples of the use of satellite data in fisheries, by the fishing fleets of various countries, exist but are concealed by commercial secrecy.

There was a programme 'An Experimental Satellite-orientated Observation Program for Commercial Fisheries' which was run for the United States West Coast fishing seasons of 1981 and 1982. The programme was sponsored by the Jet Propulsion Laboratory (JPL) (of California Institute of Technology) and the National Aeronautics and Space Administration (NASA), with the cooperation of the Ocean Services Division of the National Weather Services (NWS). Many other bodies were also involved. The data sources included conventional observations from the Fleet Numerical Oceanography Center (Navy) and the National Weather Service as well as satellite observations, principally of ocean colour (CZCS), sea surface temperature (AVHRR) and general weather circulation features (GOES). There were the operational products and the experimental products. The operational products included general weather charts, sea-surface temperature measurements, ice analyses and severe storm warnings. Transmission of the charts to the fishing fleet by radiofacsimile was available. Five-day outlooks were also provided. The experimental products included fisheries charts, depicting key isotherms (from a fine-mesh sea-surface analysis) for selected fish species, colour boundaries, convergence and upwelling areas (wind derivatives) and mixed layer depths. There were also five-day outlooks (in a voice message format) summarising fisheries-related environmental conditions, departures from normal and interpretations with respect to fishing operations.

The experimental fisheries charts were basically sea surface temperature charts but the ranges of temperatures favourable to the catching of different species of fish were indicated visually on the charts. For example, the greatest catches of salmon occur between 49 and 51°F. Most albacore are caught at temperatures between 60 and 64°F and most tropical tuna species are caught at water temperatures between 79 and 81°F. These three temperature ranges on the charts were therefore labelled as S, A and T, respectively, see Figure 7.39. New sea surface temperature analyses were made by computer every 12 hours using reports from ships (engine-room injection temperatures or bucket temperatures), satellites and bathythermograph systems. Charts were transmitted to the fishing fleet by radiofacsimile transmission. A number of other charts were also produced but we shall not discuss these here since they are not related to AVHRR data.

The original intention was that this experiment would run for a minimum of two seasons and that if it were successful – based on user evaluation – it would be carried on beyond that period with the distribution of products on a user-paying basis. However, when the fishing fleet was faced with paying for the service, in the period following the free trial period, the limited response was not enough to support an operational service. In addition to this, the ocean colour data from the CZCS ceased to be available. There is a suspicion, however, that in some countries this type of service is provided to its own fishing fleet, by a public body which is prepared to make a service of this nature available without imposing a direct charge on the consumer.

It must, of course, be remembered that what is measured by a thermal infrared scanner on a satellite is only the surface temperature, not the temperature below the surface. The idea that there is a strong correlation between sea surface temperature and the concentration of certain species of fish should, therefore, be treated with some caution. Other factors may also be involved as well. This is illustrated, for instance, by some work done in Australia (Myers and Hick 1990). A four-year project was conducted from 1981 to 1985 to examine the application of satellite-

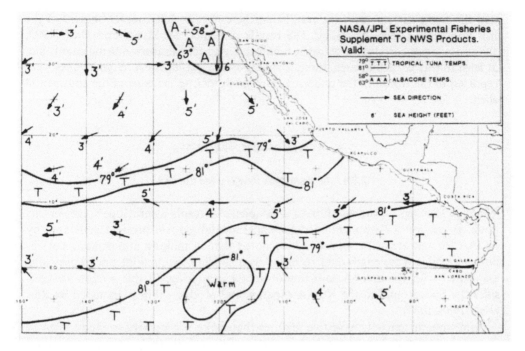

Figure 7.39 Tropical area fisheries-aid chart.

derived near-real-time sea surface temperature data (from the AVHRR) in assisting the bluefin tuna fishing industry along the southwestern coast of Australia. The key objectives of this project were to ascertain whether satellite imagery could bring economic benefits to fishermen in this part of Australia. These scientific objectives were concerned with precisely identifying the correlation, if any, between sea surface temperature measurements and fish catches and so determining the feasibility of a prediction service to the fishing industry.

Previous work on near-real-time use of satellite data in connection with fisheries (Lasker *et al.* 1981, Fiedler *et al.* 1984, Komura *et al.* 1984, Laurs *et al.* 1984), together with general biological knowledge, suggested that the project would find a strong correlation between sea surface temperatures computed from the satellite data and fish catches. Recently the use of AVHRR data, verified by ocean surveys, has provided detailed information regarding ocean movements off Western Australia (Cresswell and Pearce 1985, Godfrey and Ridgeway 1985). An important feature is the Leeuwin current, a warm current which flows south from the tropics and then flows eastwards along the south coast of Western Australia. Eddies form on the sides of this current. The distribution of southern bluefin tuna, a species of tropical origin, is not uniform within the Leeuwin current. Rather it appears that the fish congregate at its outer edge to take advantage of the better food supply. Any eddy which forms principally from this edge water may therefore reduce the concentration of fish in the current and have abnormally high concentrations of fish within itself. As the feature continues in a southerly and then easterly direction, its diminishing size and reducing temperature will further concentrate the fish. On the passage eastward, interaction with abnormal flow patterns caused by subsea anomalies which are locally termed canyons, results in a mixing of water from the eddy and the main current, with the result that there may be local replenishment of the fish numbers in these locations. The results presented by Myers and Hick (1990) showed a weak correlation between tuna catch and temperature but indicated that, at least in this particular area, there is an important spatial effect. High catches were reported in the region of the undersea canyons dissecting the edge of the continental shelf.

7.5 ATMOSPHERIC AEROSOLS

7.5.1 Aerosol path length over the sea

We have touched on the question of atmospheric aerosols at previous stages in this book. In section 4.2 we mentioned the experimental aerosol product generated by NOAA for sea areas. In section 3.3 we noted that in making atmospheric corrections to visible wavelength data recorded by a satellite the greatest uncertainty was associated with the aerosol contribution to the scattering. In this section we consider the background to the NOAA experimental product; this is discussed by Rao et al. (1988, 1989).

Atmospheric aerosols have two features that make it necessary to study them in considerable detail. The first is that, by virtue of their radiative properties, they play an important role in weather-related and climate-related phenomena. Changes in their total abundance in the vertical column (or normal extinction optical thickness) and in their physical and optical properties, namely the size distribution and absorption and scattering of radiation as determined by the refractive index of the aerosol material, and in their spatial and temporal distribution, affect the radiative transfer in the atmosphere in a manner which is dependent on the wavelength of the radiation. Secondly the aerosol concentrations in the atmosphere are highly variable, spatially and temporally. The variations in the atmospheric aerosol burden arise from natural causes and from anthropogenic causes such as industrial activities and changes in land use practices such as deforestation. Thus, when attempting to conduct quantitative studies it is seldom adequate to use some geographically typical or seasonal average of the aerosol concentration; it is the actual aerosol concentration which needs to be known in order to be able to determine the aerosol contribution to the radiative transfer (see section 3.3). Other contributions to the radiative transfer are either constant, such as the contribution from Rayleigh scattering or the effect of minor gaseous components, or can be determined from atmospheric sounding, i.e. the contribution from the water vapour concentration. The problem of determining the aerosol concentration is of particular importance if one is attempting to calculate atmospheric corrections to visible-band scanner data particularly for water quality studies (see section 7.3).

The determination of the aerosol concentration from red or near-infrared data is possible over the sea because (i) the reflectance from the sea at these wavelengths is small and its value is known, (ii) the other atmospheric effects, due to Rayleigh scattering and minor gases' contributions, are small and constant and can be calculated quite accurately and (iii) the effect of water vapour is negligible (Durkee et al. 1990). Consequently the aerosol contribution (optical path length) is both (a) a major fraction of the received signal and (b) the only unknown quantity. Therefore over a cloud-free and ice-free area of ocean with no floating biological or mineral material near the surface the aerosol optical thickness can be determined from the satellite-received signal. This is not, in general, possible over the land because (i) the reflectance of the land surface is larger than that of the sea and (ii) the value of the reflectance of the land is (usually) unknown.

Some calculations of the upwelling radiance (at the top of the atmosphere), for particular aerosol distributions, as a function of optical thickness are described by Rao *et al.* (1989), see Figure 7.40. Two aerosol size distributions were used; these are the Junge power law size distribution of the form $n(r) = Cr^{-4.5}$ and the modified gamma distribution of the form $n(r) = Ar^a \exp(-Br^g)$, where $n(r)$ is the number of aerosol particles of radius r per unit volume per unit increment in radius, and A, B, C, a and g are model-specified constants. Both partly absorbing ($m = 1.5 - i0.01$) and purely scattering ($m = 1.50$) aerosols were included in the model computations and it was assumed that the underlying surface had a Lambertian albedo of 0.015.

The aerosol retrieval algorithm currently in use at NOAA NESDIS was developed by Griggs (1975, 1983). It uses only channel-1 data and consists of a look-up table of computed radiances at the wavelength of 0.65 μm for aerosol extinction optical thicknesses ranging from 0 (purely molecular atmosphere) to 1.18 in increments of 0.118. Atmospheric aerosols are assumed to be purely scattering spherical Mie scatterers with a refractive index of 1.50 and a modified power law size distribution of the form $n(r) = C$, for $r_{min} \leq r \leq r_m$, $n(r) = C(r/r_m)^{-4.5}$, for $r_m \leq r \leq r_{max}$ and $n(r) = 0$ for $r < r_{min}$ and $r > r_{max}$. Values of 0.02 μm, 0.1 μm and 10 μm for r_{min}, r_m and r_{max} respectively have been used. The underlying surface, representing the ocean, has been assigned a Lambertian albedo of 0.015. Radiance computations have been made for the angular parameters given below:

solar zenith angle θ_0 : 42° to 84° in steps of 6°,
satellite zenith angle θ: 0° to 84° in steps of 6° and
relative azimuth angle ϕ: 140° to 180° (antisolar azimuth) in steps of 10°.

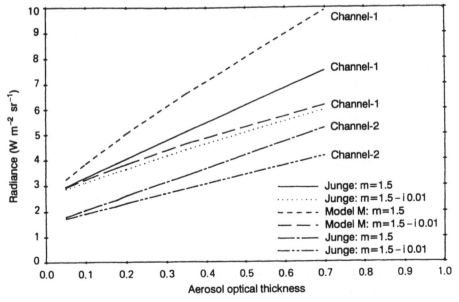

Figure 7.40 Computed values of the channel-1 and channel-2 upwelling radiance as a function of aerosol extinction optical thickness at 0.55 μm, $\theta_0 = 42°$, $\theta = 54°$ and $\phi = 180°$ (Rao *et al.* 1989).

The range of azimuth angles was governed by the fact that, in practice, only radiance measurements made in the antisolar half of the cross-track scan of the AVHRR are used to retrieve atmospheric aerosols to avoid the effects of sunglint.

Using appropriate interpolations, the value of the aerosol extinction optical thickness at 0.65 μm that would yield the same radiance as the measured value is determined from the look-up table for given values of solar zenith, satellite zenith and relative azimuth angles; the corresponding model-specified value of the aerosol extinction optical thickness at 0.50 μm, the wavelength at which atmospheric turbidity is usually expressed (see also section 7.6), is then inferred. The 0.50 μm optical thicknesses thus inferred over a period of seven days – the time required to obtain coverage of the greater part of the global oceans – are then composited and analysed, using computer-based objective analysis techniques, to generate a 100 km grid aerosol optical thickness field.

As we have already indicated, the determination of the aerosol optical thickness by this method requires to be done over a cloud-free area of the ocean. In addition to uniformity of brightness temperature and reflectance tests, a cloud screening test based on the differences between the sea surface temperatures determined using the split-window and triple-window MultiChannel Sea Surface Temperature (MCSST) retrieval algorithms (McClain et al. 1985) was implemented. It was found that cloud contamination is invariably present when this difference exceeds 5°C. The design and implementation of this test are given by McClain (1989). The work of Rao et al. (1989) was aimed at using both channel-1 and channel-2 AVHRR data, instead of just channel-1 data, as an additional or substitute cloud screening criterion.

Rao et al. (1989) considered the ratio of measured channel-1 and channel-2 albedos under a variety of sky conditions and illumination and observation geometries and observed that this ratio is appreciably greater than unity when there are no clouds present within the field of view of the instrument (see Figure 7.41). On the other hand when clouds are present this ratio tends to be very close to, or slightly less than, unity. This is also borne out by model computations.

Given that a certain set of assumptions are made in the NOAA NESDIS aerosol algorithm, there are a number of possible sources of error in the algorithm. It was felt that radiance measurements made in channel 2 could be utilised to implement consistency checks that would constitute part of model validation. However, because of the presence of moderate to strong water vapour absorption bands within the wavelength range of this channel, it is necessary to examine how changes in atmospheric precipitance water would affect the upwelling radiance. Recent modelling studies by Masuda et al. (1988a) have indicated that as the amount of precipitable water is changed from ∼0.9 cm to 2.9 cm, the upwelling radiance decreases by ∼5 to 8.5% (of the values corresponding to 0.9 cm), depending upon the various angular parameters and the aerosol model used in the computations. The feasibility of dual-channel and multichannel aerosol retrieval algorithms is presently being studied at NOAA/NESDIS.

Kaufman (1993) reported the results of sets of simultaneous measurements of the path radiance and the optical thickness from the ground at over 30 locations spread all over the world. All the measurements were performed with a single eight-channel portable sunphotometer/radiometer in the 0.44–1.03 μm range. The observations were taken for constant solar and view directions, resulting in a constant scattering angle of 120°, which resembles space observations. One set of measurements was used to develop empirical relations between the aerosol spectral optical thickness

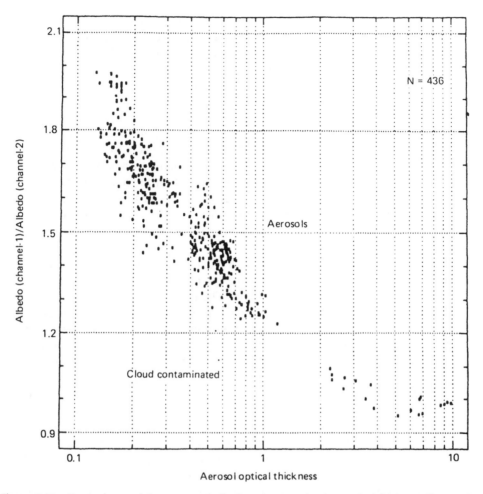

Figure 7.41 Dependence of the measured albedo ratio on extinction optical thickness (Rao *et al.* 1989).

and the scattered spectral path radiance. A second independent set was used to test these relations. It was shown that simple measurements performed from the ground can yield empirical relations that can be used to check some of the common but not validated assumptions about the particle homogeneity, sphericity, composition, and size distribution used in remote sensing models and in estimates of the radiative effects of aerosol. The introductory section of the paper by Kaufman (1993) also contains a good review of work on aerosols in relation to the analysis of remotely sensed data.

7.5.2 Dust storms, volcanic ash clouds and fires

The aerosol particles in the distribution studied in section 7.5.1 come from many causes, but undoubtedly some of those particles originate from dust storms and volcanic eruptions. However, there are special features that are associated with the observation of the high concentrations of these particles in dust storms and volcanic eruptions themselves. It is there that the airborne particles are generated before they eventually become widely dispersed. We have already noted in section 7.5.1 that the main importance of determining the aerosol optical path length is in connection with the atmospheric correction of visible wavelength remotely sensed data. However, at the very high concentrations of particles in dust storms and volcanic ash clouds there are other effects too. The propagation of thermal infrared radiation is affected and special corrections have to be made in calculating sea surface temperatures from AVHRR thermal infrared data (see section 4.2.3). In very intense volcanic ash clouds there is the problem of mechanical damage to aircraft and especially to the engines of an aircraft and so there is an important possibility for the use of satellite data, in near-real-time, to provide information to enable aircraft to take evasive flight paths. In this section we address the questions of (i) identifying dust storms and tracking the clouds of dust that they produce and (ii) identifying and tracking volcanic emission clouds, especially in the first few hours of their existence when they pose a serious threat to aircraft.

Many examples of AVHRR images of dust clouds or dust storms exist, see Figure 7.42 (in the colour section). Each year large quantities of dust are transported as dust clouds out of the deserts of the world. These movements of dust clouds can produce hazardous conditions along major transportation routes and may have climatic implications through their interaction with the solar and terrestrial radiative fields (Morales 1979, Péwé 1981). Since surface observations are generally sparse in desert regions, it is often difficult to locate the origin as well as the movement of a dust outbreak; there is a clear advantage in the use of satellite data. For example, Griggs (1975), Carlson (1979) and Norton et al. (1980) have utilised satellite measurements in the visible spectrum to monitor dust clouds as well as to estimate dust optical depth over oceanic regions. Monitoring aerosol outbreaks over land using satellite visible and near-infrared data is difficult due to the bright underlying desert surface. Shenk and Curran (1974) demonstrated the utility of the 10.5–11.5 μm window channel to track dust storms over land.

Ackerman (1989) described a technique based on the use of the differences between the channel-3 and channel-4 AVHRR data for tracking dust clouds/storms. Measurements of the refractive index of dust indicate that large differences exist at wavelengths of 3.7 and 11 μm. The imaginary part of the refractive index at 11 μm is approximately an order of magnitude greater than at 3.7 μm. Therefore, large differences may exist in the equivalent blackbody temperatures at these wavelengths in the presence of a dust layer, thus providing the basis for a method to locate and track dust outbreaks. Values of the single scattering albedo, ω_0, and asymmetry parameter, g, were calculated by Ackerman (1989) from Mie theory making various reasonable assumptions about the parameters involved, see Table 7.5. The ratio of σ_{ext} for channel-3 and channel-4 wavelengths to σ_{ext} at 0.5 μm is also given in Table 7.5. Since optical depth, τ, is a function of wavelength, results are given as a function of optical depth at a standard wavelength of 0.5 μm ($\tau_{0.5}$). This is a common wavelength for measuring atmospheric dust loading and $\tau_{0.5}$ is referred to as the turbidity. Some examples of the calculated temperature difference, $T_{B3} - T_{B4}$, are shown in Figure 7.43 as a function of turbidity for a satellite nadir viewing angle. The thick lines represent the case of a dust layer overlying a desert surface with surface albedos of 0.05 and 0.4 for the channel-4 and channel-3 spectral regions, respectively. The solid line represents the day-time case with a surface temperature of 315 K and a solar zenith angle of 43°. The broken thick line represents the desert

Table 7.5 Single scattering properties of the dust layer at 3.7 and 11 μm (Ackerman 1989)

	3.7 μm	11 μm
ω_0	0.9869	0.4162
g	0.6507	0.5038
$\sigma_\lambda/\sigma_{0.5}$	0.77	1.07

night-time case with a surface temperature of 290 K. The non-zero values of ΔT for the dust-free atmosphere result from differences in water vapour absorption and surface spectral properties at the two wavelengths. The big difference between the day-time and night-time curves is due to the fact that in the day-time the upwelling channel-3 wavelength radiation contains a significant component arising from reflected solar radiation and this, of course, is absent at night. The case of a dust layer overlying an ocean surface temperature of 298 K is represented as the thin lines in Figure 7.43. The dashed line excludes the solar component and can be considered the night-time case. Again there is a large contribution from solar reflectance during the day-time. Both over the land and over the sea the magnitude of the temperature difference is smaller at night than it is during the day.

There are several atmospheric variables which will affect satellite measured radiances at 3.7 and 11 μm and which therefore have to be taken into account in calculating $(T_{B3} - T_{B4})$. These include atmospheric temperature and moisture profiles, the microphysical properties of the dust, the vertical distribution of the dust, solar zenith angle, satellite viewing angle, and the surface radiative properties. To test the predictions of the type given in Figure 7.43 one would need turbidity (or optical depth) measurements to be made. In the absence of simultaneous satellite data and

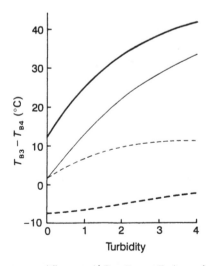

Figure 7.43 Radiative temperature difference $(\Delta T = T_{B3} - T_{B4})$ as a function of turbidity for a satellite viewing angle of 0° and a solar zenith angle of 43°. Thick lines represent a desert case while thinner lines represent an ocean case. Solid and dashed lines represent the day and night cases, respectively (Ackerman 1989).

ground-based turbidity data, i.e. ground-based measurements of the optical path in a *vertical* direction, Ackerman (1989) compared measurements of ($T_{B3} - T_{B4}$) from the AVHRR with surface visibility data (essentially measurements of the reciprocal of the optical path in a *horizontal* direction). This was done with data from the occasion of a large dust outbreak over the Saudi Arabian peninsula in late June 1979. Surface observations indicated the dust outbreak region to be cloud-free and therefore values of ΔT are not a result of cloud. Temperature differences of greater than 35°C are observed which would, based on Figure 7.43, indicate optical thickness values greater than 2. The spatial distribution of the ΔT values was found to be consistent with the evolution of the outbreak and, as might be expected, gradients of ΔT were found to delineate the boundaries of the outbreak at a greater horizontal resolution than the surface observations. Temperature differences measured by the satellite were well correlated with visibility, see Figure 7.44.

Values of the emissivity of quartz and of Sahara dust over the wavelength range of 7–17 µm for a range of particle size and zenith angle are given by Takashima and Masuda (1988).

The history of the use of meteorological satellite data to detect and monitor volcanic eruptions and their ash clouds has been summarised by Rao *et al.* (1990) and Walter (1991). Data from both geostationary and polar-orbiting satellites have been used. An extensive list of volcanic eruptions which have been studied with satellite data is given by Holasek and Rose (1991). The geostationary satellite data have the advantage that images are acquired at half-hourly intervals and therefore the temporal evolution of the eruption cloud can be followed in some considerable detail. A sequence of GOES images for the El Chichón eruption is given by Rao *et al.* (1990). El Chichón (which is situated at 17.33° N, 93.20° W) first erupted with a large explosion on 28 March 1982. Intermittent emission continued until three

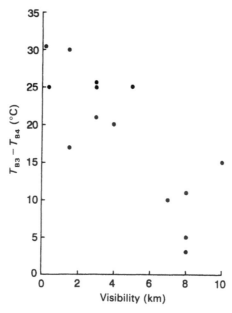

Figure 7.44   Satellite measured ΔT as a function of surface visibility for 23 June 1979 (Ackerman 1989).

further explosions occurred on 3 April and 4 April 1982. Six hours after the first of the two eruptions of 3 April 1982 the ash cloud had spread both to the northeast and to the southwest and covered an area of 86 500 km^2. The last of the eruptions was on 4 April 1982 and the complete cloud from this and the previous eruptions covered an area of 1.2 M km^2. The northeast moving part of the ash cloud was primarily tropospheric and the southeast moving part was stratospheric.

It is necessary to distinguish between the initial eruption cloud (i.e. less than a few hours) and the dispersed phased (i.e. greater than ~ 12 hours). We shall refer to the initial cloud as the eruption cloud and the later phase as the dispersed volcanic cloud. This separation is important because the compositional changes between the two phases give rise to different optical, radiative and dynamical effects. The eruption cloud consists mainly of ash particles with the possibility of large (~ 5 mm) particles and very large (a few cm) ash agglomerations, termed volcanic hail (Hobbs *et al.* 1981). The cloud is optically thick, high and probably very inhomogeneous.

Large quantities of ash are deposited relatively quickly from a volcanic eruption cloud, but the smaller particles spread out and remain in the stratosphere for a long time. We have already mentioned this in section 4.2.3 because it has to be taken into account in calculating atmospheric corrections to sea surface temperatures derived from AVHRR data. In the case of El Chichón the ash and aerosol belt circling the Earth has been studied by Robock and Matson (1983), Thomas (1983), DeLuisi (1983) and McCormick (1983).

AVHRR data were used by Tucker and Matson (1985) to determine the extent of the heavy tephra fallout in the surrounding area. Tephra samples were collected from 99 sites around El Chichón in mid-April. Three separate layers ejected by the explosions of 28 March, 3 April and 4 April were evident near the volcano. Further away from El Chichón only the 28 March and 4 April tephra could be distinguished. Ground samples indicated that the axis of maximum deposition extended approximately northeast from the volcano for the 28 March explosion and roughly east from the volcano for the 4 April explosion (Varekamp *et al.* 1984) (Figure 7.45). Tucker and Matson (1985) used AVHRR data obtained from NOAA-7 on 5 March 1982 at 14.30 hours (local time), i.e. before the eruptions, and from NOAA-6 on 13 April 1982 at 07.30 hours, i.e. after the eruptions. Both images were largely cloud-free and centred over the El Chichón area. The 13 April image coincided with the mid-April 1982 ground collection of 99 tephra depth measurements in a 70 000 km^2 area (Varekamp *et al.* 1984). Additional tephra thicknesses which were obtained from 3–7 April 1982 were reported from 30 sites west, north and east of El Chichón (Hoffer *et al.* 1982). The satellite imagery was mapped to a Mercator projection and the NDVI was generated. Laboratory spectral reflectance measurements were made for a specimen of tephra, a healthy soybean leaf and for soybean leaves lightly dusted with tephra and heavily dusted with tephra, see Figure 7.46. Deposition of tephra on the leaves results in increased reflectance (i.e. decreased pigment absorption) in channel 1 and decreased reflectance (i.e. decreased internal scattering) in channel 2 (Figure 7.46). Accordingly, the NDVI is reduced by deposition of volcanic dust on leaves versus the dust-free situation. The area for several hundred kilometres around El Chichón is heavily vegetated and thus the tephra-caused interference with the normal solar irradiance-green-leaf interaction could be determined by comparing the pre- and post-eruption NDVI images. This enabled a synoptic view to be obtained over a wide area and the ground data could then be used for the calibration of the AVHRR data in terms of tephra fall-out concentra-

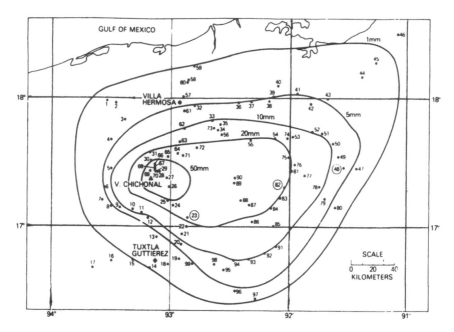

Figure 7.45 Isopach map showing the thickness of compacted ash at a density of 1.2 g cm^{-2} for tephra ejected from El Chichón volcano 28 March–4 April 1982 (Tucker and Matson 1985).

tion. The area of greatest reduction in the NDVI occurred within an 80 km radius. Sizeable areas of tephra fall-out were also evident extending in all directions from El Chichón as far as 500 km in some directions. NOAA weekly global NDVI data were analysed from April 1982 and March–April 1983 and confirmed that the reduction in the April 1982 AVHRR data was not the result of vegetation seasonality.

The 99 ground-collected tephra samples were scattered within the immediate vicinity of El Chichón. The geographical coordinates of these sampling locations were determined and the associated pre-eruption (5 March 1982) and post-eruption (13 April 1982) NDVI values were calculated. The differences between the 5 March and 13 April NDVI values were also calculated. Regression analysis was performed on the tephra depth, pre-eruption and post-eruption NDVI values, and the difference of the NDVI values. This regression relation was then used to calibrate a map of the NDVI difference (between 5 March and 13 April 1982) in terms of tephra (fall-out) concentration. Some cloud was present in each of the two AVHRR images used and so only areas free of cloud on both days were analysed for tephra deposition. The total tephra area was planimetered and determined to be >270 000 km². A combination of the satellite's capability to detect large areas of tephra deposition along with the tephra NDVI correlation might allow volcanologists to determine more accurately the quantity of material ejected from an eruption.

One important aspect of volcanic eruption clouds is the danger that they present to aircraft. The potential danger to aircraft was demonstrated during the eruptions of the Galunggung volcano (7° 15′ S, 108° 03′ E) on the Indonesian island of Java in June and July 1982, when in two separate incidents Boeing 747 aircraft on routine commercial flights experienced difficulties on encountering the volcanic eruption clouds (Prata *et al.* 1985). Engines stalled on both aircraft and emergency landings

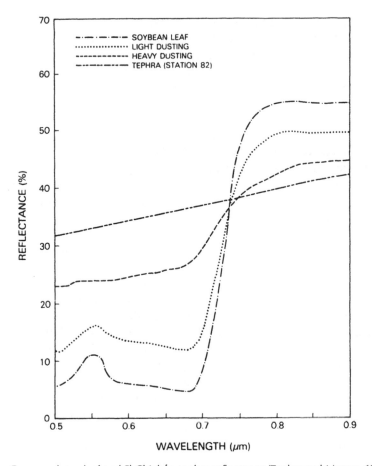

Figure 7.46   Green soybean leaf and El Chichón tephra reflectance (Tucker and Matson 1985).

were necessary. Fortunately no loss of life occurred, but the damage to the aircraft was considerable. Hanstrum and Watson (1983) have discussed these incidents in more detail. As a result of these incidents, the International Civil Aviation Organisation (ICAO) formed a study group to investigate the problems related to aviation from volcanic eruptions. One of the issues of interest is the use of satellite imagery to define the horizontal and vertical extent of volcanic clouds. Most of the larger particles are deposited within a short distance (a few hundred kilometres) from the volcano, leaving only small ash particles (micron size), volcanic gases (SO_2, HCl and CO_2) and water vapour as the major constituents of the drifting volcanic clouds. These clouds may be transported rapidly around the Earth. For instance, a volcanic cloud from the Mount Hudson eruption of 15 August 1991 was encountered by three airliners over southeastern Australia only five days later (Barton *et al.* 1992).

It is thus clear that it is very important to devise an operational scheme for use by meteorological forecasters which will give early warning of the presence of volcanic clouds. In particular it is important to be able to distinguish volcanic clouds from ice clouds and clouds containing water vapour and water droplets. It is difficult to do this from visible and near-infrared (channel-1 and channel-2) AVHRR data. This problem was addressed by Prata (1989a,b) who studied the Galunggung eruptions and Barton *et al.* (1992) who studied the Mount Hudson eruption. Prata (1989b) performed radiative transfer calculations for thermal infrared radiation for volcanic ash clouds that were assumed to consist of a mixture of ash particles, ice spheres, water droplets and ash particles coated with sulphuric acid. These calculations showed that the temperature difference $\Delta T = T_{B4} - T_{B5}$ between the channel-4 and channel-5 brightness temperatures provided a possible means for distinguishing between ice clouds and volcanic ash clouds. ΔT is usually positive for ice and water clouds (see also Volz 1973, Yamanouchi *et al.* 1987), whereas for volcanic pumice, quartz and acid droplets ΔT is negative, provided the mean particle size is less than about 3 μm. Contours of ΔT derived from AVHRR data following the Mount Galunggung volcano eruption are shown in Figure 7.47. Negative values of ΔT occur in several locations near to the volcano and in a region some distance downwind from the main vent. Large positive differences also occur (region A in Figure 7.47), where there is presumably a high proportion of entrained water vapour and (or) ice particle formation. We have already mentioned that the large particles are quickly deposited whereas the small particles remain in the stratosphere and can travel large distances. Thus in Figure 7.47, showing the initial eruption cloud, there are both positive and negative values of ΔT. However, in the case of the dispersed cloud from Mount Hudson, observed five days later and thousands of miles away from the volcano, when only the small particles remain, the negative ΔT is clearly apparent, see Figure 7.48. Negative values of ΔT were also observed in the ash cloud from the Mount Pinatubo eruption of 12 June 1991 (Potts 1993, Reynolds 1993). The question of the determination of the sizes and total masses of particles in volcanic clouds using channel-4 and channel-5 AVHRR data has been studied by Wen and Rose (1994) for the cloud of the Crater Peak/Spurr volcano, Alaska, from the eruption of 19 August 1992. They were able to estimate that the

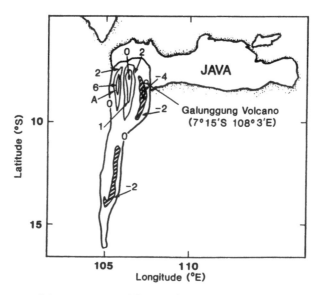

Figure 7.47 Contours of the temperature difference ΔT derived from observations of an eruption of the Mt Galunggung volcano determined from NOAA-7 AVHRR data (Prata 1989b).

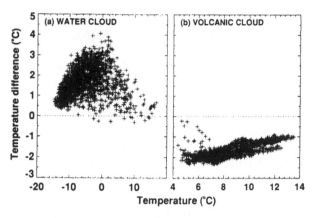

Figure 7.48 Two-dimensional scattergram ($(T_{B4} - T_{B5})$ versus T_{B4}) of (a) water clouds and (b) volcanic clouds (Barton *et al.* 1992).

mean particle radius was of the order of 2–2.5 μm and that the total amount of ash in the cloud was $(0.24-0.31) \times 10^6$ tons.

As far as potential hazards to air traffic are concerned it is the period shortly (a few hours to a few days) after the initial eruption when the volcanic cloud is most likely to be a hazard to air traffic. In an operational environment, where AVHRR imagery is available in real time, a volcanic cloud signature can be recognised by computing an image of the $T_{B4} - T_{B5}$ temperature differences. Regions of negative $T_{B4} - T_{B5}$ are indicative of volcanic clouds. It is therefore possible to use this as an operational tool to assist in the forecast of volcanic cloud hazards to air traffic. However, it is possible for negative $T_{B4} - T_{B5}$ differences to occur under certain unusual atmospheric conditions (not associated with cloudiness) and over clear arid land. Therefore these possibilities should be eliminated first. Before such a forecast is made the distribution of any active volcanoes in the region should be known and recent pressure and streamline charts consulted. An operational scheme using real-time AVHRR data is cost-effective and when combined with meteorological analyses and other data (e.g. geostationary satellite imagery) should provide a means for monitoring the movement of the volcanic cloud and avoiding air-traffic incidents like those mentioned above that occurred as a result of the eruption of the Galunggung volcano in Java in 1982.

Measurements of the effects of El Chichón volcanic ash on atmospheric haze in south Texas were made by Richardson (1984). The objective was to relate ground measurements of atmospheric optical depth and radiance values of the NOAA-7 AVHRR to increases in atmospheric haze conditions in south Texas after the El Chichón volcanic eruptions. Direct irradiation measurements were made at Weslaco, Texas. From these measurements the seasonal cosine correlation of the nadir atmospheric transmission (T) to day of year (D) before and after the eruption were found to be

$$T = 0.686 + 0.106 \cos \omega t \tag{7.5.1}$$

and

$$T = 0.617 + 0.092 \cos \omega t \tag{7.5.2}$$

respectively, where $\omega t = 2\pi D/365$. This indicates a definite reduction in the atmospheric transmittance resulting from the eruption. Richardson (1984) analysed data from the AVHRR on NOAA-7 (an afternoon satellite) for eight dates between October 1981 and July 1982; four dates were before the eruption and four dates were after the eruption. Both channel-1 and channel-2 radiances were increased by the volcanic ash. This is shown in Figure 7.49 which shows data from all five AVHRR channels along one scan line passing through the volcanic ash cloud very soon after the El Chichón eruptions. It will be seen from Figure 7.49 that the signal in channel 3 is substantially reduced by the volcanic ash cloud and that the effect on the signals in channels 4 and 5 is relatively small. This work demonstrates that the volcanic ash cloud significantly affects the channel-1 and channel-2 received radiances. In turn this means that the effects of volcanic dust cannot be ignored if one is attempting to make atmospheric corrections to channel-1 and channel-2 data, for instance if one wishes to eliminate atmospheric effects when calculating vegetation indices.

In addition to desert dust clouds and volcanic ash clouds, there are also aerosol particles arising from the smoke plumes of large fires, whether of natural origin or initiated by human actions. We have discussed such fires already (see section 6.2), but not from the point of view of the smoke aerosol. Very often the smoke plumes associated with surface fires are less significant than the clouds produced by large volcanic eruptions but, of course, there are exceptions. For example, towards the end of the Gulf War, in early 1991, a very large number of oil wells were set on fire in Kuwait and they burned for many months before they were able to be extinguished. The dispersal of the smoke from the oil fires in Kuwait was of interest for several scientific reasons, the main one being regional environmental impact, and others being hazardous air conditions for health and air traffic, possible impact on

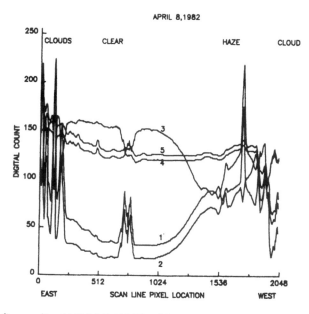

Figure 7.49 Scan line profile of NOAA-7 AVHRR, all five channels, from 8 April 1982 in the Gulf of Mexico showing the effect of dust from the El Chichón volcano eruptions on 4 April 1982 (Richardson 1984).

global climate, and an unfortunate but unprecedented opportunity to study the properties of the smoke itself and its behaviour over time. From 10 May to 14 June 1991, a field experiment was undertaken to study the physical, chemical, and radiative properties of the smoke from the oil fires. *In situ* sampling of the smoke was conducted by the University of Washington and the National Center for Atmospheric Research (NCAR) research aircraft. Satellite data from Meteosat and from the AVHRR were used in the planning of the aircraft missions, and the potential environmental impact of the smoke (Dech and Glaser 1991, Limaye *et al.* 1992). This work of Limaye *et al.* is interesting because it points out the spectral signature difference between carbon-based particles (soot) and high-quartz-content particles (desert dust, vocanic ash).

7.6 PRECIPITATION AND WATER RESOURCES

Water is one of our most precious natural resources and hydrological and water management problems now represent key areas of applied science in the current worldwide economic and ecological situations. Accordingly the need to assess and forecast water supplies has increased rapidly. Satellite remote sensing technologies have been used more and more to satisfy these urgent requirements for information. No other method is capable of providing the necessary data in near-real-time in regular temporal sequences, and on a regional, continental or global scale, to measure, monitor and model the complex hydrological processes in the environment. Several of the elements involved in the use of remote sensing in hydrology are discussed in other sections in this book, e.g. evaporation and soil moisture (section 5.6), sediment discharge and pollution (section 7.3) and temperature (section 4.2). In this section we shall be concerned only with rainfall and snow, both in terms of precipitation and run-off. In other words the discussion of the use of AVHRR in hydrology is scattered among several sections in this book.

In meteorological work one uses an integrated set of data from a variety of different sources and polar-orbiting satellite data are only one of these sources. One cannot, therefore, sensibly separate AVHRR data from other sets of satellite data or from conventional *in situ* data when studying rainfall or, more generally, all types of precipitation. Table 7.6, which is due to Barrett (1989), summarises the various current and proposed rainfall monitoring methods involving the use of satellite data. Chapter VIII-6 of the book by Rao *et al.* (1990) provides a useful summary which we shall make use of and which we shall, however, amplify in some aspects. Microwave data from Earth-observing satellites, e.g. from the SMMR (Scanning Multichannel Microwave Radiometer), can also be used in precipitation estimation (see section VII-6.3 of Rao *et al.* (1990)).

The geostationary satellites provide the most useful data for precipitation estimates; the data from the polar-orbiting satellites are of somewhat more limited use. The main reason for this is that the geostationary weather satellites, such as Meteosat or GOES, generate visible and infrared images of their whole area of coverage every 30 minutes. This is a very appropriate frequency for the generation of time-lapse sequences of developing weather systems. For most areas of the globe the polar-orbiting satellites of the TIROS-N/NOAA series of satellites provide much less frequent and more irregular coverage of any one location on the surface of the Earth. The exception to this is in the polar regions where the frequency of coverage

Table 7.6 Summary of current and proposed satellite rainfall monitoring methods (after Barrett 1989)

Method	Chief applications	Satellite	Sensor (s)	Present status
Cloud indexing	meteorology, climatology, hydrology, crop prediction, hazard monitoring, etc.	polar orbiting geostationary	visible and/or infrared	–
Climatological	crop prediction	polar orbiting	visible and/or infrared	quasi-operational
Life history	severe storm assessment, meteorological research	geostationary	infrared	–
Bispectral	meteorological research	polar orbiting geostationary	visible and infrared	–
Cloud physics	cloud research and atmospheric thermodynamics	geostationary	infrared and microwave	developmental
Passive	oceanic meteorology and climatology	polar orbiting	microwave	–
Active	cloud and rainfall research forecasting	? polar orbiting	satellite radar (active microwave)	future

by the AVHRR is much higher than at the equator and where the ground area is out of sight of any geostationary satellite. In an area away from the polar regions the frequency of coverage for AVHRR data depends on latitude and it also depends on the extent to which one is prepared to use data from large off-nadir view angles. Satellite data are widely used in the study of clouds and of precipitation for large areas in climate modelling (see section 7.9) but in this section we are concerned with the question of trying to make precipitation estimates over smaller areas.

There are two basic types of techniques for precipitation estimation, namely cloud indexing and cloud history techniques. Cloud indexing is the principal method used where only polar-orbiting satellite data are available. This technique involves characterising a cloudy area by an index number based either on imagery or on the quantitative values of the radiances, and then empirically estimating its precipitation potential. The cloud history technique is used with geostationary satellite data and it enables cloud fields to be followed throughout their life cycles. This technique will not be described in this book; details will be found, for example, in section VII-6.2.2 of Rao *et al.* (1990).

Rainfall estimates can be based on the dominant synoptic or subsynoptic weather pattern, the amount of cloud cover and the expected intensity of precipitation associated with cloud types.

Three cloud indexing techniques are described in chapter VII-6 of Rao *et al.* (1990), in which further references will be found.

i. The Follansbee technique (Follansbee 1973) This is straightforward and simple. A 24 hour rainfall estimate for an area is based on a weighted sum of the percentages of the area covered by cumulonimbus, nimbostratus and cumulus congestus clouds. Empirical rainfall intensity weights are applied to these percentages: 1.0 for cumulonimbus, 0.25 for nimbostratus and 0.02 for cumulus congestus. Visible and infrared pictures are used to determine the cloud types and the percentages of the area occupied by each type. Local climatologies can be used to adapt the weighting factors for particular areas.

ii. The Earth Satellite Corporation (ESC) technique This technique operates on a 54 km grid and provides six-hour rainfall estimates based on satellite observed cloud top temperatures and cloud areas, vertical motion from the NMC spectral model, and the time of day. The method requires four pictures a day from two polar satellites to compute estimates every six hours. When only two pictures are available, the six-hourly intermediate cloud fields are interpolated from twelve-hourly satellite data and numerical prediction diagnostic fields.

iii. The Bristol technique Methods for estimating areas of precipitation from satellite imagery have been extensively developed by the University of Bristol (Barrett *et al.* 1986, Barrett 1989). First attempts in the 1970s used manual techniques based on interpreting hard copy imagery and the determination of cloud indices translated into rainfall estimates through local specific regressions. These evolved towards interactive methods, resulting from a collaboration with NOAA, and were followed by attempts to develop an objective data analysis based upon life history or bispectral techniques. The latter enable a classification of rain/no-rain areas, while the former aim at estimating rainfall from infrared temperature thresholding, climatic information and surface station reports (Figure 7.50). This approach has also been developed by the University of Reading and at the Centre de Météorologie Spatiale at Lannion in France. The European Space Operations Centre also routinely produces a precipitation index deduced from Meteosat 11 μm channel data eight times a day (Turpeinen 1988). The latest results are from the quasi-linear inverse relation between the cumulative surface temperature on clear days with rainfall on a long-term basis (the whole rainy season or one month) established in the Sahelian region (Seguin *et al.* 1989). A more recent review is given by Liberti (1993).

One problem with all of these methods, which try and infer rainfall amounts from infrared radiances, is that there is no direct relation between cloud top temperature and precipitation. Hence these methods will only be useful when averaged over long time periods, when it is reasonable to make the assumption that the frequency of occurrence of large convective clouds can be indirectly related to likelihood of rainfall. Microwave satellite data, which come outside the scope of this book, offer the possibility of improved estimates of rainfall from satellites.

A classification scheme for clouds in AVHRR, using about 20 different classes, has been developed at the Swedish Meteorological and Hydrological Institute over a number of years. It is now operational (Karlsson 1989a, Liljas 1989) and a method has now been devised to use the results of this classification to estimate precipitation rates (Karlsson 1989b, Pylkkö and Aulamo 1991, Aulamo and Pylkkö 1993). It can be regarded as a development of the cloud inducing techniques and the initial results from its use appear to be encouraging. For the GPCP-AIP/2 (Global Precipitation Climatology Project – Algorithm Intercomparison Project) data set, a

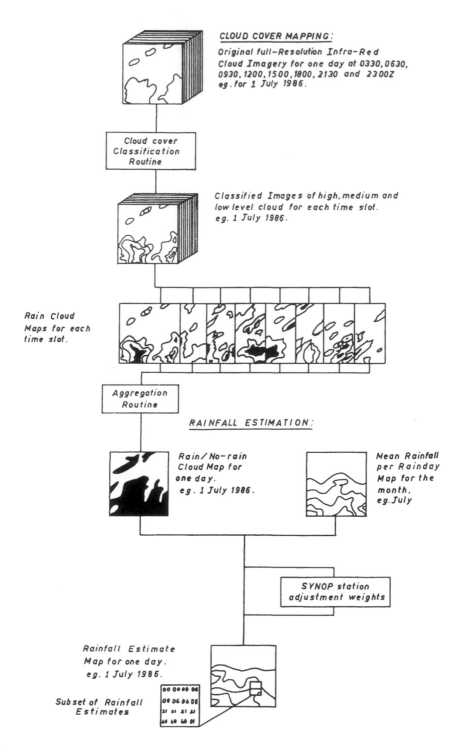

Figure 7.50   PERMIT (Polar-orbiting Effective Rainfall Monitoring Interactive Technique) production of daily rainfall estimates (Barrett 1989).

statistical analysis of the AVHRR data has been carried out in order to test the quality of the data, to study the statistical properties of the data and to help to select interesting cases from the point of view of estimating precipitation (Liberti 1991).

We did not describe the use of data from Meteosat, or other geostationary satellites, for rainfall estimation because the subject matter of this book is restricted to AVHRR data and their use. However, the use of data from geostationary satellites for the estimation of rainfall has become one of the themes of research in tropical meteorology. Most of the methods used are based on attempts to find a correlation between cloud top temperatures and/or the brightness characteristics measured by the satellite and rainfall measured on the ground by radar or rain gauge. Generally, either a threshold temperature on an infrared image or a threshold brightness on a visible image is defined. The threshold values are used to determine the active rain cloud areas which are correlated with echo radar or the measured rainfall. In addition to the references cited in the account, to which we have already referred, in section VII-6.2.2 of Rao *et al.* (1990), several further references are cited by Kakane and Imbernon (1992). These methods have been found to be complex and involve many parameters. Several studies carried out in recent years and cited by Kakane and Imbernon (1992), with the geostationary satellite Meteosat, show the existence of a quasi-linear relation between surface temperature (T_S) and rainfall (P). The reason for the use of Meteosat is that it provides images every 30 minutes. Kakane and Imbernon (1992) describe the adaptation of the method to use AVHRR data. The reasons quoted for using AVHRR instead of Meteosat included the existence of the two thermal infrared channels in the AVHRR, the good calibration of these thermal infrared channels, the possibility of making accurate atmospheric corrections to the AVHRR data to determine the surface temperature, T_S and the authors' possible improved access to AVHRR data from the direct readout ground station at Niamey in Niger. Kakane and Imbernon used rainfall data from 57 meteorological stations spread throughout Senegal and 34 AVHRR scenes from the period between 24 June and 8 October 1987. Surface temperatures were calculated using the split-window method with the set of coefficients of Lagouarde (1991), i.e.

$$T_S = 2.84 + 3.77 T_{B4} - 2.77 T_{B5} \ . \tag{7.6.1}$$

Fifteen-day compositing was used to eliminate values from cloud-covered pixels. From the 34 AVHRR images, eight composite images were obtained. This was followed by:

(1) extraction of surface temperatures over 3 × 3 pixels centred on each reference station for the resulting eight images;

(2) correlation between T_S (accumulated) values and P (accumulated); by accumulated T_S is meant the mean maximum surface temperature for the period of accumulation.

The results for the 57 stations are plotted in Figure 7.51 along with the results of a least squares fit tabulated in Table 7.7. The correlation increases during the rainfall season and shows a maximum at the end of September. The low values for June indicate the slow commencement of rainfall and the increase is due to the increase in rainfall as the rainfall season progresses. The drop in r^2 for October marks the approach of the end of the rainfall season.

This work showed that it is possible to use AVHRR data to estimate accumu-

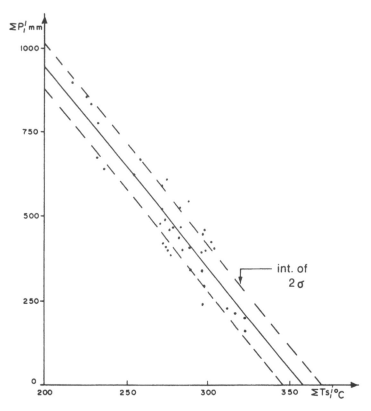

Figure 7.51 Relation between accumulated rainfall and accumulated temperature for Senegal for 15 June–30 September 1987 (Kakane and Imbernon 1992).

lated rainfall provided one restricts the data to the rainfall season. The reason why it works is that the process of evapotranspiration, following the fall of the rain, diminishes the surface temperature and the duration of this reduction will be longer following heavier rainfall.

Apart from considering remote sensing in relation to precipitation, one can also consider the role of remote sensing, and of AVHRR data in particular, to runoff prediction. There is a recent review by Haefner and Pampaloni (1992) which concen-

Table 7.7 Relation between accumulated rainfall (ΣP) and average surface temperature $(\bar{T}_s)$: $\Sigma P = a + b\bar{T}_s$ (Kakane and Imbernon 1992)

Date	a	b	r^2	σ
30 June	75.4	− 1.4	0.11	27.5
15 July	412.6	− 8.4	0.38	49.8
31 July	610.7	− 10.9	0.47	56.8
15 August	883.1	− 16.2	0.58	60.6
31 August	1325.9	− 24.9	0.73	64.1
15 September	1678.0	− 31.9	0.81	63.9
30 September	2140.6	− 41.8	0.86	65.4
15 October	2205.0	− 42.5	0.79	84.9

trates on work within Europe. This is concerned with processes in which land surfaces and inland waters are involved (i.e. excluding ground water, water pollution, oceans and coastal zones, etc.) and leading to an examination of the operational use of remote sensing. The use of remote sensing in relation to rainwater runoff is still at an embryo stage. Remotely sensed data in general, and AVHRR data in particular, are only one of many input data sources for assessments of rainwater runoff. An integrated approach combining different remote sensing systems with ground-based measurements and with the mathematical models is the most promising approach for a long-term operational monitoring system. It is essential to achieve a careful structuring and organising of the data, such as a Geographical Information System (GIS) combined with a Digital Terrain Model (DTM) where each set of data is properly geocoded (Haefner 1987, Haefner and Pampaloni 1992). Unfortunately insufficient attention is given in general to these pre-processing procedures.

Modelling the variations of a hydrological system is based on three consecutive steps (Schultz 1987, p. 426):

(a) specification of the mathematical structure of the model – deterministic or stochastic, lumped system or distributed system type, linear or non-linear, etc;

(b) model calibration, i.e. system identification by computing the model parameters for the system under consideration;

Table 7.8 Concept for surface runoff estimation of rainfall (after Schultz 1987)

1. Rainfall model, transforming remote sensor data into rainfall data.
Input Data. Life history: hydrometeorological variables or observation series from the weather services. Satellite data: IR data from polar orbiting satellites, NOAA, TIROS-N. IR data from geostationary satellites, GOES, Meteosat, GMS. Ground-based weather radar, air photos.
Model Type. Mean daily temperature-weighted cloud-cover index, bi-spectral cloud model, microwave method, linear interpolation of daily observed parameters, transfer function in convolution integrals, bi-spectral cloud model, simple regressions.

$$\downarrow$$

2. Runoff model, transforming rainfall data into surface runoff.
Input Data. Parameters from the rainfall model in spatial distributed parameters like cloud-cover index, estimation of rainfall, flood hydrographs, soil cover, soil type, snow/ice cover, drainage density, soil moisture, etc.
Model Type. Generally black box mathematical models, distributed type models for small scales.

$$\downarrow$$

3. Runoff forecast.
Monthly river runoff Flood forecast Flood and runoff forecast Daily runoff

GMS: Geostationary Meteorological Satellite.

(c) specification of the model input to be used for the simulation of the hydrological process in the hydrological system, with the aid of the mathematical model
and the model parameters identified under (a) and (b) above.

In order to be useful, remote sensing data must be transformed into a hydro-
meteorological output (e.g. rainfall or runoff). The best structure for hydrological
models based on remote sensing data is not clear at present. Models were developed
for real-time forecasting and for rainfall estimation using weather satellite data. The
models developed apply a transfer function which transforms the input (a fractional
cloud-cover index of infrared density range valid for the catchment area) into a daily
runoff indicator. The concept for surface runoff estimation of rainfall is represented
in Table 7.8 covering the three consecutive steps: rainfall model, surface runoff
model and runoff forecasting.

The use of remote sensing in monitoring snow-melt runoff has been much more
successful than in monitoring rainwater runoff. The reason is as follows. As we have
seen, the estimation of rainfall from cloud studies using remote sensing data is difficult and somewhat indirect; moreover, once the rain has fallen it soaks in or runs off
very quickly and cannot be observed directly by satellite in the place where it fell.
Snow, on the other hand, once it has fallen lies for a while in the place where it fell
so that at least the area covered by snow can be determined using AVHRR data (see
section 7.1). The fact that there is a time lapse between snowfall and run-off (at
melting) means that the forecasting of snow-melt runoff is feasible and it is now
carried out successfully in certain areas. This has already been discussed in section
7.1.

7.7 GEOLOGICAL STUDIES

The use of air photos and of high-resolution satellite data (Landsat MSS, TM,
SPOT HRV, etc.) for geological studies is well established and described in various
texts. At first sight it might seem that there was little hope of being able to use
AVHRR data for geological studies. The first reason is the low spatial resolution of
the AVHRR. The second is the fact, which applies to other sources of remotely
sensed data as well as the AVHRR, that large areas of the surface of the Earth are
thickly covered with vegetation which partially or completely obscures the geological features. However, there are some situations in which it is possible to extract
geological information from AVHRR data. We shall mention briefly some work on
(i) the identification of ancient glacier marks, (ii) the investigation of seismoactive
regions and (iii) volcanic craters and lava flows. The first of these involves using
visible and near-infrared data, while the second and third involve using thermal
infrared data.

The work of Johnston et al. (1989) on the identification of ancient glacier marks
was basically a large-area or small-scale photointerpretation exercise. Air photos
have been widely used for studying ancient glacier movements and some workers
have also used Landsat images as well (e.g. Punkari 1982, 1984, Boulton et al. 1985).
But there are problems associated with the expense of Landsat data and the need to
do complicated mosaicking of separate scenes to cover large areas. It is perhaps
rather surprising that one should simply take AVHRR (visible or near infrared)
images as photographs and apply conventional photointerpretation methods to
them. One might suppose that the spatial resolution of AVHRR was so poor that

no useful information would be obtained; nevertheless Johnston *et al.* set out to see if AVHRR data could be used to give information about ancient glaciers and were successful.

It is common knowledge that, at certain stages in the past, large areas of currently temperate regions were covered by ice. Ice, of course, moves and leaves lineations on the landscape and it is from this surviving geological evidence that one can try to reconstruct the distribution and dynamics of the ice sheets. The geological products of glaciation include both erosional features and depositional features. These are clearly visible on conventional aerial photography and in the field. They had been studied to some extent with Landsat MSS data too. However, the ice sheets covered enormous areas and piecing together hundreds, or even thousands, of air photos, or tens or hundreds of Landsat images, is far from trivial. The cost of the data and of their analysis would also be very great. Exactly the features of AVHRR data which make AVHRR attractive for studying vegetation indices at continental or global scale, i.e. its large area of coverage and its cheapness, made it attractive for this purpose too.

The area chosen for this preliminary study was Finland, see Figure 7.52; this area

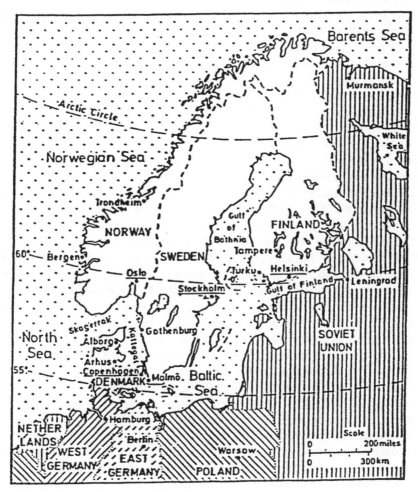

Figure 7.52   Study area for ancient glacier marks (Johnston *et al.* 1989).

was chosen because glacial markings are particularly pronounced in this region, despite the vegetation. Suitable cloud-free images were selected following a browse file search. Scenes from near- to mid-summer were chosen as these have the highest digital values and least chance of snow cover. The two scenes for which digital data were obtained were from 4 June 1980 and 16 July 1982. The raw digital data for these two scenes were geometrically rectified and contrast enhanced and filtered to enhance linear features. Principal components were also used and a colour composite involving the first three principal components was produced. Photointerpretations were made on coloured hard copy images along with the original black and white (unrectified) images. Figures 7.53–7.55 show the main results extracted from the AVHRR data. Figure 7.53 shows ice-marginal features plotted from the imagery. Relations between these features can be seen such as the major radial lineation pattern of central Finland which is bounded by the Salpausselkä moraines; these moraines represent a substantial halt in the ice retreat which occurred about 11 000 years ago. The Salpausselkäs do not represent the ultimate extent of the ice, since at their peak the Pleistocene ice sheets covered Poland and parts of Russia. Figure 7.54 shows both erosional and depositional lineations. A comparison of Figure 7.54 with Figure 7.53 shows clearly the relation between lineations and ice-marginal features. These diagrams were used to suggest general lines of ice movement and ice-lobe structure (Figure 7.55). Most of the movements are north-south, but in the west one can see a west-east flow coalescing with the north-south flow.

The spatial resolution of the AVHRR has both positive and negative aspects for this particular type of study. On the positive side the data are relatively cheap and cover a larger area in one image than any of the other commonly available polar-orbiting Earth-observing systems. The 'global' view offered by this wide coverage is particularly beneficial where continental areas are being considered, such as in this example. At such a scale the systems of glacial dynamics and ice-sheet structure can be perceived and the interrelations between structures such as moraines and lineation patterns can be examined. The major negative aspect of the low resolution of the AVHRR is the somewhat obvious difficulty encountered in identifying smaller morphological forms. This restricts the usefulness of AVHRR data for mapping more detailed morphology and was confirmed by a comparison for the Lestijarvi area between the results from the AVHRR data (Johnston *et al.* 1989) and the Landsat MSS results of Punkari (1982). The results from the two sources of data were consistent with one another in general terms, but differences were found, most likely arising from the poorer spatial resolution of the AVHRR data.

The possibility of using AVHRR thermal infrared data in the study of seismoactive regions for the prediction of earthquakes has been studied by Tronin (1996). The study of lineaments, morphological structures and neotectonic movements in seismically active regions using high-resolution (Landsat, SPOT, etc.) data in the visible and near-infrared regions of the electromagnetic spectrum is commonplace. As we have just seen, AVHRR data can also be used to some extent despite their much lower spatial resolution. However, the present example is concerned with the use of thermal infrared data. As we have seen earlier (in Chapter 4) the intensity of the thermal infrared emission from the Earth's surface depends on the emissivity and the surface temperature. The ground temperature, in its turn, depends on solar heating, surface thermal properties and meteorological conditions, as well as on processes occurring in the near-surface atmosphere and in the upper crust. It is the last of these factors, processess occurring in the upper crust, which is relevant for

Figure 7.53 Moraine complexes determined from AVHRR data (Johnston *et al.* 1989).

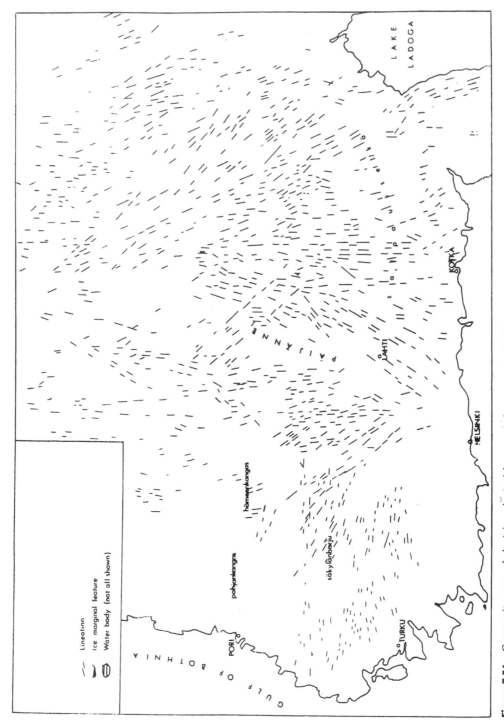

Figure 7.54 Composite map of glacial geological features (Johnston et al. 1989).

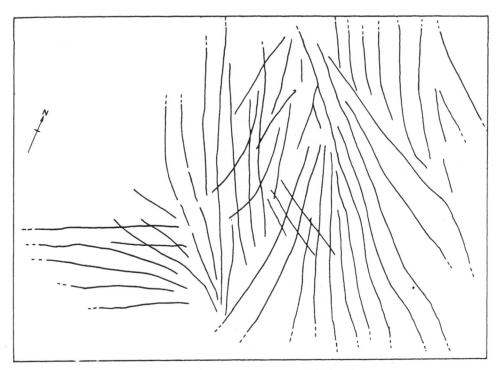

Figure 7.55   Ice directions and lobe structure (Johnston *et al.* 1989).

studies in seismoactive regions. It has been reported that pre-seismic activity alters the characteristics of the soil, including its moisture, gas content and composition. There are also numerous observations of surface and near-surface temperature changes prior to core earthquakes. For example, soil temperature anomalies of 2.5°C were measured in the zone of preparation of the Tangshan earthquake in China in 1979 (which had a magnitude of more than 7.0 on the Richter scale). Thermal changes due to stress fields have also been determined in laboratory studies of materials (for references see Tronin 1996). A change in near-surface properties can be expected to change the value of the thermal inertia and, therefore, to change the diurnal surface temperature pattern, see section 4.4. The frequent acquisition of AVHRR thermal infrared data enables the period prior to an earthquake to be monitored closely.

The examination of night-time thermal infrared images obtained over central Asia several days before the large Gazli earthquakes in 1976 and 1984, both with magnitude greater than 7.0 on the Richter scale, revealed the appearance of abnormal thermal anomalies, exceeding $100\,000\ km^2$ in area. This observation provided the incentive to study a long series of satellite images of this seismically active region of Central Asia, the Tien-Shan, Kizilkum and Karakum deserts, South Kazakhstan. The great number of cloudless days in the central Asian region and the flat relief provide favourable conditions for thermal infrared studies of the land surface. Night-time data (an hour before dawn) were chosen to avoid the effect of solar heating over the region. Both the high seismicity of the area and the favourable survey conditions form a convenient target for studying the relation between the outgoing thermal infrared radiation flux and seismicity.

The analysis of a continuous series (100–250 scenes) of night-time AVHRR data for a period of 10 years allowed the identification of a set of thermal anomalies in the central Asian seismoactive region. For example, a positive anomaly of several degrees Celsius at the foot of Kopetdag has a linear shape of 25–30 km in width and about 500 km in length. This anomaly is related to the Kopetdag Fault, the boundary structure of the first order, separating Alpine Kopetdag formations from the Turan Plate. Moreover, the anomaly coincides with the 'thermal line' of the Kopetdag hot water basin, a unique hydrogeological structure. A second anomaly of about 50 km in width and 300 km in length occurs at the foot of the Karatau Range. Spatially it coincides with the Karatau Fault, the cutting structure of the first order separating the Turan Plate from the Central Asian Folded Zone. The Karatau Fault proves to be the extension of the deep Talas-Fergana Fault, which in many respects controls the geodynamics of the region. These anomalies are stable. However, some stable anomalies contain components with variable area and intensity. Such a variable component is classified as a non-stable anomaly and various examples of non-stable anomalies were found.

The previous studies of Gorny et al. (1988) on the Gazli earthquakes of 1976 and 1984, both with magnitudes larger than 7.0, had shown that there was a possible correlation between the non-stationary thermal anomaly in the point of intersection of the Tamdy-Tokrauss and the Karatau Faults and the seismic activity in the region of Gazli. The examination of night-time AVHRR thermal infrared images taken over this region several days before the events revealed the appearance of this anomaly. After the main Gazli earthquake of 19 March 1984, there was established a background distribution of the outgoing infrared radiation flux. In the image from 29 July the background distribution of the terrestrial outgoing radiation flux was observed. In the images from 30 July, 1 August and 2 August there was noted the appearance of an anomaly over the Tamdy-Tokrauss Fault. In the images from 3 August and 4 August the anomaly had disappeared and there was again observed in this region only the background distribution. On 5 August 1984 there occurred a 4.3 earthquake in the area of Gazli within the zone of the Tamdy-Tokrauss fault, with its epicentre at 40° 20′ N, 63° 35′ W. After the earthquake, between 6 August and 10 August there was registered in the region another appearance and development of the thermal anomaly at the point of intersection of the same faults. The maximum area of the anomaly was observed in the data from 7 August and 8 August. Beginning on 10 August 1984, the distribution of the outgoing infrared radiation flux reverted to the background state. There followed, on 14 August 1984, 5.3 and 4.9 earthquakes in the Gazli area that were genetically connected with the Tamdy-Tokrauss fault. In other words in each case what was observed was the appearance of an anomaly which lasted for a few days and then disappeared and the disappearance was followed, a few days later, by an earthquake. Having described these observations, Tronin (1996) then attempted to quantify and model the origins of thermal anomalies associated with seismic activity. We should, perhaps, make the point that studying earthquakes in desert regions on a historical basis and establishing correlations between thermal anomalies and earthquakes is not quite the same thing as making predictions of earthquakes, not just in desert areas but in areas that are highly developed and densely inhabited (such as in California or Japan).

Another potential geological application of AVHRR data is in studying volcanic craters and lava flows. Studies using Landsat TM short-wavelength infrared bands 5 and 7 have shown potential for measuring the temperature, radiant flux and emittance of volcano hotspots (Rothery *et al.* 1988, Glaze *et al.* 1989). While it is interesting to try to use AVHRR data in this context, the spatial resolution is a problem. Since the sizes of the craters and lava flows are often smaller than the IFOV of the AVHRR the mixed pixel method developed by Matson and Dozier (1981) and described previously in section 6.2 can be used in this context too. This method was applied to the study of two Italian volcanoes, Stromboli and Vulcano, by Oppenheimer (1989). Consider, for instance, a pixel including a lava lake with surface temperature T_h surrounded by ground at temperature T_b. The channel-3 and channel-4 or channel-5 pixel-integrated temperatures should lie between T_b and T_h but differ because of the wavelength dependence of the Planck function. From them, both T_h and p, the hot fraction of the pixel, can be calculated, if one assumes a value for T_b. Figure 7.56 shows a curve relating target temperature, background temperature and pixel fraction, p, where the channel-3 and channel-4 observed temperatures were 48.3°C and 18.9°C, respectively (a real example from AVHRR data from Mount Etna, Italy). However, there are problems with this approach, particularly in volcanic situations:

(i)   A two-component surface temperature field is unrealistic with such large pixels. An AVHRR IFOV with side ~ 1 km straddling a volcano will sample ground of a range of altitudes, aspects and surface cover, so selection of a background temperature is not trivial. It is also unreasonable to model the hot volcanic surface (e.g. lava body or fumarole field) by a single temperature. The simplest thermal model of lava surfaces comprises two temperatures, those of glowing cracks and cooler crust.

(ii)   Even if we are willing to accept a two-component temperature distribution, it is readily seen in Figure 7.56 that uncertainty in T_b is magnified greatly in deriving the dual-band target temperature. A very small, very hot source may not produce a resolvable anomaly in channel 4, in which case the dual-band calculation becomes invalid.

(iii) The AVHRR instantaneous field of view is not well constrained; generally pixels are 'blurred' both across and along scan lines (see Figure 3.8). It is

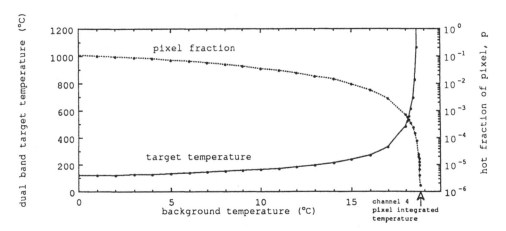

Figure 7.56 Example of dual-band target temperature, T_h, and fraction of IFOV calculated for various values of T_b (Oppenheimer 1989).

important to account for this in calculations that involve areas and in the summation of radiant flux for two or more areas. This is particularly relevant when comparison of multitemporal data sets is undertaken.

(iv) Conditions in a volcano's crater fluctuate rapidly. During the field work on Stromboli by Oppenheimer (1989) incandescent scoria were frequently erupted from intracrater boccas, some of which were filled with lava. Radiant temperatures of these vents (measured from crater rims at distances of about 50 m) fluctuated rapidly due to the activity of lava inside, and the amount of volcanogenic fumes in the field of view. Such temporal variation brings into question the quantitative value of an isolated satellite 'snapshot' of volcanic activity.

Oppenheimer was able to observe a thermal anomaly in channel-3 AVHRR data for Stromboli but only in one of the four scenes used (although nearby Mount Etna, also in a strombolian phase at the time, showed a clear channel-3 anomaly in three AVHRR scenes); this supports the idea that what one sees depends on the instant of the satellite scene capture. As expected, the geothermal flux at Vulcano is too small to detect (Oppenheimer and Rothery 1989). Thermal imagery of Mount Erebus from NOAA-6 has been studied by Wiesnet and d'Aguanno (1982). The question of remote sensing of geothermal fluxes, using AVHRR data, in areas of high geothermal, but not volcanic, activity has been studied by Mackay (1992); however, the size of the AVHRR field of view coupled with the relatively small heat fluxes made it difficult to produce conclusive results (Cracknell et al. 1989).

7.8 SUNGLINT AND NEAR-SURFACE WINDSPEEDS

Sunglint is a phenomenon that has been widely neglected or ignored in the use of satellite remote sensing techniques to study the oceans. This section is concerned with a description of the effect of near-surface winds on sunglint patterns and of the established theory of Cox and Munk for establishing a quantitative relation between near-surface windspeed and the intensity distribution in an area of sunglint. The results from some of our own calculations (Cracknell et al. 1987, 1988) based on using this theory with AVHRR data from the Mediterranean will be presented. Correlations between near-surface windspeeds and the warming of the surface layer, as manifested in thermal infrared channel data from the AVHRR, will also be discussed. One of the best descriptions of sunglint is to be found in the report by Fett (1977) which, however, is not very easy to obtain. The present section is based on the write up of a summer school lecture on the subject (Cracknell 1990).

In most applications of scanner data one is either dealing with diffusely reflected radiation, in the visible or near-infrared bands, or with emitted radiation, in the thermal infrared bands. In such cases sunglint, that is the direct specular reflection of sunlight at the surface of the sea, is something that people usually regard as a nuisance and they try to avoid it. Thus, for example, in the NOAA aerosol algorithm the data from the half of the scan line towards the Sun is not processed in case it may contain sunglint. Also on the Nimbus-7 satellite there was special provision made to introduce a tilt with a mirror so as deliberately to avoid observing any areas of sunglint. That was reasonable enough in the context of what CZCS was concerned with, i.e. studying ocean colour and water quality. In the present section we take a different view and try positively to use sunglint to obtain some useful oceanographic or meteorological information (see also La Violette *et al.* 1990). Basically, it is possible to study the area and intensity of a region of sunglint and to extract information about the surface roughness, i.e. waveheight, and therefore the near-surface windspeed.

Suppose that the surface of the sea is perfectly flat, i.e. it behaves as a plane mirror. Then if the satellite's scanner (the AVHRR for instance) is looking in the right direction for specular reflection there will be a very bright signal, see Figure 7.57. The area of this bright image will be very small; in terms of an AVHRR image it would occupy a rather small number of pixels (about 100 perhaps). There is, however, one complication for which Figure 7.57 does not allow. The geometry of Figure 7.57 applies to a camera in which the whole scene is imaged simultaneously. With a scanner such as the AVHRR a complete scene is imaged over a period of several minutes, during which time the satellite travels a large distance. The details of the geometry of the imaging process become much more complicated and the general consequence is that the sunglint pattern becomes extended in a direction parallel to the satellite's motion, which is approximately north-south (or vice versa).

One can obviously just look manually through an archive of satellite data and search for examples of scenes with sunglint data. However, this is tedious. A computer program for the automatic identification of areas within AVHRR scenes at which sunglint is to be expected (if clouds are absent) has been described by Cracknell *et al.* (1987). The European Space Agency's operation at ESRIN, Frascati, in Italy routinely includes a test for sunglint along with its tests for clouds in the preparation of its AVHRR quicklook products (see section 2.5.3).

So far we have assumed that the surface of the sea is perfectly smooth and occasionally that may be very nearly true. However, waves are nearly always present and we need to examine that situation in some detail. Suppose that the sea surface is not smooth but has a single wave present, see Figure 7.58. In this figure the scale of the waves, relative to the IFOV (instantaneous field of view) of the scanner is wrong. Consider a wavecrest (or trough) in the IFOV; the original geometry still survives because the tangent plane is horizontal. For parts of the

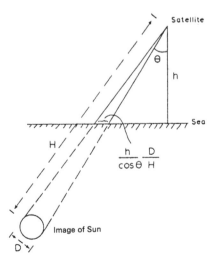

Figure 7.57 Sketch to illustrate specular reflection from the Sun at a smooth surface (not to scale).

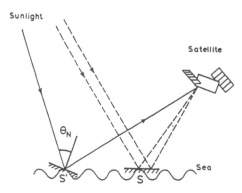

Figure 7.58 Sketch to illustrate specular reflection from the Sun at a surface with the shape of a sinusoidal wave.

wave near to a trough or crest, where the inclination of the tangent plane is quite small, the specular reflection conditions will still lead to the reflected ray reaching the scanner on the satellite. But for other parts of the wave, where the inclination of the tangent plane is greater, the geometrical conditions required for sunglint no longer apply. Thus, the intensity received from the area S is reduced. The greater the amplitude (height) of the waves then, qualitatively, the greater the reduction in the sunglint intensity that can be expected. However, if we consider an area outside S then there will be some part of the wave for which the tangent is now sloping at an angle for which it is in just the right orientation to satisfy the conditions for specular reflection. Thus, whereas for a calm sea we would get no specular reflection except from the small area S, now we get some specular reflection from an area outside S. Thus we see, in qualitative terms, that if the sea is not calm the sunglint will be (a) enlarged in area and (b) reduced in intensity. The theory of Cox and Munk (1954, 1956), which was developed long before anyone could reasonably foresee the vast amount of relevant satellite data that would become available, is concerned with obtaining a quantitative relation between the sunglint pattern and the near-surface windspeed. In this theory one considers a piece of wave (not in S) from which specular reflection occurs; we define the following:

θ = zenith $\quad\big\}$ of reflected ray relative to the Sun,
α = azimuth

β = angle of incidence = angle of reflection,

ξ = solar zenith angle at the ocean surface, and

θ_N = zenith angle of the normal at the point on the wave at which reflection occurs.

Then, from the appropriate spherical trigonometry, one finds that

$$\cos \theta_N = \frac{\cos \theta + \cos \xi}{2 \cos \beta}.$$ (7.8.1)

Cox and Munk showed empirically that for uniform ocean surface roughness there was a Gaussian distribution of wave slopes with a distribution function

$$P(\theta_N, V) = \frac{1}{2\pi\sigma^2}\left(\exp -\frac{\tan^2 \theta_N}{\sigma^2}\right)$$ (7.8.2)

where V is the near-surface windspeed. The standard deviation was found to be related (empirically) to the windspeed V by

$$\sigma^2 = 0.00512V + 0.003 \tag{7.8.3}$$

where V is expressed in m s^{-1}. Then we have to relate the satellite-received radiance, $L_s(\lambda)$, to the probability distribution of the slopes, i.e. of the θ_N. $L_s(\lambda)$ can be written in the form

$$L_s(\lambda) = L_w(\lambda) + L_{sk}(\lambda) + L_g(\lambda) \tag{7.8.4}$$

where

$L_w(\lambda)$ = diffuse radiance reflected from the water surface,
$L_{sk}(\lambda)$ = diffuse sky radiance, and
$L_g(\lambda)$ = sunglint radiance.

If we consider an area that is far away from the sunglint area then only $L_w(\lambda)$ and $L_{sk}(\lambda)$ are present. Therefore we take $L_s(\lambda)$ from the satellite data for the sunglint area and we assume that we can take $L_w(\lambda)$ and $L_{sk}(\lambda)$ from the satellite data from elsewhere in the scene and thus determine $L_g(\lambda)$. This does assume that $L_w(\lambda)$ and $L_{sk}(\lambda)$ do not vary significantly between the sunglint area and elsewhere. Then we are able to make use of the expression

$$\frac{L_g(\lambda) \cos^4 \theta_N \cos \theta}{L_s(\lambda) r(\omega)} = P(\theta_N, V) \tag{7.8.5}$$

where $r(\omega)$ is the Fresnel reflectivity, to determine $P(\theta_N, V)$. Thus for a given pixel in the sunglint area we find $L_g(\lambda)$ and $L_s(\lambda)$ from the satellite data, we do all the geometry to find θ_N for this pixel and then use equation (7.8.5) to determine $P(\theta_N, V)$. Then equation (7.8.2) can be used to determine σ from $P(\theta_N, V)$ and then, from this value of σ, equation (7.8.3) can be used to determine V, the windspeed corresponding to this pixel. Thus it is possible to determine a pixel-by-pixel map of the near-surface windspeed throughout the sunglint area.

An example of some results obtained by Cracknell *et al.* (1988) for 34 points in the sunglint area in the data from 15 June 1982 are shown in Table 7.9. The weighted meteorological data values were obtained from windspeed data for the Mediterranean for that day from the UK Meteorological Office. Two points should be noted. First, the Meteorological Office's windspeeds were not given at exactly the locations used in the sunglint-derived windspeed calculations and some interpolation had to be done to account for this. Secondly, there was a time difference, of the order of 60–90 minutes, between the acquisition of the meteorological data and of the satellite data. The differences between the satellite-derived and Meteorological Office-derived windspeeds in Table 7.9 vary from $+1.0$ to -0.8 m s^{-1}. This agreement is quite close, but it is not clear how much of the error is attributable to the satellite-derived values and how much is attributable to the Meteorological Office-derived values.

What we have described so far has assumed a uniform windfield. However, this is often not the case in practice. It is common to find dark areas within a sunglint pattern and there are various important cases to consider.

(a) There may be dark areas within the region near the centre of the sunglint pattern. These correspond to calm areas and are illustrated schematically in Figure 7.59 which is taken from some calculations of McClain and Strong (1969). This

Table 7.9 Windspeeds (Cracknell *et al.* 1988)

No.	Weighted meteorological data ($m\ s^{-1}$)	Computed ($m\ s^{-1}$)	Difference ($m\ s^{-1}$)
1	4.7	4.0	0.7
2	4.3	4.1	0.2
3	4.9	4.1	0.8
4	4.9	4.1	0.8
5	4.9	4.2	0.7
6	5.0	4.2	0.8
7	5.0	4.8	0.2
8	5.0	4.4	0.6
9	5.0	4.5	0.5
10	5.0	4.8	0.2
11	5.0	5.0	0.0
12	4.8	5.2	− 0.4
13	4.6	4.6	0.0
14	4.2	4.9	− 0.7
15	4.2	4.9	− 0.7
16	4.7	4.9	− 0.2
17	4.7	4.7	0.0
18	5.0	4.7	0.3
19	4.7	4.9	− 0.2
20	4.6	5.4	− 0.8
21	4.8	5.4	− 0.6
22	5.1	5.7	− 0.6
23	4.6	3.7	0.9
24	5.0	4.4	0.6
25	5.0	4.7	0.3
26	4.9	5.2	− 0.3
27	4.9	4.4	0.5
28	5.1	4.3	0.8
29	4.9	5.3	− 0.4
30	4.8	5.2	− 0.4
31	6.2	5.2	1.0
32	5.0	4.4	0.6
33	5.5	4.5	1.0
34	5.0	4.8	0.2

figure is based on calculations of the sunglint intensity for a uniform windspeed of 5 $m\ s^{-1}$ except in the shaded area, which would appear dark on an image, for which a windspeed of zero was assumed.

An example of such a feature in an AVHRR image is shown in Figure 7.60. The large dark patch SE of Mallorca and the smaller dark patches north of Mallorca and Menorca in Figure 7.60(a) (visible) and Figure 7.60(b) (near-infrared) are all indicative of calm areas. In Figure 7.60(d) (thermal infrared) these calm areas show up as exhibiting local heating of the surface layer (see Robinson *et al.* 1984, Ramp *et al.* 1991 for example) and fit quite exactly with the areas shown in channels 1 and 2 (Figures 7.60(a) and (b)). This is printed, as band 4 in Figure 7.60(d) with dark corresponding to a high intensity of radiation and light, e.g. the clouds in the north-

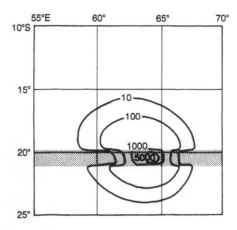

Figure 7.59 Theoretical sea-surface sunglint for a composite pattern. Isopleths are relative reflected intensity per 10^4 steradian incident flux; satellite height 722 km (McClain and Strong 1969).

(a)

Figure 7.60 NOAA-7 AVHRR images 1348 GMT on 1 July 1982 (a) channel 1.

(b)

Figure 7.60　(b) Channel 2.

west corner of the scene, corresponding to a low intensity of radiation. For channel-3 data (3.5–3.9 μm) shown in Figure 7.60(c) the sunglint area is dominated by reflected radiation at this wavelength; this is printed, like Figure 7.60(d), with dark corresponding to a high intensity of radiation and light corresponding to a low intensity. The calm areas noted from the other bands reflect less radiation to the scanner (as in Figures 7.60(a) and (b)) and hence appear light in Figure 7.60(c) too; they are also very well co-registered with the channel-1 and channel-2 data shown in Figures 7.60(a) and (b).

(b) There may by dark areas well away from the PSP (primary specular point) and in the large sunglint area of low intensity indicating calm conditions as illustrated schematically in Figure 7.61. This, like Figure 7.59, is taken from McClain and Strong (1969) and is based on a uniform windspeed of 5 m s^{-1} everywhere except in the shaded area, for which a windspeed of zero was assumed.

(c) If, however, the windspeed becomes quite large, then the sunglint pattern becomes very much reduced in intensity. An example is shown in Figure 7.62 where a high wind rushing out through the Straits of Gibraltar from the Mediterranean destroys the sunglint in an area to the west of the Straits, see, for example, Scorer

(c)

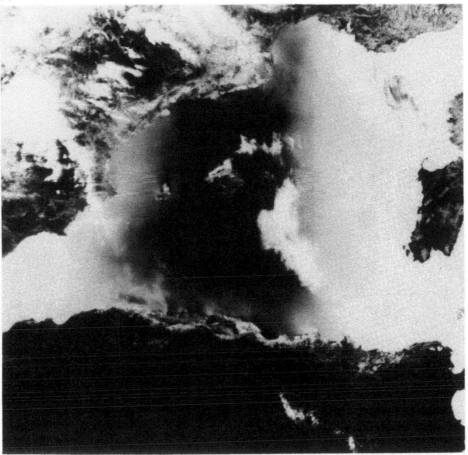

Figure 7.60   (c) Channel 3.

(1986). According to Strong and Ruff (1970) this occurs at windspeeds of about 10 m s^{-1}. That is, dark patches may occur or, if the windspeed is large over the whole area, the whole sunglint pattern will disappear completely. Thus with dark patches, particularly in the extended sunglint area, one needs to be careful because they may arise from very calm conditions or they may arise from rather rough sea conditions and high windspeeds; ambiguities may need to be resolved by reference to other data.

7.9 CLIMATOLOGY

7.9.1 Global data sets

A very good summary of the applications of Earth-observing satellite data in climatology is given in section 3 of a recent review by Saunders and Seguin (1992) and a large number of references will be found there. However, that is concerned with data from all satellite systems; in this section we shall be more selective because this book is only concerned with AVHRR data.

(d)

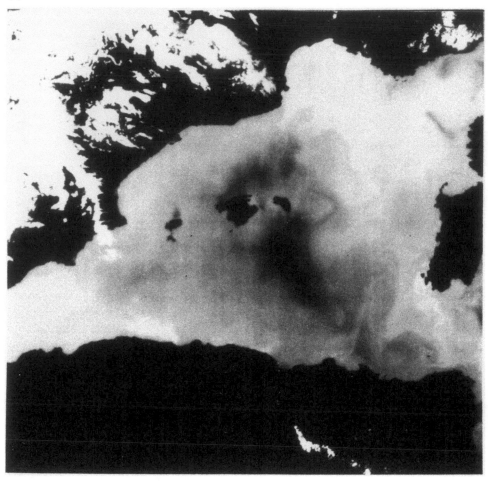

Figure 7.60   (d) Channel 4.

Earth-orbiting artificial satellites have been in existence for about 30 years and the AVHRR instruments have been gathering data for about 17 years. The archives of data that are available, both on a local scale and on a global scale, enable daily, monthly and annual means of parameters such as sea surface temperature, land-surface properties and cloudiness to be derived. After several years of data have been collected, long-term changes in the climate and the interannual variability of the parameters can be investigated. Sea surface temperature is one example where high-quality data have now been available for over ten years allowing long-term trends and variability in SST to be studied (Strong 1989). Sea surface temperature is one obvious example of a surface climatological parameter and its retrieval from AVHRR data was described above in section 4.2. Several examples of land-surface parameters can also be mentioned.

The availability of infrared radiances from satellites over land at a local scale has made case studies possible into the relation between the land-surface radiative temperature and topography. Some were based on AVHRR night-time data, although

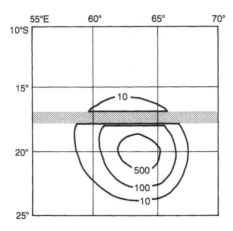

Figure 7.61 Theoretical sea-surface sunglint for a composite pattern, with isopleths and satellite height as in Figure 7.59 (McClain and Strong 1969).

some have been based on HCMM or Landsat TM data. There has been a considerable amount of work on surface radiation balance components. Work in this area has progressed significantly in the shortwave region, while very few studies have been published for the longwave region. Incident solar radiation at the surface has been computed in two ways, by using the direct approach based on regressions between satellite-based radiance data and ground-based measurements with pyranometers, and the more general approach, based on physical modelling, using radiative transfer equations.

Figure 7.62 Effect of wind through the Straits of Gibraltar, NOAA-7 AVHRR image from 28 June 1982.

The estimation and mapping of albedo from space has also made significant progress, but in contrast only a few preliminary attempts of using satellite imagers or sounders, for retrieving atmospheric longwave radiation for climate studies have been presented. Work on the estimation of surface heat fluxes is an important contribution to climate modelling work. The work performed within the TELLUS Project between 1976 and 1980 by Working Group II of EARSeL is summarised by Saunders and Seguin (1992); it involved the development of models, the derivation of simple algorithms and field campaigns.

Clouds are crucially important in models of the Earth's climate because of their direct effect on the radiative fluxes in the atmosphere which drive the general circulation. For this reason a global climatology of cloud cover and type, covering many annual cycles, is an important requirement for climate models. Before the advent of satellites, cloud climatologies were based primarily on surface and aircraft observations at single locations. This is obviously not ideal since the spatial coverage is poor in the northern hemisphere and almost non-existent in the southern hemisphere. In addition surface based observations cannot view the cloud top which is the part of the cloud primarily affecting the radiative fluxes by reflecting solar radiation and emitting thermal radiation. Satellite radiometers view the cloud tops and are able to produce a global data set of spatially averaged cloud parameters, their only drawback being that they cannot detect cloud below the upper layer. Rainfall monitoring from satellite data is another major contribution to climate studies, but we have already discussed this in section 7.6.

Global change is a significant issue in both the scientific and non-scientific community. Increasing greenhouse gas concentrations, stratospheric ozone depletion, tropical deforestation and reductions in biological diversity have provoked widespread concern. Some of these derive from local processes which have global implications (deforestation); others are related to global-scale processes with local implications (stratospheric ozone depletion). Understanding global-scale processes and phenomena which modify the Earth's environment has become a critical requirement of the scientific community (International Geosphere-Biosphere Program 1990). A new discipline, Earth system science, has emerged as a response to the need for improved understanding of the Earth as a linked system (Ehrlich *et al.* 1994).

A major need identified by the Earth system science community is for improved global data sets for describing and modelling Earth processes. Traditional Earth science disciplines have generated global products which have limited use for integrated studies of the Earth.

The global change scientific community has identified the AVHRR 1 km data set as an important database for land-cover mapping and land processes modelling at continental and global scales (International Geosphere-Biosphere Program 1992). The IGBP identified several land-cover variables that can be directly estimated using AVHRR data. These include: vegetation indices, the Earth's albedo, solar radiation flux at the surface, evapotranspiration and surface temperature. Vegetation indices are required for the estimation of leaf area index, photosynthetic capacity and primary productivity. A large portion of the multitemporal studies conducted to date describe vegetation dynamics using NDVI. Quantitative correlation of spectral data to land-cover parameters, however, is still rare in the literature. We have already seen in this book many examples of the use of AVHRR in applications that could, in principle, be operated at a global scale. Table 2 of the review by Ehrlich *et al.* (1994) gives a number of additional references that we have not cited in this book. AVHRR 1 km imagery will be used to produce global 1 km Normalised Difference Vegetation Index (NDVI) composites (Deering *et al.* 1975) at weekly or bi-weekly intervals (International Geosphere-Biosphere Program 1992). NDVI composites are derived by retaining only the highest NDVI within a multi-date NDVI data set on a pixel-by-pixel basis (Holben 1986). The 1 km NDVI composite will then be used to produce 1 km derivatives such as a characterisation of the global land surface and to estimate land-cover characteristics directly (International Geosphere-Biosphere Program 1992). These 1 km derivative products will benefit a number of global change projects such as the International Global Atmospheric Chemistry project (IGAC), Biospheric Aspects of the Hydrological Cycle (BAHC) and Global Change and Terrestrial Ecosystems (GCTE), as well as environmental monitoring research at regional and continental scales. The AVHRR 1 km data set will also constitute a test case for the sensors scheduled to be placed on board the Earth Observation System (EOS) space station such as the Moderate-Resolution Imaging Spectrometer (MODIS), which, by the beginning of the next century, will provide improved spatial and spectral coverage of the Earth.

We shall attempt to summarise the contributions of AVHRR data to studies of the climate. The archives of AVHRR data which are available therefore only span periods that are very short in terms of climatological studies and the timescales of climate change. One might be tempted to suppose, therefore, that the meteorological satellites have little to offer to climatological studies, but that would be a mistake.

7.9.2 Climate modelling

A large fraction of current climatological research is concerned with using numerical climatological models to attempt to predict (or calculate or determine) the effect on the Earth's climate of various changes that may be supposed to occur either as a result of natural events or as a result of human activities. Various aspects of the role of satellite data in such work have been considered by Vaughan and Cracknell (1994). Climate modelling involves the use of general circulation models (GCMs) which are very similar to the GCMs which are now widely used in numerical weather prediction. These models require the input of boundary condition data over a regular grid of points and a very important role for meteorological satellites is in providing input data for climatological calculations. The parameters required include quantities such as albedo, radiation (energy) fluxes, surface temperatures, and the extent of cloud cover, the extent of ice and snow cover and the type of land-surface cover (including its temporal variation).

It is not our purpose to give an extensive account of climate modelling (see, for example, Trenberth 1992). Our task is much more restricted, namely to consider the extent to which AVHRR data can contribute to climate studies. In many ways this question has already been addressed in earlier sections of this book. The AVHRR can be used to determine certain geophysical parameters. If these parameters are relevant to climate studies and to climate modelling then that identifies a possible use of AVHRR data for climatological work. We are not, in this section, suddenly going to identify new geophysical parameters to be extracted from AVHRR data.

There are a number of problems in using AVHRR data in climatological studies. They are not primarily problems of interpretation of AVHRR data. Some of them are related to the logistics of handling the large data sets involved. Others are concerned with the long-term consistency of the data; in other words one needs to be sure that time-dependent features observed in a long-term data set correspond to real environmental changes and do not arise from artefacts of the data set. Consideration of the stability of the observing platform and of the instrumentation on board the satellite are both of great importance. For a long data set which spans the operational lifetime of several satellites there is also the consideration that the performance of the instruments carried on each of the different satellites should be consistent. Let us consider these factors in turn. The variation in the time of day or night of the closest AVHRR pass can be quite large. For instance McGregor and Gorman (1994) show that over the seven-year period from 1982 for the vicinity of Wellington (New Zealand) and taking data from five different satellites in the series there can be as much as $3\frac{1}{2}$ hours time difference in the time of the closest pass. This arises from two factors, the first is the effect of phasing, associated with the 102 minute spread of satellite pass time, and the second is a slow drift over the operational lifetime of the satellite. The question of the changes in the calibration of the AVHRR throughout its lifetime was studied in some detail in section 2.2.6, both in relation to channel 1 and channel 2, which have no on-board calibration, and also in relation to channels 3, 4 and 5, for which there is on-board calibration. This has led to quite a lot of work being done on post-launch calibration of the AVHRR instruments and the need both to allow for the drift in the behaviour of any one AVHRR instrument and to allow for differences among the calibration of different instruments in the series.

7.9.3   Energy fluxes

We mentioned in section 1.5 the Earth Radiation Budget Experiment (ERBE) which was flown on NOAA-9 and NOAA-10. This was by no means the first attempt to make measurement of the Earth's radiation budget from satellites. Before the existence of Earth-orbiting satellites, the Earth radiation budget (ERB) was estimated by using radiative transfer calculations. From 1959 onwards there was a whole succession of instruments flown on satellites for the purpose of making Earth radiation budget measurements. A summary is given, for example, at the start of section IX-1.1 of the book by Rao et al. (1990). The computation of the surface energy budget over the sea by means of AVHRR-derived SST and ECMWF analysis fields has been described by Pagano et al. (1991) We have also mentioned, in section 2.4.2, the NOAA operational Heat Budget products which are, essentially, long-term (seasonal) averages with rather low spatial resolution. As the precision of climate models improves so there is a need for albedo and radiation flux data of higher spatial resolution and with allowance for temporal variation. There have been various energy flux studies of specific areas, for example in a part of France (Taconet and Vidal-Madjar 1988), the Alps (Mannstein 1989), Iceland (Cracknell et al. 1989) and polar regions (Lutz and Bauer 1989).

   One experiment that it is important to mention in connection with energy flux studies is the Hydrologic Atmospheric Pilot EXperiment-MOdélisation du BILan HYdrique (HAPEX-MOBILHY) (André et al. 1986, Phulpin et al. 1989). One of the main objectives of this experiment is to develop latent heat flux retrieval methods, especially in connection with GCM studies, and to study how satellite data may be used in this context. This experiment, in which French, British and American teams were involved, took place in southwestern France from September 1985 to January 1987. During an intensive observation period from May to July 1986 *in situ* and remote sensing data were collected. Many studies have been undertaken and several have already yielded encouraging results (André *et al.* 1988). The remote sensing data included aircraft measurements and satellite measurements (from Meteosat, AVHRR, Landsat and SPOT).

An important aspect of this experiment was the development of new parametrisation schemes for the surface energy budget in which some coefficients depend on the soil type, the vegetation canopy and the moisture content. Most of these coefficients have been calibrated with *in situ* measurements. The ultimate goal is to use parameters derived from satellite data to map these coefficients within the HAPEX-MOBILHY area and then to propose a general method that can be applied in temperate regions. Phulpin *et al.* (1989) concentrated on identifying the vegetation type and on determining the NDVI and the temporal variations of the NDVI.

Dalu and Liberti (1989) have studied the problem of trying to use AVHRR data in the estimation of radiative heat flux at the sea surface, since the sea surface radiative flux (SSRF) components are important input boundary conditions in climate models. It would be rather laborious and expensive to obtain *in situ* measurements and, in any case, the fluxes are calculated with empirical formulae, using conventional meteorological data (e.g. Henderson-Sellers 1986). For this reason it is useful to estimate the surface radiative fluxes from satellite data, which in general can give the desired spatial and temporal resolution (Schmetz *et al.* 1986, Masuda and Takashima 1990b, Ranson *et al.* 1991).

In the evaluation of the net SSRF from remotely sensed data, the main problem is the complex relation between the measured upwelling radiances and the SSRF components. It is simpler to parametrise each component of the net SSRF as a function of the information that can be derived from remotely sensed data. The AVHRR can give information on the sea surface temperature, the columnar water vapour content and the aerosol optical thickness. The surface fluxes were calculated by Dalu and Liberti (1989) with the radiative transfer model LOWTRAN-6 for different atmospheric and geometrical conditions. On the basis of the results of these numerical simulations, each component of the net SSRF was parametrised as a function of the physical variables that can be retrieved from the AVHRR data. The simulations were made using different combinations of the profiles contained in the model (tropics, mid-latitude summer, etc.). The type of aerosol was fixed (maritime), and the internal LOWTRAN Mie database was used for the phase function.

We consider first a cloudless sky. The net radiative flux Φ_n at the surface for clear sky conditions is

$$\Phi_n = \Phi_{0D}(1 - R_0) + \Phi_{0s}(1 - A_0) + \Phi_L(1 - A_L) - \Phi_S \qquad (7.9.1)$$

where Φ_{0D} and Φ_{0s} are the shortwave direct and scattered fluxes respectively, R_0 and A_0 are the sea surface broadband shortwave reflectivity and albedo, Φ_L is the longwave downwelling flux emitted by the atmosphere, A_L is the broadband longwave reflectivity and Φ_S is the radiative flux emitted by the surface. The direct flux is given by

$$\Phi_{0D} = \Phi_\infty \, t_0(\theta_0, \tau_a, w) \cos \theta_0 \qquad (7.9.2)$$

where $t_0(\theta_0, \tau_a, w)$ is the weighted atmospheric transmittance obtained from the ratio Φ_{0D}/Φ_∞ and θ_0 is the solar zenith angle. The transmittance $t_0(\theta_0, \tau_a, w)$, calculated as described above, was parametrised by Dalu and Liberti as

$$t_0(\theta_0, \tau_a, w) = \exp[(-f + vw + a\tau_a)(\sec \theta_0)^\gamma] \tag{7.9.3}$$

where the values of the coefficients are $f = 0.195 \pm 0.005$, $v = 0.0038 \pm 0.0002$, $a = 0.863 \pm 0.006$ and $\gamma = 0.75 \pm 0.09$. The coefficient γ is mainly related to the bending of the path for the variation of the atmospheric refractive index with height. The flux of the scattered radiation was parametrised as

$$\Phi_{0S} = \Phi_{0D} F(w, \tau_a)\exp[(0.36\tau_a + 0.36)(\sec \theta_0 - 1)] \tag{7.9.4}$$

where

$$F(w, \tau_a) = 0.033 + 0.62\tau_a + (1.6\tau_a - 0.2)10^{-4}\, w^2 . \tag{7.9.5}$$

Values of the sea surface broadband reflectivity R_0 as a function of the solar zenith angle are given by Henderson-Sellers (1986), while values of the albedo A_0 as a function of solar zenith angle and atmospheric transmittance are tabulated by Payne (1972). The sea surface radiative flux is given by

$$\Phi_S = \varepsilon\sigma T_S^4 \tag{7.9.6}$$

where $\sigma = 5.67 \times 10^{-8}$ W m^{-2} K^{-4} is the Stefan–Boltzmann constant and ε is the emissivity. $\varepsilon = 0.972$ was taken as the value of the emissivity so that the reflectivity is then $A_L = 1 - \varepsilon = 0.028$.

The downwelling thermal infrared radiances are mainly dependent on the vertical temperature profile of the bottom layer of the atmosphere. The average radiative temperature T_a of the atmosphere is then correlated to the air temperature at the surface, which is assumed, over the oceans, to be equal to T_s. With this hypothesis, the thermal infrared flux was parametrised by Dalu and Liberti as proportional to σT_S^4, i.e.

$$\Phi_L = (F_f + F_w w - F_a \tau_a)\sigma T_S^4 \tag{7.9.7}$$

where the values of the coefficients are $F_f = 0.806 \pm 0.007$, $F_w = 0.00046 \pm 0.00008$ and $F_a = 0.0063 \pm 0.0003$, calculated to compensate for the bias introduced by replacing T_a with T_S. Using these equations, with a variation of T_S from 294 to 297 K, τ_a from 0.2 to 0.4 and w from 10 to 20 kg m^{-2}, Dalu and Liberti (1989) obtained the variations in the sea surface radiative flux given in Table 7.10.

The presence of clouds prevents a direct estimate of the sea surface temperature, T_S. However this quantity only changes slowly in time and is relatively uniform in space. So it is possible to have a satisfactory estimate of its value by using a temporal and spatial interpolation. Sensitivity studies show that, when clouds are present,

Table 7.10 Variations of the fluxes (W m^{-2}) for given variations of the parameters in a mid-latitude summer atmosphere

		$\Delta\Phi_{0D}$	$\Delta\Phi_{0S}$	$\Delta\Phi_L$	$\Delta\Phi_S$
ΔT_S	(297–294) (K)	0	0	14	17
$\Delta \tau_a$	(0.4–0.2)	−135	105	−3	0
Δw	(20–10) (kg m^{-2})	−36	0.3	5	0

w and τ_a are less important and average climatological values can be used. A rough evaluation of the surface fluxes can be obtained with a parametrisation of the net sea surface radiative flux components as functions of the cloud albedo, the cloud top temperature, the climatological values of w and τ_a, the interpolated value of T_S and the geometrical variables, which are known.

Neglecting multiple reflections between cloud and surface, the radiative net flux for a total overcast situation is

$$\Phi_n = \Phi_{0D} t_c (1 - A_0) + \Phi_{ac}(1 - A_L) - \Phi_S \tag{7.9.8}$$

where t_c is the ratio of the flux calculated when cloud is present to the direct flux Φ_{0D} calculated for clear sky conditions. This is equivalent to a transmittance and can be parametrised as a function of the cloud albedo (Slingo 1989, Stephens et al. 1984). Φ_{ac} can be expressed as

$$\Phi_{ac} = A(zb)\Phi_L + B(zb)\Phi_c . \tag{7.9.9}$$

The coefficient $A(zb)$ is the ratio of the flux due to the atmosphere between the cloud and the surface and the flux Φ_L is given by equation (7.9.7). The flux Φ_c is $\varepsilon_c \sigma T_c^4$, where ε_c is the cloud emissivity and T_c is the cloud bottom temperature. The coefficient $B(zb)$ is related to the absorption and diffusion of the atmosphere between the cloud and the surface. The coefficients $A(zb)$ and $B(zb)$ were estimated separately, by calculating the fluxes with and without clouds, for average atmospheric conditions, and values are given in the paper by Dalu and Liberti (1989).

We now consider the question of the information content of the AVHRR in relation to this theory and these calculations. The question of the determination of the sea surface temperature in the absence of cloud has already been discussed at considerable length in section 4.2. The question of the determination of aerosol concentrations has also been discussed in section 7.5.

The intensity of the scattered radiation increases with decreasing wavelength. If two channels with sufficiently different wavelength are available, the ratio between the measured albedo changes as a function of the aerosol load. The following empirical relation between the aerosol optical thickness and the ratio A_2/A_1 of the AVHRR channel-2 and channel-1 albedo can be obtained from the results published by Rao et al. (1989):

$$\tau_a = 8.4(A_2/A_1)^2 - 7.6(A_2/A_1) + 1.7 . \tag{7.9.10}$$

The values of the albedos A_1 and A_2 can be determined from the AVHRR channel-1 and channel-2 data using the calibration coefficients obtained from the pre-launch calibration (see section 2.2.2). The night-time aerosol optical thickness can be obtained from the AVHRR thermal infrared channel data (McClain 1989).

A quadratic relation was found between the atmospheric correction $(T_S - T_{B4})$ and the brightness temperature difference $(T_{B4} - T_{B5})$ in the two AVHRR-2 channels 4 and 5 (Weinreb and Hill 1980). Since the quadratic term is relatively small, this quadratic relation can be replaced with a linear one optimised for the average water vapour amount typical of the region and season under examination. The intercept in the split window linear algorithm is then due to the linearisation of the quadratic equation (Dalu et al. 1985). The best choice of the coefficients for the Mediterranean Sea, taking into account also the secondary effects, is (Dalu and Liberti 1989)

$$T_S - T_{B4} = 2.68(T_{B4} - T_{B5}) - 0.4 . \tag{7.9.11}$$

The same coefficients were found by McMillin and Crosby (1984), and were obtained from the comparison of a large amount of AVHRR and *in situ* data.

It is rather difficult to evaluate the atmospheric water vapour content by remote sensing techniques because of its large variability in space and time. Over the sea, however, for the natural equilibrium created between evaporation and diffusion, it can be assumed that the relative humidity at the surface is never too different from the climatic average of 80 per cent (Prabhakara *et al.* 1979). In this case the columnar water vapour content is directly proportional to the difference $(T_{B4} - T_{B5})$ (Dalu 1986), i.e.

$$w = 19.6(T_{B4} - T_{B5}) \cos \theta .\tag{7.9.12}$$

The errors expected for the estimation of the variables are: 0.5 K for T_S, 0.08 for τ_a, and 4 kg m^{-2} for w. The corresponding errors in the derived fluxes are 26 W m^{-2} for the shortwave component, 6 W m^{-2} for the atmospheric component and 4 W m^{-2} for the surface component.

Multispectral methods to classify the clouds with AVHRR data have already been mentioned in section 3.2. From such a classification, some radiative properties of the clouds can be inferred from climatology. From the AVHRR channel-1 albedo measurements it is possible to evaluate the liquid and the ice water path, which can be related to the cloud emissivity and to the cloud geometrical thickness. For stratiform clouds, the cloud thickness can be directly derived from the measured albedo and the solar zenith angle (Feigelson 1984). Channel 4 gives a good evaluation of the cloud top temperature, since the atmospheric absorption above the cloud can be neglected. The cloud top height can be calculated from the measured brightness temperature, starting from T_S with a climatological temperature lapse rate.

From the above discussion we see that it seems to be possible to derive a rough estimate of the cloud thickness, transmittance, emissivity and top temperature from AVHRR data and to obtain an estimate of the surface fluxes, even in cloudy conditions. However the derivation of some of the cloud parameters is rather qualitative and it is difficult to evaluate the associated errors. Some improvement in the quality of the estimated fluxes in cloudy conditions is expected to be possible if AVHRR data are integrated with microwave radiometric data. Some progress is also expected from improved radiative transfer and cloud models (see also Schmetz 1989).

7.9.4 Cloud cover

Information on cloud cover is available from the visible and near-infrared channels of a scanner on board a satellite in a far more complete and comprehensive form than could possibly be obtained any other way. We have also already mentioned the question of cloud classification in section 3.2. One of the main problems in extracting information on cloud cover from AVHRR data for climatological studies is an organisational and data management problem.

The advantages of a cloud climatology derived from satellite data have led to the implementation of the first project of the World Climate Research Programme known as the International Satellite Cloud Climatology Project, ISCCP (Schiffer and Rossow 1983, 1985, Rossow and Schiffer 1991, Weare 1992). The aim of the ISCCP, which commenced data collection in July 1983, is to bring together different radiance measurements of clouds form both geostationary and polar-orbiting satellites. The calibrated radiances are combined into one common data set, on which a cloud analysis algorithm is applied to retrieve fractional cloud cover to within 5 per cent and cloud top height to within 1 km. These cloud parameters are averaged over 250 km by 250 km boxes and sampled every three hours. This cloud climatology data set will have even greater value when combined with the coincident top-of-atmosphere radiation data such as that recorded by the Earth Radiation Budget Experiment (ERBE) organised by NASA. The need to monitor the calibration of the AVHRR in flight and to obtain reliable post-launch calibration coefficients for ISCCP was emphasised by Brest and Rossow (1992), see also section 2.2.6.

One particularly important feature of the ISCCP is its polar component which provides one of the major applications of Antarctic AVHRR data (Turner 1989). This has involved the production of the first high-latitude data sets of cloud properties for use in the validation of the cloud fields derived by general circulation models. Although very sophisticated automatic cloud detection schemes have been developed for use in mid-latitudes many of these break down in the polar regions because of the similar temperature and albedo of the ice surface and the cloud tops. More elaborate schemes are therefore being developed for the particular conditions at high latitude (Raschke 1987, Raschke et al. 1992). The techniques being employed often make use of the small differences between the brightness temperatures of the infrared channels that are found over ice and water droplet clouds (Yamanouchi et al. 1987). Textural classification has not been used to the same extent as in the fields of land and agricultural applications. However, Ebert (1987, 1992) used pattern recognition techniques in the cloud discrimination scheme and this technique will certainly find further application over the next few years. Day-time channel 3 is one of the most important channels for use in the polar regions as it can discriminate water droplet cloud from the ice and snow covered surface (see, for example, chapter 4 of Scorer 1986). Although little use has been made of the data in a quantitative sense, they have great value when the imagery is interpreted visually. In the long periods when no solar radiation is received the channel-3 data are very difficult to use objectively because of calibration difficulties. Radiances at this wavelength are so small below 0°C that the brightness temperatures have large errors which limit their use in computing interchannel differences.

7.9.5 Sea, snow and ice

It is somewhat easier to deal with the sea, snow and ice than with the land-surface areas. First, it is necessary to map the areas covered by snow and ice and, of course, to take into account the variations with the seasons. The AVHRR provides data of sufficient spatial resolution and also with sufficient frequency for this purpose (see section 7.1, also Eckhardt 1989). The AVHRR data, from various spectral channels, can also be used to characterise the type of ice or state of the snow; once this is done, the known albedo and emissivity values of the appropriate class can be used in calculations of the surface energy fluxes. For the sea, its reflectivity and emissivity are known and so radiant energy fluxes can readily be calculated. In addition to the albedo at visible and near-infrared wavelengths and the emissivity at thermal infrared wavelengths, the surface temperature is also an important parameter for climate model calculations. For the sea it is possible to obtain reliable sea surface temperatures over all the world's oceans from satellite-flown thermal infrared scanners. The data can be provided at the regularly spaced grid points used in climatological models. We have already seen in chapter 4 that the AVHRR is particularly suitable for this purpose because the data are well calibrated and global coverage is readily available. We have also described in considerable detail in chapter 4 how the AVHRR data have to be handled and processed to determine sea surface temperatures accurately. Strong (1989) analysed global mean sea surface temperatures obtained from satellite data for the period from 1982 to 1988. The results of this analysis indicated a gradual but significant warming of about 0.1°C per annum which is about twice as large as that observed in sea surface temperature data from conventional sources (ships and buoys). The study of AHVRR GAC data for sea surface temperature maps from 1981 to 1989 for monitoring long-term changes in the Atlantic Ocean off West Africa has been descibed by Van Camp et al. (1991). The annual surface temperature cycle of lakes has been shown to respond to medium-term climate changes and a case study of Lough Neagh in Northern Ireland has been studied by Brown et al. (1991).

7.9.6 Land surfaces

As climatic research has developed, and in particular as numerical climate modelling has become more sophisticated, it has become very important to have information regarding the characteristics and spatial distribution of the Earth's land cover. The description of the land surface (other than that covered by snow or ice) and the specification of its properties over a regular grid on a global basis for climatological modelling is far more difficult than for sea or for areas of snow or ice. For unvegetated surfaces such as bare rock, desert sand or salt flats the areas are relatively easy to map and the properties of the surfaces are fairly constant. For vegetated surfaces, however, it is much more difficult because the surface properties are highly variable, both spatially and temporally. It is also possible to address the question of attempting to detect climatic change as manifested by changes in the extent of arid and semi-arid areas (Tucker et al. 1994).

Capabilities to inventory and map land-cover conditions and to monitor change are required for, among other things, modelling the global carbon and hydrological cycles, studying land-surface–climate interactions and establishing rates of tropical deforestation. Global land process research heretofore has had to rely upon simple interpretations of gross land-cover and surface properties, such as biomass, albedo, surface roughness and canopy resistance, at low spatial resolution. The Matthews land-cover and natural vegetation (Matthews 1983, 1985) and the Olson and Watts (1982) major world ecosystems global databases are the most common sources of land-cover and surface parameter data. These databases have, respectively, 1° by 1° and 0.5° by 0.5° spatial resolution. Higher resolution data with greater precision for classification are clearly required (International Geosphere-Biosphere Program 1990) and it is also necessary to allow for temporal variations (both seasonal and longer-term variations) in the land-cover and surface parameter values. Thus data from various satellites, including AVHRR data, can be used to obtain information of this type. The International Satellite Land-Surface Climatology Project (ISLSCP) was set up a number of years ago as part of the World Climate Research Programme (WCRP) to improve and use techniques for retrieving land-surface parameters from the standpoint of climate change research and also to study the climatic impact on the land surface (soil, vegetation, ice cover) through the use of satellite data analysis. A review of the earlier days of the ISLSCP and a large set of references is given by Kondratyev (1989). The work of the ISLSCP has done much to contribute to current understanding of how ground properties can be derived quantitatively from satellite data. This work has involved a number of very intensive field experiments in several parts of the world (see e.g. Rasool and Bolle 1984, Becker *et al.* 1988, Sellers *et al.* 1988, Sellers and Hall 1992). ISLSCP draws on data from a whole range of instruments flown on satellites and the details of the ISLSCP are more appropriate to a book on climate modelling than to the present book. In our discussion of the determination of land-surface information for climate studies we shall concentrate on the use of data from the AVHRR.

We consider one example of the use of AVHRR data which relates to a rather small area (in global terms) and which also actually was undertaken in relation to a mesoscale numerical weather prediction model rather than a climate model. This example relates to the albedo. It will be recalled that the surface albedo is the reflectance of the surface, integrated over all visible and near-infrared wavelengths and over all angles in the upward hemisphere. The surface albedo varies with underlying surface type (e.g. coniferous forest, crops, urban development), snow cover, surface wetness and solar zenith angle. Surface albedo values are required for numerical weather prediction models or climatological models averaged over each model grid square over land. For instance the mesoscale forecasting model developed by the UK Meteorological Office to cover the British Isles assumed a surface albedo of 18% for all the grid squares containing land. Saunders (1990) addressed the question of trying to replace this constant value with measured albedos which take into account the different surface types averaged over each grid square and the seasonal variations in albedo. Satellites provide the only realistic way to obtain such data. Saunders studied data from eight AVHRR passes over the British Isles and near-continent for relatively cloud-free conditions between the months of April and October. The retrieved surface albedos were presented as an average over a 15 km grid square. The surface albedos of the agricultural areas of the British Isles showed a marked increase from 17 to 22% between April and July. In contrast urban areas and rougher hilly terrain (e.g. the Scottish Highlands) gave lower surface albedos of ~13% which remained more constant throughout the period between April and October.

The example which we have just considered illustrates the great variability, both in space and in time, of one of the properties of a vegetated surface. Therefore it will be appreciated that it is extremely difficult to obtain reliable information on global land cover, which is required in the study of climate change. Townshend *et al.* (1991) argued that conventional maps of global vegetation cover are of limited accuracy and value. Considerable variations exist in the published data, see Figure 7.63. The total area estimates in (a) vary because of the different estimates of the unvegetated areas (deserts, ice caps). Even where there is correspondence between different authors for particular categories, such as agricultural land, this can arise simply because the estimates are derived from the same source. The variations of these global compilations arise from a number of reasons. First, they rely on compilation from numerous separate sources, since direct ground observations using an agreed system of classification have never been directly attempted at a global scale. Integrating local studies into regional ones and thence into global maps raises problems, if only because the original maps were often produced at different dates. A greater

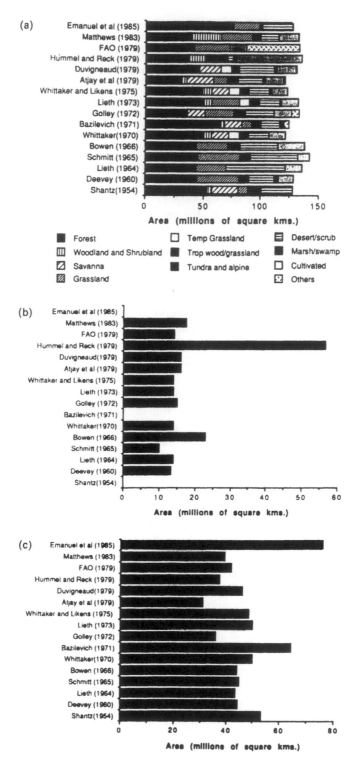

Figure 7.63 Variations in estimates of the global extent of cover types for (a) ten main cover types, (b) cultivated land and (c) forest land (Townshend *et al.* 1991). The references in the figure are all cited in the paper by Townshend *et al.* (1991).

problem is that widely varying criteria are used in classification. Consequently the compilers of the global estimates have had to reclassify maps into a common scheme based on inherently incompatible classifications. Even if the same approach to classification has been adopted, the size of the specific parameters used to separate classes can vary. Hence the proportions occupied by classes with the same label in Figure 7.63 are usually not directly comparable. Not only does the total area occupied by different classes vary substantially between authors but the detailed spatial distribution often varies substantially even where the total global estimates of a cover type are similar. Major attempts have been made to synthesise the current knowledge to generate digital global databases; although they represented improvements on previous knowledge, the points we have just noted inevitably cast doubt on their internal consistency.

The absence of satisfactory land-cover maps at very broad scales has encouraged the use of remote sensing data to provide this information. Selecting the instrument with the most appropriate spatial resolution is important (Townshend and Justice 1989). To determine a global land-cover database using Landsat MSS or TM or SPOT data would involve an enormous cost in terms of purchase of data and processing time. It would also yield a database of far higher spatial resolution than the grid spacing currently used in numerical climate modelling. It would also involve further enormous costs to update it or to take into account seasonal variations. AVHRR data on the other hand are of only moderate, but still quite adequate, spatial resolution and the AVHRR data are collected more frequently, with virtually the entire globe being imaged twice each day. The high frequency of coverage enhances the likelihood that cloud-free observations can be obtained for specific temporal windows, and makes it possible to monitor change in land-cover conditions over short periods, such as a growing season. Moreover, the moderate resolution of the data makes it feasible to collect, store and process continental or global data sets.

Preliminary land-cover maps have been derived at continental scales by Tucker *et al.* (1985a) for Africa and by Townshend *et al.* (1987) for South America using the NOAA GVI product with pixel sizes of approximately 15–20 km. Initial work on continental land-cover classification used a measure of the mean annual NDVI value and seasonality (Tucker *et al.* 1985a), as represented by the first two principal components of an annual set of AVHRR data. Subsequently, multiple images from different dates throughout the year have been used in attempting land-cover classifications (Townshend *et al.* 1987, 1989b) and considerable improvements in accuracy have been found as the number of images was increased.

Despite the encouraging results from these preliminary investigations, the resultant maps derived from satellite data have substantial limitations. Most attempts have used the NOAA GVI product, which has a number of significant deficiencies. One of these is its very coarse resolution of approximately 20 km which we have already mentioned. The GVI product also has the problem that off-nadir pixels, which suffer markedly from atmospheric effects, are preferentially sampled, especially in higher latitudes, in generating the 20 km pixels from the finer resolution data (Goward *et al.* 1991); also in the GVI the original DN values are used rather than the reflectance or radiance, which yields distorted values of the NDVI (see section 5.4). Additional limitations include the fact that analyses have relied on only a single year's data, and consequently interannual variability in vegetation response to climatic conditions can lead to errors in the inferred distribution of vegetation types especially in semi-arid and seasonally flooded areas. Also, boundaries between cover types with similar phenologies such as those between the moist savannas, tropical rain forest, and degraded tropical rain forest in Africa are unreliable, while wetlands are not well distinguished from other vegetation types on these maps.

In view of these problems Townshend *et al.* (1991) addressed the question of alternative data sources to the AVHRR. Although they discussed other existing instruments (including passive microwave scanners, see also section 5.8) they concluded that AVHRR data are likely to remain important for global vegetation studies for some time to come. However, operational provision of global land-cover information in the future will require better data sets in terms of spectral, radiometric, temporal, and spatial properties. Only recently, with the expansion of the number of HRPT ground stations worldwide, have spatially extensive data sets at the highest nominal resolution (1.1 km) started to become available on a continuing basis for major land areas.

The value of the AVHRR data arises primarily from their multitemporal use, rather than immediately for classification purposes. The number of spectral bands in the AVHRR is too small for a sensitive multispectral classification of a single image to be very successful. However, while different types of vegetation may have similar values of the channel-1 and channel-2 reflectances (and therefore of the NDVI) at certain times of the year, they may have very different values at other times of the year. Thus a multitemporal approach using data from throughout the year is essential for satisfactory discrimination between most cover classes, when using the NDVI (see, e.g. Thomas and Henderson-Sellers 1987, Henderson-Sellers 1989).

A special issue of the *International Journal of Remote Sensing* was devoted to 'Global data sets for the land from the AVHRR' (*International Journal of Remote Sensing*, volume 15, number 17, 20 November 1994) and we cannot attempt to summarise adequately here all that is in that special issue. We simply note a few salient points. Townshend (1994) identified four main data processing streams in the creation of useful global and regional data sets from the raw AVHRR data, see Figure 7.64. In the first a global data set at relatively coarse resolution, the Global Vegetation Index (GVI) product, was created by NOAA using GAC data, see section 2.4.2. Notwithstanding the limitations of this data set, its creation was an important precursor to the creation of all subsequent large-area AVHRR data sets. In the second data processing stream, improved data sets were created as the result of efforts by the group led by C. J. Tucker at the NASA Goddard Space Flight Center. Although these were created for all the main land masses of the Earth, they have been available only by continent at their full resolution of about 8 km. Research efforts to improve these and similar data sets continue in various laboratories. In the third data processing stream efforts are currently under way to create medium-resolution data sets based on GAC data on an operational basis. In the fourth data processing stream the full potential of the data, in terms of spatial resolution, is being realised through the creation of a global 1 km data set. The main data sets which have been produced or which are under production are summarised in Table 7.11.

The GVI product, from the first data processing stream, has already been described in outline in section 2.4.2 where references are also cited (see also Townshend 1994). As part of the International Geosphere-Biosphere Program (IGBP), the GVI data set is being made more widely available through the Diskette Project (International Geosphere-Biosphere Program 1990). To overcome some of the limitations of the GVI data sets, plans are currently being formulated by NOAA to

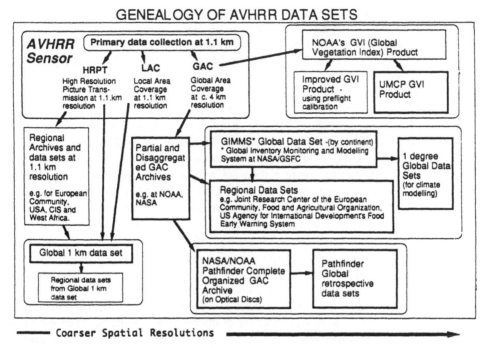

Figure 7.64 Coverage of global data sets from the AVHRR (Townshend 1994).

Table 7.11 AVHRR global data sets (archives of AVHRR data requiring substantial processing before usage are not included) (Townshend 1994)

Name of data set	Agency	Bin size	Coverage	Frequency	Availability
Global Vegetation Index (GVI)	NOAA	15–20 km	Global	Weekly and two weekly	1982–present
Diskette project	NCAR	15–20 km	Africa and Global	Monthly	1982–89
Modified GVI	UMCP	15–20 km	Global	Biweekly	1982–90
GIMMS product	NASA/GSFC	7–8 km	Global by continent	Monthly	1982–present
GIMMS derived 1° product	NASA/GSFC	1 degree	Global	Monthly	1982–present*
Land pathfinder	NASA/NOAA	8 km	Global	Ten days and daily	1982–present*
TREES	CEC/ISPRA	1.1 km and 4 km	Tropics	Daily	1982–present
1 km product	EDC/IGBP/ NASA/ NOAA/ESA	1.1 km	Global (by 1995)	Ten days	March 1992– September 1994

* To be available by 1995. At the time of writing 1987–88 only available.

produce a second generation GVI product and this may be generated retrospectively back to 1982 (Townshend *et al.* 1991). A data set in which the calibration coefficients calculated before launch for channels 1 and 2 for the different AVHRR sensors is currently available. Another revised form of the GVI product is being prepared at the University of Maryland (Goward *et al.* 1993, 1994).

The second data processing stream involves an improved product at higher spatial resolution which has been under production by the Global Inventory Monitoring and Modeling Systems (GIMMS) group at the NASA Goddard Space Flight Center since 1984 (Townshend and Justice 1986, Los *et al.* 1994, Sellers *et al.* 1994, Tucker *et al.* 1994). GAC data are reprojected on to an equal area projection and resampled by continent to create a data set with a spatial resolution typically of about 7.6 km, this being the size of the basic IFOV for GAC data at view angles of about 35° off nadir. Thus near-nadir pixels are somewhat over-sampled and pixels from higher view angles are under-sampled. The raw pixel values are converted using available calibration data, which are limited in amount and quality. A simple procedure for cloud detection is applied by thresholding channel-5 brightness temperatures. The NDVI is then calculated for each pixel, with zeros being included where clouds have been flagged. Composites are subsequently generated by selecting the highest NDVI value for a 15-day period of each pixel for all continents except Africa, where the period is 10 days because of operational requirements of the Food and Agricultural Organization (FAO) of the United Nations. Data are currently being composited by continent. The resultant product is therefore substantially different from the NOAA GVI product in terms of its spatial resolution, projection, cloud clearing procedures and the values used to calculate the NDVI.

An ongoing research programme is being conducted by members of GIMMS to produce improved global data sets. These include the application of algorithms to correct for stratospheric aerosols associated with volcanic eruptions. A further development is the creation of data sets for global climate modelling with much reduced spatial resolution of only 1° using the GIMMS data sets (Los *et al.* 1994, Sellers *et al.* 1994); they have been applied for global land-cover mapping (Defries and Townshend 1994).

The third data processing stream involves the operational generation of medium-resolution data sets from 1990 onwards. Experiences with the GVI and GIMMS products are providing the basis of the NASA/NOAA-sponsored AVHRR Land Pathfinder data set, which is being created to act as a precursor for the international Earth Observing System (EOS). The Pathfinder project has focused on a number of data sets with global coverage that have been collected for a number of years. Current plans are to generate a land global data set based on GAC AVHRR data at 8 km spatial resolution retrospectively to 1982. Plans for the creation of this land data set are outlined by James and Kalluri (1994) and further information is given by Maiden and Greco (1994).

A related product is also being created at the Joint Research Centre (JRC) of the Commission of the European Communities (CEC) at Ispra, in Italy, for global climate and environmental studies. Much of this work has concentrated on pre-processing of long-term data sets, with particular emphasis placed on the continent of Africa (Belward *et al.* 1991). These data have been used to produce the Normalised Difference Vegetation Index, channel-2 reflectance, channel-3 and channel-4 brightness temperatures, an approximate surface temperature and a cloud probability image on a daily basis from July 1981 to January 1990 for the whole African continent at a resolution of 5 km. These data will be made available to the international research community through the Centre for Earth Observation (CEO) initiative of the CEC and the European Space Agency (ESA). The data sets have already provided new information concerning inter-annual and intra-annual variations in vegetation fire dynamics for Africa and have been used to derive forest seasonality information through the JRC's thematic projects TRopical Ecosystem and Environment observations by Satellites (TREES) (Malingreau et al. 1993) and Fire In global Resource and Environmental monitoring (FIRE) (Malingreau and Belward 1994).

The fourth data processing stream is concerned with constructing a data set at the full HRPT or LAC spatial resolution with an IFOV of 1.1 km. In the early days complete global coverage with 1.1 km resolution data could not be obtained because the tape-recorders on board the satellite can only collect partial coverage and data from local ground receiving stations have remained largely in local archives. But increasing awareness of the value of AVHRR data and the relatively modest costs of setting up ground receiving stations to receive HRPT data have led to the collection and local archiving of data for the majority of the Earth's land-surface. As a result of the activities of various groups (for details see Eidenshink and Faundeen 1994), the availability of data at 1 km resolution was identified as an important priority for a number of the Core Projects of the International Geosphere-Biosphere Program. The feasibility of meeting these requirements was provided by the production of the data sets of the whole conterminous United States by the Earth Resources Observation System (EROS) Data Center of the United States Geological Survey (Loveland et al. 1991, Eidenshink 1992). The scientific case for the data set and its required characteristics is described in a report published by the International Geosphere-Biosphere Program (1992) and by Townshend et al. (1994). The first stages in the creation of the data set are outlined by Buongiorno et al. (1993) and Eidenshink and Faundeen (1994). The principal objective of the EROS Data Center's programme was to define and evaluate the potential for using AVHRR 1 km digital imagery and multisource data (such as broad-scale climate, terrain, ecoregions) to characterise global land cover.

The EROS Data Center has direct reception capabilities for AVHRR HRPT data covering most of North America. By involving a number of organisations, access to a large number of HRPT direct readout stations enables full-resolution coverage of nearly all the land areas of the Earth to be obtained, see Figure 7.65 and also Table 2.28. These organisations include NOAA, NASA, USGS, the European Space Agency (ESA), the Australian Commonwealth Scientific and Industrial Research Organisation (CSIRO) and the Satellite Meteorological Centre (SMC) of China, and the project officially began on 1 April 1992 with 23 ground stations distributed worldwide. For those areas of land surface not covered by these stations, or if any station goes temporarily out of action, the on-board tape-recorders were scheduled to collect data to fill in the gaps. The only exception is that of Antarctica where complete land coverage is not always available. Simple Automatic Picture Transmission (APT) receivers have long been used on the Antarctic bases as an aid to short-period weather forecasting, monitoring of sea ice conditions for marine navigation and to help in aircraft operations. Given the restrictions of the on-board tape-recording facilities, the direct reception of High Resolution Picture Transmission (HRPT) data therefore provides the only means of obtaining large amounts of full-resolution imagery and, despite the harsh environmental conditions, tracking receivers and HRPT ground stations are now installed at coastal bases in Antarc-

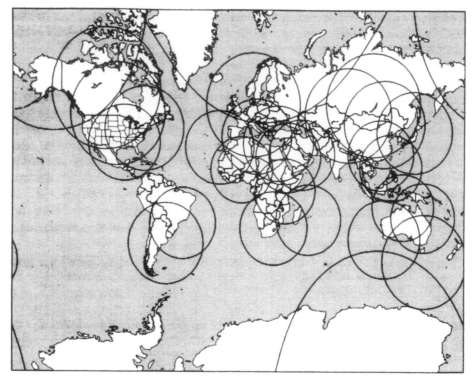

Figure 7.65   Coverage areas of the HRPT ground station network (Eidenshink and Faundeen 1994).

tica, see the Appendix. With the drop in the price of simple HRPT receivers and of the computing systems needed to process the data, it is anticipated that a number of other receivers will be installed over the next few years. However, since the major uses of the full spatial resolution data sets are related to surface vegetation cover, the lack of complete coverage of Antarctica is currently not regarded as a major limitation.

Each of the major participants and their affiliates is committed to gather, manage and share the data. Their responsibilities include routine collection of all the high resolution AVHRR data within their coverage region, quality control of the raw data, providing the data in a standard format and delivery of the data on a timely basis. These responsibilities are intended to complement, rather than interfere with, normal ground station operations and commitments to other programmes. AVHRR data reception activities are integrated with georegistration, product generation (such as greenness maps and land-cover classifications) and archiving systems. Because spatially extensive 1 km data sets possessing high temporal resolution have been largely unavailable, capabilities to use such data for regional land-cover characterisation have not been well explored. Information on data set distribution is available from Customer Services, EDC, Sioux Falls, South Dakota, 57198, USA, telephone: . . . 1-605-594-6507. All of the three previous various data processing streams have involved the reprocessing of AVHRR data back to the early 1980s, taking advantage of globally archived data. For the 1 km product such retrospective processing will not be possible because of the absence of global archives at this resolution.

One of the important future sources of land-cover information is expected to be the MODIS-N sensor, the nadir-pointing version of the Moderate Resolution Imaging Spectrometer, of the EOS programme (see Table 7.12). This instrument will be a substantial improvement as a global sensor of land cover compared with the AVHRR in terms of the number of spectral bands, their location, calibration, spatial resolution and geometrical registration. Additional spectral bands will permit improved atmospheric correction. However, the instrument will be no better in terms of the total swath width and the frequency of imaging. Following the placing of MODIS-N in orbit in the late 1990s, it is inevitable that longer term comparisons will have to rely on data from earlier sensors such as the AVHRR. Obtaining consistent relations between results from these different sensors will be an essential, if difficult, process for the longer term monitoring of land cover (Townshend *et al.* 1991).

Table 7.12 Comparison of the principal sensor characteristics of MODIS-N and AVHRR relevant for land-cover characterisation (Townshend *et al.* 1991)

	AVHRR	MODIS-N	
		Centre	Band width
Spectral bands for land-cover applications	580–680 nm	470 nm	20 nm
	725–1100 nm	555 nm	20 nm
	1580–1750 nm[a]	659 nm	20 nm
		865 nm	40 nm
		1240 nm	20 nm
		1640 nm	20 nm
		2130 nm	50 nm
	3 thermal bands	9 thermal bands	
IFOV (nadir)	1.1 km	500 m	
		250 m (659 and 865 nm)	
Swath width	2700 km	2330 km	
Calibration	absent	lunar	
Radiometric quantization	10 bit[b]	12 bit	
Global frequency	1–2 days	1–2 days	
View angle	55.4°	55°	
Tilt capability	none	none	

[a] Proposed future spectral band.
[b] Plans exist to increase the quantisation to 12 bits from NOAA-K onwards.

AVHRR HRPT ground stations with a digital archive

The following tables provide information on the location of AVHRR HRPT stations with a known digital archive. Information is provided on the acquisition and archiving policies of these stations. Addresses and telephone/fax numbers are also given. The information provided is believed to have been correct at the time of original compilation (April 1991). Inevitably changes in many of the details of this listing will occur with time. (International Geosphere-Biosphere Program 1992).

SITE NAME	SITE LOCATION	ADDRESS	ACQUISITION				ARCHIVE			COMMENTS
			SATELLITE NUMBER	OVERPASS TAKEN D = DAY N = NIGHT	FREQUENCY D = DAILY S = SPORADIC	DATA ARCHIVED R = RAW P = PROCESSED	MONTHS RETAINED 999 = PERMANENT	OPEN CATALOGUE? Y, N, U	DATA RETRIEVABLE? Y, N, U	
ANTARCTICA, SYOWA	LAT: 69°E LONG: 39°34'E	NATIONAL INSTITUTE OF POLAR RESEARCH TAKASHI, YAMANOUCHI 1-9-10 KAGA, ITABASHI-KU TOKYO, 173 JAPAN PHONE: +81 3962-4711 FAX:: +81 3962-2529 TELEX: 272 3 515 POLRSCJ	6-7 9-12	D	D	R + P	999	Y	Y	ARCHIVE SINCE 1/1/87
ANTARCTICA, TERRANOVA BAY	LAT: 74°24'S LONG: 164°07'E	ROBERTO CERVELLATI PNRA ENEA-CRE CASACCIA PAS S.P. AN GUILLARESE 301 00060 S.M. di GALERIA, ROMA, ITALY FAX: +39 6 3048 4893	10-12	D + N	D	R	999	Y	Y	GOOD DATA DURING SUMMER MONTHS ONLY
ANTARCTICA, O'HIGGINS	LAT: 63°19'S LONG: 57°54'W	KLAUS REINIGGER DLR 8031 OBERPFAFFENHOFEN GERMANY FAX: +49 815328 1443	9-12	D	S	R	999	Y	Y	FOUR SHORT PERIOD CAMPAIGNS PER YEAR. 4 TO 6 WEEKS EACH
ANTARCTICA, MCMURDO STATION AGENCY: ANTARCTIC RESEARCH CENTER SCRIPPS OCEANOGRAPHIC INSTITUTE	LAT: 77°50'56"S LONG: 166°39'36"E	ROBERT WHIRTNER SCRIPPS INST. OCEANOGRAPHY ANTARCTIC RESEARCH CENTER OCEAN RES. DIV., A-014 LAJOLLA, CA 92093, USA PHONE: (619) 534-3785 FAX: (619) 534-7383	6-11	D	D	R	999	Y	Y	8 × 10 REDUCED QUICK LOOK, ARCHIVE SINCE 10/85 ALL BANDS
ANTARCTICA, PALMER STATION AGENCY: RESEARCH ANTARCTIC CENTER, SCRIPPS OCEANOGRAPHIC INSTITUTE	LAT: 64°46'30'S LONG: 64°04'00"W	ROBERT WHIRTNER SCRIPPS INST. OCEANOGRAPHY ANTARCTIC RESEARCH CENTER OCEAN RES. DIV., A-014 LAJOLLA, CA 92093, USA PHONE: (619) 534-3785 FAX: (619) 534-7383	6-11	D	D	R	999	Y	Y	8 × 10 REDUCED QUICK LOOK, ARCHIVE SINCE 6/89. ALL BANDS
ARGENTINA, BUENOS AIRES	LAT: 34°24'S LONG: 58°18'W	SALVADOR ALAIMO SERVICIO METEROLOGICO NACIONAL 25 DE MAYO 658 BUENOS AIRES, ARGENTINA	11	D	D	R	UNCERTAIN	UNCERTAIN	UNCERTAIN	DATA ARCHIVED IN A RAW TELEMETRY FORMAT-REQUIRE CONVERSION TO CCT

SITE NAME	SITE LOCATION	ADDRESS	SATELLITE NUMBER	ACQUISITION OVERPASS TAKEN D = DAY N = NIGHT	FREQUENCY D = DAILY S = SPORADIC	DATA ARCHIVED R = RAW P = PROCESSED	ARCHIVE MONTHS RETAINED 999 = PERMANENT	OPEN CATALOGUE? Y, N, U	DATA RETRIEVABLE? Y, N, U	COMMENTS
AUSTRALIA, ALICE SPRINGS	LAT: 23 45 S LONG: 133 52 E	ROBIN BUCKLEY AUSTRALIAN CENTRE FOR REMOTE SENSING (ACRES), P.O. BOX 28 BELCONNEN, ACT 2616, AUSTRALIA PHONE: +61 6 252 4402 FAX: +61 6 251 6326 TELEX: AA 61510 ACRES C/AUSTRALIA, PUB: TELEMEMO, O/ACRES, U/N: ROBIN BUCKLEY	10-12	D	D	R	999	Y	Y	ARCHIVE SINCE 1/2/89
AUSTRALIA, HOBART, TASMANIA	LAT: 42 48 S LONG: 147 18 E	DR CARL HILSSON CSIRO MARINE LABORATORIES PO BOX 1538 HOBART TASMANIA 7001 AUSTRALIA PHONE: +61 (002)206251 FAX: +61 (002)240530 TELEX: AA 57182	7-12	D, N	D	R	999	Y	Y	PAUL TILDESLEY PHONE: 6102 206251 FAX: 6102 240 530 NOAA 7-11 ARCHIVE SINCE 4/86 ALL BANDS
AUSTRALIA, ASPENDALE	LAT: 38 S LONG: 145 E	ALEXANDER C DILLEY CSIRO DIV. OF ATMOS RESEARCH PRIVATE BAG NO 1, MORDIALLOC VICTORIA 3195, AUSTRALIA PHONE: +61 (03)586-7675 FAX: +61 (03)586-7600 TELEX: AA34463	7-12	D	D	R	999	Y	Y	ALL BANDS ARCHIVE SINCE 3/84
AUSTRALIA, PERTH	LAT: 32 0S LONG: 115 53 E	MR HENRY HOUGHTON WESTERN AUSTRALIA SATELLITE TECHNOLOGY AND APPLICATIONS CONSORTIUM (WASTAC) REMOTE SENSING APPLICATIONS CENTRE, 8TH FLOOR, JARDINE HOUSE, 184 ST. GEORGE'S TERRACE PERTH, WESTERN AUSTRALIA 6000 PHONE: 619-323-1520 FAX: 619-321-8576	7-12	D	D	R	999	Y	Y HARD COPY	ARCHIVE SINCE 1/1/81. HARD COPY CATALOG ONLY

| SITE NAME | SITE LOCATION | ADDRESS | SATELLITE NUMBER | ACQUISITION | | DATA ARCHIVED R = RAW P = PROCESSED | ARCHIVE | | | COMMENTS |
				OVERPASS TAKEN D = DAY N = NIGHT	FREQUENCY D = DAILY S = SPORADIC		MONTHS RETAINED 999 = PERMANENT	OPEN CATALOGUE? Y, N, U	DATA RETRIEVABLE? Y, N, U	
AUSTRALIA, TOWNSVILLE	LAT: 19°18'S LONG: 146°48'E	JOHN LILLEYMAN NORTH-EAST AUSTRALIAN SATELLITE IMAGERY SYSTEM DEPT. OF ELECTRICAL AND ELECTRONIC ENGINEERING JAMES COOK UNIV. OF NORTH QUEENSLAND, TOWNSVILLE QUEENSLAND, 4811 AUSTRALIA PHONE: +61-77-814379 FAX: +61-77-251348 TELEX: 47009	9-12	D + N	D	R	12	Y	Y	ACQUIRE 1-2 PASSES WEEKLY PRE-12/89. ALL PASSES POST-12/89 ARCHIVE SINCE 1/4/88 DIGITAL BROWSE
BANGLADESH, DHAKKA	LAT: 23°45'N LONG: 90°18'E	A. M. CHOUDHURY BANGLADESH SPACE RESEARCH AND REMOTE SENSING ORGANIZATION (SPARRSO) AGARGAN, SHER-E-BANGLA NAGAR, DHAKKA-1207 PHONE: 327335, TELEX: 642 215 SRS BJ	6-12	D + N	D	R + P	999	N	N	ARCHIVE HARDCOPY SINCE 1982, SELECT DIGITAL SINCE 1985
BRAZIL, CACHOEIRA PAULISTA	LAT: 22°45'S LONG: 45°00'W	SERGIO DE PAULA PEREIA INSTITUTO DE PESQUISAS ESPACIAS (DEPARTMENT OF OPERATIONS) RODOVIA PRESIDENTE DUTRA KM 40/SP. CP 001 12630 CACHOEIRA PAULISTA, SAO PAULO, BRASIL PHONE: +55 125 61 1377 FAX: +55 125 61 2088	9-12	D + N	S	R	999	N	Y	ARCHIVE SINCE 1/6/89: 1 PASS JUNE TO OCT. N-9/11 CH. 1-3: 3 PASSES WEEKLY NOV to MAY N-9/11 CH. 1-2
CANADA, DOWNSVIEW, ONTARIO	LAT: 43°46'N LONG: 79°28'W	MR. A. ALDUNATE ATMOSPHERIC ENVIRONMENT SERVICE SATELLITE DATA LAB. 4905 DUFFERIN STREET DOWNSVIEW, ONTARIO CANADA M3H 5T4 PHONE: 416-739-4896 FAX: 416-739-4521	9-11	D	D	R + P	10 YEARS	Y	U	DAILY PASS OVER EASTERN CANADA IS SENT TO CCRS FOR ARCHIVE. LIMITED ARCHIVE FOR RESEARCH

SITE NAME	SITE LOCATION	ADDRESS	ACQUISITION			ARCHIVE				COMMENTS
			SATELLITE NUMBER	OVERPASS TAKEN D = DAY N = NIGHT	FREQUENCY D = DAILY S = SPORADIC	DATA ARCHIVED R = RAW P = PROCESSED	MONTHS RETAINED 999 = PERMANENT	OPEN CATALOGUE? Y, N, U	DATA RETRIEVABLE? Y, N, U	
CANADA, EDMONTON, ALBERTA	LAT: 53°30'N LONG: 113°30'W	RON GOODSON ATMOSPHERIC ENVIRONMENT SERVICE, WESTERN REGION FORECAST OPERATIONS EDMONTON, ALBERTA CANADA PHONE: 403-468-7910 FAX: 403-468-7916	10 + 11		D	R + P	10 DAYS OR 24 MO. HARDCOPY	N	?	10 DAY ROTATING FOR DIGITAL 1 YEAR HARDCOPY
CANADA, PRINCE ALBERT, SASKATCHEWAN	LAT: 53°12'N LONG: 105°55'W	IAN PRESS MR. LEON BRONSTEIN CANADA CENTER FOR REMOTE SENSING, 2464 SHEFFIELD RD, OTTAWA, ONTARIO K1A 0Y7 CANADA PHONE: 613-991-5505 FAX: 613-952-9089 TELEX: 053-3589	6-12	D	D	R	999	Y	Y	ARCHIVE 7/83
CANARY ISLANDS, MASPALOMAS	LAT: 27°46'N LONG: 15°37'W	MANOLO SOZA ESTACION ESPACIAL DE MASPALOMAS LAS PALOMAS DE GRAN CANARIA P.O. BOX 29, MASPALOMAS, SPAIN PHONE: +34 28 761 876 FAX: +34 928 766 956	9-12	D	D	R	999	Y	Y	ARCHIVE SINCE 1/7/86 AT GRAN CANARIA, SPAIN
CHILE, SANTIAGO	LAT: 33°18'S LONG: 70°24'W	RAMON L. MOLINA CENTRO DE ESTUDIOS ESPACIALES, UNIVERSIDAD DE CHILE CASILLA 411-3 SANTIAGO, CHILE PHONE: +56 2 721 816 FAX: +56 2 844 1003	7-12	D + N	D	R	999	Y	N	ARCHIVE SINCE 1/11/83
CZECHOSLOVAKIA, PRAGUE	LAT: 50°04'N LONG: 14°27'E	MARTIN SETVAK & STEPAN KYJOVSKY CZECH HYDROMETEOROLOGICAL INSTITUT, PRAGUE CHMI, OBSERVATORY LIBUS NA SABATCE 17 143 06 PRAGUE 4, CZECHOSLOVAKIA PHONE: +42 2 468380 FAX: +42 2 40|0800	6-12			R		Y	Y	SELECTED DATA FOR METEOROLOGICAL RESEARCH

SITE NAME	SITE LOCATION	ADDRESS	SATELLITE NUMBER	ACQUISITION OVERPASS TAKEN D = DAY N = NIGHT	FREQUENCY D = DAILY S = SPORADIC	DATA ARCHIVED R = RAW P = PROCESSED	ARCHIVE MONTHS RETAINED 999 = PERMANENT	OPEN CATALOGUE? Y, N, U	DATA RETRIEVABLE? Y, N, U	COMMENTS
FRANCE, LANNION	LAT: 48°45'N LONG: 03°28'W	PASCAL BRUNEL CENTRE DE METEOROLOGIE SPATIALE, BP 147 22302 LANNION CEDEX, FRANCE PHONE: +33 96 05 67 45 FAX: +33 96 05 67 37	6-12	D + N	S	R + P	25 MONTH ROLLING	Y	Y	ACQUIRE DATA WITH EQUATOR CROSSING 20W-40E ASC. OR 150W-140E DES.
GERMANY, BERLIN	LAT: 52°28'N LONG: 13°12'E	FREE INSTITUTE OF BERLIN INST. OF METEOROLOGY, FB 24/WE7 ATTN: DIRK KOSLOWSKY DIETRICH-SCHAEFER-WEG 6 D-1000, BERLIN 41 FEDERAL REPUBLIC OF GERMANY FAX: +49 30 838 3874	6-11	D + N	D	R + P	999	N	Y	ARCHIVE SINCE 1/9/88
GERMANY, BREMERHAVEN	LAT: 53°31'N LONG: 08°34'E	ALFRED WEGENER INSTITUTE FOR POLAR AND SEA RESEARCH AM HANDELSHAFEN 12 D-2850 BREMERHAVEN, GERMANY PHONE: +49 471 483 1523 FAX: +49 471 483 1425 TELEX: 238 695 POLAR D ALFRED.WEGENER OR PIPOR.OFFICE	9-12		S	R + P		N	Y	SHORT PERIOD SCIENTIFIC CAMPAIGNS SINCE 1/4/88
GERMANY, HAMBURG	LAT: 53°34'N LONG: 09°58'E	RAINER FUDLOGG BSH, BERNHARD NOCHT STR. 78 D-2000 HAMBURG 36 POSTFACH 301220 D-2000 HAMBURG 36 PHONE: +49 40 3190 5125 FAX: +49 40 3190 5150 TELEX: 2 11 138 BSH HH D	9-12		D	R		Y	U	ARCHIVE SINCE 12/89
GERMANY, OBERPFAFFENHOFEN (MUNICH)	LAT: 48°30'N LONG: 11°09'E	WALTER EBKE DLR 8031 OBERPFAFFENHOFEN GERMANY PHONE: +49 8153 28 1187 FAX: +49 8513 28 1443	6-12	D + N	D	R	999	Y	Y	PRE-1983 LIMITED ARCHIVE ONE PASS DAILY 1983-86

SITE NAME	SITE LOCATION	ADDRESS	SATELLITE NUMBER	ACQUISITION		ARCHIVE				COMMENTS
				OVERPASS TAKEN D = DAY N = NIGHT	FREQUENCY D = DAILY S = SPORADIC	DATA ARCHIVED R = RAW P = PROCESSED	MONTHS RETAINED 999 = PERMANENT	OPEN CATALOGUE? Y, N, U	DATA RETRIEVABLE? Y, N, U	
INDIA, SHADNAGAR, HYDERABAD	LAT: 17°07'N LONG: 78°20'E	PROF. B. L. DEEKSHATULU DIRECTOR, NRSA BALANAGAR HYDERABAD 500 037 INDIA	11	D + N	D	R	999			DATA ARCHIVED ON HDDT SINCE 31/1/87
ITALY, ROMA	LAT: 41°26'N LONG: 12°16'E	PAOLO PAGANO SERVIZIO METEOROLOGICO DELL' AERONAUTICA PIAZZALE DEGLI ARCHIVI, 34, 00144 ROMA, ITALY PHONE: +39 6 5996 410 FAX: +39 6 5920 994	9-12	D	D	R		N	N	TIP EVERY PASS, SOMETIMES AVHRR. ARCHIVE SINCE 1/11/89
ITALY, SCANZANO	LAT: 37°54'N LONG: 13°21'E	SANDRO AND DEPISCOPO, ENNIO RICCOTTILLI, TELESPAZIO VIA TIBURTINA 965 00156 ROMA, ITALY PHONE: +39 6 4069 3375	9-12	D	D	R		Y	U	START ACQUISITIONS 1/1/91
INDONESIA, JAPAN	LAT: 6°S LONG: 106°E	TEJASUKMANA BAMBANG SBSCL LAPAN J1 KALISARI PEKAYON JAKARTA TIMUR, INDONESIA PHONE: +62 21 8710065 FAX: +62 21 8710786	6-12	D	D	R	999	N	Y	ARCHIVE INDONESIA AREA SINCE 1/1/1981
JAPAN, KIYOSE	LAT: 35°45'N LONG: 139°31'E	SEI-ICHI SAITOH METEOROLOGICAL INFORMATION CENTER, JAPAN WEATHER ASSOCIATION 5, 4-CHOMR, KOUJIMACHI, CHIYODA-KU TOKYO, JAPAN PHONE: +81 3 238 0480 FAX: +81 3 262 9549	6-12	D	D	R + P	999	Y	Y	ARCHIVE SINCE 1/3/87 (JAPAN ISLAND)

SITE NAME	SITE LOCATION	ADDRESS	ACQUISITION				ARCHIVE			COMMENTS
			SATELLITE NUMBER	OVERPASS TAKEN D = DAY N = NIGHT	FREQUENCY D = DAILY S = SPORADIC	DATA ARCHIVED R = RAW P = PROCESSED	MONTHS RETAINED 999 = PERMANENT	OPEN CATALOGUE? Y, N, U	DATA RETRIEVABLE? Y, N, U	
JAPAN, TOKYO, TOKAI UNIVERSITY	LAT: 32°49'N LONG: 130°52'E	HARUHISA SHIMODA TOKAI UNIVERSITY RESEARCH AND INFORMATION CENTER (TRIC) 2-28-4 TOMIGAY, SHIBUYA-KU TOKYO 151, JAPAN PHONE: +81 03 481 0611 FAX: +81 03 481 0610	6-12	D	D	R + P	999	N	Y	ARCHIVE SINCE 1/4/87
JAPAN, TOKYO, TOKYO UNIVERSITY	LAT: 36°N LONG: 140°E	MIKIO TAKAGI INSTITUTE OF INDUSTRIAL SCIENCE UNIVERSITY OF TOKYO 7-22-1, ROPPONGI MINATO-KU (CNPR) TOKYO 106, JAPAN PHONE: +81 3 479 0289 FAX: +81 3 402 6226	6-12	D	D	R	999	Y	Y	ARCHIVE SINCE 26/1/83
JAPAN, SENDAI	LAT: 38°15'N LONG: 140°50'E	DR. HIROSHI KAWAMURA EARTH OBSERVING SATELLITE CENTER, CENTER FOR ATMOS. AND OCEANIC VAR, FACILITY OF SCIENCE, TOHOKU UNIVERSITY, SENDAI 980, JAPAN PHONE: +81 22 222 1800 X3346 FAX: +81 22 268 2179	9-12	D	D	R + P	999	Y	Y	ARCHIVE SINCE 1/4/88 PROCESSED CH 2+4 OF TOHOKU AREA
KOREA, SEOUL: KMS	LAT: 37°36'N LONG: 127°00'E	CHOI, HEE SEUNG SATELLITE METEOROLOGY DEPARTMENT, KOREA, METEOROLOGICAL SERVICE 1, SONGWOL-DONG, CHONGNO-GU SEOUL 110-101, KOREA PHONE: +82 2 737 0011 FAX: +82 2 737 0325 TELEX: 27276 KMSSEL K	10-12	D + N	D	P	999	Y	Y	ARCHIVE SINCE 20/6/89 OF GOOD QUALITY

SITE NAME	SITE LOCATION	ADDRESS	ACQUISITION				ARCHIVE			
			SATELLITE NUMBER	OVERPASS TAKEN D = DAY N = NIGHT	FREQUENCY D = DAILY S = SPORADIC	DATA ARCHIVED R = RAW P = PROCESSED	MONTHS RETAINED 999 = PERMANENT	OPEN CATALOGUE? Y, N, U	DATA RETRIEVABLE? Y, N, U	COMMENTS
KOREA, SEOUL NATIONAL UNIVERSITY	LAT: 37°30′N LONG: 127°E	CGYBGM HIBG YUI REMOTE SENSING AND IMAGE PROCESSING LAB. DEPARTMENT OF OCEANOGRAPHY SEOUL NATIONAL UNIVERSITY SEOUL 151-742, KOREA PHONE: +82 2 880 6747 FAX: +82 2 882 4216 TELEX: 29664 SNUROK K	9-12	D	D	R	999	Y	Y	ARCHIVE SINCE 25/11/89 ALL ACQUIRED PASSES
LA REUNION	LAT: 20°52′S LONG: 55°28′E	MICHEL PETIT CENTRE ORSTROM 2051 AVENUE DU VAL DE MONTERRAND 34032 MONTPELIER, FRANCE PHONE: +33 67 617 445 FAX: +33 67 547 800	9-12	D	D	R	999	Y	Y	ARCHIVE SINCE 15/12/89 AT LA REUNION. FRANCE ONE IMAGE DAILY
NIGER, NIAMEY	LAT: 13°31′N LONG: 02°04′E	NIAMEY FIELD OFFICE CENTRE AGRHYMET FAX: 011 227 732435 TELEX: 011 982 5448 NI	10-12	D	D	R + P	999	Y	Y OR FRASCATI	ARCHIVE SINCE 1/5/89 AT NIAMEY (NDVI)
NETHERLANDS, DE BILT	LAT: 52°N LONG: 05°E	JAN R. BIJMA P.O. BOX 201 3730 KONINGIN WILHELMINALAAN 10, DE BILT, NETHERLANDS PHONE: +31 30 206 435 FAX: +31 30 210 407 TELEX: 47096 KNMI NL	6-12	D	D	P	999	Y	Y	ARCHIVE SINCE 1/2/90 (CLOUD-FREE)
NEW ZEALAND, WELLINGTON	LAT: 41°12′S LONG: 174°30′E	DEPT. OF SCIENTIFIC AND INDUSTRIAL RESEARCH (D SIR) PRIVATE BAG PALMEISTON NORTH NEW ZEALAND FAX: 063-66-664	10-12	D	D	P GEO-LOCATED	999	N	N	ARCHIVE SELECTED ACQUISITION SINCE 1/10/81

SITE NAME	SITE LOCATION	ADDRESS	SATELLITE NUMBER	ACQUISITION		DATA ARCHIVED R = RAW P = PROCESSED	ARCHIVE			COMMENTS
				OVERPASS TAKEN D = DAY N = NIGHT	FREQUENCY D = DAILY S = SPORADIC		MONTHS RETAINED 999 = PERMANENT	OPEN CATALOGUE? Y, N, U	DATA RETRIEVABLE? Y, N, U	
NORWAY, TROMSØ	LAT: 69°39'N LONG: 18°56'E	ROLF-TERJE ENOKSEN STATION MANAGER TROMSØ TELEMETRY STATION P.O. BOX 387 N-9001 TROMSØ, NORWAY	6-12	D + N	D	R	999	Y	Y	DATA ACQUISITION SINCE 1/84. DATA ARE SENT TO ESA-EARTHNET FOR ARCHIVE PRE-90=2 PASSED DAILY. POST-90= ALL PASSES ON OPTICAL DISC.
POLAND, KRAKOW	LAT: 50°N LONG: 20°E	LESLAW BARANSKI INST. OF MET. AND WATER MANAGEMENT (SATELLITE CENTER) KRAKOW, PIOTRA BOROWEGO STR, 14 30-215 KRAKOW, POLAND PHONE:+48 12 113 844 FAX:+48 12 116 798	9-12	D	D	R	999	N	Y	ARCHIVE SINCE 23/4/87 GOOD QUALITY DATA
SOUTH AFRICA, HARTEBEESTHOEK	LAT: 25°53'S LONG: 27°42'E	TIMOTHY BOYLE, MARAIS IKE SATELLITE APPLICATIONS CENTRE HARTEBEESTHOEK MIKOMTER/CSIR P.O. BOX 395 0001 PRETORIA, SOUTH AFRICA PHONE:+27 11 642 4693 FAX:+27 11 642 2446	6-12	D + N	D	R	999	Y	Y	ARCHIVE SINCE 1/11/84 AT HARTEBEESTHOEK
SWITZERLAND, BERNE	LAT: 46°10'N LONG: 06°00'E	MICHAEL BAUMGARTNER GEOGRAPHIUSCHES INSTITUT DER UNIVERSITAT BERN HALLERSTRASSE 12 3012 BERNE, SWITZERLAND PHONE:+41 31 658 020 FAX:+41 31 658 511	6-12	D + N	D	R	999	Y	Y	ARCHIVE SINCE 1/1/81 CH. 2&4 SOMETIMES CH. 3
TAIWAN, KEELUNG	LAT: 25°07'N LONG: 121°43'E	CHI-YUAN LIN TAIWAN FISHERIES RESEARCH INSTITUTE, 199 HOU-IN ROAD KEELUNG 20220, TAIWAN PHONE:+886 2 462 2101 FAX:+886 2 462 9388	10-12	D + N	D	R	999	N	?	ARCHIVE SINCE 1/4/89 GOOD QUALITY

SITE NAME	SITE LOCATION	ADDRESS	SATELLITE NUMBER	ACQUISITION		DATA ARCHIVED R = RAW P = PROCESSED	ARCHIVE			COMMENTS
				OVERPASS TAKEN D = DAY N = NIGHT	FREQUENCY D = DAILY S = SPORADIC		MONTHS RETAINED 999 = PERMANENT	OPEN CATALOGUE? Y, N, U	DATA RETRIEVABLE? Y, N, U	
TAIWAN, TAIPEI	LAT: 25°N LONG: 121°E	TAI-CHUNG YEN METEOROLOGICAL SATELLITE CENTER, CENTRAL WEATHER BUREAU, 64 KUNG YUAN ROAD TAIPEI, TAIWAN PHONE: +886 2 371 3181 X701 FAX: +886 2 331 5915	6-12	D + N	D	R	999	N	N	ACQUISITION START 28/1/81
UNITED KINGDOM, DUNDEE	LAT: 56°27'N LONG: 2°30'W	PETER BAYLIS NERC SATELLITE STATION UNIVERSITY OF DUNDEE DUNDEE DD1 4HN, SCOTLAND PHONE: +44 382 23181 X4406 FAX: +44 382 202 830 TELEX: 76293 ULDUND G	6-12	D + N	D	R	999	Y	Y	DATA ARCHIVED SINCE 6/11/78
UNITED KINGDOM, LASHAM	LAT: 51°11'N LONG: 01°02'W	ANTHONY MILES SPACE DEPARTMENT, Y60H BUILDING, ROYAL AEROSPACE ESTABLISHMENT FARNBOROUGH, HAMPSHIRE GU14 6TD, ENGLAND PHONE: +44 252 2441 X5787 FAX: +44 252 375 329	6-12	D	D	R	2 WEEK 1991 = LONG TERM	N	N	2 WEEK ROLLING START ARCHIVE IN 1991
USA, AUSTIN, TX	LAT: 30°16'N LONG: 97°45'W	MELBA CRAWFORD, DIRECTOR CENTER FOR SPACE RESEARCH UNIVERSITY OF TEXAS AT AUSTIN AUSTIN, TX 78712, USA PHONE: (512) 471-3070 FAX: (512) 471-8727	10, 11	D	D	R	999	Y	Y	
USA, BATON ROUGE, LA	LAT: 30°24'N LONG: 91°10'W	DR. OSCAR HUH LOUISIANA STATE UNIVERSITY COASTAL STUDIES INSTITUTE 3RD FLOOR, OLD GEOLOGY BLDG. HOWE-RUSSELL GEOSCIENCE COMPLEX, BATON ROUGE, LA 70803-7527, USA PHONE: (504) 388-2952 FAX: (504) 388-2520	9, 10, 11	D + N	D	R	999	Y	Y	ARCHIVE SINCE 28/6/88

SITE NAME	SITE LOCATION	ADDRESS	ACQUISITION				ARCHIVE			COMMENTS
			SATELLITE NUMBER	OVERPASS TAKEN D = DAY N = NIGHT	FREQUENCY D = DAILY S = SPORADIC	DATA ARCHIVED R = RAW P = PROCESSED	MONTHS RETAINED 999 = PERMANENT	OPEN CATALOGUE? Y, N, U	DATA RETRIEVABLE? Y, N, U	
USA, FAIRBANKS, AK	LAT: 64°58'N LONG: 147°30'W	NOAA/NESDIS/NCDC SATELLITE DATA SERVICES DIV. ROOM 100 PRINCETON EXECUTIVE SQUARE 5627 ALLENTOWN ROAD WASHINGTON, DC 20233, USA PHONE: (301) 763-8400 ALASKA PHONE: (907) 474-7487 FAX: (301) 763-8443	6-11	D + N	D	R	999	Y	Y	DATA UPLINKED VIA DOMSAT TO NOAA/NESDIS IN SUITLAND, MARYLAND FOR ARCHIVE
USA, UH SNUG HARBOR, HAWAII	LAT: 21°19'N LONG: 157°15'W	PIERRE FLAMENT DEPARTMENT OF OCEANOGRAPHY UNIVERSITY OF HAWAII AT MANOA 1000 POPE ROAD HONOLULU, HI 96822, USA	9-11		D	P	12 ON-LINE SST TO CD	Y	Y	ARCHIVE SINCE 1/6/90
USA, SIOUX FALLS, SD	LAT: 43°32'N LONG: 96°44'W	EROS DATA CENTER CUSTOMER SERVICES SIOUX FALLS, SD 57198, USA PHONE: (605) 594-6507 FAX: (605) 594-7589	9-11	D	D	R	999	Y	Y	ACQUISITION AND ARCHIVE SINCE 5/87 BROWSE IMAGES ON MICROFICHE DEVELOPING DIGITAL BROWSE
USA, STENNIS SPACE CENTER, MS	LAT: 30.12'N LONG: 89.33'W	ED ARTHUR NAVAL OCEAN & ATMOSPHERIC LAB. (NOARL) CODE 321 STENNIS SPACE CENTER, MS 39529-5004, USA PHONE: (601) 688-5265 FAX: (601) 688-4149	10-11		S	P	12	Y	Y	ARCHIVE = 1 YEAR ROLLING
USA, WALLOPS ISLAND, VA	LAT: 37°52'N LONG: 75°27'W	NOAA/NESDIS/NCDC SATELLITE DATA SERVICES DIV. ROOM 100 PRINCETON EXECUTIVE SQUARE 5627 ALLENTOWN ROAD WASHINGTON, DC 20233, USA PHONE: (301) 763-8400 FAX: (301) 763-8443	6-11	D + N	D	R	999	Y	Y	DATA UPLINKED VIA DOMSAT TO NOAA/NESDIS IN SUITLAND, MARYLAND FOR ARCHIVE

References

Abel, P. (1990) Prelaunch calibration of the NOAA-11 AVHRR visible and near ir channels. *Remote Sensing of Environment*, **31**, 227–229.

Abel, P., Guenther, B., Galimore, R. N. and Cooper, J. W. (1993) Calibration results for NOAA-11 AVHRR channels 1 and 2 from congruent path aircraft observations. *Journal of Atmospheric and Oceanic Technology*, **10**, 493–508.

Achard, F. and Blasco, F. (1990) Analysis of vegetation seasonal evolution and mapping of forest cover in West Africa with the use of NOAA AVHRR HRPT data. *Photogrammetric Engineering and Remote Sensing*, **56**, 1359–1365.

Ackerman, S. A. (1989) Using the radiative temperature difference at 3.7 and 11 μm to track dust outbreaks. *Remote Sensing of Environment*, **27**, 129–133.

Ackleson, S. G. and Holligan, P. M. (1989) AVHRR observations of a Gulf of Maine Coccolithophore bloom. *Photogrammetric Engineering and Remote Sensing*, **55**, 473–474.

Agarwal, V. K. and Ashajayanthi, A. V. (1982) Simulation study of multispectral estimation of sea-surface temperature from infrared observations. *Remote Sensing of Environment*, **12**, 371–380.

Ali, A., Quadir, D. A. and Huh, O. K. (1987) Agricultural, hydrologic and oceanographic studies in Bangladesh with NOAA AVHRR data. *International Journal of Remote Sensing*, **8**, 917–925.

Ali, A., Quadir, D. A. and Huh, O. K. (1989) Study of river flood hydrology in Bangladesh with AVHRR data. *International Journal of Remote Sensing*, **10**, 1873–1891.

Allam, R. J. (1986) On a slight mis-registration of AVHRR channels 3 and 4. *International Journal of Remote Sensing*, **7**, 887–894.

Allam, R. J., Brown, P. B., Dibben, P., Smith, F. and Standley, A. P. (1989) Imagery products from AVHRR for operational meteorology: the Autosat-2 project in the UK Meteorological Office. Proceedings of 4th AVHRR Data Users' Meeting, Rothenburg, F.R. Germany, 5–8 September 1989. EUM P 06 (Darmstadt – Eberstadt: EUMETSAT), pp. 5–10.

Al-Taee, M. A., Cracknell, A. P. and Vaughan, R. A. (1993) A two-look technique applied to sea surface temperature retrieval from NOAA AVHRR data, *Towards Operational Applications*. Proceedings of the 19th Annual Conference of the Remote Sensing Society, Chester, UK, September 1993 (Nottingham: Remote Sensing Society), pp. 92–97.

Andersen, T. (1982) Operational snowmapping by satellites. In *Hydrological aspects of alpine and high-mountain areas*, edited by J. W. Glen (Wallingford: IAHS), pp. 149–154.

Anding, D. and Kauth, R. (1972) Reply to the comment by G. A. Maul and M. Sidran. *Remote Sensing of Environment*, **2**, 171–173.

André, J. C., Goutorbe, J. P. and Perrier, A. (1986) HAPEX-MOBILHY, A hydrologic atmospheric experiment for the study of water budget and evaporation flux at the climate scale. *Bulletin of the American Meteorological Society*, **67**, 138–144.

André, J. C., Goutorbe, J. P., Perrier, A., Becker, F., Bessemoulin, P., Bougeault, P., Brunet, Y., Brutsaert, W., Carlson, T., Cuenca, R., Gash, J., Gelpe, J., Hildebrand, P., Laguoard, J. P., Lloyd, C., Mahrt, L., Mascart, P., Mazaudier, C., Noilhan, J., Ottlé, O., Payen, M., Phulpin, T., Stull, R., Shuttleworth, J., Schmugge, T., Tagonet, O., Tarrieu, C., Thepenier, R. M., Vallencogne, C., Vidal-Madjar, D. and Neill, A. (1988) HAPEX-MOBILHY: first results from the Special Observing Period. *Annalae Geophysicae*, **6**, 447–492.

Antoine, J. Y., Derrien, M., Gaillard, O., Le Borgne, P., Le Goas, C. and Marsouin, A. (1991) Operational restitution of small scale SST from AVHRR data over the N.E. Atlantic and the Mediterranean Sea. Proceedings of 5th AVHRR Data Users' Meeting, Tromsø, Norway, 25–28 June 1991. EUM P 09 (Darmstadt – Eberstadt: EUMETSAT), pp. 301–306.

Antoine, J. Y., Derrien, M., Harang, L., Le Borgne, P., Le Gleau, H. and Le Goas, C. (1992) Errors at large satellite zenith angles on AVHRR derived sea surface temperatures. *International Journal of Remote Sensing*, **13**, 1797–1804.

Arino, O., Dedieu, G. and Deschamps, P. Y. (1992) Determination of land surface spectral reflectances using Meteosat and NOAA/AVHRR shortwave channel data. *International Journal of Remote Sensing*, **13**, 2263–2287.

Arino, O., Viovy, N. and Belward, A. S. (1991) Vegetation dynamics of West Africa classified using AVHRR NDVI time series. Proceedings of 5th AVHRR Data Users' Meeting, Tromsø, Norway, 25–28 June 1991. EUM P 09 (Darmstadt – Eberstadt: EUMETSAT), pp. 441–447.

Asrar, G., Kanematsu, E. T., Jackson, R. D. and Pinter, P. J. (1985) Estimation of total above ground phytomass production using remotely sensed data. *Remote Sensing of Environment*, **17**, 211–220.

Aulamo, H. and Pylkkö, P. (1993) Deriving the intensity of rain from AVHRR data and vertical temperature fields. Proceedings of 6th AVHRR Data Users' Meeting, Belgirate, Italy, 29 June–2 July 1993. EUM P 12 (Darmstadt – Eberstadt: EUMETSAT), pp. 231–236.

Bachmann, M. and Bendix, J. (1991) Geocorrection of NOAA/AVHRR imagery. An algorithm based on orbital parameters and one ground control point. Proceedings of 5th AVHRR Data Users' Meeting, Tromsø, Norway, 25–28 June 1991. EUM P 09 (Darmstadt – Eberstadt: EUMETSAT), pp. 11–16.

Bachmann, M. and Bendix, J. (1992) An improved algorithm for NOAA-AVHRR image referencing. *International Journal of Remote Sensing*, **13**, 3205–3215.

Baldwin, D. and Emery, W. J. (1995) Spacecraft attitude variations of NOAA-11 inferred from AVHRR imagery. *International Journal of Remote Sensing*, **16**, 531–548.

Barale, V. and Schlittenhardt, P. M. (1993) *Ocean colour: theory and applications in a decade of CZCS experience* (Dordrecht: Kluwer).

Baranski, L. A. (1992) ATMOSATLAND PC software for the processing and application of NOAA/HRPT data for environmental studies. *International Journal of Remote Sensing*, **13**, 291–301.

Baranski, L. A. and Mrugalski, J. (1991) AVHRR – useful source of hydrological parameters. Proceedings of 5th AVHRR Data Users' Meeting, Tromsø, Norway, 25–28 June 1991. EUM P 09 (Darmstadt – Eberstadt: EUMETSAT), pp. 129–134.

Barber, D. G. and Richard, P. R. (1992) Use of AVHRR imagery in arctic marine mammal research. *International Journal of Remote Sensing*, **13**, 167–175.

Barker, H. W. and Davies, J. A. (1992) Cumulus cloud radiative properties and the characteristics of satellite radiance wavenumber spectra. *Remote Sensing of Environment*, **42**, 51–64.

Barnes, J. C. and Bowley, C. J. (1968) Snow cover distribution as mapped from satellite photographs. *Water Resources Research*, **4**, 257–272.

Barnes, J. C. and Smallwood, M. D. (1975) Synopsis of current satellite snow mapping techniques, with emphasis on the application of near-infrared data. In *Operational application of satellite snowcover observations*, edited by A. Rango. NASA-SP-391 (Washington, DC: Scientific and Technical Information Office, NASA, pp. 199–214.

Barnes, J. C. and Smallwood, M. (1982) TIROS-N series direct readout services user's guide. (Washington, DC: US Department of Commerce, NOAA/NESDIS).

Barrett, E. C. (1974) *Climatology from satellites* (London: Methuen).

Barrett, E. C. (1989) Satellite remote sensing of rainfall. In *Applications of remote sensing to agrometeorology*, edited by F. Toselli (Dordrecht: Kluwer), pp. 305–326.

Barrett, E. C., D'Souza, G. and Power, C. H. (1986) Bristol techniques for the use of satellite data in rain cloud and rainfall monitoring. *Journal of the British Interplanetary Society*, **39**, 517–526.

Barrett, E. C., Kidd, C. and Xu, H. (1991) Snow monitoring by AVHRR-HRPT and DMSP SSM/I data analyses. Proceedings of 5th AVHRR Data Users' Meeting, Tromsø, Norway, 25–28 June 1991. EUM P 09 (Darmstadt – Eberstadt: EUMETSAT), pp. 103–108.

Bartlett, D. S., Johnson, R. W., Hardisky, M. A. and Klemas, V. (1986) Assessing impacts of off-nadir observation on remote sensing of vegetation: use of the Suits model. *International Journal of Remote Sensing*, **7**, 247–264.

Bartlett, D. S., Whiting, G. J. and Hartman, J. M. (1990) Use of vegetation indices to estimate intercepted solar radiation and net carbon dioxide exchange of a grass canopy. *Remote Sensing of Environment*, **30**, 115–128.

Barton, I. J. (1989) Digitization effects in AVHRR and MCSST data. *Remote Sensing of Environment*, **29**, 87–89.

Barton, I. J. and Bathols, J. M. (1989) Monitoring floods with AVHRR. *Remote Sensing of Environment*, **30**, 89–94.

Barton, I. J., Prata, A. J., Watterson, I. G. and Young, S. A. (1992) Identification of the Mount Hudson volcanic cloud over S.E. Australia. *Geophysical Research Letters*, **19**, 1211–1214.

Barton, I. J. and Takashima, T. (1986) An AVHRR investigation of surface emissivity near Lake Eyre, Australia. *Remote Sensing of Environment*, **20**, 153–163.

Baumgartner, M. F. (1987) Schneeschmelz-Abflussimulationen basierend auf Schneeflachenbestimmungen mit digitalen Landsat-MSS und NOAA/AVHRR Daten. (Snow-melt run-off simulations based on snow area determination with digital Landsat-MSS and NOAA/AVHRR data). *Remote Sensing Series* (University of Zurich, Department of Geography), **11**.

Baumgartner, M. F. and Apfl., G. (1993) Alpine snow cover analysis system. Proceedings of 6th AVHRR Data Users' Meeting, Belgirate, Italy, 29 June–2 July 1993. EUM P 12 (Darmstadt – Eberstadt: EUMETSAT), pp. 355–363.

Baylis, P. E. (1981) Guide to the design and specification of a primary user receiving station for meteorological and oceanographic satellite data. In *Remote sensing in meteorology, oceanography and hydrology*, edited by A. P. Cracknell (Chichester: Ellis Horwood), pp. 81–96.

Baylis, P. E. (1983) University of Dundee satellite data reception and archiving facility. In *Remote sensing applications in marine science and technology*, edited by A. P. Cracknell (Dordrecht: Reidel), pp. 29–34.

Baylis, P., Brush, J. and Parmiggiani, F. (1989) The AVHRR receiving and processing system for the Italian Antarctic research program. Proceedings of 4th AVHRR Data Users' Meeting, Rothenburg, F.R. Germany, 5–8 September 1989. EUM P 06 (Darmstadt – Eberstadt: EUMETSAT), pp. 41–44.

Becker, F. (1981) Angular reflectivity and emissivity of natural media in the thermal infrared bands. In *Signatures spectrales d'objets en télédétection*. Colloque ISPRS, Avignon, France, 1981, edited by G. Guyot and M. Verbrugghe (Versailles: INRA), pp. 57–72.

Becker, F. (1982) Absolute sea surface temperature measurement by remote sensing and atmospheric corrections using differential radiometry. In *Processes in marine remote sensing*, edited by F. J. Vernberg and F. P. Diemer (Columbia: University of South Carolina Press), pp. 151–174.

Becker, F. (1987) The impact of spectral emissivity on the measurement of land surface temperature from a satellite. *International Journal of Remote Sensing*, **8**, 1509–1522.

Becker, F., Bolle, H.-J. and Rowntree, P. R. (1988) The International Satellite Land-Surface Climatology Project. ISLSCP Report 10, ISLSCP Secretariat.

Becker, F. and Choudhury, B. J. (1988) Relative sensitivity of Normalized Difference Vegetation Index (NDVI) and Microwave Polarization Difference Index (MPDI) for vegetation and desertification monitoring. *Remote Sensing of Environment*, **24**, 297–311.

Becker, F. and Li, Z. H. (1990) Towards a local split window method over land surfaces. *International Journal of Remote Sensing*, **11**, 369–393.

Bellec, B. and Le Gléau, H. (1992) The multispectral colour-composite technique: an improved method to display meteorological satellite imagery. *International Journal of Remote Sensing*, **13**, 1981–1998.

Belward, A. S. (1992) Spatial attributes of AVHRR imagery for environmental monitoring. *International Journal of Remote Sensing*, **13**, 193–208.

Belward, A. S., Grégoire, J.-M., D'Souza, G., Trigg, S., Hawkes, M., Burstet, J.-M., Serça, D., Tireford, J.-L., Charlot, J.-M. and Vuattoux, R. (1993) *In situ*, real-time fire detection using NOAA/AVHRR data. Proceedings of 6th AVHRR Data Users' Meeting, Belgirate, Italy, 29 June–2 July 1993. EUM P 12 (Darmstadt – Eberstadt: EUMETSAT), pp. 333–339.

Belward, A. S. and Lambin, E. (1990) Limitations to the identification of spatial structures from AVHRR data. *International Journal of Remote Sensing*, **11**, 921–927.

Belward, A. S., Malingreau, J. P. and Verstraete, M. M. (1991) Long term AVHRR data sets for global environmental studies. Proceedings of 5th AVHRR Data Users' Meeting, Tromsø, Norway, 25–28 June 1991. EUM P 09 (Darmstadt – Eberstadt: EUMETSAT), pp. 195–202.

Berger, F. H. (1989) Cloud classes derived from AVHRR data. Proceedings of 4th AVHRR Data Users' Meeting, Rothenburg, F.R. Germany, 5–8 September 1989. EUM P 06 (Darmstadt – Eberstadt: EUMETSAT), pp. 65–68.

Berie, G. and Cornillon, P. (1981) Remote sensing, a tool for managing the marine environment: eight case studies. Marine Technical Report 77, Department of Ocean Engineering, University of Rhode Island, Rhode Island, USA.

Bernstein, R. L. (1982) Sea surface temperature estimation using the NOAA-6 satellite Advanced Very High Resolution Radiometer. *Journal of Geophysical Research*, **87**, 9455–9465.

Blanchet, D. (1988) Variations of the local failure pressure with depth through first-year and multi-year ice. *Journal of Offshore Mechanics and Arctic Engineering*, **110**, 159–168.

Bocoum, O. (1991) Calibrage des canaux 1 & 2 de NOAA-AVHRR pour une comparaison annuelle des données NDVI. Proceedings of 5th AVHRR Data Users' Meeting, Tromsø, Norway, 25–28 June 1991. EUM P 09 (Darmstadt – Eberstadt: EUMETSAT), pp. 35–40.

Böhm, E., Marullo, S. and Santoleri, R. (1991) AVHRR visible-IR detection of diurnal warming events in the western Mediterranean Sea. *International Journal of Remote Sensing*, **12**, 695–701.

Bordes, P., Brunel, P. and Marsouin, A. (1992) Automatic adjustment of AVHRR navigation. *Journal of Atmospheric and Oceanic Technology*, **9**, 15–27.

Boulton, G. S., Smith, G. D., Jones, A. S. and Newsome, J. (1985) Glacial geology and glaciology of the last mid-latitude ice sheets. *Journal of the Geological Society of London*, **142**, 1–8.

Bowers, S. A. and Hanks, R. J. (1965) Reflection of radiant energy from soils. *Soil Science*, **100**, 130–138.

Bradbury, P. A. (1989) Regional crop condition monitoring using AVHRR NDVI: a Spanish case study. Proceedings of 4th AVHRR Data Users' Meeting, Rothenburg, F.R. Germany, 5–8 September 1989. EUM P 06 (Darmstadt – Eberstadt: EUMETSAT), pp. 309–312.

Brakke, T. W. and Otterman, J. (1990) Canopy bidirectional reflectance dependence of leaf orientation. *International Journal of Remote Sensing*, **11**, 1023–1032.

Breaker, L. C. (1990) Estimating and removing sensor-induced correlation from Advanced Very High Resolution Radiometer satellite data. *Journal of Geophysical Research*, **95**, 9701–9711.

Brest, C. L. and Rossow, W. B. (1992) Radiometric calibration and monitoring of NOAA AVHRR data for ISCCP. *International Journal of Remote Sensing*, **13**, 235–273.

Bromwich, D. H. (1992) A satellite case study of a katabatic surge along the Transantarctic Mountains. *International Journal of Remote Sensing*, **13**, 55–66.

Brown, J. W., Brown, O. B. and Evans, R. H. (1993) Calibration of Advanced Very High Resolution Radiometer infrared channels: a new approach to nonlinear correction. *Journal of Geophysical Research*, **98**, 18257–18268.

Brown, O. B., Brown, J. W. and Evans, R. H. (1985) Calibration of Advanced Very High Resolution Radiometer infrared observations. *Journal of Geophysical Research*, **90**, 11667–11677.

Brown, R., Wooster, M., Blake, B., Sear, C., Hutchinson, P. and Stromme, T. (1993) Use of real time AVHRR imagery to assess fishery conditions off Namibia. Proceedings of 6th AVHRR Data Users' Meeting, Belgirate, Italy, 29 June–2 July 1993. EUM P 12 (Darmstadt – Eberstadt: EUMETSAT), pp. 385–392.

Brown, S., Mason, I. and Harris, A. (1991) Lake surface temperatures from AVHRR for climate research. Proceedings of 5th AVHRR Data Users' Meeting, Tromsø, Norway, 25–28 June 1991. EUM P 09 (Darmstadt – Eberstadt: EUMETSAT), pp. 211–216.

Brunel, P. and Marsouin, A. (1987) An operational method using ARGOS orbital elements for navigation of AVHRR imagery. *International Journal of Remote Sensing*, **8**, 569–578.

Brunel, P. and Marsouin, A. (1989) Navigation of AVHRR images using Argos or TBUS orbital elements. Proceedings of 4th AVHRR Data Users' Meeting, Rothenburg, F.R. Germany, 5–8 September 1989. EUM P 06 (Darmstadt – Eberstadt: EUMETSAT), pp. 11–15.

Brunel, P., Marsouin, A. and Bordes, P. (1991) Automatic adjustment of AVHRR navigation. Proceedings of 5th AVHRR Data Users' Meeting, Tromsø, Norway, 25–28 June 1991. EUM P 09 (Darmstadt – Eberstadt: EUMETSAT), pp. 41–46.

Brush, R. J. H. (1988) The navigation of AVHRR imagery. *International Journal of Remote Sensing*, **9**, 1491–1502.

Brush, R. J. H. (1993) Anomalous effect of sunglint on the AVHRR in the NOAA-12 satellite. *International Journal of Remote Sensing*, **14**, 629–634.

Bryceson, K. P. (1989) The use of Landsat MSS data to determine the distribution of locust eggbeds in the Riverina region of New South Wales, Australia. *International Journal of Remote Sensing*, **10**, 1749–1762.

Bunting, J. T. and D'Entremont, R. P. (1982) Improved cloud detection utilizing Defense Meteorological Satellite Program near infrared measurements. Technical report AFGL-TR-82-0027, Air Force Geophysics Laboratory (LYS), Hanscom Air Force Base, Massachusetts 01731, USA.

Buongiorno, A., Pittella, G. and Eidenshink, J. (1993) 'Global Land 1-km AVHRR Data Set' project: the status of data collection archiving and processing after the first year of experience. Proceedings of 6th AVHRR Data Users' Meeting, Belgirate, Italy, 29 June–2 July 1993. EUM P 12 (Darmstadt – Eberstadt: EUMETSAT), pp. 485–491.

Burns, B. A., Viehoff, T. and Schmidt, M. (1989) Development of digital algorithms for sea ice monitoring with AVHRR imagery. Proceedings of 4th AVHRR Data Users' Meeting, Rothenburg, F.R. Germany, 5–8 September 1989. EUM P 06 (Darmstadt – Eberstadt: EUMETSAT), pp. 241–244.

Buttleman, K. (1988) Oceanographic features shown in NOAA-10 image. *Photogrammetric Engineering and Remote Sensing*, **54**, 1692.

Cabot, F., Dedieu, G. and Maisongrande, P. (1993) Surface albedo from space over HAPEX Sahel sites. Proceedings of 6th AVHRR Data Users' Meeting, Belgirate, Italy, 29 June–2 July 1993. EUM P 12 (Darmstadt – Eberstadt: EUMETSAT), pp. 51–57.

Cahoon, D. R., Stocks, B. J., Levine, J. S., Cofer, W. R. and Chung, C. C. (1992) Evaluation of a technique for satellite-derived area estimation of forest fires. *Journal of Geophysical Research*, **97**, 3805–3814.

Callison, R. D. and Cracknell, A. P. (1984) Atmospheric correction to AVHRR brightness temperatures for waters around Britain. *International Journal of Remote Sensing*, **5**, 185–198.

Carey, R. M. and Pritchard, J. (1989) Flooding in Bangladesh. *Photogrammetric Engineering and Remote Sensing*, **55**, 974.

Carlson, T. N. (1979) Atmospheric turbidities in Saharan dust outbreaks as determined by analysis of satellite brightness data. *Monthly Weather Review*, **107**, 322–335.

Carlson, T. N. and Prospero, J. M. (1972) The large-scale movement of Saharan air outbreaks over the northern equatorial Atlantic. *Journal of Applied Meteorology*, **11**, 283–297.

Caselles, V. and Delegido, J. (1987) A simple model to estimate the daily value of the regional maximum evapotranspiration from satellite temperature and albedo images. *International Journal of Remote Sensing*, **8**, 1151–1162.

Caselles, V., Delegido, J., Sobrino, J. A. and Hurtado, E. (1992) Evaluation of the maximum evapotranspiration over the La Mancha region, Spain, using NOAA AVHRR data. *International Journal of Remote Sensing*, **13**, 939–946.

Caselles, V. and Sobrino, J. A. (1989) Determination of frosts in orange groves from NOAA-9 AVHRR data. *Remote Sensing of Environment*, **29**, 135–146.

Castagne, N., Le Borgne, P., Le Vourch, J. and Olry, J.-P. (1986) Operational measurement of sea surface temperatures at CMS Lannion from NOAA-7 AVHRR data. *International Journal of Remote Sensing*, **7**, 953–984.

Che, N. and Price, J. C. (1992) Survey of radiometric calibration results and methods for visible and near infrared channels of NOAA-7, -9 and -11 AVHRRs. *Remote Sensing of Environment*, **41**, 19–27.

Chedin, A., Scott, N. A. and Benoir, A. (1982) A single-channel, double-viewing angle method for sea surface temperature determination from coincident METEOSAT and TIROS-N radiometric measurements. *Journal of Applied Meteorology*, **21**, 613–618.

Cheevasuvit, F., Taconet, O., Vidal-Madjar, D. and Maitre, H. (1985) Thermal structure of an agricultural region as seen by NOAA-7 AVHRR. *Remote Sensing of Environment*, **17**, 153–163.

Chiesa, C. C. and Tyler, W. A. (1994) Beyond cubic convolution: ERIM restoration for remotely-sensed imagery. *Earth Observation Magazine*, February 1994, 40–44.

Choudhury, B. J. (1987) Relationships between vegetation indices, radiation absorption and net photosynthesis evaluated by a sensitivity analysis. *Remote Sensing of Environment*, **22**, 209–233.

Choudhury, B. J. (1989a) Monitoring global land surface using Nimbus-7 37 GHz data. Theory and examples. *International Journal of Remote Sensing*, **10**, 1579–1605.

Choudhury, B. J. (1989b) Estimating evaporation and carbon assimilation using infrared temperature data; vistas in modelling. In *Theory and applications of optical remote sensing*, edited by G. Asrar (New York: Wiley), pp. 628–690.

Choudhury, B. J. (1990) Monitoring arid lands using AVHRR-observed visible reflectance and SMMR-37GHz polarization difference. *International Journal of Remote Sensing*, **10**, 1949–1956.

Choudhury, B. J. and DiGirolamo, N. E. (1994) Relations between SMMR 37 GHz polarization difference and rainfall and atmospheric carbon dioxide concentration. *International Journal of Remote Sensing*, **15**, 3643–3661.

Choudhury, B. J. and Tucker, C. J. (1987a) Monitoring global vegetation using Nimbus-7 37 GHz data: some empirical relations. *International Journal of Remote Sensing*, **8**, 1085–1090.

Choudhury, B. J. and Tucker, C. J. (1987b) Satellite observed seasonal and inter-annual variation of vegetation over the Kalahari, the Great Victoria Desert and the Great Sandy Desert: 1979–1984. *Remote Sensing of Environment*, **23**, 233–241.

Choudhury, B. J., Tucker, C. J., Golus, R. E. and Newcomb, W. W. (1987) Monitoring vegetation using Nimbus-7 scanning multi-channel microwave radiometer's data. *International Journal of Remote Sensing*, **8**, 533–538.

Chouret, A. (1974) Les effets de la sécheresse actuelle en Afrique sur le niveau du lac Tchad. *Cahiers ORSTOM: Série Hydrologie*, **XI**, No. 1, pp. 35–46.

Christensen, F. T., Lu, Q.-M. and Pedersen, L. T. (1994) Multi-year sea ice mapping by thermal infrared radiometry. *International Journal of Remote Sensing*, **15**, 1229–1250.

Cicone, R. C. and Metzler, M. D. (1984) Comparison of Landsat MSS, Nimbus-7 CZCS, and NOAA-7 AVHRR features for land-use analysis. *Remote Sensing of Environment*, **14**, 257–265.

Cihlar, J., St.-Laurent, L. and Dyer, J. A. (1991) Relation between the Normalized Difference Vegetation Index and ecological variables. *Remote Sensing of Environment*, **35**, 279–298.

Clark, J. R. and La Violette, P. E. (1981) Detecting the movement of oceanic fronts using registered TIROS-N imagery. *Geophysical Research Letters*, **8**, 229–232.

Clevers, J. G. P. W. (1989) The application of a weighted infrared-red vegetation index for estimating leaf area index by correcting for soil moisture. *Remote Sensing of Environment*, **29**, 25–37.

Coakley, J. A. and Baldwin, D. G. (1984) Towards the objective analysis of clouds from satellite imagery data. *Journal of Climate and Applied Meteorology*, **23**, 1065–1099.

Coakley, J. A. and Bretherton, F. P. (1982) Cloud cover from high-resolution scanner data; detecting and allowing for partially filled fields of view. *Journal of Geophysical Research*, **87**, 4917–4932.

Coll, C., Valor, E., Schmugge, T. J. and Casselles, V. (1993) A procedure for estimating the land surface emissivity difference in the AVHRR channels 4 and 5. Application to the Valencian area Spain. Proceedings of 6th AVHRR Data Users' Meeting, Belgirate, Italy, 29 June–2 July 1993. EUM P 12 (Darmstadt – Eberstadt: EUMETSAT), pp. 73–78.

Collier, P., Runacres, A. M. E. and McClatchey, J. (1989) Mapping very low surface temperature in the Scottish Highlands using NOAA AVHRR data. *International Journal of Remote Sensing*, **10**, 1519–1529.

Collins, M. J. (1989) Synoptic ice motion from AVHRR imagery; automated measurements versus wind-driven theory. *Remote Sensing of Environment*, **29**, 79–85.

Colwell, J. N. (1974) Vegetation canopy reflectance. *Remote Sensing of Environment*, **3**, 175–183.

Colwell, R. N. (1983) *Manual of remote sensing* (Falls Church, Virginia: American Society of Photogrammetry).

Condit, H. R. (1970) The spectral reflectance of American soils. *Photogrammetric Engineering*, **36**, 955–966.

Constans, J., Fraysse, G., Leger, G. and Roux, J. (1984) The continuous measurement of the temperature of the upper layer of the sea – a practical approach. *International Journal of Remote Sensing*, **5**, 103–114.

Cooper, D. I. and Asrar, G. (1989) Evaluating atmospheric correction models for retrieving surface temperatures from AVHRR over a tallgrass prairie. *Remote Sensing of Environment*, **27**, 93–102.

Coulson, K. L. and Reynolds, D. W. (1971) The spectral reflectance of natural surfaces. *Journal of Applied Meteorology*, **10**, 1285–1295.

Cox, C. and Munk, W. (1954) Measurement of the roughness of the sea surface from photographs of the sun glitter. *Journal of the Optical Society of America*, **44**, 838–850.

Cox, C. and Munk, W. (1956) Slopes of the sea surface deduced from photographs of sun glitter. *Bulletin of the Scripps Institute of Oceanography*, **6**, 401–488.

Cox, G. F. N. and Weeks, W. F. (1988) Numerical simulations of the profile properties of undeformed first-year sea ice during the growth season. *Journal of Geophysical Research*, **93**, 12449–12460.

Cracknell, A. P. (1981) *Remote sensing in meteorology, oceanography and hydrology* (Chichester: Ellis Horwood).

Cracknell, A. P. (1990) Sunglint and the study of near-surface windspeeds over the oceans. In *Microwave remote sensing for oceanographic and marine weather-forecast models*, edited by R. A. Vaughan (Dordrecht: Kluwer), pp. 125–139.

Cracknell, A. P. (1991) Rapid remote recognition of habitat changes. *Preventive Veterinary Medicine*, **11**, 315–323.

Cracknell, A. P. (1993) A method for the correction of sea surface temperatures derived from satellite thermal infrared data in an area of sunglint. *International Journal of Remote Sensing*, **14**, 3–8.

Cracknell, A. P. and Hayes, L. W. B. (1991) *Introduction to remote sensing* (London: Taylor & Francis).

Cracknell, A. P. and Huang, W. G. (1988) Surface currents off the west coast of Ireland studied from satellite images. *International Journal of Remote Sensing*, **9**, 439–449.

Cracknell, A. P., Khattak, S. and Vaughan, R. A. (1987) The automatic identification of areas of sunglint in AVHRR data. *Advances in digital image processing*, Proceedings of the Annual Conference of the Remote Sensing Society, Nottingham, UK, September 1987 (Nottingham: Remote Sensing Society), pp. 403–415.

Cracknell, A. P., Khattak, S. and Vaughan, R. A. (1988) Sunglint in AVHRR data and the determination of near surface windspeeds. *International Archives of Photogrammetry and Remote Sensing*, **27**(B7), 118–124.

Cracknell, A. P., Mackay, G. and Vaughan, R. A. (1989) Estimation of ground heat flux using AVHRR data for Iceland. Proceedings of 4th AVHRR Data Users' Meeting, Rothenburg, F.R. Germany, 5–8 September 1989. EUM P 06 (Darmstadt – Eberstadt: EUMETSAT), pp. 317–322.

Cracknell, A. P., Muirhead, K., Callison, R. D. and Campbell, N. A. (1983) Satellite remote sensing, environmental monitoring and the offshore oil industries. In Proceedings of an EARSeL/ESA Symposium on Remote Sensing Applications for Environmental Studies, Brussels, Belgium, 26–29 April 1983 (Paris: European Space Agency), pp. 163–171.

Cracknell, A. P. and Paithoonwattanakij, K. (1989a) Fast automatic registration for satellite imagery by using heuristic planning. *IEE Proceedings*, **136F**, 221–225.

Cracknell, A. P. and Paithoonwattanakij, K. (1989b) Pixel and sub-pixel accuracy in geometrical correction of AVHRR imagery. *International Journal of Remote Sensing*, **10**, 661–667.

Cracknell, A. P. and Xue, Y. (1996) Thermal inertia determination from space – a tutorial review. *International Journal of Remote Sensing*, **17**, 431–461.

Craig, R. G. and Labovitz, M. L. (1980) Sources of variation in Landsat autocorrelation. Proceedings of the 14th International Symposium on Remote Sensing of Environment, San José, Costa Rica (Ann Arbor, Michigan: Environmental Research Institute of Michigan), pp. 1755–1767.

Crane, R. G. and Anderson, M. R. (1984) Satellite discrimination of snow/cloud surfaces. *International Journal of Remote Sensing*, **5**, 213–223.

Cresswell, G. and Pearce, A. (1985) Ocean circulation off Western Australia and the Leeuwin current. CSIRO Division of Oceanography Information Service, Sheet 16-3, CSIRO Division of Oceanography, Hobart, Tasmania, Australia.

Crippen, R. E. (1988) The dangers of underestimating the importance of data adjustments in band ratioing. *International Journal of Remote Sensing*, **9**, 767–776.

Crippen, R. E. (1990) Calculating the vegetation index faster. *Remote Sensing of Environ-*

ment, **34**, 71–73.

Croft, T. A. (1977) *Nocturnal images of the earth from space* (Order Number 68197) (Reston, Virginia: US Geological Survey).

Croft, T. A. (1978) Night-time images of the Earth from space. *Scientific American*, **239**, No. 1, 68–79.

Cross, A. (1990) NOAA/AVHRR as a data source for a tropical forest GIS: Deforestation in Amazonia. Proceedings of the International Geoscience and Remote Sensing Symposium, Washington, DC, USA, 21–25 May 1990 (New York: IEEE), pp. 223–226.

Cross, A. M. (1992) Monitoring marine oil pollution using AVHRR data: observations off the coast of Kuwait and Saudi Arabia during January 1991. *International Journal of Remote Sensing*, **13**, 781–788.

Cross, A. M., Settle, J. J., Drake, N. A. and Paivinen, R. T. M. (1991) Subpixel measurement of tropical forest cover using AVHRR data. *International Journal of Remote Sensing*, **12**, 1119–1129.

Csiszár, I., Fejes, E., Kerényi, J. and Rimóczi-Paál., A. (1993) Dependence of the daily amplitude of the active surface temperature on the vegetation index and the radiative balance components. Proceedings of 6th AVHRR Data Users' Meeting, Belgirate, Italy, 29 June–2 July 1993. EUM P 12 (Darmstadt – Eberstadt: EUMETSAT), pp. 67–72.

Curran, P. J. (1983) Multispectral remote sensing for estimation of green leaf area index. *Philosophical Transactions of the Royal Society of London*, **309A**, 257–270.

Dabrowska-Zielinska, K. (1989) Estimates of evapotranspiration using AVHRR data corrected for absorption by atmospheric water vapour. Proceedings of 4th AVHRR Data Users' Meeting, Rothenburg, F.R. Germany, 5–8 September 1989. EUM P 06 (Darmstadt – Eberstadt: EUMETSAT), pp. 185–192.

Dabrowska-Zielinska, K. and Grusczcynska, M. (1993) NOAA AVHRR applied for monitoring of soil water condition. Proceedings of 6th AVHRR Data Users' Meeting, Belgirate, Italy, 29 June–2 July 1993. EUM P 12 (Darmstadt – Eberstadt: EUMETSAT), pp. 151–160.

D'Aguanno, J. (1986) Use of AVHRR data for studying katabatic winds in Antarctica. *International Journal of Remote Sensing*, **7**, 703–713.

Dalu, G. (1985) Emittance effect on the remotely sensed sea surface temperature. *International Journal of Remote Sensing*, **6**, 733–740.

Dalu, G. (1986) Satellite remote sensing of atmospheric water vapour. *International Journal of Remote Sensing*, **7**, 1089–1097.

Dalu, G. and Liberti, G. L. (1989) Parameterization of the sea surface radiative fluxes as functions of the information content of the AVHRR-2 data. Proceedings of 4th AVHRR Data Users' Meeting, Rothenburg, F.R. Germany, 5–8 September 1989. EUM P 06 (Darmstadt – Eberstadt: EUMETSAT), pp. 197–203.

Dalu, G. and Viola, A. (1987) An improved calibration scheme for AVHRR-2. *International Journal of Remote Sensing*, **8**, 1501–1508.

Dalu, G., Viola, A. and Marullo, S. (1985) Sea surface temperature from AVHRR-2 data. *Il Nuovo Cimento*, **8C**, 793–804.

Dancette, C. (1983) Besoins en eau du mil au Sénégal: adaptation en zone semi-aride tropicale, *Agronomie Tropicale*, **4**, 267–280.

Daughtry, C. S. T., Gallo, K. D. and Bauer, M. E. (1982) Spectral estimates of solar radiation intercepted by corn canopies. AgRISTARS Technical Report sr-PZ-04236 Purdue University, West Lafayette, IN.

Dave, J. V. (1978) Extensive data sets of the diffuse radiation in realistic atmospheric models with aerosols and common absorbing gases. *Solar Energy*, **21**, 361–369.

Dave, J. V. (1981) Transfer of visible radiation in the atmosphere. Palo Alto Scientific Report No. G320-3411, IBM Corporation, Palo Alto Scientific Center, 1530 Page Mill Road, Palo Alto, California 94304.

Dech, S. W. (1989) The use of NOAA-AVHRR satellite data for monitoring sea ice motion in the Greenland sea. Proceedings of 4th AVHRR Data Users' Meeting, Rothenburg, F.R. Germany, 5–8 September 1989. EUM P 06 (Darmstadt – Eberstadt: EUMETSAT), pp. 323–327.

Dech, S. W. and Glaser, R. (1991) Ecological effects of burning oilfields in Kuwait derived from AVHRR data. Proceedings of 5th AVHRR Data Users' Meeting, Tromsø, Norway, 25–28 June 1991. EUM P 09 (Darmstadt – Eberstadt: EUMETSAT), pp. 449–453.

Dech, S. W. and Glaser, R. (1992) Burning oilwells in Kuwait – smoke plume monitoring and effects on vegetation derived from AVHRR data. *International Journal of Remote Sensing*, **13**, 3243–3249.

De Cotiis, A. G., Jandhyala, V. and Goodrum, G. (1991) The Image Library and Browse System (ILABS). *Palaeography, Palaeoclimatology, Palaeoecology (Global and Planetary Change Section)*, **90**, 281–285.

Deering, D. W., Rouse, J. W., Haas, R. H. and Schell, J. A. (1975) Measuring 'forage production' of grazing units from Landsat MSS data. Proceedings of the 10th International Symposium on Remote Sensing of Environment (Ann Arbor, Michigan: Environmental Research Institute of Michigan), volume II, pp. 1169–1178.

Defries, R. S. and Townshend, J. R. G. (1994) NDVI-derived land cover classifications at a global scale. *International Journal of Remote Sensing*, **15**, 3567–3586.

DeLuisi, J. (1983) Measurements of the El Chichón dust cloud from Mauna Loa observatory. Proceedings of the 7th Annual Climate Diagnostics Workshop, October 17–22, 1982 (National Centre for Atmospheric Research Boulder, CO), pp. 383–385.

Derrien, M., Engel, F., Farki, B., Fdhil, M., Frayssinet, P., Le Gléau, H. and Sairouni, A. (1993) Vegetation description with NOAA-11/AVHRR. Proceedings of 6th AVHRR Data Users' Meeting, Belgirate, Italy, 29 June–2 July 1993. EUM P 12 (Darmstadt – Eberstadt: EUMETSAT), pp. 101–108.

Derrien, M., Farki, B., Le Gléau, H. and Sairouni, A. (1992) Vegetation cover mapping over France using NOAA-11/AVHRR. *International Journal of Remote Sensing*, **13**, 1787–1795.

Derrien, M. and Le Gléau, H. (1993) Some numerical cloud studies derived from the operational AVHRR processing at Météo-France. Proceedings of 6th AVHRR Data Users' Meeting, Belgirate, Italy, 29 June–2 July 1993. EUM P 12 (Darmstadt – Eberstadt: EUMETSAT), pp. 403–409.

Derrien, M., Le Gléau, H., Harang, L., Noyalet, A. and Piriou, J. L. (1989) An automatic cloud classification using AVHRR at night. Proceedings of 4th AVHRR Data Users' Meeting, Rothenburg, F.R. Germany, 5–8 September 1989. EUM P 06 (Darmstadt – Eberstadt: EUMETSAT), pp. 69–72.

Deschamps, P. Y. and Phulpin, T. (1980) Atmospheric correction of infrared measurements of sea surface temperature using channels at 3.7, 11 and 12 μm. *Boundary Layer Meteorology*, **18**, 131–143.

Desjardins, R., Gray, J. and Bonn, F. (1990) Atmospheric corrections for remotely-sensed thermal data in a cool humid temperature zone. *International Journal of Remote Sensing*, **11**, 1369–1389.

Dey, B. and Feldman, U. (1990) Observations of winter polynyas and fractures using NOAA AVHRR TIR images and Nimbus-7 SMMR sea ice concentration charts. *Remote Sensing of Environment*, **30**, 141–149.

Diallo, O., Diouf, A., Hanan, N. P., Ndiaye, A. and Prévost, Y. (1991) AVHRR monitoring of savanna primary production in Senegal, West Africa: 1987–1988. *International Journal of Remote Sensing*, **12**, 1259–1279.

Dirks, R. W. J. and Spitzer, D. (1987) On the radiation transfer in the sea, including fluorescence and stratification effects. *Limnology and Oceanography*, **32**, 942–953.

Djavadi, D. and Cracknell, A. P. (1986) Cloud cover and the monitoring of strawburning using AVHRR data. *International Journal of Remote Sensing*, **7**, 949–951.

Domain, F., Citeau, J. and Noel, J. (1980) Sea surface temperatures studied by Meteosat data along the coast of Senegal and Mauritania. *Coastal and marine applications of remote sensing*, Proceedings of the 6th Annual Conference of the Remote Sensing Society, University of Dundee, UK, 18–19 December 1979, edited by A. P. Cracknell (Reading: Remote Sensing Society), pp. 59–67.

Dousset, B., Flament, P. and Bernstein, R. (1993) Los Angeles fires seen from space. *EOS, Transactions, American Geophysical Union*, **74**, 33–38.

Dozier, J. (1981) A method for satellite identification of surface temperature fields of subpixel resolution. *Remote Sensing of Environment*, **11**, 221–229.

Dozier, J. (1984) Snow reflectance from Landsat-4 Thematic Mapper. *IEEE Transactions on Geoscience and Remote Sensing*, **22**, 323–328.

D'Souza, G., Belward, A. S. and Malingreau, J.-P. (1996) *Advances in the use of NOAA AVHRR data for land applications* (Dordrecht: Kluwer).

D'Souza, G. and Hiederer, R. (1989) Monitoring rangeland vegetation in the Sahel using Landsat MSS and NOAA AVHRR. Proceedings of 4th AVHRR Data Users' Meeting, Rothenburg, F.R. Germany, 5–8 September 1989. EUM P 06 (Darmstadt – Eberstadt: EUMETSAT), pp. 111–116.

Dudhia, A. (1989) Noise characteristics of the AVHRR infrared channels. *International Journal of Remote Sensing*, **10**, 637–644.

Duggin, M. J. (1985) Factors limiting the discrimination and quantification of terrestrial features using remotely-sensed radiance. *International Journal of Remote Sensing*, **6**, 3–21.

Durkee, P. A. Pfeil, F., Frost, E. and Shema, R. (1990) Global analysis of aerosol particle characteristics and implications for effects on climate. *Atmospheric Environment*, **25A**, 2457–2471.

Dwivedi, R. M. and Narain, A. (1987) Remote sensing of phytoplankton, an attempt from the Landsat Thematic Mapper. *International Journal of Remote Sensing*, **8**, 1563–1569.

Eales, P. (1989) Automatic linear feature matching for satellite image navigation. Proceedings of 4th AVHRR Data Users' Meeting, Rothenburg, F.R. Germany, 5–8 September 1989. EUM P 06 (Darmstadt – Eberstadt: EUMETSAT), pp. 21–26.

Ebert, E. E. (1987) A pattern recognition algorithm for distinguishing surface and cloud types in the polar regions. *Journal of Climate and Applied Meteorology*, **26**, 1412–1427.

Ebert, E. E. (1989) Analysis of polar clouds from satellite imagery using pattern recognition and a statistical cloud analysis scheme. *Journal of Applied Meteorology*, **28**, 382–399.

Ebert, E. E. (1992) Pattern recognition analysis of polar clouds during summer and winter. *International Journal of Remote Sensing*, **13**, 97–109.

Eck, T. F. and Kalb, V. L. (1991) Cloud-screening for Africa using a geographically and seasonally variable infrared threshold. *International Journal of Remote Sensing*, **12**, 1205–1221.

Eckhardt, M. (1989) Polar sea ice distribution derived from AVHRR data. Proceedings of 4th AVHRR Data Users' Meeting, Rothenburg, F.R. Germany, 5–8 September 1989. EUM P 06 (Darmstadt – Eberstadt: EUMETSAT), pp. 205–208.

Eckhardt, M., Gallas, J. and Tonn, W. (1992) Sea ice distribution in the Greenland and Barents Seas based on satellite information for the period 1966–89. *International Journal of Remote Sensing*, **13**, 23–35.

Ehrlich, D., Estes, J. E. and Singh, A. (1994) Applications of NOAA-AVHRR 1 km data for environmental monitoring: a review. *International Journal of Remote Sensing*, **15**, 145–161.

Eidenshink, J. C. (1992) The 1990 conterminous US AVHRR data set. *Photogrammetric Engineering and Remote Sensing*, **58**, 809–813.

Eidenshink, J. C. and Faundeen, J. L. (1994) The 1 km AVHRR global data set: first stages in implementation. *International Journal of Remote Sensing*, **15**, 3443–3462.

Ellingsen, E. (1989) The Tromsø telemetry station: present capabilities and future plans. *International Journal of Remote Sensing*, **10**, 613–616.

Emery, W. J., Brown, J. and Nowak, Z. P. (1989) AVHRR image navigation: summary and review. *Photogrammetric Engineering and Remote Sensing*, **55**, 1175–1183.

Emery, W. J., Fowler, C. W., Hawkins, J. and Preller, R. H. (1991) Fram Strait satellite image-derived ice motions. *Journal of Geophysical Research*, **96**, 4751–4768.

Emery, W. J. and Ikeda, M. (1984) A comparison of geometric correction methods for AVHRR imagery. *Canadian Journal of Remote Sensing*, **10**, 46–56.

Emery, W. J. and Schluessel, P. (1989) Global differences between skin and bulk sea surface temperatures. *EOS Transactions of the American Geophysical Union*, **70**, 210–211.

Emery, W. J., Thomas, A. C., Collins, M. J., Crawford, W. R. and Mackas, D. L. (1986) An objective method for computing advective surface velocities from sequential infrared satellite images. *Journal of Geophysical Research*, **91**, 12865–12878.

Engelstad, M., Sengupta, S. K., Lee, T. and Welch, R. M. (1992) Automated detection of jet contrails using the AVHRR split window. *International Journal of Remote Sensing*, **13**, 1391–1412.

England, C. F. and Hunt, G. E. (1985) A bispectral method for the automatic determination of parameters for use in imaging satellite cloud retrievals. *International Journal of Remote Sensing*, **6**, 1545–1553.

EUMETSAT (1993) *EUMETSAT Directory of Meteorological Satellite Applications*. EUM BR 03 (Darmstadt – Eberstadt: EUMETSAT).

Eyre, J. R. (1986) Estimation of sea surface temperature from AVHRR data; comments on the paper by Singh *et al.* (1985). *International Journal of Remote Sensing*, **7**, 465–469.

Eyre, J. R., Brownscombe, J. L. and Allam, R. J. (1984) Detection of fog at night using Advanced Very High Resolution Radiometer (AVHRR) imagery. *Meteorological Magazine*, **113**, 266–271.

Faanes, T. (1991) A multispectral algorithm for snow mapping using NOAA AVHRR. Proceedings of 5th AVHRR Data Users' Meeting, Tromsø, Norway, 25–28 June 1991. EUM P 09 (Darmstadt – Eberstadt: EUMETSAT), pp. 455–461.

Faizoun, A. and Dedieu, G. (1993) Atmospheric effects on NOAA/AVHRR shortwave measurements: Sensitivity study and use of atmospheric climatologies to correct AVHRR time series. Proceedings of 6th AVHRR Data Users' Meeting, Belgirate, Italy, 29 June–2 July 1993. EUM P 12 (Darmstadt – Eberstadt: EUMETSAT), pp. 161–168.

Feigelson, E. M. (1984) *Radiation in a cloudy atmosphere* (Dordrecht: Reidel).

Ferrare, R. A., Fraser, R. S. and Kaufman, Y. J. (1990) Satellite measurements of large scale air pollution: measurements of forest fire smoke. *Journal of Geophysical Research*, **95**, 9911–9925.

Fett, R. W. (1977) *Navy tactical applications guide*, volume I. *Techniques and applications of image analysis* (Park Ridge, Illinois: Walter A. Bohan).

Fiedler, P. C. and Laurs, R. M. (1990) Variability of the Columbia River plume observed in visible and infrared imagery. *International Journal of Remote Sensing*, **11**, 999–1010.

Fiedler, P. C., Smith, G. B. and Laurs, R. M. (1984) Fisheries applications of satellite data in the eastern North Pacific. *Marine Fisheries Review*, **46**, 1–13.

Fischer, J. C. (1989) Advanced Medium Resolution Imaging Radiometer (AMRIR). Proceedings of 4th AVHRR Data Users' Meeting, Rothenburg, F.R. Germany, 5–8 September 1989. EUM P 06 (Darmstadt – Eberstadt: EUMETSAT), pp. 279–283.

Fiúza, A. F. G. (1992) The measurement of sea surface temperature from satellites. In *Space oceanography*, edited by A. P. Cracknell (Singapore: World Scientific), pp. 197–279.

Flannigan, M. D. (1985) Forest fire monitoring using the NOAA satellite series. MSc thesis, Department of Atmospheric Science, Colorado State University, Fort Collins, Colorado.

Flannigan, M. D. and Vonder Haar, T. H. (1986) Forest fire monitoring using NOAA satellite AVHRR. *Canadian Journal of Forest Research*, **16**, 975–982.

Flasse, S. P., Verstraete, M. M. and Meyer, D. J. (1993) Inverting a bidirectional reflectance model to remove directional effects in AVHRR data. Proceedings of 6th AVHRR Data Users' Meeting, Belgirate, Italy, 29 June–2 July, 1993. EUM P 12 (Darmstadt – Eberstadt: EUMETSAT), pp. 79–85.

Foerster, J. W. (1993) Northeast north Pacific Ocean: surface current pattern shifts during the spring. *Remote Sensing of Environment*, **43**, 149–159.

Follansbee, W. A. (1973) Estimation of average daily rainfall from satellite cloud photographs. NOAA Technical Memorandum NESS 44 (Washington, DC: Department of Commerce), 39 pp.

Foote, R. and Draper, L. T. (1980) TIROS-N Advanced Very High Resolution Radiometer (AVHRR). *Coastal and marine applications of remote sensing*, Proceedings of the 6th Annual Conference of the Remote Sensing Society, University of Dundee, UK, 18–19 December 1979, edited by A. P. Cracknell (Reading: Remote Sensing Society), pp. 25–35.

Forkert, T., Strauss, B. and Wendling, P. (1993) A new algorithm for the automated detection of jet contrails from NOAA-AVHRR satellite images. Proceedings of 6th AVHRR Data Users' Meeting, Belgirate, Italy, 29 June–2 July 1993. EUM P 12 (Darmstadt – Eberstadt: EUMETSAT), pp. 513–519.

Foster, J. L. and Hall, D. K. (1991) Observations of snow and ice features during the polar winter using moonlight as a source of illumination. *Remote Sensing of Environment*, **37**, 77–88.

França, G. B. and Cracknell, A. P. (1994) Retrieval of land and sea surface temperature using NOAA-11 AVHRR data in northeastern Brazil. *International Journal of Remote Sensing*, **15**, 1695–1712.

França, G. B. and Cracknell, A. P. (1995) A simple cloud masking approach using NOAA AVHRR daytime data for tropical areas. *International Journal of Remote Sensing*, **16**, 1697–1705.

Fraser, R. S. and Kaufman, Y. J. (1985) The relative importance of scattering and absorption in remote sensing. *IEEE Transactions in Geoscience and Remote Sensing*, **23**, 625–633.

Frouin, R. and Gautier, C. (1987) Calibration of NOAA-7 AVHRR, GOES-5, and GOES-6 VISSR/VAS solar channels. *Remote Sensing of Environment*, **22**, 73–101.

Frouin, R., Gautier, C. and Morcrette, J. J. (1988) Downward longwave irradiance at the ocean surface from satellite data: methodology and in situ validation. *Journal of Geophysical Research*, **93**, 597–619.

Fung, I. K., Prentice, E., Matthews, J., Lerner, J. and Russell, G. (1983) Three-dimensional tracer study of atmospheric CO_2: response to seasonal exchanges with terrestrial biosphere. *Journal of Geophysical Research*, **88**, 1281–1294.

Fusco, L. (1989) The European coordinate TIROS network: operational experience and future developments. Proceedings of 4th AVHRR Data Users' Meeting, Rothenburg, F.R. Germany, 5–8 September 1989. EUM P 06 (Darmstadt – Eberstadt: EUMETSAT), pp. 289–297.

Fusco, L. (1991) CEO (Committee on Earth Observation) AVHRR work plan: an international effort for the coordination of AVHRR related activities. Proceedings of 5th AVHRR Data Users' Meeting, Tromsø, Norway, 25–28 June 1991. EUM P 09 (Darmstadt – Eberstadt: EUMETSAT), pp. 95–100.

Fusco, L., Muirhead, K. and Tobiss, G. (1989) Earthnet's coordination scheme for AVHRR data. *International Journal of Remote Sensing*, **10**, 625–636.

Gagliardini, D. A., Karszenbaum, H., Legeckis, R. and Klemas, V. (1984) Application of Landsat MSS, NOAA/TIROS AVHRR, and Nimbus CZCS to study the La Plata River and its interaction with the ocean. *Remote Sensing of Environment*, **15**, 21–36.

Galladet, T. C. and Simpson, J. J. (1991) Automated cloud screening of AVHRR imagery using split-and-merge clustering. *Remote Sensing of Environment*, **38**, 77–121.

Gallegos, S. C., Hawkins, J. D. and Chiu Fu Cheng (1993) A new automated method of cloud masking for Advanced Very High Resolution Radiometer full-resolution data over the ocean. *Journal of Geophysical Research*, **98**, 8505–8516.

Gallo, K. P. and Daughtry, C. S. T. (1987) Differences in vegetation indices for simulated Landsat-5 MSS and TM, NOAA-9 AVHRR, and SPOT-1 sensor systems. *Remote Sensing of Environment*, **23**, 439–452.

Gallo, K. P. and Eidenshink, J. C. (1988) Differences in visible and near-ir responses, and derived vegetation indices, for the NOAA-9 and NOAA-10 AVHRRs: a case study. *Photogrammetric Engineering and Remote Sensing*, **54**, 485–490.

Gallo, K. P. and Flesch, T. K. (1989) Large-area crop monitoring with the NOAA AVHRR: estimating the silking stage of corn development. *Remote Sensing of Environment*, **27**, 73–80.

Garand, L. (1988) Automated recognition of oceanic cloud patterns. Part I: Methodology and application to cloud climatology. *Journal of Climate*, **1**, 20–39.

Garand, L. and Weinman, J. A. (1986) A structural-stochastic model for the analysis and synthesis of cloud images. *Journal of Climate and Applied Meteorology*, **25**, 1052–1068.

Gastellu-Etchegorry, J. P. and Boely, T. (1988) Methodology for an operational monitoring of remotely-sensed sea surface temperatures in Indonesia. *International Journal of Remote Sensing*, **9**, 423–438.

Gaston, A. (1981) Vegetation of Chad, northeast and southeast of Lake Chad. PhD thesis, University of Paris XII, 338 pp.

Gerson, D. J. and Rosenfeld, A. (1975) Automatic sea ice detection in satellite pictures. *Remote Sensing of Environment*, **4**, 187–198.

Gervin, J. C., Kerber, A. G., Witt, R. G., Lu, Y. C. and Sekhon, R. (1985) Comparison of level I land cover classification accuracy for MSS and AVHRR data. *International Journal of Remote Sensing*, **6**, 47–57.

Gesell, G. (1989) An algorithm for snow and ice detection using AVHRR data: an extension to the APOLLO software package. *International Journal of Remote Sensing*, **10**, 897–905.

Gesell, G., König, T., Mannstein, H. and Kriebel, K. T. (1993) SHARK-APOLLO quantitative satellite data analysis based on ESRIN/SHARP and DLR/APOLLO. Proceedings of 6th AVHRR Data Users' Meeting, Belgirate, Italy, 29 June–2 July 1993. EUM P 12 (Darmstadt – Eberstadt: EUMETSAT), pp. 583–587.

Glaze, L., Francis, P. W. and Rothery, D. A. (1989) Measuring thermal budgets of active volcanoes by satellite remote sensing. *Nature*, **338**, 144–146.

Glenn, S. M., Forristall, G. Z., Cornillon, P. and Milkowski, G. (1990) Observations of Gulf Stream Ring 83-E and their interpretation using feature models. *Journal of Geophysical Research*, **95**, 13043–13063.

Godfrey, J. S. and Ridgeway, K. R. (1985) The large scale environment of the poleward flowing Leeuwin current, Western Australia: longshore steric height, gradients, wind stresses and geostrophic flow. *Journal of Physical Oceanography*, **15**, 489–495.

Goel, N. S. and Reynolds, N. E. (1989) Bidirectional canopy reflectance and its relationship to vegetation characteristics. *International Journal of Remote Sensing*, **10**, 107–132.

Gond, V., Loudjani, P., Cabot, F. and Viovy, N. (1993) Classification of West African vegetation based upon temporal evolution of corrected NDVI data. Proceedings of 6th AVHRR Data Users' Meeting, Belgirate, Italy, 29 June–2 July 1993. EUM P 12 (Darmstadt – Eberstadt: EUMETSAT), pp. 341–347.

Gordon, H. R., Brown, J. W., Brown, O. B., Evans, R. H. and Clark, D. K. (1983a) Nimbus-7 CZCS: reduction in its radiometric sensitivity with time. *Applied Optics*, **22**, 3929–3931.

Gordon, H. R., Clark, D. K., Brown, J. W., Brown, O. B., Evans, R. H. and Broenkow, W. W. (1983b) Phytoplankton pigment concentrations in the Middle Atlantic Bight: comparison of ship determinations and CZCS estimates. *Applied Optics*, **22**, 20–36.

Gordon, H. R. and Morel, A. Y. (1983) *Remote assessment of ocean color for interpretation of satellite visible imagery* (New York: Springer).

Gorman, A. J. and McGregor, J. (1994) Some considerations for using AVHRR data in climatological studies: II Instrument performance. *International Journal of Remote Sensing*, **15**, 549–565.

Gorny, V. I., Salman, A. G., Tronin, A. A. and Shilin, B. V. (1988) The Earth's outgoing IR radiation as an indicator of seismic activity. *Proceedings of the Academy of Sciences of the USSR*, **301**, 67–69.

Goward, S. N., Cruickshanks, G. D. and Hope, A. S. (1985) Observed relation between thermal emission and reflected spectral radiance of a complex vegetated landscape. *Remote Sensing of Environment*, **18**, 137–146.

Goward, S. N., Dye, D. G., Turner, S. and Yang, J. (1993) Objective assessment of the NOAA Global Vegetation Index data product. *International Journal of Remote Sensing*, **14**, 3365–3394.

Goward, S. N. and Hope, A. S. (1989) Evapotranspiration from combined reflected solar and emitted terrestrial radiation: preliminary FIFE results from AVHRR data. *Advances in Space Research*, **9**, 239–249.

Goward, S. N., Markham, B., Dye, D. G., Dulaney, W. and Yang, J. (1991) Normalized Difference Vegetation Index measurements from the Advanced Very High Resolution Radiometer. *Remote Sensing of Environment*, **35**, 257–277.

Goward, S. N., Turner, S., Dye, D. G. and Liang, S. (1994) The University of Maryland improved Global Vegetation Index Product. *International Journal of Remote Sensing*, **15**, 3365–3395.

Greegor, D. and Norwine, J. (1981) A gradient model of vegetation and climate utilizing NOAA satellite imagery. Phase 1: Texas transect. Report No. E82-10107; NASA-CR-167454, 65 pp.

Grégoire, J. M., Belward, A. S. and Kennedy, P. (1993) Dynamiques de saturation du signal dans la bande 3 du senseur AVHRR: handicap majeur ou source d'information pour la surveillance de l'environnement en milieu soudano-guinéen d'Afrique de l'Ouest? *International Journal of Remote Sensing*, **14**, 2079–2095.

Grenfell, T. C. and Maykut, G. A. (1977) The optical properties of ice and snow in the Arctic Basin. *Journal of Glaciology*, **18**, 445–464.

Grenfell, T. C. and Perovich, D. K. (1984) Spectral albedos of sea ice in the Beaufort Sea. *Journal of Geophysical Research*, **89**, 3573–3580.

Griggs, M. (1975) Measurements of atmospheric aerosol optical depth over water using ERTS-1 data. *Journal of the Air Pollution Control Association*, **25**, 622–626.

Griggs, M. (1983) Satellite measurements of tropospheric aerosols. *Advances in Space Research*, **2**, 109–118.

Griggs, M. (1985) A method to correct satellite measurements of sea surface temperature for the effects of atmospheric aerosols. *Journal of Geophysical Research*, **90**, 12951–12959.

Gruber, A. (1977) Determination of the Earth–atmosphere radiation budget from NOAA satellite data. NOAA Technical Report NESS 76 (Washington, DC: Department of Commerce), 28 pp.

Gruber, A. and Winston, J. S. (1978) Earth–atmosphere radiative heating based on NOAA scanning radiometer measurements. *Bulletin of the American Meteorological Society*, **59**, 1570–1573.

Gu, Z. Q., Duncan, C. N., Renshaw, E., Mugglestone, M. A., Cowan, C. F. N. and Grant, P. M. (1989) Comparison of techniques for measuring cloud texture in remote sensed satellite meteorological image data. *IEE Proceedings*, **136F**, 236–248.

Gupta, R. K. (1992a) Processing error reduction factors in the generation of geometrically corrected NOAA/AVHRR vegetation index images. *International Journal of Remote Sensing*, **13**, 515–526.

Gupta, R. K. (1992b) District level NOAA/AVHRR vegetation indices and scan angle effects. *International Journal of Remote Sensing*, **13**, 715–735.

Gutman, G. (1988) A simple method for estimating monthly mean albedo of land surfaces from AVHRR data. *Journal of Applied Meteorology*, **27**, 973–988.

Gutman, G. (1989) On the relationship between monthly mean and maximum-value composite normalized vegetation indices. *International Journal of Remote Sensing*, **10**, 1317–1325.

Gutman, G. G. (1991) Vegetation indices from AVHRR: an update and future prospects. *Remote Sensing of Environment*, **35**, 121–136.

Gutman, G. G. (1992) Satellite daytime image classification for global studies of Earth's surface parameters from polar orbiters. *International Journal of Remote Sensing*, **13**, 209–234.

Gutman, G., Tarpley, D. and Ohring, G. (1987) Cloud screening for determination of land surface characteristics in a reduced resolution satellite data set. *International Journal of Remote Sensing*, **8**, 859–870.

Haefner, H. (1987) Fernerkundung und Geographie – Thematische, methodische und technische Perspektiven (Remote sensing and geography – thematic, methodological and technical perspectives). *Erdkunde, Archiv für Wissenschaftliche Geographie*, **41**, 169–182.

Haefner, H. and Pampaloni, P. (1992) Water resources. *International Journal of Remote Sensing*, **13**, 1277–1303.

Hamnes, H., Solbakk, S. A., Henriksen, E., Finnseth, A., Steinbakk, L. and Lunde, N. H. (1991) Near real time AVHRR production chain at the Tromsø satellite station. Proceedings of 5th AVHRR Data Users' Meeting, Tromsø, Norway, 25–28 June 1991. EUM P 09 (Darmstadt – Eberstadt: EUMETSAT), pp. 3–10.

Hanan, N. P., Prince, S. D. and Hiernaux, P. H. Y. (1991) Spectral modelling of multi-component landscapes in the Sahel. *International Journal of Remote Sensing*, **12**, 1243–1258.

Hansen, B. U. (1989) Monitoring AVHRR derived vegetation indices and biomass production in southern Greenland. Proceedings of 4th AVHRR Data Users' Meeting, Rothenburg, F.R. Germany, 5–8 September 1989. EUM P 06 (Darmstadt – Eberstadt: EUMETSAT), pp. 137–140.

Hanstrum, B. N. and Watson, A. S. (1983) A case study of two eruptions of Mount Galunggung and an investigation of volcanic eruption cloud characteristics using remote sensing techniques. *Australian Meteorological Magazine*, **31**, 171–177.

Harris, A. R., Brown, S. J. and Mason, I. M. (1994) The effect of windspeed on sea surface temperature retrieval from space. *Geophysical Research Letters*, **21**, 1715–1718.

Harris, A. R. and Mason, I. M. (1989) Lake area measurement using AVHRR: a case study. *International Journal of Remote Sensing*, **10**, 885–895.

Harris, A. R. and Mason, I. M. (1991) An extension to the split-window technique. Proceedings of 5th AVHRR Data Users' Meeting, Tromsø, Norway, 25–28 June 1991. EUM P 09 (Darmstadt – Eberstadt: EUMETSAT), pp. 253–258.

Harris, A. R. and Mason, I. M. (1992) An extension to the split-window technique giving improved atmospheric correction and total water vapour. *International Journal of Remote Sensing*, **13**, 881–892.

Harrison, A. R. and Lucas, R. M. (1989) Multi-spectral classification of snow using NOAA AVHRR imagery. *International Journal of Remote Sensing*, **10**, 907–916.

Hastings, D. A. and Emery, W. J. (1992) The Advanced Very High Resolution Radiometer (AVHRR): A brief reference guide. *Photogrammetric Engineering and Remote Sensing*, **58**, 1183–1188.

Hastings, D., Matson, M. and Horvitz, A. H. (1988a) AVHRR. Monitoring giant icebergs. *Photogrammetric Engineering and Remote Sensing*, **54**, 663–665.

Hastings, D., Matson, M. and Horvitz, A. H. (1988b) AVHRR Catalog. *Photogrammetric Engineering and Remote Sensing*, **54**, 1469–1470.

Hastings, D., Matson, M. and Horvitz, A. H. (1988c) AVHRR. Yellowstone Park fires: smoke and hot spots. *Photogrammetric Engineering and Remote Sensing*, **54**, 1629.

Hastings, D., Matson, M. and Horvitz, A. H. (1989a) AVHRR. Monitoring global aerosol movements. *Photogrammetric Engineering and Remote Sensing*, **55**, 424–425.

Hastings, D., Matson, M. and Horvitz, A. H. (1989b) Ship 'tracks' detected on AVHRR imagery. *Photogrammetric Engineering and Remote Sensing*, **55**, 558.

Hastings, D., Matson, M. and Horvitz, A. H. (1989c) AVHRR. *Photogrammetric Engineering and Remote Sensing*, **55**, 1422.

Hayes, L. (1985) The current use of TIROS-N series of meteorological satellites for land-cover studies. *International Journal of Remote Sensing*, **6**, 35–45.

Hayes, L. and Cracknell, A. P. (1984) Vegetation depiction by AVHRR – a Scottish sampling. *Satellite remote sensing – review and preview*. Proceedings of the 10th anniversary International Conference of the Remote Sensing Society, University of Reading, UK, 18–21 September 1984 (Nottingham: Remote Sensing Society), pp. 181–190.

Hayes, R. M. (1980) Operational use of remote sensing during the Campeche Bay oil well blowout. Proceedings of the 14th International Symposium on Remote Sensing of Environment, San José, Costa Rica, 23–30 April 1980 (Ann Arbor, Michigan: Environmental Research Institute of Michigan), volume 2, pp. 1187–1196.

Henderson-Sellers, A. (1984) *Satellite sensing of a cloudy atmosphere: observing the third planet* (London: Taylor & Francis).

Henderson-Sellers, A. (1986) Calculating the surface energy balance for lake and reservoir modeling: a review. *Reviews of Geophysics*, **24**, 625–649.

Henderson-Sellers, A. (1989) Climate, models and geography. In *Remodelling geography*, edited by B. MacMillan (Oxford: Blackwell), pp. 117–146.

Hepplewhite, C. L. (1989) Remote observation of the sea surface and atmosphere: the oceanic skin effect. *International Journal of Remote Sensing*, **10**, 801–810.

Hielkema, J. U., Prince, S. D. and Astle, W. L. (1986) Rainfall and vegetation monitoring in the savanna zone of the Democratic Republic of Sudan using the NOAA Advanced Very High Resolution Radiometer. *International Journal of Remote Sensing*, **7**, 1499–1513.

Ho, D. and Asem, A. (1986) NOAA AVHRR image referencing. *International Journal of Remote Sensing*, **7**, 895–904.

Ho, D., Asem, A. and Deschamps, P. Y. (1986) Atmospheric correction for the sea surface temperature using NOAA-7 AVHRR and METEOSAT-2 infrared data. *International Journal of Remote Sensing*, **7**, 1323–1333.

Hobbs P. V., Radke, L. F., Eltgroth, M. W. and Hegg, D. A. (1981) Airborne studies of the emissions from the eruptions of Mount St Helens. *Science, New York*, **211**, 816–818.

Hobbs, T. J. (1995) The use of NOAA AVHRR NDVI data to assess herbage production in the arid rangelands of Central Australia. *International Journal of Remote Sensing*, **16**, 1289–1302.

Hoffer, J. M., Gomez, F. and Muels, P. (1982) Eruption of El Chichón volcano, Chiapas, Mexico, 28 March to 7 April 1982. *Science, New York*, **218**, 1307–1308.

Hoffer, R. M. (1975) Computer-aided analysis of SKYLAB Multispectral scanner data in mountainous terrain for land use, forestry, water resources, and geologic applications. Final report on NASA contract No. NAS9-13380. SKYLAB EREP project 398. LARS information note 121275, Laboratory for Applications of Remote Sensing, Purdue University, West Lafayette, Indiana, USA.

Holasek, R. E. and Rose, W. I. (1991) Anatomy of 1986 Augustine volcano eruptions as recorded by multispectral image processing of digital AVHRR weather satellite data. *Bulletin of Volcanology*, **53**, 420–435.

Holben, B. N. (1986) Characteristics of maximum-value composite images from temporal AVHRR data. *International Journal of Remote Sensing*, **7**, 1417–1434.

Holben, B. N., Eck, T. F. and Fraser, R. S. (1991) Temporal and spatial variability of aerosol optical depth in the Sahel region in relation to vegetation remote sensing. *International Journal of Remote Sensing*, **12**, 1147–1163.

Holben, B. and Fraser, R. S. (1984) Red and near-infrared sensor response to off-nadir viewing. *International Journal of Remote Sensing*, **5**, 145–160.

Holben, B. N., Kaufman, Y. J. and Kendall, J. D. (1990) NOAA-11 AVHRR visible and near-IR inflight calibration. *International Journal of Remote Sensing*, **11**, 1511–1519.

Holben, B., Kimes, D. and Fraser, R. S. (1986) Directional reflectance response in AVHRR red and near-IR bands for three cover types and varying atmospheric conditions. *Remote Sensing of Environment*, **19**, 213–236.

Holligan, P. M., Viollier, M., Harbour, D. S., Camus, P. and Champagne-Philippe, M. (1983)

Satellite and ship studies of coccolithophore production along a continental shelf edge. *Nature*, **304**, 339–342.

Holyer, R. J. (1984) A two-satellite method for measurement of sea surface temperature. *International Journal of Remote Sensing*, **5**, 115–131.

Hope, A. S. (1988) Estimation of wheat canopy resistance using combined remotely sensed spectral reflectance and thermal observations. *Remote Sensing of Environment*, **24**, 369–383.

Huete, A. R. (1988) A soil-adjusted vegetation index (SAVI). *Remote Sensing of Environment*, **25**, 295–309.

Huete, A. R., Hua, G., Qi, J., Chehbouni, A. and van Leeuwen, W. J. D. (1992) Normalization of multidirectional red and NIR reflectances with the SAVI. *Remote Sensing of Environment*, **41**, 143–154.

Huete, A. R., Post, D. F. and Jackson, R. D. (1984) Soil spectral effects on 4-space vegetation discrimination. *Remote Sensing of Environment*, **15**, 155–165.

Huete, A. R. and Tucker, C. J. (1991) Investigation of soil influences in AVHRR red and near-infrared vegetation index imagery. *International Journal of Remote Sensing*, **12**, 1223–1242.

Hughes, M. F. (1993) Data collection and direct broadcast from NOAA satellites: user services. Proceedings of 6th AVHRR Data Users' Meeting, Belgirate, Italy, 29 June–2 July, 1993. EUM P 12 (Darmstadt – Eberstadt: EUMETSAT), pp. 431–434.

Huh, O. K. and DiRosa, D. (1981) Analysis and interpretation of TIROS-N AVHRR infrared imagery, western Gulf of Mexico. *Remote Sensing of Environment*, **11**, 371–383.

Huh, O. K. and Shim, T. (1987) Satellite observations of surface temperatures and flow patterns, Sea of Japan and East China Sea, late March 1979. *Remote Sensing of Environment*, **22**, 379–393.

Hunt, G. R., Salisbury, J. W. and Bunting, J. T. (1974) Distinction between snow and cloud in DMSP imagery. Report to the Fourth Air Force Cold Regions Laboratory/American Weather Service Satellite Working Group.

Hutchinson, C. F. (1991) Uses of satellite data for famine early warning in sub-Saharan Africa. *International Journal of Remote Sensing*, **12**, 1405–1421.

Idso, S. B., Schmugge, T. J., Jackson, R. D. and Reginato, R. J. (1975) The utility of surface temperature measurements for the remote sensing of soil water status. *Journal of Geophysical Research*, **80**, 3044–3049.

Ikeda, M., Johannessen, J. A., Lygre, K. and Sandven, S. (1989) A process study of mesoscale meanders and eddies in the Norwegian Coastal Current. *Journal of Physical Oceanography*, **19**, 20–35.

International Geosphere-Biosphere Program (1990) A study of global change. The initial core projects. Report No. 12 (Stockholm: IGBP Secretariat).

International Geosphere-Biosphere Program (1992) A study of global change. Improved global data for land applications. Report No. 20 (Stockholm: IGBP Secretariat).

ITT (1980) Advanced Very High Resolution Radiometer (Model 2) for the TIROS 'N' spacecraft. Alignment and calibration data book, AVHRR/2, Flight Model SN 202, ITT Aerospace/Optical Division, Fort Wayne, Indiana.

Iverson, L. R., Cook, E. A. and Graham, R. L. (1989) A technique for extrapolating and validating forest cover across large regions. Calibrating AVHRR data with TM data. *International Journal of Remote Sensing*, **10**, 1805–1812.

Jaakkola, S. (1990) Managing data for the monitoring of tropical forest cover: the Global Resource Information Database approach. *Photogrammetric Engineering and Remote Sensing*, **56**, 1355–1357.

Jackson, R. D. (1983) Spectral indices in *n*-space. *Remote Sensing of Environment*, **13**, 1401–1429.

Jackson, R. D., Pinter, P. J., Reginato, R. J. and Idso, S. B. (1980) Hand-held radiometry. Agricultural Reviews and Manuals W-19, US Department of Agriculture, Science and

Education Administration, Oakland, California.

Jackson, R. D., Reginato, R. J. and Idso, S. B. (1977) Wheat canopy temperature: a practical tool for evaluation of water requirements. *Water Resources Research*, **13**, 651–656.

Jacobowitz, H. (1991) Estimation of the Earth's radiation budget from narrowband NOAA satellite observations. *Palaeography, Palaeoclimatology, Palaeoecology (Global and Planetary Change Section)*, **90**, 11–16.

Jaeger, J. C. (1953) Conduction of heat in a solid with periodic boundary conditions, with an application to the surface temperature of the Moon. *Proceedings of the Cambridge Philosophical Society*, **49**, 355–359.

James, M. E. and Kalluri, S. N. V. (1994) The Pathfinder AVHRR land data set: an improved coarse resolution data set for terrestrial monitoring. *International Journal of Remote Sensing*, **15**, 3347–3363.

Johannessen, J. A., Johannessen, O. M. and Haughan, P. M. (1989a) Remote sensing and model simulation studies of the Norwegian coastal current during the algal bloom in May 1988. *International Journal of Remote Sensing*, **10**, 1893–1906.

Johannessen, J. A., Johannessen, O. M., Svendsen, E., Shuchman, R., Manley, T., Campbell, W. J., Josberger, E. G., Sandven, S., Gascard, J. C., Olaussen, T., Davidson, K. and Van Leer, J. (1987a) Mesoscale eddies in the Fram Strait marginal ice zone during the 1983 and 1984 Marginal Ice Zone Experiments. *Journal of Geophysical Research*, **92**, 6754–6772.

Johannessen, O. M., Johannessen, J. A., Svendsen, E., Shuchman, R. A., Campbell, W. J. and Josberger, E. (1987b) Ice-edge eddies in the Fram Strait marginal ice zone. *Science, New York*, **236**, 427–429.

Johannessen, J. A., Svendsen, E., Sandven, S., Johannessen, O. M. and Lygre, K. (1989b) Three-dimensional structure of mesoscale eddies in the Norwegian Coastal Current. *Journal of Physical Oceanography*, **19**, 3–19.

Johnson, G. E., Van Dijk, A. and Sakamoto, C. M. (1987) The use of AVHRR data in operational agricultural assessment in Africa. *Geocarto International*, **2**, 41–60.

Johnston, A. C., Cracknell, A. P., Vaughan, R. A., Boulton, G. S. and Clark, C. (1989) Identification of ancient glacier marks using AVHRR imagery. *International Journal of Remote Sensing*, **10**, 917–929.

Jönsson, L. (1989) Flow studies of the northern part of the Strait of Öresund by means of remote sensing. Proceedings of 4th AVHRR Data Users' Meeting, Rothenburg, F.R. Germany, 5–8 September 1989. EUM P 06 (Darmstadt – Eberstadt: EUMETSAT), pp. 229–234.

Jönsson, L. (1993) Determination of flow velocities using NOAA-AVHRR data. Proceedings of 6th AVHRR Data Users' Meeting, Belgirate, Italy, 29 June–2 July 1993. EUM P 12 (Darmstadt – Eberstadt: EUMETSAT), pp. 375–383.

Justice, C. O., Dugdale, G., Townshend, J. R. G., Narracott, A. S. and Kumar, M. (1991a) Synergism between NOAA-AVHRR and Meteosat data for studying vegetation development in semi-arid West Africa. *International Journal of Remote Sensing*, **12**, 1349–1368.

Justice, C. O., Eck, T. F., Tanré, D. and Holben, B. N. (1991b) The effect of water vapour on the normalized difference vegetation index derived for the Sahelian region from NOAA AVHRR data. *International Journal of Remote Sensing*, **12**, 1165–1187.

Justice, C. O., Townshend, J. R. G. and Choudhury, B. J. (1989) Comparison of AVHRR and SMMR data for monitoring vegetation phenology on a continental scale. *International Journal of Remote Sensing*, **10**, 1607–1632.

Justice, C. O., Townshend, J. R. G., Holben, B. N. and Tucker, C. J. (1985) Analysis of the phenology of global vegetation using meteorological satellite data. *International Journal of Remote Sensing*, **6**, 1271–1318.

Justice, C. O., Townshend, J. R. G. and Kalb, V. L. (1991c) Representation of vegetation by continental data sets derived from NOAA-AVHRR data. *International Journal of Remote Sensing*, **12**, 999–1021.

Kahle, A. B. (1977) A simple thermal model of the Earth's surface for geologic mapping by remote sensing. *Journal of Geophysical Research*, **82**, 1673–1680

Kahle, A. B., Schieldge, J. P., Adams, M. J., Alley, R. E. and Levine, C. J. (1981) Geological application of thermal inertia imaging using HCMM data. JPL Publication 81-55, JPL Quarterly Technical Review, Jet Propulsion Laboratory, Pasadena.

Kakane, V. C. K. and Imbernon, J. (1992) Estimation of rainfall in Senegal using the satellite NOAA-9/AVHRR. *International Journal of Remote Sensing*, **13**, 2059–2068.

Kamachi, M. (1989) Advective surface velocities derived from sequential images for rotational flow field: limitations and applications of maximum cross-correlation method with rotational registration. *Journal of Geophysical Research*, **94**, 18227–18233.

Kanemasu, E. T., Heilman, J. L., Bagley, J. O. and Powers, W. L. (1977) Using Landsat data to estimate evapotranspiration of winter wheat. *Environmental Management*, **1**, 515–520.

Karlsson, K. G. (1989a) Development of an operational cloud classification model. *International Journal of Remote Sensing*, **10**, 687–693.

Karlsson, K. G. (1989b) Multispectrally derived cloud and precipitation information on a mesoscale grid. Proceedings of 4th AVHRR Data Users' Meeting, Rothenburg, F.R. Germany, 5–8 September 1989. EUM P 06 (Darmstadt – Eberstadt: EUMETSAT), pp. 91–95.

Karlsson, K. G. (1993) Comparison of operational AVHRR-based cloud analyses with surface observations. Proceedings of 6th AVHRR Data Users' Meeting, Belgirate, Italy, 29 June–2 July 1993. EUM P 12 (Darmstadt – Eberstadt: EUMETSAT), pp. 223–229.

Kaufman, Y. J. (1987) The effect of subpixel clouds on remote sensing. *International Journal of Remote Sensing*, **8**, 839–857.

Kaufman, Y. J. (1993) Aerosol optical thickness and atmospheric path radiance. *Journal of Geophysical Research*, **98**, 2677–2692.

Kaufman, Y. J. and Holben, B. N. (1993) Calibration of the AVHRR visible and near-ir bands by atmospheric scattering, ocean glint and desert reflection. *International Journal of Remote Sensing*, **14**, 21–52.

Kelly, K. A. (1985) Separating clouds from ocean in infrared images. *Remote Sensing of Environment*, **17**, 67–83.

Kennedy, P. J. (1989) Monitoring the phenology of Tunisian grazing lands. *International Journal of Remote Sensing*, **10**, 835–845.

Kennedy, P. J., Belward, A. S. and Grégoire, J.-M. (1994) An improved approach to fire monitoring in West Africa using AVHRR data. *International Journal of Remote Sensing*, **15**, 2235–2255.

Kerber, A. G. and Schutt, J. B. (1986) Utility of AVHRR channels 3 and 4 in land-cover mapping. *Photogrammetric Engineering and Remote Sensing*, **52**, 1877–1883.

Kerr, Y. H., Imbernon, J., Dedieu, G., Hautecoeur, O., Lagouarde, J. P. and Seguin, B. (1989) NOAA AVHRR and its uses for rainfall and evapotranspiration monitoring. *International Journal of Remote Sensing*, **10**, 847–854.

Kerr, Y. H. and Lagouarde, J. P. (1989) On the derivation of land surface temperature from AVHRR data. Proceedings of 4th AVHRR Data Users' Meeting, Rothenburg, F.R. Germany, 5–8 September 1989. EUM P 06 (Darmstadt – Eberstadt: EUMETSAT), pp. 157–160.

Kerr, Y. H., Lagouarde, J. P. and Imbernon, J. (1992) Accurate land surface temperature retrieval from AVHRR data with use of an improved split window algorithm. *Remote Sensing of Environment*, **41**, 197–209.

Key, J. and Haefliger, M. (1992) Arctic ice surface temperature retrieval from AVHRR thermal channels. *Journal of Geophysical Research*, **97**, 5885–5893.

Key, J., Maslanik, J. A. and Schweiger, A. J. (1989) Classification of merged AVHRR and SMMR Arctic data with neural networks. *Photogrammetric Engineering and Remote Sensing*, **55**, 1331–1338.

Khattak, S., Vaughan, R. A. and Cracknell, A. P. (1991) Sunglint and its observation in AVHRR data. *Remote Sensing of Environment*, **37**, 101–116.

Khazenie, N. and Richardson, K. A. (1991) Classification of cloud types based on spatial textural measures using NOAA-AVHRR data. *International Geoscience and Remote Sensing Symposium on Remote Sensing: Global Monitoring for Earth Management*, **3**, 1701–1705.

Khorram, S. (1981) Water quality mapping from Landsat digital data. *International Journal of Remote Sensing*, **2**, 145–153.

Khosraviani, G. and Cracknell, A. P. (1984) Atmospheric effects in multiple-look observations from space. Proceedings of the 18th International Symposium on Remote Sensing of Environment, Paris, France, 1–5 October 1984 (Ann Arbor, Michigan: Environmental Research Institute of Michigan), pp. 751–757.

Khosraviani, G. and Cracknell, A. P. (1987) A two-look technique for studying atmospheric effects in optical scanner data for the ocean. *International Journal of Remote Sensing*, **8**, 291–308.

Kidder, S. Q. and Wu, H. T. (1987) A multispectral study of the St. Louis area under snow-covered conditions using NOAA-7 AVHRR data. *Remote Sensing of Environment*, **22**, 159–172.

Kidwell, K. B. (1984) NOAA polar orbiter data (TIROS-N, NOAA-6, NOAA-7 and NOAA-8) user's guide. National Oceanic and Atmospheric Administration, National Environmental Satellite, Data, and Information Service, Washington, DC.

Kidwell, K. B. (1990) Global vegetation index user's guide. National Oceanic and Atmospheric Administration, National Environmental Satellite, Data, and Information Service, Washington, DC.

Kidwell, K. B. (1991) NOAA polar orbiter data (TIROS-N, NOAA-6, NOAA-7, NOAA-8, NOAA-9, NOAA-10, NOAA-11 and NOAA-12) user's guide. National Oceanic and Atmospheric Administration, National Environmental Satellite, Data, and Information Service, Washington, DC.

Kikuchi, T., Satow, K., Ohata, T., Yamanouchi, T. and Nishio, F. (1992) Wind and temperature regime in Mizuho Plateau, East Antarctica. *International Journal of Remote Sensing*, **13**, 67–79.

Killmayer, A. (1993) Drought impact information system. Proceedings of 6th AVHRR Data Users' Meeting, Belgirate, Italy, 29 June–2 July 1993. EUM P 12 (Darmstadt – Eberstadt: EUMETSAT), pp. 349–354.

Kimball, A. W. (1980) Technical operating report SSC data evaluation. Westinghouse Electric Corporation, Defense and Electronic Systems Center, Baltimore, Maryland, USA.

Kimes, D. S. and Holben, B. N. (1992) Extracting spectral albedo from NOAA-9 AVHRR multiple view data using an atmospheric correction procedure and an expert system. *International Journal of Remote Sensing*, **13**, 275–289.

Kimes, D. S., Newcomb, W. W., Tucker, C. J., Zonneveld, I. S., Van Wijngaarden, W., De Leeuw, J. and Epema, G. F. (1985) Directional reflectance factor distributions for cover types of Northern Africa. *Remote Sensing of Environment*, **18**, 1–19.

Kimes, D. S. and Sellers, P. J. (1985) Inferring hemispherical reflectance of the Earth's surface for global energy budgets from remotely sensed nadir or directional radiance values. *Remote Sensing of Environment*, **18**, 205–223.

Kimes, D. S., Sellers, P. J. and Diner, D. J. (1987) Extraction of spectral hemispherical reflectance (albedo) of surfaces from nadir and directional reflectance data. *International Journal of Remote Sensing*, **8**, 1727–1746.

Kimes, D. S., Smith, J. A. and Ranson, K. J. (1980) Vegetation reflectance measurements as a function of solar zenith angle. *Photogrammetric Engineering and Remote Sensing*, **46**, 1563–1573.

King, C. (1989) Programme to evaluate the potential of NOAA-AVHRR for monitoring crops and assessing yield (Agriculture Project – JRC Ispra). Proceedings of 4th AVHRR

Data Users' Meeting, Rothenburg, F.R. Germany, 5–8 September 1989. EUM P 06 (Darmstadt – Eberstadt: EUMETSAT), pp. 145–151.

Klemas, V. and Philpott, W. D. (1981) Drift and dispersion studies of ocean-dumping waste using Landsat imagery and current drogues. *Photogrammetric Engineering and Remote Sensing*, **47**, 533–542.

Kloster, K. (1989) Using TBUS orbital elements for AVHRR image gridding. *International Journal of Remote Sensing*, **10**, 653–659.

Kloster, K. (1991) Evaluation of computed sea-ice concentrations from MSS and AVHRR in the Barents Sea. Proceedings of 5th AVHRR Data Users' Meeting, Tromsø, Norway, 25–28 June 1991. EUM P 09 (Darmstadt – Eberstadt: EUMETSAT), pp. 117–122.

Knipling, E. B. (1970) Physical and physiological basis for the reflectance of visible and near-infrared radiation from vegetation. *Remote Sensing of Environment*, **1**, 155–159.

Koepke, P. (1989) Removal of atmospheric effects from AVHRR solar channel clear-sky albedoes. Proceedings of 4th AVHRR Data Users' Meeting, Rothenburg, F.R. Germany, 5–8 September 1989. EUM P 06 (Darmstadt – Eberstadt: EUMETSAT), pp. 267–270.

Kogan, F. N. (1990) Remote sensing of weather impacts on vegetation in non-homogeneous areas. *International Journal of Remote Sensing*, **11**, 1405–1419.

Komura, F., Yoda, H., Kubo, Y. and Fukahata, S. (1984) Development of image processing system for fishery remote sensing. *Hitachi Review*, **33**, 103–108.

Kondratyev, K. Ya. (1989) *The International Satellite Land-Surface Climatology Project (ISLSCP): state-of-the-art and prospect (review)* (Moscow: Centre for International Projects, USSR State Committee for Environmental Protection).

Koslowsky, D. (1989) On-line processing of NDVI maps. Proceedings of 4th AVHRR Data Users' Meeting, Rothenburg, F.R. Germany, 5–8 September 1989. EUM P 06 (Darmstadt – Eberstadt: EUMETSAT), pp. 161–164.

Krasnopolsky, V. M. and Breaker, L. C. (1994) The problem of AVHRR image navigation revisited. *International Journal of Remote Sensing*, **15**, 979–1008.

Kriebel, K. T. (1976) On the variability of the reflected radiation field due to differing distributions of the irradiation. *Remote Sensing of Environment*, **4**, 257–264.

Kriebel, K. T. (1989) Cloud liquid water path derived from AVHRR data using APOLLO. *International Journal of Remote Sensing*, **10**, 723–729.

Kukla, G. and Robinson, D. (1981) Accuracy of operational snow and ice charts. Proceedings of the 1981 International Geoscience and Remote Sensing Symposium, volume 2 (New York: IEEE), pp. 974–987.

Kummer, D. M. (1992) Remote sensing and tropical deforestation: a cautionary note from the Philippines. *Photogrammetric Engineering and Remote Sensing*, **58**, 1469–1471.

Labovitz, M. L., Toll, D. L. and Kennard, R. E. (1980) Preliminary evidence for the influence of physiography and scale upon the autocorrelation function of remotely sensed data, NASA TM 82064, Goddard Space Flight Center, Greenbelt, MD.

Labs, D. and Neckel, H. (1967) The absolute radiation intensity of the centre of the Sun disc in the spectral range 3288–12480 Å. *Zeitschrift für Astrophysik*, **65**, 133–155.

Labs, D. and Neckel, H. (1968) The radiation of the solar photosphere. *Zeitschrift für Astrophysik*, **69**, 1–73.

Labs, D. and Neckel, H. (1970) Transformation of the absolute solar radiation data into the International Practical Temperature Scale of 1968. *Solar Physics*, **15**, 79–87.

Lagouarde, J.-P. (1991) Use of NOAA AVHRR data combined with an agrometeorological model for evaporation mapping. *International Journal of Remote Sensing*, **12**, 1853–1864.

Langaas, S. and Andersen, H. S. (1991) Semi-automatic bispectral approach to NOAA AVHRR based detection of savanna fires. Proceedings of 5th AVHRR Data Users' Meeting, Tromsø, Norway, 25–28 June 1991. EUM P 09 (Darmstadt – Eberstadt: EUMETSAT), pp. 399–404.

Langley, S. P. (1881) The bolometer and radiant energy. *Proceedings of the American Academy of Arts and Sciences*, **8**, 342–358.

Lasker, R., Peláez, J. and Laurs, R. M. (1981) The use of satellite infrared imagery for describing ocean processes in relation to spawning of the Northern Anchovy (*Engraulis mordax*). *Remote Sensing of Environment*, **11**, 439–453.

Lauritson, L., Nelson, G. J. and Porto, F. W. (1979) Data extraction and calibration of TIROS-N/NOAA radiometers. NOAA Technical Memorandum NESS 107. US Department of Commerce, National Oceanic and Atmospheric Administration, National Environmental Satellite, Data and Information Service, Washington DC 20233. See also Planet (1988).

Laurs, R. M., Fiedler, P. C. and Montgomery, D. R. (1984) Albacore tuna catch distributions relative to environmental features observed from satellites. *Deep Sea Research*, **31**, 1085–1099.

La Violette, P. E. and Holyer, R. J. (1988) Noise and temperature gradients in multichannel sea surface temperature imagery of the ocean. *Remote Sensing of Environment*, **25**, 231–241.

La Violette, P. E., Johnson, D. R. and Brooks, D. A. (1990) Sun-glitter photographs of Georges Bank and the Gulf of Maine from the Space Shuttle. *Oceanography*, (April), 43–49.

Le Comte, D. (1989) Using AVHRR for early warning of famine in Africa. *Photogrammetric Engineering and Remote Sensing*, **55**, 168–169.

Lee, A. J. and Ramster, J. W. (1979) Atlas of the Seas around the British Isles. Ministry of Agriculture, Fisheries and Food, Fisheries Research Technical Report 20 (London: HMSO).

Lee, J., Weger, R. C., Sengupta, A. K. and Welch, R. M. (1990) A neural network approach to cloud classification. *IEEE Transactions on Geoscience and Remote Sensing*, **28**, 846–855.

Legeckis, R. (1975) Application of synchronous meteorological satellite data to the study of time dependent sea surface temperature along the boundary of the Gulf Stream. *Geophysical Research Letters*, **2**, 435–438.

Legeckis, R. (1979) A survey of world wide SST fronts detected by environmental satellites. *Journal of Geophysical Research*, **83**, 4501–4522.

Legeckis, R., Legg, E. and Limeburner, R. (1980) Comparison of polar and geostationary satellite infrared observations of sea surface temperatures in the Gulf of Maine. *Remote Sensing of Environment*, **9**, 339–350.

Legeckis, R. and Pritchard, J. (1976) Algorithm for correcting the VHRR imagery for geometric distortions due to the Earth curvature, Earth rotation and spacecraft roll attitude errors. NOAA Technical Memorandum NESS 77, NOAA, National Environmental Satellite Service, Washington, DC.

Legg, C. A. (1989) The potential of meteorological satellite data for monitoring flooding disasters. Proceedings of 4th AVHRR Data Users' Meeting, Rothenburg, F.R. Germany, 5–8 September 1989. EUM P 06 (Darmstadt – Eberstadt: EUMETSAT), pp. 165–169.

Le Gléau, H., Derrien, M., Harang, L., Lavanant, L. and Noyalet, A. (1989) Operational cloud mask using the AVHRR of NOAA-11. Proceedings of 4th AVHRR Data Users' Meeting, Rothenburg, F.R. Germany, 5–8 September 1989. EUM P 06 (Darmstadt – Eberstadt: EUMETSAT), pp. 85–89.

LeSchack, L. A. (1981) Correlation of under-ice roughness with satellite and airborne thermal infrared data. Proceedings of the 6th International Conference on Port and Ocean Engineering under Arctic Conditions, Québec City, Québec, Canada, 1981, volume 1, pp. 156–165.

Li, Z. R. and Becker, F. (1991) Determination of land surface temperature and emissivity from AVHRR data. Proceedings of 5th AVHRR Data Users' Meeting, Tromsø, Norway, 25–28 June 1991. EUM P 09 (Darmstadt – Eberstadt: EUMETSAT), pp. 405–410.

Li, Z. R. and Becker, F. (1993) Feasibility of land surface temperature and emissivity determination from AVHRR data. *Remote Sensing of Environment*, **43**, 67–85.

Li, Z. R. and McDonnell, M. J. (1988) Atmospheric correction of thermal images. *Internation-*

al Journal of Remote Sensing, **9**, 107–121.

Liberti, G. L. (1991) GPCP-AIP/2 properties of AVHRR data. Preliminary results. Proceedings of 5th AVHRR Data Users' Meeting, Tromsø, Norway, 25–28 June 1991. EUM P 09 (Darmstadt – Eberstadt: EUMETSAT), pp. 235–240.

Liberti, G. L. (1993) Precipitation estimation with AVHRR data: a review. Proceedings of 6th AVHRR Data Users' Meeting, Belgirate, Italy, 29 June–2 July 1993. EUM P 12 (Darmstadt – Eberstadt: EUMETSAT), pp. 31–37

Liljas, E. (1987) Multispectral classification of cloud, fog and haze. In *Remote sensing applications in meteorology and climatology*, edited by R. A. Vaughan (Dordrecht: Reidel), pp. 301–319.

Liljas, E. (1989) Experience of an operational cloud classification method. Proceedings of 4th AVHRR Data Users' Meeting, Rothenburg, F.R. Germany, 5–8 September 1989. EUM P 06 (Darmstadt – Eberstadt: EUMETSAT), pp. 73–78.

Liljas, E. (1991) Automated cloud classification from AVHRR data. Problems and solutions. Proceedings of 5th AVHRR Data Users' Meeting, Tromsø, Norway, 25–28 June 1991. EUM P 09 (Darmstadt – Eberstadt: EUMETSAT), pp. 327–332.

Lillesand, T. M., Meisner, D. E., Downs, A. L. and Deuell, R. L. (1982) Use of GOES and TIROS/NOAA satellite data for snow-cover mapping. *Photogrammetric Engineering and Remote Sensing*, **48**, 251–259.

Limaye, S. S., Ackerman, S. A., Fry, P. M., Isa, M., Ali, H., Ali, G., Wright, A. and Rangno, A. (1992) Satellite monitoring of smoke from the Kuwait oil fires. *Journal of Geophysical Research*, **97**, 14551–14563.

Lin, X. and Coakley, J. A. (1993) Retrieval of properties for semitransparent clouds from multispectral infrared imagery. *Journal of Geophysical Research*, **98**, 18501–18514.

Linthicum, K. J., Bailey, C. L., Davies, F. G. and Tucker, C. J. (1987) Detection of Rift Valley fever viral activity in Kenya by satellite remote sensing imagery. *Science, New York*, **235**, 1656–1659.

Liu, Z. K., Zheng, S. Y. and Zeng, Q. Z. (1986) Snow survey from meteorological satellite images in the Qilian Mountain Basin in northwest China. *International Journal of Remote Sensing*, **7**, 1335–1340.

Llewellyn-Jones, D. T., Minnett, P. J., Saunders, R. W. and Zavody, A. M. (1984) Satellite multi-channel infrared measurements of sea surface temperature of the N.E. Atlantic Ocean using AVHRR/2. *Quarterly Journal of the Royal Meteorological Society*, **110**, 613–631.

Lloyd, D. (1989a) A phenological description of Iberian vegetation using short wave vegetation index imagery. *International Journal of Remote Sensing*, **10**, 827–833.

Lloyd, D. (1989b) Mapping phytophenology using shortwave vegetation index imagery. Proceedings of 4th AVHRR Data Users' Meeting, Rothenburg, F.R. Germany, 5–8 September 1989. EUM P 06 (Darmstadt – Eberstadt: EUMETSAT), pp. 131–136.

Lønseth, L. and Bern, T.-I. (1991) Operational use of AVHRR products to monitor the Norwegian coastal current. Proceedings of 5th AVHRR Data Users' Meeting, Tromsø, Norway, 25–28 June 1991. EUM P 09 (Darmstadt – Eberstadt: EUMETSAT), pp. 313–319.

López, S., González, F., Llop, R. and Cuevas, J. M. (1991) An evaluation of the utility of NOAA AVHRR images for monitoring forest fire risk in Spain. *International Journal of Remote Sensing*, **12**, 1841–1851.

Los, S. O., Justice, C. O. and Tucker, C. J. (1994) A global 1° by 1° NDVI data set for climate studies derived from the GIMMS continental NDVI data. *International Journal of Remote Sensing*, **15**, 3493–3518.

Loudjani, P., Saugier, B., Saint, G. and Menaut, J. C. (1993) Estimation of regional net primary production using coarse resolution satellite data. A case study: West Africa. Proceedings of 6th AVHRR Data Users' Meeting, Belgirate, Italy, 29 June–2 July 1993. EUM P 12 (Darmstadt – Eberstadt: EUMETSAT), pp. 125–134.

Loveland, T. R., Merchant, J. W., Ohlen, D. O. and Brown, J. F. (1991) Development of a land-cover characteristics database for the conterminous US. *Photogrammetric Engineering and Remote Sensing*, **57**, 1453–1463.

Lutz, H. J. and Bauer, P. (1989) Determination of the surface radiation budget in polar regions. Proceedings of 4th AVHRR Data Users' Meeting, Rothenburg, F.R. Germany, 5–8 September 1989. EUM P 06 (Darmstadt – Eberstadt: EUMETSAT), pp. 255–258.

Lynn, D. W. (1986) Monotemporal, multitemporal, and multidate thermal infrared data acquisition from satellites for soil and surface-material survey. *International Journal of Remote Sensing*, **7**, 213–231.

Lyon, J. G., Bedford, K. W., Yen, C. C. J., Lee, D. H. and Mark, D. J. (1988) Determinations of suspended sediment concentrations from multiple day Landsat and AVHRR data. *Remote Sensing of Environment*, **25**, 107–115.

Ma Ai Nai and Xue Yong (1990) A study of remote sensing information model of soil moisture. Proceedings of 11th Asian Conference on Remote Sensing, Guangzhou, China, 15–21 November 1990 (Beijing: International Academic Publishers), pp. ACRSp-11-1– ACRSp-11-5.

McClain, E. P. (1981) Multiple atmospheric window techniques for satellite derived sea surface temperatures. In *Oceanography from space*, edited by J. F. R. Gower (New York: Plenum), pp. 73–85.

McClain, E. P. (1989) Global sea surface temperatures and cloud clearing for aerosol optical depth estimates. *International Journal of Remote Sensing*, **10**, 763–769.

McClain, E. P., Pichel, W. G. and Walton, C. C. (1985) Comparative performance of AVHRR-based multichannel sea surface temperatures. *Journal of Geophysical Research*, **90**, 11587–11601.

McClain, E. P. and Strong, A. E. (1969) On anomalous dark patches in satellite-viewed sunglint areas. *Monthly Weather Review*, **97**, 875–844.

McClatchey, J. (1992) The use of climatological observations as ground truth for distributions of minimum temperature derived from AVHRR data. *International Journal of Remote Sensing*, **13**, 155–163.

McClimans, T. A. (1989) Activities at the Norwegian Hydrotechnical Laboratory. *International Journal of Remote Sensing*, **10**, 617–623.

McConaghy, D. C. (1980) Measuring sea surface temperature from satellites: A ground truth approach. *Remote Sensing of Environment*, **10**, 307–310.

McCormick, M. P. (1983) Aircraft and ground-based lidar measurements of El Chichón stratospheric aerosols. Proceedings of the 7th Annual Climate Diagnostics Workshop, 18–22 October 1983 (Boulder, Colorado: National Center for Atmospheric Research), pp. 386–389.

McGinnis, D. F., Wiesnet, D. R. and Pritchard, J. A. (1975) Determination of snow depth and snow extent from NOAA-2 satellite very high resolution data. *Water Resources Research*, **11**, 897–902.

McGregor, J. and Gorman, A. J. (1994) Some considerations for using AVHRR data in climatological studies: I. Orbital characteristics of NOAA satellites. *International Journal of Remote Sensing*, **15**, 537–548.

Mackay, G. (1992) Surface energy balance modelling using NOAA-AVHRR sensor data applied to the study of regional thermal effects and geothermal activity, Iceland. PhD thesis, Dundee University.

Mackay, G., Cracknell, A. P. and Vaughan, R. A. (1991) Ground heat flux and apparent thermal inertia imagery used to study the regional heat flow pattern and geothermal activity in Iceland. Proceedings of 5th AVHRR Data Users' Meeting, Tromsø, Norway, 25–28 June 1991. EUM P 09 (Darmstadt – Eberstadt: EUMETSAT), pp. 273–278.

McKenzie, R. L. and Nisbet, R. M. (1982) Applicability of satellite-derived sea-surface temperatures in the Fiji region. *Remote Sensing of Environment*, **12**, 349–361.

McMillin, L. M. (1975) Estimation of sea surface temperature from two infrared window measurements with different absorption. *Journal of Geophysical Research*, **80**, 5113–5117.

McMillin, L. M. and Crosby, D. S. (1984) Theory and validation of the multiple window sea surface temperature technique. *Journal of Geophysical Research*, **89**, 3655–3661.

Maiden, M. and Greco, S. (1994) NASA's Pathfinder data set program: land surface parameters. *International Journal of Remote Sensing*, **15**, 3333–3345.

Maitre, H. and Wu, Y. (1987) Improving dynamic programming to solve image registration. *Pattern Recognition*, **20**, 442–462.

Major, D. J., Baret, F. and Guyot, G. (1990) A ratio vegetation index adjusted for soil brightness. *International Journal of Remote Sensing*, **11**, 727–740.

Malingreau, J. P. (1984) Remote sensing and disaster monitoring: a review of applications in Indonesia. Proceedings of 18th International Symposium on Remote Sensing of Environment, Paris, France, 1–4 October 1984 (Ann Arbor, Michigan: Environmental Research Institute of Michigan), pp. 283–297.

Malingreau, J.-P. (1986) Global vegetation dynamics: satellite observations over Asia. *International Journal of Remote Sensing*, **7**, 1121–1146.

Malingreau, J. P., Achard, F., D'Souza, G., Stibig, H. J., D'Souza, J., Estreguil, C. and Eva, H. (1993) AVHRR for global tropical forest monitoring: the lessons of the TREES project. Proceedings of 6th AVHRR Data Users' Meeting, Belgirate, Italy, 29 June–2 July 1993. EUM P 12 (Darmstadt – Eberstadt: EUMETSAT), pp. 323–332.

Malingreau, J. P. and Belward A. S. (1989) Vegetation monitoring using AVHRR data at different resolutions. Proceedings of 4th AVHRR Data Users' Meeting, Rothenburg, F. R. Germany, 5–8 September 1989. EUM P 06 (Darmstadt – Eberstadt: EUMETSAT), pp. 141–144.

Malingreau, J. P. and Belward, A. S. (1992) Scale considerations in vegetation monitoring using AVHRR data. *International Journal of Remote Sensing*, **13**, 2289–2307.

Malingreau, J. P. and Belward, A. S. (1994) Recent activities in the European Community for the creation and analysis of global AVHRR data sets. *International Journal of Remote Sensing*, **15**, 3397–3416.

Malingreau, J. P., Laporte, N. and Gregoire, J. M. (1990) Exceptional fire events in the tropics. Southern Guinée January 1987. *International Journal of Remote Sensing*, **11**, 2121–2123.

Malingreau, J. P., Stephens, G. and Fellows, L. (1985) Remote sensing of forest fires: Kalimantan and North Borneo in 1982–3. *Ambio*, **14**, 314–315.

Malingreau, J. P. and Tucker, C. J. (1987) The contribution of AVHRR data for measuring and understanding global processes: large-scale deforestation in the Amazon Basin. Proceedings of IGARSS'87, Ann Arbor, Michigan, USA, 18–21 May 1987, IEEE 87CH2434-9 (New York: IEEE), pp. 484–489.

Malingreau, J. P. and Tucker, C. J. (1988) Large scale deforestation in the southeastern Amazon Basin. *Ambio*, **17**, 49–55.

Malingreau, J. P. and Tucker, C. J. (1990) Ranching in the Amazon Basin. Large-scale changes observed by AVHRR. *International Journal of Remote Sensing*, **11**, 187–189.

Malingreau, J. P., Tucker, C. J. and Laporte, N. (1989) AVHRR for monitoring global tropical deforestation. *International Journal of Remote Sensing*, **10**, 855–867.

Malkevich, M. S. and Gorodetsky, A. K. (1988) Determination of ocean surface temperature taking account of atmospheric effects by measurements of angular IR-radiation distribution of the 'ocean–atmosphere' system made from the satellite 'Cosmos-1151'. *Remote Sensing Reviews*, **3**, 137–161.

Mannstein, H. (1989) Energy budget components in the Alps from AVHRR data. Proceedings of 4th AVHRR Data Users' Meeting, Rothenburg, F.R. Germany, 5–8 September 1989. EUM P 06 (Darmstadt – Eberstadt: EUMETSAT), pp. 259–262.

Mannstein, H. and Gesell, G. (1991) Deconvolution of AVHRR data. Proceedings of 5th AVHRR Data Users' Meeting, Tromsø, Norway, 25–28 June 1991. EUM P 09 (Darmstadt – Eberstadt: EUMETSAT), pp. 53–58.

Marelli, L. (1992) Acquisition, processing and distribution of experimental satellite data. *International Journal of Remote Sensing*, **13**, 1125–1136.

Markham, B. L. (1985) The Landsat sensors' spatial responses. *IEEE Transactions on Geoscience and Remote Sensing*, **23**, 864–875.

Marsh, S. E., Walsh, J. L., Lee, C. T., Beck, L. R. and Hutchinson, C. F. (1992) Comparison of multi-temporal NOAA-AVHRR and SPOT-XS satellite data for mapping land-cover dynamics in the west African Sahel. *International Journal of Remote Sensing*, **13**, 2997–3016.

Marsouin, A. and Brunel, P. (1991) Nagivation of AVHRR images using ARGOS or TBUS bulletins. *International Journal of Remote Sensing*, **12**, 1575–1592.

Martinec, J., Rango, A. and Mayor, E. (1983) The snowmelt-runoff model (SRM) users manual (Washington, DC: National Aeronautics and Space Administration), Publication **1100**.

Maselli, F., Conese, C., Petkov, L. and Gilabert, M.-A. (1992) Use of NOAA-AVHRR NDVI data for environmental monitoring and crop forecasting in the Sahel. Preliminary results. *International Journal of Remote Sensing*, **13**, 2743–2749.

Maslanik, J. A., Key, J. R. and Barry, R. G. (1989) Merging AVHRR and SMMR data for remote sensing of ice and cloud in polar regions. *International Journal of Remote Sensing*, **10**, 1691–1696.

Mason, I. M., Rapley, C. G., Street-Perrott, F. A. and Guskowska, M. A. J. (1985) *ERS-1 observations for climate research*. ESA SP-233 (Paris: European Space Agency), pp. 235–241.

Masuda, K. and Takashima, T. (1990a) Deriving cirrus information using the visible and near-IR channels of the future NOAA-AVHRR radiometer. *Remote Sensing of Environment*, **31**, 65–81.

Masuda, K. and Takashima, T. (1990b) Sensitivity of shortwave radiation absorbed in the ocean to cirrus parameters. *Remote Sensing of Environment*, **33**, 75–86.

Masuda, K., Takashima, T. and Rao, C. R. N. (1988a) Remote sensing of atmospheric aerosols over the oceans using multispectral radiances measured with the Advanced Very High Resolution Radiometer (AVHRR) onboard the NOAA meteorological satellites. In *Aerosols and climate*, edited by P. V. Hobbs and M. P. McCormick (Hampton, Virginia: A. Deepak), pp. 39–49.

Masuda, K., Takashima, T. and Takayama, Y. (1988b), Emissivity of pure and sea waters for the model sea surface in the infrared window regions. *Remote Sensing of Environment*, **24**, 313–329.

Matson, M. and Dozier, J. (1981) Identification of subresolution high temperature sources using a thermal IR sensor. *Photogrammetric Engineering and Remote Sensing*, **47**, 1311–1318.

Matson, M. and Holben, B. N. (1987) Satellite detection of tropical burning in Brazil. *International Journal of Remote Sensing*, **8**, 509–516.

Matson, M., Schneider, S. R., Aldridge, B. and Satchwell, B. (1984) Fire detection using the NOAA-Series satellites. NOAA Technical Report NESDIS 7, Washington, DC.

Matson, M., Stephens, G. and Robinson, J. (1987) Fire detection using data from the NOAA-N satellites. *International Journal of Remote Sensing*, **8**, 961–970.

Matthews, E. (1983) Global vegetation and land use: new high resolution data bases for limited studies. *Journal of Climatology and Applied Meteorology*, **22**, 474–487.

Matthews, E. (1985) *Atlas of archived vegetation, land use and seasonal albedo data sets*, NASA Technical Memorandum 86199, Washington, DC.

Maul, G. A. (1981) Applications of GOES visible-infrared data to quantify mesoscale ocean surface temperature. *Journal of Geophysical Research*, **86**, 8009–8021.

Maul, G. A. (1983) Zenith angle effects in multichannel infrared sea surface remote sensing. *Remote Sensing of Environment*, **13**, 439–451.

Maul, G. A. and Sidran, M. (1972) Comment on 'Estimation of sea surface temperature from space' by D. Anding and R. Kauth. *Remote Sensing of Environment*, **2**, 165–169.

May, D. A., Stowe, L. L., Hawkins, J. D. and McClain, E. P. (1992) A correction for Saharan dust effects on satellite sea surface temperature measurements. *Journal of Geophysical Research*, **97**, 3611–3619.

Melinotte, J. M. and Arino, O. (1993) AVHRR CD-Browser Ionia. Proceedings of 6th AVHRR Data Users' Meeting, Belgirate, Italy, 29 June–2 July 1993. EUM P 12 (Darmstadt – Eberstadt: EUMETSAT), pp. 589–592.

Merson, R. H. (1989) An AVHRR mosaic image of Antarctica. *International Journal of Remote Sensing*, **10**, 669–674.

Miller, J. R., Wu, Jiyou, Boyer, M. G., Belanger, M. and Hare, E. W. (1991) Seasonal patterns in leaf reflectance red-edge characteristics. *International Journal of Remote Sensing*, **12**, 1509–1523.

Miller, N., Chine, E. and Howard, S. (1983) Evaluation of AVHRR data to develop fire fuels information as an input to IAMS. Final Report (Internal Report of the US Geological Survey).

Minnett, P. J. (1986) On the use of synthetic 12 μm data in a split-window retrieval of sea surface temperature from AVHRR measurements. *International Journal of Remote Sensing*, **7**, 1887–1891.

Minnett, P. J. (1989) On the *in situ* validation of AVHRR measurements of sea-surface temperature. Proceedings of 4th AVHRR Data Users' Meeting, Rothenburg, F.R. Germany, 5–8 September 1989. EUM P 06 (Darmstadt – Eberstadt: EUMETSAT), pp. 209–214.

Minnett, P. J. and Saunders, R. W. (1989) Validation of spaceborne radiometers: coordinated ship and aircraft measurements. Proceedings of 4th AVHRR Data Users' Meeting, Rothenburg, F.R. Germany, 5–8 September 1989. EUM P 06 (Darmstadt – Eberstadt: EUMETSAT), pp. 329–334.

Mitchell, R. M. (1989) Atmospheric correction of vegetation indices using split pass AVHRR imagery. Proceedings of 4th AVHRR Data Users' Meeting, Rothenburg, F.R. Germany, 5–8 September 1989. EUM P 06 (Darmstadt – Eberstadt: EUMETSAT), pp. 153–156.

Moctezuma, M., Maitre, H. and Parmiggiani, F. (1993) Automatic tracking of ice floes on AVHRR images sequences. Proceedings of 6th AVHRR Data Users' Meeting, Belgirate, Italy, 29 June–2 July 1993. EUM P 12 (Darmstadt – Eberstadt: EUMETSAT), pp. 541–546.

Morales, C. (Editor) (1979) *Saharan dust* (Chichester: Wiley).

Morassutti, M. P. (1992) Component reflectance scheme for DMSP-derived sea ice reflectances in the Arctic Basin. *International Journal of Remote Sensing*, **13**, 647–662.

Morel, A. and Prieur, L. (1977) Analysis of variations in ocean color. *Limnology and Oceanography*, **22**, 709–722.

Moulin, S. and Fischer, A. (1993) Simulation of the temporal variations of NOAA/AVHRR reflectances. Coupling of functional model and satellite data. Proceedings of 6th AVHRR Data Users' Meeting, Belgirate, Italy, 29 June–2 July 1993. EUM P 12 (Darmstadt – Eberstadt: EUMETSAT), pp. 277–283.

Muirhead, K. and Cracknell, A. P. (1984a) Identification of gas flares in the North Sea using satellite data. *International Journal of Remote Sensing*, **5**, 199–212.

Muirhead, K. and Cracknell, A. P. (1984b) Gas flares and forest fires – the potential of AVHRR Band 3. *Satellite remote sensing – review and preview*. Proceedings of the 10th Anniversary International Conference, University of Reading, UK, 18–21 September 1984 (Nottingham: Remote Sensing Society), pp. 411–419.

Muirhead, K. and Cracknell, A. P. (1985) Straw burning over Great Britain detected by AVHRR. *International Journal of Remote Sensing*, **6**, 827–833.

Muirhead, K. and Cracknell, A. P. (1986), see Saull (1986).

Muirhead, K. and Malkawi, O. (1989) Automatic classification of AVHRR images. Proceedings of 4th AVHRR Data Users' Meeting, Rothenburg, F.R. Germany, 5–8 September 1989. EUM P 06 (Darmstadt – Eberstadt: EUMETSAT), pp. 31–34.

Myers, D. G. and Hick, P. T. (1990) An application of satellite-derived sea surface temperature data to the Australian fishing industry in near-real time. *International Journal of Remote Sensing*, **11**, 2103–2112.

Narendra Nath, A. N., Rao, M. V. and Rao, K. H. (1993) Observed high temperatures in the sunglint area over the North Indian Ocean. *International Journal of Remote Sensing*, **14**, 849–853.

Neckel, H. and Labs, D. (1984) The solar radiation between 3300 and 12500 Å. *Solar Physics*, **90**, 205–208.

Needham, B. H. (1988) Next generation AVHRR: AMRIR. *Photogrammetric Engineering and Remote Sensing*, **54**, 1333–1335.

Negré, T., Imbernon, J., Guinot, J. P., Seguin, B., Berges, J. C. and Guillot, B. (1988) Estimation et suivi de la pluviométrie au Sénégal par satellite Meteosat. *Agronomie Tropicale*, **43**, 279–288.

Nelson, R. (1989) Regression and ratio estimators in integrated AVHRR and MSS data. *Remote Sensing of Environment*, **30**, 201–216.

Nelson, R. and Holben, B. (1986) Identifying deforestation in Brazil using multiresolution satellite data. *International Journal of Remote Sensing*, **7**, 429–448.

Nelson, R., Horning, N. and Stone, T. A. (1987) Determining the rate of forest conversion in Mato Grosso, Brazil, using Landsat MSS and AVHRR data. *International Journal of Remote Sensing*, **8**, 1767–1784.

Nemani, R. R. and Running, S. W. (1989) Estimation of surface resistance to evapotranspiration from NDVI and thermal-IR AVHRR data. *Journal of Climate and Applied Meteorology*, **28**, 276–294.

Nikitin, P. A. (1991) Satellite-based monitoring of sea ice. Proceedings of 5th AVHRR Data Users' Meeting, Tromsø, Norway, 25–28 June 1991. EUM P 09 (Darmstadt – Eberstadt: EUMETSAT), pp. 411–414.

Ninnis, R. M., Emery, W. J. and Collins, M. J. (1986) Automated extraction of pack ice motion from Advanced Very High Resolution Radiometer imagery. *Journal of Geophysical Research*, **91**, 10725–10734.

Njoku, E. G. and Patel, I. R. (1986) Observations of the seasonal variability of soil moisture and vegetation cover over Africa using satellite microwave radiometry. Proceedings of the ISLSCP Conference, Rome, Italy, 2–6 December 1985. SP-248, (Paris: European Space Agency), pp. 349–353.

Norton, C., Mosher, F. R., Hinton, B., Martin, D. W., Santek, D. and Kuhlow, W. (1980) A model for calculating desert aerosol turbidity over the oceans from geostationary satellite data. *Journal of Applied Meteorology*, **19**, 633–644.

Norwine, J. and Greegor, D. H. (1983) Vegetation classification based on Advanced Very High Resolution Radiometer (AVHRR) satellite imagery. *Remote Sensing of Environment*, **13**, 69–87.

Nykjaer, L. (1993) Coastal upwelling off Northwest Africa observed with AVHRR, meteorological and biological data. Proceedings of 6th AVHRR Data Users' Meeting, Belgirate, Italy, 29 June–2 July, 1993. EUM P 12 (Darmstadt – Eberstadt: EUMETSAT), pp. 367–373.

O'Brien, D. M. and Turner, P. J. (1992) Navigation of coastal AVHRR images. *International Journal of Remote Sensing*, **13**, 509–514.

O'Brien, H. W. and Munis, R. H. (1975) Red and near-infrared spectral reflectance of snow. Research Report 332, Cold Regions Research and Engineering Laboratory, US Army Corps of Engineers, Hanover, NH, 18 pp.

Ødegaard, H. A. and Østrem, G. (1977) Application of satellite data for snow mapping. Landsat-2 Contract No. 29020: Hydrological investigations in Norway, Report No. 9, Norwegian Water Resources and Electricity Board.

Ohring, G. and Gruber, A. (1983) Satellite radiation observations and climate theory. *Advances in Geophysics*, **25**, 237–304.

Olesen, F. S. (1987) Vertical sounding from satellite. In *Remote sensing applications in meteorology and climatology*, edited by R. A. Vaughan (Dordrecht: Reidel), pp. 155–172.

Olesen, F. S. (1992) Vertical sounding of atmosphere and imaging radiometer. In *Space oceanography*, edited by A. P. Cracknell (Singapore: World Scientific), pp. 127–138.

Olesen, F. S. and Grassl, H. (1985) Cloud detection and classification over oceans at night with NOAA-7. *International Journal of Remote Sensing*, **8**, 1435–1444.

Olesen, F. S., Kind, O. and Messinger, N. (1993) Determination of the diurnal change of land surface temperature. Proceedings of 6th AVHRR Data Users' Meeting, Belgirate, Italy, 29 June–2 July 1993. EUM P 12 (Darmstadt – Eberstadt: EUMETSAT), pp. 59–65.

Olson, J. S. and Watts, J. A. (1982) Major world ecosystem complexes map, scale = 1 : 30,000,000. Oak Ridge National Laboratory, Oak Ridge, Tennessee.

Oppenheimer, C. (1989) AVHRR volcano hotspot monitoring. Proceedings of 4th AVHRR Data Users' Meeting, Rothenburg, F.R. Germany, 5–8 September 1989. EUM P 06 (Darmstadt – Eberstadt: EUMETSAT), pp. 335–338.

Oppenheimer, C. M. M. and Rothery, D. A. (1989) Infrared remote sensing of hot volcanoes assessed by field observations. *Remote sensing for operational applications*, Technical Contents of the 15th Annual Conference of the Remote Sensing Society, University of Bristol, UK, 13–15 September 1989 (Nottingham: Remote Sensing Society), pp. 317–322.

Ormsby, J. P., Choudhury, B. J. and Owe, M. (1987) Vegetation spatial variability and its effect on vegetation indices. *International Journal of Remote Sensing*, **8**, 1301–1306.

Østrem, G. (1981) The use of remote sensing in hydrology in Norway. In *Remote sensing in meteorology, oceanography and hydrology*, edited by A. P. Cracknell (Chichester: Ellis Horwood), pp. 258–269.

Østrem, G., Andersen, T., Odegaard, H. and Del Llano, R. (1979) Snow and ice mapping: Norwegian examples for run-off prediction. Technical Conference on Use of Data from Meteorological Satellites. SP-143 (Paris: European Space Agency), pp. 177–181.

Øvergård, A. (1989) Near realtime processing and distribution of AVHRR products covering the Arctic. Proceedings of 4th AVHRR Data Users' Meeting, Rothenburg, F.R. Germany, 5–8 September 1989. EUM P 06 (Darmstadt – Eberstadt: EUMETSAT), pp. 35–39.

Pagano, P., De Leonibus, L. and Schiarini, S. (1991) Use of AVHRR data in the air–sea interactions. Proceedings of 5th AVHRR Data Users' Meeting, Tromsø, Norway, 25–28 June 1991. EUM P 09 (Darmstadt – Eberstadt: EUMETSAT), pp. 243–252.

Pairman, D. and Kittler, J. (1986) Clustering algorithms for use with images of clouds. *International Journal of Remote Sensing*, **7**, 855–866.

Paithoonwattanakij, K. (1989) Automatic pattern recognition techniques for geometrical correction on satellite data. PhD thesis, Dundee University.

Paivinen, R., Witt, R. and Cross, A. (1990) Application of NOAA/AVHRR data for tropical forest cover mapping. *Global natural resource monitoring and assessment: preparing for the 21st century*, Proceedings of the International Conference Venice, Italy, 23–26 September 1989 (Maryland: American Society of Photogrammetry and Remote Sensing), pp. 1359–1367.

Paltridge, G. W. and Barber, J. (1988) Monitoring grassland dryness and fire potential in Australia with NOAA/AVHRR data. *Remote Sensing of Environment*, **25**, 381–394.

Paltridge, G. W. and Mitchell, R. M. (1990) Atmospheric and viewing angle correction of vegetation indices and grassland fuel moisture content derived from NOAA/AVHRR. *Remote Sensing of Environment*, **31**, 121–135.

Parikh, J. (1977) A comparative study of cloud classification techniques. *Remote Sensing of Environment*, **6**, 67–81.

Pathak, P. N. (1982) Comparison of sea-surface temperature observations from TIROS-N and ships in the North Indian Ocean during MONEX (May–July 1979). *Remote Sensing of Environment*, **12**, 363–369.

Patt, F. S. and Gregg, W. W. (1994) Exact closed-form geolocation algorithm for Earth survey sensors. *International Journal of Remote Sensing*, **15**, 3719–3734.

Payne, R. E. (1972) Albedo of the sea surface. *Journal of Atmospheric Science*, **29**, 959–970.

Pearce, A. F., Prata, A. J. and Manning, C. R. (1989) Comparison of NOAA/AVHRR-2 sea surface temperatures with surface measurements in coastal waters. *International Journal of Remote Sensing*, **10**, 37–52.

Pereira, M. C. (1988) Deteccão, Monitoramento e Analise de Alguns Efeitos Ambientais de Quemadas no Amazonia Atraves da Utilização de Imagens dos Satelites NOAA e Landsat, e Davos Aeronave. INPE-4503-tdl/325, Inst Nacional de Pesquisas Espacias, 12.201 São José dos Campos, SP, Brazil.

Perry, C. R. and Lautenschlager, L. F. (1984) Functional equivalence of spectral vegetation indices. *Remote Sensing of Environment*, **14**, 169–182.

Peterson, D. L., Spanner, M. A., Running, S. W. and Teuber, K. B. (1987) Relationship of Thematic Mapper Simulator data to leaf area index of temperate coniferous forest. *Remote Sensing of Environment*, **22**, 323–341.

Péwé, T. L. (Editor) (1981) *Desert dust: origin, characteristics and effect on man.* Geological Society of America, Special Paper 186.

Philipson, W. R. and Teng, W. L. (1988) Operational interpretation of AVHRR vegetation indices for world crop information. *Photogrammetric Engineering and Remote Sensing*, **54**, 55–59.

Phulpin, T., Jullien, J P. and Lasselin, D. (1989) AVHRR data processing to study the surface canopies in temperate regions: first results of HAPEX-MOBILHY. *International Journal of Remote Sensing*, **10**, 869–884.

Pinter, P. J., Zipoli, G., Maracchi, G. and Reginato, R J. (1987) Influence of topography and sensor view angles on NIR/red ratio and greenness vegetation indices of wheat. *International Journal of Remote Sensing*, **8**, 953–957.

Pinty, B. and Verstraete, M. M. (1992) GEMI: A non-linear index to monitor global vegetation from satellites. *Vegetatio*, **101**, 15–20.

Pinty, B., Verstraete, M. M. and Dickinson, R. E. (1989) A physically-based model for the NDVI. Proceedings of 4th AVHRR Data Users' Meeting, Rothenburg, F.R. Germany, 5–8 September 1989. EUM P 06 (Darmstadt – Eberstadt: EUMETSAT), pp. 105–109.

Pittella, G. and Bamford, C. (1989) The Earthnet 'SHARK' system for TIROS data acquisition processing and archive. Proceedings of 4th AVHRR Data Users' Meeting, Rothenburg, F.R. Germany, 5–8 September 1989. EUM P 06 (Darmstadt – Eberstadt: EUMETSAT), pp. 45–49.

Planet, W. G. (Editor) (1988) Data extraction and calibration of TIROS-N/NOAA radiometers. NOAA Technical Memorandum NESS 107 – Rev. 1. US Department of Commerce, National Oceanic and Atmospheric Administration, National Environmental Satellite, Data and Information Service, Washington, DC 20233. See also Lauritson *et al.* (1979).

Pollock, R. B. and Kanemasu, E. T. (1979) Estimating leaf-area index of wheat with Landsat data. *Remote Sensing of Environment*, **8**, 307–312.

Popp, T. (1993) Atmospheric correction of satellite images in the solar spectral range. Proceedings of 6th AVHRR Data Users' Meeting, Belgirate, Italy, 29 June–2 July 1993. EUM P 12 (Darmstadt – Eberstadt: EUMETSAT), pp. 87–93.

Potts, R. J. (1993) Satellite observations of Mt Pinatubo ash clouds. *Australian Meteorological Magazine*, **42**, 59–68.

Prabhakara, C., Dalu, G. and Kunde, V. G. (1975) Estimation of sea surface temperature from remote sensing in the 11 and 13 μm window region. *Journal of Geophysical Research*, **79**, 5039–5044.

Prabhakara, C., Dalu, G., Lo, R. C. and Nath, N. R. (1979) Remote sensing of seasonal distribution of precipitable water vapour over the oceans and the inference of boundary layer structure. *Monthly Weather Review*, **107**, 1388–1401.

Prangsma, G. J. and Roozekrans, J. N. (1989) Using NOAA AVHRR imagery in assessing water quality parameters. *International Journal of Remote Sensing*, **10**, 811–818.

Prata, A. J. (1989a) Observations of volcanic ash clouds in the 10–12 μm window using AVHRR/2 data. *International Journal of Remote Sensing*, **10**, 751–761.

Prata, A. J. (1989b) Infrared radiative transfer calculations for volcanic ash clouds. *Geophysical Research Letters*, **16**, 1293–1296.

Prata, A. J. (1990) Satellite-derived evaporation from Lake Eyre, South Australia. *International Journal of Remote Sensing*, **11**, 2051–2068.

Prata, A. J., Cechet, R. P., Barton, I. J. and Llewellyn-Jones, D. T. (1989) The Along Track Scanning Radiometer for ERS-1 – scan geometry and data simulation. Proceedings of 4th AVHRR Data Users' Meeting, Rothenburg, F.R. Germany, 5–8 September 1989. EUM P 06 (Darmstadt – Eberstadt: EUMETSAT), pp. 285–287.

Prata, A. J. and Platt, C. M. R. (1991) Land surface temperature measurements from the AVHRR. Proceedings of 5th AVHRR Data Users' Meeting, Tromsø, Norway, 25–28 June 1991. EUM P 09 (Darmstadt – Eberstadt: EUMETSAT), pp. 433–438.

Prata, A. J. and Wells, J. B. (1990) A satellite sensor image of the Leeuwin current, Western Australia. *International Journal of Remote Sensing*, **11**, 173–180.

Prata, A. J., Wells, J. B. and Ivanac, M. W. (1985) A 'satellite's eye view' of volcanoes on the Lesser Sunda islands. *Weather*, **40**, 245–250.

Pratt, D. A. and Ellyet, C. D. (1979) The thermal inertia approach to mapping of soil moisture and geology. *Remote Sensing of Environment*, **8**, 151–168.

Price, J. C. (1977) Thermal inertia mapping: a new view of the Earth. *Journal of Geophysical Research*, **82**, 2582–2590.

Price, J. C. (1983) Estimating surface temperatures from satellite thermal infrared data – a simple formulation for the atmospheric effect. *Remote Sensing of Environment*, **13**, 353–361.

Price, J. C. (1987a) Calibration of satellite radiometers and the comparison of vegetation indices. *Remote Sensing of Environment*, **21**, 15–27.

Price, J. C. (1987b) Radiometric calibration of satellite sensors in the visible and near infrared: history and outlook. *Remote Sensing of Environment*, **22**, 3–9.

Price, J. C. (1991) Timing of NOAA afternoon passes. *International Journal of Remote Sensing*, **12**, 193–198.

Price, J. C. (1992) Estimating vegetation amount from visible and near infrared reflectances. *Remote Sensing of Environment*, **41**, 29–34.

Prince, S. D. (1991a) Satellite remote sensing of primary production: comparison of results for Sahelian grasslands 1981–1988. *International Journal of Remote Sensing*, **12**, 1301–1311.

Prince, S. D. (1991b) A model of regional primary production for use with coarse resolution satellite data. *International Journal of Remote Sensing*, **12**, 1313–1330.

Prospero, J. M. and Carlson, T. N. (1972) Vertical and areal distribution of Saharan dust over the western equatorial North Atlantic Ocean. *Journal of Geophysical Research*, **77**, 5255–5265.

Punkari, M. (1982) Glacial geomorphology and dynamics in the eastern parts of the Baltic Shield interpreted using Landsat imagery. *Photogrammetric Journal of Finland*, **9**, 77–93.

Punkari, M. (1984) The relations between glacial dynamics and tills in the eastern part of the Baltic Shield. *Striae*, **20**, 43–51.

Puyou-Lascassies, P., Husson, A. and Jeanjean, H. (1993) SEAMEO-France project. Methodology and training in monitoring deforestation using satellite data for sustainable forest management. Proceedings of 6th AVHRR Data Users' Meeting, Belgirate, Italy, 29 June–2 July 1993. EUM P 12 (Darmstadt – Eberstadt: EUMETSAT), pp. 303–312.

Pylkkö, P. and Aulamo, H. (1991) Using AVHRR data to estimate rainfall in northern latitudes. Proceedings of 5th AVHRR Data Users' Meeting, Tromsø, Norway, 25–28 June 1991. EUM P 09 (Darmstadt – Eberstadt: EUMETSAT), pp. 333–337.

Quarmby, N. A., Townshend, J. R. G., Settle, J. J., White, K. H., Milnes, M., Hindle, T. L. and Silleos, N. (1992) Linear mixture modelling applied to AVHRR data for crop area estimation. *International Journal of Remote Sensing*, **13**, 415–425.

Ramp, S. R., Garwood, R. W., Davis, C. O. and Snow, R. L. (1991) Surface heating and patchiness in the coastal ocean off central California during a wind relaxation event. *Journal of Geophysical Research*, **96**, 14947–14957.

Randerson, J. T. and Simpson, J. J. (1993) Recurrent patterns in surface thermal fronts associated with cold filaments along the west coast of North America. *Remote Sensing of Environment*, **46**, 146–163.

Rango, A. (1975) Employment of satellite snowcover observations for improving seasonal runoff estimates. NASA Special Publication 391 (Washington, DC: NASA), pp. 157–174.

Rango, A. and Peterson, R. (Editors) (1980) *Operational applications of satellite snowcover observations*. NASA-CP-2116, Goddard Space Flight Center, Greenbelt, Maryland, 301 pp.

Ranson, K. J., Irons, J. R. and Daughtry, C. S. T. (1991) Surface albedo from bidirectional reflectance. *Remote Sensing of Environment*, **35**, 201–211.

Rao, C. R. N. (1987) Pre-launch calibration of channels 1 and 2 of the Advanced Very High Resolution Radiometer. NOAA Technical Report NESDIS 36 (Washington DC: US Department of Commerce).

Rao, C. R. N. (1992) Aerosol radiative corrections to the retrieval of sea surface temperatures from infrared radiances measured by the Advanced Very High Resolution Radiometer (AVHRR). *International Journal of Remote Sensing*, **13**, 1757–1769.

Rao, C. R. N. (1993a) Nonlinearity corrections for the thermal infrared channels of the Advanced Very High Resolution Radiometer: assessment and recommendations. NOAA Technical Report NESDIS 69 (Washington DC: US Department of Commerce).

Rao, C. R. N. (1993b) Degradation of the visible and near infrared channels of the Advanced Very High Resolution Radiometer on the NOAA-9 spacecraft: assessment and recommendations for corrections. NOAA Technical Report NESDIS 70 (Washington DC: US Department of Commerce).

Rao, C. R. N. and Chen, J. (1993) Calibration of the visible and near-infrared channels of the Advanced Very High Resolution Radiometer (AVHRR) after launch. *Recent Advances in Sensors, Radiometric Calibration and Processing of Remotely-Sensed Data*, 14–16 April 1993, Orlando, Florida (Washington, DC: Society of Photo-Optical Engineers), pp. 56–66.

Rao, C. R. N. and Chen, J. (1994) Post-launch calibration of the visible and near-infrared channels of the Advanced Very High Resolution Radiometer on NOAA-7, -9 and -11 spacecraft. NOAA Technical Report NESDIS 78 (Washington, DC: US Department of Commerce).

Rao, C. R. N. and Chen. J. (1995) Inter-satellite calibration linkages for the visible and near-infrared channels of the Advanced Very High Resolution Radiometer on the NOAA-7, -9, and -11 spacecraft. *International Journal of Remote Sensing*, **16**, 1931–1942.

Rao, C. R. N., Stowe, L. L. and McClain, E. P. (1989) Remote sensing of aerosols over the oceans using AVHRR data: theory, practice and applications. *International Journal of Remote Sensing*, **10**, 743–749.

Rao, C. R. N., Stowe, L. L., McClain, E. P. and Sapper, J. (1988) Development and application of aerosol remote sensing with AVHRR data from the NOAA satellites. In *Aerosol and climate*, edited by P. V. Hobbs and M. P. McCormick (Hampton, Virginia: A. Deepak), pp. 69–80.

Rao, P. K., Holmes, S. J., Anderson, R. K., Winston, J. S. and Lehr, P. E. (1990) *Weather satellites: systems, data and environmental applications* (Boston, Massachusetts: American Meteorological Society).

Rapley, C. G., Guskowska, M. A. J., Cudlip, W. and Mason, I. M. (1987) An exploratory study of inland water and land altimetry using Seasat data. ESA Contract Report 6483/85/NL/BI, European Space Agency, Paris.

Raschke, E. (1987) Report of the International Satellite Cloud Climatology Project (ISCCP)

Workshop on Cloud Algorithms in the Polar Regions. Tokyo, Japan, 19–21 August 1986.

Raschke, E., Bauer, P. and Lutz, H. J. (1992) Remote sensing of clouds and surface radiation budget over polar regions. *International Journal of Remote Sensing*, **13**, 13–22.

Rasmussen, M. S. (1992) Assessment of millet yields and production in northern Burkina Faso using integrated NDVI from the AVHRR. *International Journal of Remote Sensing*, **13**, 3431–3442.

Rasool, S. I. and Bolle, H. J. (1984) ISLSCP: International Satellite Land-Surface Climatology Project. *Bulletin of the American Meteorological Society*, **65**, 143–144.

Rees, W. G. and Squire, V. A. (1989) Technological limitations to satellite glaciology. *International Journal of Remote Sensing*, **10**, 7–22.

Reiff, J. (1988) The contribution of satellite information to operational weather forecasting. Achievements and objectives in the 1990s. *International Journal of Remote Sensing*, **9**, 1675–1680.

Reutter, H. and Olesen, F.-S. (1991) Atmospheric correction of AVHRR IR-data with TOVS soundings. Proceedings of 5th AVHRR Data Users' Meeting, Tromsø, Norway, 25–28 June 1991. EUM P 09 (Darmstadt – Eberstadt: EUMETSAT), pp. 143–147.

Reynolds, R. W. (1993) Impact of Mount Pinatubo aerosols on satellite-derived sea surface temperatures. *Journal of Climate*, **6**, 768–774.

Richardson, A. J. (1984) El Chichón volcanic ash effects on atmospheric haze measured by NOAA-7 AVHRR data. *Remote Sensing of Environment*, **16**, 157–164.

Richardson, A. J. and Wiegand, C. L. (1977) Distinguishing vegetation from soil background information. *Photogrammetric Engineering and Remote Sensing*, **43**, 1541–1552.

Richardson, A. J. and Wiegand, C. L. (1991) Comparison of two models for simulating the soil-vegetation composite reflectance of a developing cotton canopy. *International Journal of Remote Sensing*, **11**, 447–459.

Rimóoczi-Paál, A. (1993) Radiation balance investigations using NOAA AVHRR data. Proceedings of 6th AVHRR Data Users' Meeting, Belgirate, Italy, 29 June–2 July 1993. EUM P 12 (Darmstadt – Eberstadt: EUMETSAT), pp. 493–498.

Robinson, I. S. (1985) *Satellite oceanography: an introduction for oceanographers and remote-sensing scientists* (Chichester: Ellis Horwood).

Robinson, I. S. and Ward, N. (1989) Comparison between satellite and ship measurements of sea surface temperature in the north-east Atlantic Ocean. *International Journal of Remote Sensing*, **10**, 787–799.

Robinson, I. S., Wells, N. C. and Charnock, H. (1984) The sea surface thermal boundary layer and its relevance to the measurement of sea surface temperature by airborne and space-borne radiometers. *International Journal of Remote Sensing*, **5**, 19–45.

Robinson, J. M. (1991) Fire from space: global fire evaluation using infrared remote sensing. *International Journal of Remote Sensing*, **12**, 3–24.

Robock, A. and Matson, M. (1983) Circumglobal transport of the El Chichón volcanic dust cloud. *Science, New York*, **221**, 195–197.

Rogers, D. J. (1991) Satellite imagery, tsetse and trypanosomiasis in Africa. *Preventative Veterinary Medicine*, **11**, 201–220.

Rogers, D. J. and Randolph, S. E. (1991) Mortality rates and population density of tsetse flies correlated with satellite imagery. *Nature*, **351** 739–741.

Rogers, D. J. and Randolph, S. E. (1993) Distribution of tsetse and ticks in Africa: past, present and future. *Parasitology Today*, **9**, 266–271.

Rogers, D. J. and Williams, B. G. (1993) Monitoring trypanosomiasis in space and time. *Parasitology*, **106**, S77–S92.

Rogers, D. J. and Williams, B. G. (1994) Tsetse distribution in Africa: seeing the wood *and* the trees. In *Large-scale ecology and conservation biology*, edited by P. J. Edwards, R. May and N. R. Webb (Oxford: Blackwell), pp. 247–271.

Roozekrans, J. N. (1989) Operational AVHRR products of the North Sea. Proceedings of 4th

AVHRR Data Users' Meeting, Rothenburg, F.R. Germany, 5–8 September 1989. EUM P 06 (Darmstadt – Eberstadt: EUMETSAT), pp. 235–239.

Roozekrans, J. N. (1993) The monitoring of desertification processes in Spain using NOAA-AVHRR data. Proceedings of 6th AVHRR Data Users' Meeting, Belgirate, Italy, 29 June–2 July 1993. EUM P 12 (Darmstadt – Eberstadt: EUMETSAT), pp. 313–322.

Rosema, A. (1986a) Results of the Group Agromet Monitoring Project (GAMP). *ESA Journal*, **10**, 17–41.

Rosema, A. (1986b) GAMP methodology: integrated mapping of rainfall, evapotranspiration, germination, biomass development and thermal inertia, based on Meteosat and conventional meteorological data. Proceedings of the ISLSCP Conference, Rome, Italy, 2–6 December 1985. SP-248 (Paris: European Space Agency), pp. 549–557.

Rossini, P., Taddei, R. and Terpessi, C. (1993) Crop monitoring and forecasting in south Italy with NDVI/AVHRR data. Proceedings of 6th AVHRR Data Users' Meeting, Belgirate, Italy, 29 June – 2 July 1993. EUM P 12 (Darmstadt – Eberstadt: EUMETSAT), pp. 547–549.

Rossow, W. B., Mosher, F., Kinsella, E., Arking, A., Desbois, M., Harrison, E., Minnis, P., Ruprecht, E., Seze, G., Simmer, C. and Smith, E. (1985) ISCCP cloud algorithm intercomparison. *Journal of Climate and Applied Meteorology*, **24**, 877–903.

Rossow, W. B. and Schiffer, R. A. (1991) ISCCP cloud data products. *Bulletin of the American Meteorological Society*, **72**, 2–20.

Rothery, D., Francis, P. W. and Wood, C. A. (1988) Volcano monitoring using short wavelength infrared data from satellites. *Journal of Geophysical Research*, **93**, 7993–8008.

Roujean, J. L., Leroy, M., Podaire, A. and Deschamps, P. Y. (1992) Evidence of surface reflectance bidirectional effects from a NOAA/AVHRR multi-temporal data set. *International Journal of Remote Sensing*, **13**, 685–698.

Rouse, J. W., Haas, R. H., Schell, J. A. and Deering, D. W. (1973) Monitoring vegetation systems in the Great Plains with ERTS. 3rd ERTS Symposium, volume I.

Running, S. W. and Nemani, R. R. (1988) Relating seasonal patterns of the AVHRR vegetation index to simulated photosynthesis and transpiration of forests in different climates. *Remote Sensing of Environment*, **24**, 347–367.

Sathyendranath, S. and Morel, A. (1983) Light emerging from the sea – interpretation and uses in remote sensing. In *Remote sensing applications in marine science and technology*, edited by A. P. Cracknell (Dordrecht: Reidel), pp. 323–357.

Saull, R. J. (1986) Strawburning over Great Britain detected by AVHRR: a comment. *International Journal of Remote Sensing*, **7**, 169–172; includes Reply by K. Muirhead and A. P. Cracknell.

Saunders, R. W. (1986) An automated scheme for the removal of cloud contamination from AVHRR radiances over western Europe. *International Journal of Remote Sensing*, **7**, 867–886.

Saunders, R. W. (1989) Modelled atmospheric transmittances for the AVHRR channels. Proceedings of 4th AVHRR Data Users' Meeting, Rothenburg, F.R. Germany, 5–8 September 1989. EUM P 06 (Darmstadt – Eberstadt: EUMETSAT), pp. 247–253.

Saunders, R. W. (1990) The determination of broad band surface albedo from AVHRR visible and near-infrared radiances. *International Journal of Remote Sensing*, **11**, 49–67.

Saunders, R. W. and Kriebel, K. T. (1988) An improved method for detecting clear sky and cloudy radiances from AVHRR data. *International Journal of Remote Sensing*, **9**, 123–150; errata, *ibid.*, **9**, 1393–1394.

Saunders, R. W. and Seguin, B. (1992) Meteorology and climatology. *International Journal of Remote Sensing*, **13**, 1231–1259.

Saunders, R. W., Ward, N. R., England, C. F. and Hunt, G. E. (1981) Sea surface temperature measurements around the UK derived from Meteosat and TIROS-N data. *Matching Remote Sensing Technologies and their Applications*, Proceedings of an International

Conference, London, UK, December 1981 (Nottingham: Remote Sensing Society), pp. 191–198.

Scharfen, G., Barry, R. G., Robinson, D. A., Kukla, G. J. and Serreze, M. C. (1987) Large-scale patterns of snow melt on Arctic sea ice from meteorological satellite imagery. *Annals of Glaciology*, **9**, 200–205.

Schiffer, R. A. and Rossow, W. B. (1983) The International Satellite Cloud Climatology Project (ISCCP): The first project of the World Climate Research Programme. *Bulletin of the American Meteorological Society*, **64**, 779–784.

Schiffer, R. A. and Rossow, W. B. (1985) ISCCP global radiance data set. A new resource for climate research. *Bulletin of the American Meteorological Society*, **66**, 1498–1505.

Schluessel, P. (1989) Satellite-derived low-level atmospheric water vapour content from synergy of AVHRR and HIRS. *International Journal of Remote Sensing*, **10**, 705–721.

Schluessel, P., Emery, W. J., Grassl, H. and Mammen, T. H. (1990) On the bulk–skin temperature difference and its impact on satellite remote sensing of sea surface temperature. *Journal of Geophysical Research*, **95**, 13341–13356.

Schluessel, P., Shin, H.-Y, Emery, W. J. and Grassl, H. (1987) Comparison of satellite derived sea surface temperatures with *in situ* skin measurements. *Journal of Geophysical Research*, **92**, 2859–2874.

Schmetz, P., Schmetz, J. and Raschke, E. (1986) Estimation of daytime downward longwave radiation at the surface from satellite and grid point data. *Theoretical and Applied Climatology*, **37**, 136–149.

Schneider, S. R. (1980) The NOAA/NESS programme for operational snowcover mapping: preparing for the 1980s. In *Operational applications of satellite snowcover observations*, edited by A. Rango and P. Peterson, NASA-CP-2116, Goddard Space Flight Center, Greenbelt, Maryland, pp. 21–40.

Schneider, S. R., McGinnis, D. F. and Stephens, G. (1985) Monitoring Africa's Lake Chad basin with Landsat and NOAA satellite data. *International Journal of Remote Sensing*, **6**, 59–73.

Schneider, S. and Matson, M. (1977) Satellite observations of snowcover in the Sierra Nevadas during the Great California Drought. *Remote Sensing of Environment*, **4**, 327–334.

Schultz, G. A. (1987) Parameter determination and input estimation in rainfall-runoff modelling based on remote sensing techniques. In *Water for the future: Hydrology in Perspective*, edited by O. C. Rodda and N. C. Matalas (Wallingford, Oxfordshire: International Association of Hydrological Sciences, Institute of Hydrology), Publication **164**, 425–438.

Schulz, J. (1993) Latent heat flux at the air–sea interface from a combination of SSM/I with AVHRR data. Proceedings of 6th AVHRR Data Users' Meeting, Belgirate, Italy, 29 June–2 July 1993. EUM P 12 (Darmstadt – Eberstadt: EUMETSAT), pp. 179–185.

Schwalb, A. (1978) The TIROS-N/NOAA A–G satellite series, NOAA Technical Memorandum NESS 95. United States Department of Commerce, Washington DC (revised 1982).

Schwalb, A. (1982) Modified version of the TIROS-N/NOAA A–G satellite series (NOAA E–J) Advanced TIROS-N (ATN). NOAA Technical Memorandum NESS 116. United States Department of Commerce, Washington, DC.

Scorer, R. S. (1986) *Cloud investigation by satellites* (Chichester: Ellis Horwood).

Scorer, R. S. (1987) Ship trails. *International Journal of Remote Sensing*, **8**, 687–688.

Scorer, R. S. (1989) Cloud reflectance variations in channel-3. *International Journal of Remote Sensing*, **10**, 675–686.

Seguin, B., Assad, E., Freteaud, J. P., Imbernon, J., Kerr, Y. H. and Lagouarde, J. P. (1989) Use of meteorological satellites for water balance monitoring in Sahelian regions. *International Journal of Remote Sensing*, **10**, 1101–1117.

Seguin, B., Lagouarde, J. P. and Savane, M. (1991) The assessment of regional crop water conditions from meteorological satellite thermal infrared data. *Remote Sensing of*

Environment, **35**, 141–148.

Seidel, K., Burkart, U., Baumann, R., Martinec, J., Haefner, H. and Itten, K. (1989) Satellite data for evaluation of snow reserves and runoff forecasts. Hydrology and Water Resources Symposium on Comparisons in Australian Hydrology (Christchurch, New Zealand: University of Canterbury), pp. 24–27.

Seljelv, L.-G. and Enoksen, R. T. (1993) Near real-time services. A presentation of Tromsø satellite station infrastructure; capabilities and examples. Proceedings of 6th AVHRR Data Users' Meeting, Belgirate, Italy, 29 June–2 July, 1993. EUM P 12 (Darmstadt – Eberstadt: EUMETSAT), pp. 499–503.

Sellers, P. J. (1985) Canopy reflectance, photosynthesis and transpiration. *International Journal of Remote Sensing*, **6**, 1335–1371.

Sellers, P. J. (1987) Canopy reflectance, photosynthesis and transpiration. II. The role of biophysics in the linearity of their interdependence. *Remote Sensing of Environment*, **21**, 143–183.

Sellers, P. J. and Hall, F. G. (1992) FIFE in 1992: results, scientific gains and future directions. *Journal of Geophysical Research*, **97**, 19091–19109.

Sellers, P. J., Hall, F. G., Asrar, G., Strebel, D. E. and Murphy, R. E. (1988) The first ISLSCP Field Experiment (FIFE). *Bulletin of the American Meteorological Society*, **69**, 22–27.

Sellers, P. J., Tucker, C. J., Collatz, G. J., Los, S. O., Justice, C. O., Dazlich, D. A. and Randall, D. A. (1994) A global 1° by 1° NDVI data set for climate studies. Part 2: The generation of global fields of terrestrial biophysical parameters from the NDVI. *International Journal of Remote Sensing*, **15**, 3519–3545.

Serafini, Y. V. (1987) Estimation of the evapotranspiration using surface and satellite data. *International Journal of Remote Sensing*, **8**, 1547–1562.

Setzer, A. W. and Malingreau, J. P. (1993) Temporal variation in the detection limit of fires in AVHRR's ch. 3. Proceedings of 6th AVHRR Data Users' Meeting, Belgirate, Italy, 29 June–2 July 1993. EUM P 12 (Darmstadt – Eberstadt: EUMETSAT), pp. 575–579.

Setzer, A. W., Pereira, M. C., Pereira, A. C. and Almeida, S. A. O. (1988) Relatorio de Atividades do Projeto IBDF-INPE 'SEQE' – Ano 1987. INPE-4534-RPE/565, Inst. Nacional de Pesquisas Espacias, 12.201, São José dos Campos, SP, Brazil.

Setzer, A. W. and Verstraete, M. M. (1994) Fire and glint in AVHRR's channel 3: a possible reason for the non-saturation mystery. *International Journal of Remote Sensing*, **15**, 711–718.

Sharman, M. (1989) Monitoring agriculture in Europe using AVHRR: management and use of a large multi-temporal dataset. Proceedings of 4th AVHRR Data Users' Meeting, Rothenburg, F.R. Germany, 5–8 September 1989. EUM P 06 (Darmstadt – Eberstadt: EUMETSAT), pp. 175–178.

Sharman, M. J. and Millot, M. (1993) Comparing time profiles: problems in monitoring vegetation condition. Proceedings of 6th AVHRR Data Users' Meeting, Belgirate, Italy, 29 June–2 July 1993. EUM P 12 (Darmstadt – Eberstadt: EUMETSAT), pp. 261–268.

Sheng, Y., Xiao, Q. and Chen, W. (1993) Method of applying AVHRR data to monitor flood disaster. Proceedings of 6th AVHRR Data Users' Meeting, Belgirate, Italy, 29 June–2 July 1993. EUM P 12 (Darmstadt – Eberstadt: EUMETSAT), pp. 293–302.

Shenk, W. E. and Curran, R. J. (1974) The detection of dust storms over land and water with satellite visible and infrared measurements. *Monthly Weather Review*, **102**, 830–837.

Sheres, D. and Kenyon, K. E. (1990a) An eddy, coastal jets and incoming swell all interacting near Pt Conception, California. *International Journal of Remote Sensing*, **11**, 5–25.

Sheres, D. and Kenyon, K. E. (1990b) Swell refraction by the Pt Conception, California, eddy. *International Journal of Remote Sensing*, **11**, 27–40.

Shibayama, M., Wiegand, C. L. and Richardson, A. J. (1986) Diurnal patterns of bidirectional vegetation indices for wheat canopies. *International Journal of Remote Sensing*, **7**, 233–246.

Short, N. M. and Stuart, L. M. (1982) The Heat Capacity Mapping Mission (HCMM) anthology. NASA SP-465, NASA Scientific and Technical Information Branch, Washington, DC.

Sidran, M. (1980) Infrared sensing of sea surface temperature from space. *Remote Sensing of Environment*, **10**, 101–114.

Simonett, D. S. and Coiner, J. C. (1971) Susceptibility of environments to low resolution imaging for land-use mapping. Proceedings of the 7th International Symposium on Remote Sensing of Environment (Ann Arbor, Michigan: Environmental Research Institute of Michigan), pp. 373–394.

Simpson, J. H. (1981) Sea surface fronts and temperatures. In *Remote sensing in meteorology, oceanography and hydrology*, edited by A. P. Cracknell (Chichester: Ellis Horwood), pp. 295–311.

Simpson, J. J. (1990) On the accurate detection and enhancement of oceanic features observed in satellite data. *Remote Sensing of Environment*, **33**, 17–33.

Simpson, J. J. and Gobat, J. I. (1994) Robust velocity estimates, stream functions and simulated Lagrangian drifters from sequential spacecraft data. *IEEE Transactions on Geoscience and Remote Sensing*, **32**, 479–493.

Simpson, J. J. and Gobat, J. I. (1996) Improved cloud detection for daytime AVHRR scenes over land. *Remote Sensing of Environment*, **55**, 21–49.

Simpson, J. J. and Humphrey, C. (1990) An automated cloud screening algorithm for daytime Advanced Very High Resolution Radiometer imagery. *Journal of Geophysical Research*, **95**, 13459–13481.

Simpson, J. J. and Keller, R. H. (1995) An improved fuzzy logic segmentation of sea ice, clouds, and ocean in remotely sensed Arctic imagery. *Remote Sensing of Environment*, **54**, 290–312.

Simpson, J. J. and Yhann, S. R. (1994) Reduction of noise in AVHRR channel 3 data with minimum distortion. *IEEE Transactions on Geoscience and Remote Sensing*, **32**, 315–328.

Singh, S. M. (1984) Removal of atmospheric effects on a pixel by pixel basis from the thermal infrared data from instruments on satellites. The Advanced Very High Resolution Radiometer (AVHRR). *International Journal of Remote Sensing*, **5**, 161–183; erratum, *ibid.*, **5**, 618.

Singh, S. M. (1986) Vegetation index and possibility of complementary parameters from AVHRR/2. *International Journal of Remote Sensing*, **7**, 295–300.

Singh, S. M. (1988a) Simulation of solar zenith angle effect on Global Vegetation Index (GVI) data. *International Journal of Remote Sensing*, **9**, 237–248.

Singh, S. M. (1988b) Lowest order correction for solar zenith angle to Global Vegetation Index (GVI) data. *International Journal of Remote Sensing*, **9**, 1565–1572.

Singh, S. M. (1989) Lowest-order correction to GVI data for solar zenith angle effect. *International Journal of Remote Sensing*, **10**, 819–825.

Singh, S. M. (1991) Near real time atmospheric correction algorithm. Proceedings of 5th AVHRR Data Users' Meeting, Tromsø, Norway, 25–28 June 1991. EUM P 09 (Darmstadt – Eberstadt: EUMETSAT), pp. 71–75.

Singh, S. M. (1992a) Fast atmospheric correction algorithm. *International Journal of Remote Sensing*, **13**, 933–938.

Singh, S. M. (1992b) Vegetation dynamics, CO_2 cycle and the El Niño phenomenon. *International Journal of Remote Sensing*, **13**, 2069–2077.

Singh, S. M. and Cracknell, A. P. (1986) The estimation of atmospheric effects for SPOT using AVHRR channel-1 data. *International Journal of Remote Sensing*, **7**, 361–377.

Singh, S. M., Cracknell, A. P. and Fiúza, A. F. G. (1985) The estimation of atmospheric corrections to one-channel (11 μm) data from the AVHRR; simulation using AVHRR/2. *International Journal of Remote Sensing*, **6**, 927–945.

Singh, S. M., Cracknell, A. P. and Fiúza, A. F. G. (1986) Estimation of sea surface temperature from AVHRR data: reply to some comments by J. R. Eyre. *International Journal of Remote Sensing*, **7**, 1191–1196.

Singh, S. M. and Saull, R. J. (1988) The effect of atmospheric correction on the interpretation of multitemporal AVHRR-derived vegetation index dynamics. *Remote Sensing of Environment*, **25**, 37–51.

Singh, S. M. and Warren, D. E. (1983) Sea surface temperatures from infrared measurements. In *Remote sensing applications in marine science and technology*, edited by A. P. Cracknell (Dordrecht: Reidel), pp. 231–262.

Slingo, A. (1989) A GCM parameterization for the shortwave radiative properties of water clouds. *Journal of Atmospheric Science*, **46**, 1419–1427.

Smith, R. C., Eppley, R. W. and Baker, K. S. (1982) Correlation of primary production as measured aboard ship in southern California coastal waters and as estimated from satellite chlorophyll images. *Marine Biology*, **66**, 281–288.

Smith, R. C. G. and Choudhury, B. J. (1990) On the correlation of indices of vegetation and surface temperature over south-eastern Australia. *International Journal of Remote Sensing*, **11**, 2113–2120.

Snijders, F. L. (1989) Operational monitoring of environmental conditions relevant to crop production and desert locust plague prevention using NOAA-AVHRR data. Proceedings of 4th AVHRR Data Users' Meeting, Rothenburg, F.R. Germany, 5–8 September 1989. EUM P 06 (Darmstadt – Eberstadt: EUMETSAT), pp. 179–184.

Sobrino, J. A. and Caselles, V. (1991) A methodology for obtaining the crop temperature from NOAA-9 AVHRR data. *International Journal of Remote Sensing*, **12**, 2461–2475.

Sobrino, J. A., Caselles, V. and Becker, F. (1989) Significance of the remotely sensed thermal infrared measurements obtained over a citrus orchard. *ISPRS Journal of Photogrammetry and Remote Sensing*, **44**, 343–354.

Sobrino, J. A., Li, Z. L., Stoll, M. P. and Becker, F. (1993) Determination of the surface temperature from ATSR data. Proceedings of the 25th International Symposium on Remote Sensing and Global Environmental Change. Tools for Sustainable Development, Graz, Austria, 4–8 April 1993 (Ann Arbor, Michigan: Environmental Research Institute of Michigan), volume II, pp. 99–109.

Søgaard, H. (1983) Snow mapping in Greenland based on multitemporal satellite data. In *Hydrological applications of remote sensing and remote data transmission*, edited by A. I. Johnson (Wallingford, Oxfordshire: International Association of Hydrological Sciences, Institute of Hydrology), Publication 145, pp. 491–498.

Søgaard, H. (1986) Snow mapping in western Greenland. Proceedings of the European Association of Remote Sensing Laboratories Symposium on Europe from Space. SP-258 (Paris: European Space Agency), pp. 383–393.

Sørensen, K., Lindell, T. and Røed, L. P. (1991) Combination of AVHRR and high resolution data for water quality monitoring. Proceedings of 5th AVHRR Data Users' Meeting, Tromsø, Norway, 25–28 June 1991. EUM P 09 (Darmstadt – Eberstadt: EUMETSAT), pp. 175–180.

Soufflet, V., Tanré, D., Begue, A., Podaire, A. and Deschamps, P. Y. (1991) Atmospheric effect on NOAA AVHRR data over Sahelian regions. *International Journal of Remote Sensing*, **12**, 1189–1203.

Soukeras, S. and Hayes, L. W. B. (1991) On-line access to the record of the University of Dundee AVHRR archive. *International Journal of Remote Sensing*, **12**, 183–191.

Sousa, F. M. and Fiúza, A. (1989) Recurrence of upwelling filaments off northern Portugal as revealed by satellite imagery. Proceedings of 4th AVHRR Data Users' Meeting, Rothenburg, F.R. Germany, 5–8 September 1989. EUM P 06 (Darmstadt – Eberstadt: EUMETSAT), pp. 219–223.

Spanner, M. A., Pierce, L. L., Running, S. W. and Peterson, D. L. (1990) The seasonality of AVHRR data of temperate coniferous forests: relationship with leaf area index. *Remote Sensing of Environment*, **33**, 97–112.

Sparkman, J. K. (1989) NOAA polar orbiting sensor systems: today and tomorrow. *International Journal of Remote Sensing*, **10**, 609–612.

Spitzer, D., Laane, R. and Roozekrans, J. N. (1990) Pollution monitoring of the North Sea using NOAA/AVHRR imagery. *International Journal of Remote Sensing*, 11, 967–977.

Steffen, K. and Lewis, J. E. (1988) Surface temperatures and sea ice typing for northern Baffin Bay. *International Journal of Remote Sensing*, 9, 409–422.

Stephens, G. (1995) Cover. Saharan dust outbreak. *International Journal of Remote Sensing*, 16, 2299–2300.

Stephens, G. L., Ackerman, S. and Smith, E. A. (1984) A shortwave parameterization revised to improve cloud absorption. *Journal of Atmospheric Science*, 41, 687–690.

Stewart, R. H. (1985) *Methods of satellite oceanography* (Berkeley: University of California Press).

Steyn-Ross, D. A., Steyn-Ross, M. L. and Clift, S. (1992) Radiance calibrations for Advanced Very High Resolution Radiometer infrared channels. *Journal of Geophysical Research*, 97, 5551–5568.

Stoner, E. R. and Baumgardner, M. F. (1981) Characteristic variations in reflectance of surface soils. *Soil Science Society of America Journal*, 45, 1161–1165.

Street-Perrott, F. A. and Harrison, S. P. (1985) Lake levels and climate reconstruction. In *Paleoclimate analysis and modelling*, edited by A. D. Hecht (New York: Wiley), pp. 291–340.

Stringer, W. J., Dean, K. G., Guritz, R. M., Garbeil, H. M., Groves, J. E. and Ahlnaes, K. (1992) Detection of petroleum spilled from the *MV Exxon Valdez*. *International Journal of Remote Sensing*, 13, 799–824.

Strong, A. E. (1989) Greater global warming revealed by satellite derived sea surface temperature trends. *Nature*, 338, 642–645.

Strong, A. E. and Eadie, B. J. (1978) Satellite observations of calcium carbonate precipitations in the Great Lakes. *Limnology and Oceanography*, 23, 877–887.

Strong, A. E., McClain, E. P. and McGinnis, D. F. (1971) Detection of thawing snow and ice packs through the combined use of visible and near-infrared measurements from Earth satellites. *Monthly Weather Review*, 99, 828–830.

Strong, A. E. and Ruff, I. S. (1970) Utilizing satellite-observed solar reflections from the sea surface as an indicator of surface wind speed. *Remote Sensing of Environment*, 1, 181–185.

Sturm, B. (1981) The atmospheric correction of remotely sensed data and the quantitative determination of suspended matter in marine water surface layers. In *Remote sensing in meteorology, oceanography and hydrology*, edited by A. P. Cracknell (Chichester: Ellis Horwood), pp. 163–197.

Sturm, B. (1983) Selected topics of Coastal Zone Color Scanner (CZCS) data evaluation. In *Remote sensing applications in marine science and technology*, edited by A. P. Cracknell (Dordrecht: Reidel), pp. 137–167.

Suits, G. H. (1972a) The calculation of the directional reflectance of a vegetative canopy. *Remote Sensing of Environment*, 2, 117–125.

Suits, G. H. (1972b) The cause of azimuthal variations in directional reflectance of vegetative canopies. *Remote Sensing of Environment*, 22, 175–182.

Suits, G., Malila, W. and Weller, T. (1988) Procedures for using signals from one sensor as substitutes for signals of another. *Remote Sensing of Environment*, 25, 395–408.

Summers, R. J. (1989) Educator's Guide for Building and Operating Environmental Satellite Receiving Stations. NOAA Technical Report NESDIS 44 (Washington, DC: US Department of Commerce).

Svendsen, E., Kloster, K., Farrelly, B. A., Johannessen, O. M., Johannessen, J., Campbell, W. J., Gloersen, P., Cavalieri, D. J. and Mätzler, C. (1983) Norwegian remote sensing experiment: evaluation of the Nimbus-7 Scanning Multichannel Microwave Radiometer for sea ice research. *Journal of Geophysical Research*, 88, 2781–2791.

Swithinbank, C. (1973) Higher resolution satellite pictures. *Polar Record*, 16, 739–742.

Taconet, O. and Vidal-Madjar, D. (1988) Application of a flux algorithm to a field-satellite campaign over vegetated area. *Remote Sensing of Environment*, 26, 227–239.

Takashima, T. and Masuda, K. (1988) Averaged emissivities of quartz and sahara dust powders in the infrared region. *Remote Sensing of Environment*, **26**, 301–302.

Tanaka, S., Sugimura, T., Nishimura, T., Ninomiya, Y. and Hatakeyama, Y. (1982) Compilation of the Kuroshio Current vector map from NOAA-6/AVHRR data and consideration of oceanic eddies and the short period fluctuation of the Kuroshio. *Journal of the Remote Sensing Society of Japan*, **2**, 11–30.

Tarpley, D. (1993) Extraction of climate parameters from operational polar orbiting spacecraft. Proceedings of 6th AVHRR Data Users' Meeting, Belgirate, Italy, 29 June–2 July 1993. EUM P 12 (Darmstadt – Eberstadt: EUMETSAT), pp. 3–8.

Tarpley, J. D., Schneider, S. R. and Money, R. L. (1984) Global vegetation indices from the NOAA-7 meteorological satellite. *Journal of Climate and Applied Meteorology*, **23**, 491–494.

Teillet, P. M. (1992) An algorithm for the radiometric and atmospheric correction of AVHRR data in the solar reflective channels. *Remote Sensing of Environment*, **41**, 185–195.

Teillet, P. M., Slater, P. N., Ding, Y., Santer, R. P., Jackson, R. D. and Moran, M. S. (1990) Three methods for the absolute calibration of the NOAA AVHRR sensors in-flight. *Remote Sensing of Environment*, **31**, 105–120.

Teng, W. L. (1990) AVHRR monitoring of US crops during the 1988 drought. *Photogrammetric Engineering and Remote Sensing*, **56**, 1143–1146.

Thekaekara, M. P. (1974) Extraterrestrial solar spectrum. *Applied Optics*, **13**, 518–522.

Thekaekara, M. P., Kruger, R. and Duncan, C. H. (1969) Solar irradiance measurements from a research aircraft. *Applied Optics*, **8**, 1713–1732

Thiermann, V. and Ruprecht, E. (1992) A method for the detection of clouds using AVHRR infrared observations. *International Journal of Remote Sensing*, **10**, 1829–1841.

Thomas, G. and Henderson-Sellers, A. (1987) Evaluation of satellite derived land cover characteristics for global climate modelling. *Climate Change*, **11**, 313–348.

Thomas, G. E. (1983) Satellite measurements of the El Chichón stratospheric cloud. Proceedings of the 7th Annual Climate Diagnostics Workshop, 17–22 October 1982 (Boulder, Colorado: National Center for Atmospheric Research, pp. 390–393.

Thompson, D. R. and Wehmanen, O. A. (1979) Using Landsat digital data to detect moisture stress. *Photogrammetric Engineering and Remote Sensing*, **45**, 201–207.

Thomson, A. G. and Milner, C. (1989) Population densities of sheep related to Landsat Thematic Mapper radiance. *International Journal of Remote Sensing*, **10**, 1907–1912.

Tobiss, G. and Muirhead, K. (1989) Earthnet's AVHRR catalogue service. Proceedings of 4th AVHRR Data Users' Meeting, Rothenburg, F.R. Germany, 5–8 September 1989. EUM P 06 (Darmstadt – Eberstadt: EUMETSAT), pp. 339–342.

Tomasi, C., Vitale, V. and Gasperoni, L. (1993) A simulation study of NDVI dependence features on atmospheric water vapour, aerosols and ozone. Proceedings of 6th AVHRR Data Users' Meeting, Belgirate, Italy, 29 June–2 July 1993. EUM P 12 (Darmstadt – Eberstadt: EUMETSAT), pp. 135–141.

Torlegård, A. K. I. (1986) Some photogrammetric experiments with digital image processing. *Photogrammetric Record*, **12**, 175–186.

Townshend, J. R. G. (1994) Global data sets for land applications from the Advanced Very High Resolution Radiometer. *International Journal of Remote Sensing*, **15**, 3319–3332.

Townshend, J. R. G., Choudhury, B. J., Giddings, L., Justice, C. O., Prince, S. D. and Tucker, C. J. (1989a) Comparison of data from the Scanning Multifrequency Microwave Radiometer (SMMR) with data from the Advanced Very High Resolution Radiometer (AVHRR) for terrestrial environmental monitoring: an overview. *International Journal of Remote Sensing*, **10**, 1687–1690.

Townshend, J. R. G. and Justice, C. O. (1986) Analysis of the dynamics of African vegetation using the normalized difference vegetation index. *International Journal of Remote Sensing*, **7**, 1435–1445.

Townshend, J. R. G. and Justice, C. O. (1988) Selecting the spatial resolution of satellite sensors required for global monitoring of land transformations. *International Journal of Remote Sensing*, **9**, 187–236.

Townshend, J. R. G., Justice, C. O., Choudhury, B. J., Tucker, C. J., Kalb, V. T. and Goff, T. E. (1989b) A comparison of SMMR and AVHRR data for continental land cover characterization. *International Journal of Remote Sensing*, **10**, 1633–1642.

Townshend, J. R. G., Justice, C. O., Gurney, C. and McManus, J. (1992) The impact of misregistration on change detection. *IEEE Transactions on Geoscience and Remote Sensing*, **30**, 1054–1060.

Townshend, J. R. G., Justice, C. O. and Kalb, V. T. (1987) Characterization and classification of South American land cover types using satellite data. *International Journal of Remote Sensing*, **8**, 1189–1207.

Townshend, J., Justice, C., Li, W., Gurney, C. and McManus, J. (1991) Global land cover classification by remote sensing: present capabilities and future possibilities. *Remote Sensing of Environment*, **35**, 243–255.

Townshend, J. R. G., Justice, C. O., Skole, D., Malingreau, J.-P., Cihlar, J., Teillet, P., Sadowski, F. and Ruttenberg, S. (1994) The 1 km AVHRR global data set: needs of the International Geosphere Biosphere Programme. *International Journal of Remote Sensing*, **15**, 3417–3441.

Townshend, J. R. G. and Tucker, C. J. (1984) Objective assessment of Advanced Very High Resolution Radiometer data for land cover mapping. *International Journal of Remote Sensing*, **5**, 497–504.

Trenberth, K. E. (1992) *Climate system modeling* (Cambridge: Cambridge University Press).

Trenchard, M. C. and Artley, J. A. (1981a) Application of thermal model for pan evaporation to the hydrology of a defined medium, the sponge, AgRISTARS Report No. FC-L1-04192; JSC-17440, Johnson Space Center, NASA, Houston, Texas.

Trenchard, M. C. and Artley, J. A. (1981b) Sponge, an application of pan evaporation estimates to indicate relative moisture conditions, *Agronomy Abstracts*, 1981, Annual Meeting of the American Society of Agronomy, Atlanta, Georgia.

Tronin, A. A. (1996) Satellite thermal survey – a new tool for the study of seismoactive regions. *International Journal of Remote Sensing*, **17**, 1439–1455.

Tucker, C. J. (1979) Red and photographic infrared linear combinations for monitoring vegetation. *Remote Sensing of Environment*, **8**, 127–150.

Tucker, C. J. (1980) Remote sensing of leaf water content in the near infrared. *Remote Sensing of Environment*, **10**, 23–32.

Tucker, C. J. (1989) Comparing SMMR and AVHRR data for drought monitoring. *International Journal of Remote Sensing*, **10**, 1663–1672.

Tucker, C. J. (1992) Relating SMMR 37 GHz polarization difference to precipitation and atmospheric carbon dioxide concentration: a reappraisal. *International Journal of Remote Sensing*, **13**, 177–191.

Tucker, C. J., Fung, I. Y., Keeling, C. D. and Gammon, R. H. (1986) Relationship between atmospheric CO_2 variations and a satellite-derived vegetation index. *Nature*, **319**, 195–199.

Tucker, C. J., Gatlin, J. A. and Schneider, S. R. (1984a) Monitoring vegetation in the Nile Delta with NOAA-6 and NOAA-7 AVHRR imagery. *Photogrammetric Engineering and Remote Sensing*, **50**, 53–61.

Tucker, C. J., Hielkema, J. H. and Roffey, J. (1985a) The potential of satellite remote sensing of ecological conditions for survey and forecasting desert-locust activity. *International Journal of Remote Sensing*, **6**, 127–138.

Tucker, C. J., Holben, B. N. and Goff, T. E. (1984b) Intensive forest clearing in Rondonia, Brazil, as detected by satellite remote sensing. *Remote Sensing of Environment*, **15**, 255–261.

Tucker, C. J. and Matson, M. (1985) Determination of volcanic dust deposition from El Chichón using ground and satellite data. *International Journal of Remote Sensing*, **6**,

619–627.

Tucker, C. J., Newcomb, W. W. and Dregne, H. E. (1994) AVHRR data sets for determination of desert spatial extent. *International Journal of Remote Sensing*, **15**, 3547–3565.

Tucker, C. J. and Sellers, P. J. (1986) Satellite remote sensing of primary production. *International Journal of Remote Sensing*, **7**, 1395–1416.

Tucker, C. J., Townshend, J. R. G. and Goff, T. E. (1985b) African land cover classification using satellite data. *Science, New York*, **227**, 233–250.

Turner, J. (1989) The use of AVHRR in the Antarctic. Proceedings of 4th AVHRR Data Users' Meeting, Rothenburg, F.R. Germany, 5–8 September 1989. EUM P 06 (Darmstadt – Eberstadt: EUMETSAT), pp. 59–64.

Turner, J. and Row, M. (1989) Mesoscale vortices in the British Antarctic Territory. In *Polar and Arctic Lows*, edited by P. F. Twitchell, E. A. Rasmussen and K. L. Davidson (Hampton, Virginia: A. Deepak), pp. 347–356.

Turner, J. and Warren, D. E. (1988) The structure of sub-synoptic-scale vortices in polar airstreams from AVHRR and TOVS data. Proceedings of the 2nd Conference on Polar Meteorology and Oceanography, Madison, USA, 29–31 March 1988 (Boston, Massachusetts: American Meteorological Society), pp. 126–128.

Turner, J. and Warren, D. E. (1989) Cloud track winds in the polar regions from sequences of AVHRR images. *International Journal of Remote Sensing*, **10**, 695–703.

Turpeinen, O. M. (1988) Results from the rainfall estimation studies based on Meteosat data. *Annals of Meteorology (Neue Folge)*, **25**, 655–658.

Tyler, M. A. and Stumpf, R. P. (1989) Feasibility of using satellites for detection of kinetics of small phytoplankton blooms in estuaries: tidal and migrational effects. *Remote Sensing of Environment*, **27**, 233–250.

Uspensky, A. B. and Scherbina, G. I. (1993) Land surface temperature and emissivity estimation from combined AHVRR/TOVS data. Proceedings of 6th AVHRR Data Users' Meeting, Belgirate, Italy, 29 June–2 July 1993. EUM P 12 (Darmstadt – Eberstadt: EUMETSAT), pp. 109–115.

Uspensky, A. B. and Sutosky, V. V. (1991) Towards the remote sensing of land surface temperature from AVHRR data. Proceedings of 5th AVHRR Data Users' Meeting, Tromsø, Norway, 25–28 June 1991. EUM P 09 (Darmstadt – Eberstadt: EUMETSAT), pp. 367–374.

Valley, S. L. (Editor) (1965) *Handbook of geophysics and space physics*. Air Force Cambridge Laboratories, Office of Aerospace Research, United States Air Force.

Van Camp, L., Nykjaer, L. and Schlittenhardt, P. (1991) Daily sea surface temperature maps from 1981 to 1989 for monitoring long term changes in the Atlantic Ocean off West Africa. Proceedings of 5th AVHRR Data Users' Meeting, Tromsø, Norway, 25–28 June 1991. EUM P 09 (Darmstadt – Eberstadt: EUMETSAT), pp. 223–228.

Van Camp, L. and Schlittenhardt, P. (1989) Design and use of a professional AVHRR software system operating within the ERDAS image processing environment. Proceedings of 4th AVHRR Data Users' Meeting, Rothenburg, F.R. Germany, 5–8 September 1989. EUM P 06 (Darmstadt – Eberstadt: EUMETSAT), pp. 17–20.

Van Dijk, A. (1989) Smoothing vegetation index profiles: an alternative method for reducing radiometric disturbance in NOAA/AVHRR data. Proceedings of 4th AVHRR Data Users' Meeting, Rothenburg, F.R. Germany, 5–8 September 1989. EUM P 06 (Darmstadt – Eberstadt: EUMETSAT), pp. 117–124.

Van Dijk, A., Callis, S. L., Sakamoto, C. M. and Decker, W. L. (1987) Smoothing Index Profiles: an alternative method for reducing radiometric disturbance in NOAA/AVHRR data. *Photogrammetric Engineering and Remote Sensing*, **53**, 1059–1067.

Varekamp, J. C., Luhr, J. F. and Prestegaard, K. L. (1984) The 1982 eruptions of El Chichón Volcano (Chiapas, México): character of the eruptions, ash-fall deposits and gas phase. *Journal of Volcanology and Geothermal Research*, **23**, 39–68.

Vastano, A. C. and Borders, S. E. (1984) Sea surface motion over an anticyclonic eddy on the

Oyashio Front. *Remote Sensing of Environment*, **16**, 87–90.

Vastano, A. C., Borders, S. and Wittenberg, R. (1985) Sea surface flow estimation with infrared and visible imagery. *Journal of Atmospheric and Oceanic Technology*, **2**, 401–403.

Vaughan, R. A. and Cracknell, A. P. (1994) *Remote sensing and global climate change* (Berlin: Springer).

Vaughan, R. A. and Downey, I. D. (1988) Circulation patterns in AVHRR imagery. *International Journal of Remote Sensing*, **9**, 597–600.

Vermote, E. and Kaufman, Y. J. (1995) Absolute calibration of AVHRR visible and near infrared channels using ocean and cloud views. *International Journal of Remote Sensing*, **16**, 2317–2340.

Verstraete, M. M., LePrieur, C., De Brisis, S. and Pinty, B. (1993) GEMI: A new index to estimate the continental fractional vegetation cover. Proceedings of 6th AVHRR Data Users' Meeting, Belgirate, Italy, 29 June–2 July 1993. EUM P 12 (Darmstadt – Eberstadt: EUMETSAT), pp. 143–149.

Victorov, S. V. (1996) *Regional satellite oceanography* (London: Taylor & Francis).

Vidal, A. (1991) Atmospheric and emissivity correction of land surface temperature measured from satellite using ground measurements or satellite data. *International Journal of Remote Sensing*, **12**, 2449–2460.

Vidal, A. and Perrier, A. (1989) Analysis of a simplified relation for estimating daily evapotranspiration from satellite thermal IR data. *International Journal of Remote Sensing*, **10**, 1327–1337.

Viehoff, T. (1989) Mesoscale variability of sea surface temperature in the North Atlantic. *International Journal of Remote Sensing*, **10**, 771–785.

Viehoff, T. (1990) A shipborne AVHRR-HRPT receiving and image processing system for polar research. *International Journal of Remote Sensing*, **11**, 877–886.

Vignolles, C., Puyou Lascassies, P. and Gay, M. (1993) Setting of a method of coarse resolution signal deconvolution in order to establish profiles of vegetation index evolution measurement of the time effect. Proceedings of 6th AVHRR Data Users' Meeting, Belgirate, Italy, 29 June–2 July 1993. EUM P 12 (Darmstadt – Eberstadt: EUMETSAT), pp. 285–292.

Viola, A. and Böhm, E. (1991) Satellite- and air-derived sea surface temperature measurements during TEMPO. Proceedings of 5th AVHRR Data Users' Meeting, Tromsø, Norway, 25–28 June 1991. EUM P 09 (Darmstadt – Eberstadt: EUMETSAT), pp. 187–191.

Viollier, M., Tanré, D. and Deschamps, P. J. (1980) An algorithm for remote sensing of water color from space. *Boundary-Layer Meteorology*, **18**, 247–267.

Viovy, N., Arino, O. and Belward, A. S. (1992) The Best Index Slope Extraction (BISE): a method for reducing noise in NDVI time-series. *International Journal of Remote Sensing*, **13**, 1585–1590.

Vogelmann, J. E. (1990) Comparison between two vegetation indices for measuring different types of forest damage in the north-eastern United States. *International Journal of Remote Sensing*, **11**, 2281–2297.

Volz, F. E. (1973) Infrared optical constants of ammonium sulfate, Sahara dust, volcanic pumice and flyash. *Applied Optics*, **12**, 564–568.

Vukovich, F. M., Toll, D. L. and Murphy, R. E. (1987) Surface temperature and albedo relationships in Senegal derived from NOAA-7 satellite data. *Remote Sensing of Environment*, **22**, 413–421.

Wald, L., Sèze, G. and Desbois, M. (1991) Automatic cloud screening in NOAA-AVHRR day-time imagery. Proceedings of 5th AVHRR Data Users' Meeting, Tromsø, Norway, 25–28 June 1991. EUM P 09 (Darmstadt – Eberstadt: EUMETSAT), pp. 89–93.

Walsh, S. J. (1987) Comparison of NOAA AVHRR data to meteorological drought indices. *Photogrammetric Engineering and Remote Sensing*, **53**, 1069–1074.

Walter, R. G. (1991) The use of AVHRR for monitoring northern hemisphere snow and ice and for volcanic ash. *Proceedings of 5th AVHRR Data Users' Meeting, Tromsø, Norway, 25–28 June 1991.* EUM P 09 (Darmstadt – Eberstadt: EUMETSAT), pp. 291–299.

Walton, A. E. (1985) Satellite measurement of sea surface temperature in the presence of volcanic aerosols. *Journal of Climate and Applied Climatology,* **24,** 501–507.

Walton, C. C. (1988) Nonlinear multichannel algorithms for estimating sea surface temperature with AVHRR satellite data. *Journal of Applied Meteorology,* **27,** 115–124.

Wan, Zheng-Ming (1985) Land surface temperature meaurement from space. PhD dissertation, Department of Geography, University of California, Santa Barbara.

Wannamaker, B. (1984) An evaluation of digitized APT data from the TIROS-N/NOAA-A,-J series of meteorological satellites. *International Journal of Remote Sensing,* **5,** 133–144.

Wardley, N. W. (1984) Vegetation index variability as a function of viewing angle. *International Journal of Remote Sensing,* **5,** 861–870.

Warren, D. (1989) AVHRR channel-3 noise and methods for its removal. *International Journal of Remote Sensing,* **10,** 645–651.

Watson, K. (1973) Periodic heating of a layer over a semi infinite solid. *Journal of Geophysical Research,* **78,** 5904–5910.

Watson, K. (1975) Geological application of thermal infrared images. *Proceedings of the IEEE,* **63,** 128–137.

Watson, K. (1982) Regional thermal-inertia mapping from an experimental satellite. *Geophysics,* **47,** 1681–1687.

Watson, K. and Hummer-Miller, S. (1981) A simple algorithm to estimate the effective regional atmospheric parameters for thermal inertia mapping. *Remote Sensing of Environment,* **11,** 455–462.

Watson, K., Hummer-Miller, S. and Sawatzky, D. L. (1982) Registration of Heat-Capacity Mapping Mission day and night images. *Photogrammetric Engineering and Remote Sensing,* **48,** 263–268.

Weare, B. C. (1992) A comparison of ISCCP C1 cloud amounts with those derived from high resolution AVHRR images. *International Journal of Remote Sensing,* **13,** 1965–1980.

Weinreb, M. P., Hamilton, G., Brown, S. and Koczor, R. J. (1990) Nonlinearity corrections in calibration of Advanced Very High Resolution Radiometer infrared channels. *Journal of Geophysical Research,* **95,** 7381–7388.

Weinreb, M. P. and Hill, M. L. (1980) Calculation of atmospheric radiances and brightness temperatures in infrared window channels of satellite radiometers. NOAA Technical Report NESS 80, US Department of Commerce, Washington, DC.

Welch, R. M., Navar, M. S. and Sengupta, S. K. (1989) The effect of spatial resolution upon texture-based cloud field classifications. *Journal of Geophysical Research,* **94,** 14767–14781.

Wen, S. and Rose, W. I. (1994) Retrieval of sizes and total masses of particles in volcanic clouds using AVHRR bands 4 and 5. *Journal of Geophysical Research,* **99,** 5421–5431.

Wick, G. A., Emery, W. J. and Schluessel, P. (1992) A comprehensive comparison between satellite measured skin and multichannel sea surface temperature. *Journal of Geophysical Research,* **97,** 5569–5595.

Wiesnet, D. R. and d'Aguanno, J. (1982) Thermal imagery of Mount Erebus from the NOAA-6 satellite. *Antarctic Journal of the US,* **17,** 32–34.

Williams, J. B., Trigg, S. N., Anderson, I., Sear, C. B., Wooster, M. and Navarro, P. (1993) Monitoring land use change: the potential of real time AVHRR information for range land management. *Proceedings of 6th AVHRR Data Users' Meeting, Belgirate, Italy, 29 June–2 July 1993.* EUM P 12 (Darmstadt – Eberstadt: EUMETSAT), pp. 411–416.

Williamson, H. D. (1988) Evaluation of middle and thermal infrared radiance in indices used to estimate GLAI. *International Journal of Remote Sensing,* **9,** 275–283.

Winiger, M., Heeb, M., Nejedly, G. and Roesselet, C. (1989) Regional boundary layer airflow patterns derived from digital NOAA-AVHRR data. *International Journal of Remote*

Sensing, **10**, 731–741.

Wood, B. L., Beck, L. R., Washino, R. K., Palchick, S. M. and Sebesta, P. D. (1991) Spectral and spatial characterization of rice field mosquito habitat. *International Journal of Remote Sensing*, **12**, 621–626.

Woodcock, C. E. and Strahler, A. H. (1987) The factor of scale in remote sensing. *Remote Sensing of Environment*, **21**, 311–332.

Woodwell, G. M., Houghton, R. A., Stone, T. A., Nelson, R. F. and Kovalick, W. (1987) Deforestation in the tropics: new measurements in the Amazon Basin using Landsat and NOAA/AVHRR imagery. *Journal of Geophysical Research*, **92**, 2157–2163.

World Climate Programme (1984) Cloud analysis algorithm intercomparison, World Climate Programme Report, WCP-73. ISCCP.

Wylie, B. K., Harrington, J. A., Prince, S. D. and Denda, I. (1991) Satellite and ground-based pasture production assessment in Niger: 1986–1988. *International Journal of Remote Sensing*, **12**, 1281–1300.

Xue, Y. (1989) Theoretical models of thermal inertia and monitoring soil moisture by automatic recognition of remote sensing data. MSc thesis, Peking University (in Chinese).

Xue, Y. and Cracknell, A. P. (1992) Thermal inertia mapping: from research to operation. Proceedings of the 18th Annual Conference of the Remote Sensing Society, University of Dundee, UK, 15–17 September 1992, edited by A. P. Cracknell and R. A. Vaughan (Nottingham: Remote Sensing Society), pp. 471–480.

Xue, Y. and Cracknell, A. P. (1993) Advanced thermal inertia modelling and its application: modelling the emissivity of the ground. Proceedings of the 25th International Symposium on Remote Sensing and Global Environmental Change, Graz, Austria, 4–8 April 1993 (Ann Arbor, Michigan: Environmental Research Institute of Michigan), volume II, pp. 121–122.

Xue, Y. and Cracknell, A. P. (1995a) Operational bi-angle approach to retrieve the Earth surface albedo from AVHRR data in the visible band. *International Journal of Remote Sensing*, **16**, 417–429.

Xue, Y. and Cracknell, A. P. (1995b) Advanced thermal inertia modelling. *International Journal of Remote Sensing*, **16**, 431–446.

Yamanouchi, T. and Kawaguchi, S. (1992) Cloud distribution in the Antarctic from AVHRR data and radiation measurements at the surface. *International Journal of Remote Sensing*, **13**, 111–127.

Yamanouchi, T., Suzuki, K., and Kawaguchi, S. (1987) Detection of clouds in Antarctica from infrared multispectral data of AVHRR. *Journal of the Meteorological Society of Japan*, **65**, 949–962.

Yates, H. W., Tarpley, J. D., Schneider, S. R., McGinnis, D. F. and Scofield, R. A. (1984) The role of meteorological satellites in agricultural remote sensing. *Remote Sensing of Environment*, **14**, 219–233.

Yokoyama, R. and Tanba, S. (1991) Estimation of sea surface temperature via AVHRR of NOAA-9 – comparison with fixed buoy data. *International Journal of Remote Sensing*, **12**, 2513–2528.

Zadoks, J. C., Chang, T. T. and Konzak, C. F. (1974) A decimal code for the growth stages of cereals. *Weed Research*, **14**, 415–421.

Zheng, Q., Klemas, V. and Huang, N. E. (1984) Dynamics of the slope water off New England and its influence on the Gulf Stream as inferred from satellite IR data. *Remote Sensing of Environment*, **15**, 135–153.

Index